SOIL-SPECIFIC FARMING
PRECISION AGRICULTURE

Advances in Soil Science

Series Editors: Rattan Lal and B. A. Stewart

Published Titles

Advances in Soil Science

SOIL-SPECIFIC FARMING
PRECISION AGRICULTURE

Edited by

Rattan Lal
The Ohio State University
Columbus, USA

B. A. Stewart
West Texas A&M University
Canyon, USA

CRC Press
Taylor & Francis Group
Boca Raton London New York

CRC Press is an imprint of the
Taylor & Francis Group, an **informa** business

CRC Press
Taylor & Francis Group
6000 Broken Sound Parkway NW, Suite 300
Boca Raton, FL 33487-2742

First issued in paperback 2021

Version Date: 20150813

ISBN 13: 978-1-03-209844-9 (pbk)
ISBN 13: 978-1-4822-4533-2 (hbk)

Library of Congress Cataloging-in-Publication Data

Soil-specific farming : precision agriculture / editors: Rattan Lal, B. A. Stewart.
        pages cm. -- (Advances in soil science)
     Includes bibliographical references and index.
     ISBN 978-1-4822-4533-2 (hardcover : alk. paper) 1. Precision farming. I. Lal, R., editor. II. Stewart, B. A. (Bobby Alton), 1932- editor. III. Title: Precision agriculture. IV. Series: Advances in soil science (Boca Raton, Fla.)

S494.5.P73S65 2016
338.1'4--dc23   2015027742

Visit the Taylor & Francis Web site at
http://www.taylorandfrancis.com

and the CRC Press Web site at
http://www.crcpress.com

Contents

Preface

Faced with challenges of resource scarcity (water, nutrient, energy) and environmental degradation (nonpoint source pollution, gaseous emissions), it is important to identify and adopt innovative farming systems. Precision agriculture (PA), also called soil-specific farming or satellite farming, is a strategy of sustainable intensification of agroecosystems. The latter implies producing more from less land, water, agrochemicals, energy, and other finite resources. It also implies producing more with less of an ecological footprint, such as emission of greenhouse gases (e.g., N_2O, CH_4), depletion of soil organic carbon, and loss of topsoil by erosion. While initially developed for soil-specific application of fertilizers, the concept of PA can be adopted to address a range of inputs and farm operations such as tillage, irrigation, pesticides and herbicides, developing genotypes, and soil quality and variability. The strategy is to optimize resource use while addressing spatial heterogeneity in soil characteristics and site-specific abiotic and biotic stresses that limit agronomic production and aggravate environmental degradation.

PA technology utilizes Global Positioning System (GPS) and appropriate design support tools to target soil and crop management according to spatial and temporal variations in soil/site characteristics that constrain agronomic productivity, reduce use efficiency of inputs, impair environment quality, and jeopardize sustainability. The goal is to develop and strengthen the database on soil properties, terrain characteristics, and climate parameters with regard to spatial variability over the landscape, and use specific technologies to target these constraints.

Whereas the usefulness of PA technologies to sustainable intensification and prudential use of resources is widely recognized, adoption of these concepts has been slow even in developed countries. Thus, increasing adoption of PA necessitates addressing constraints related to farm size, technology, and policy. Can PA technology be made scale-neutral so that concepts are used both for large-scale commercial farmers as well as for smallholder agriculturalists of the tropics and subtropics? Is basic soil data available at the level of field-scale to adapt PA concepts to issues such as pest and weed management, microirrigation, depth of seeding and placement of fertilizer, and cluster planting? How can the essential inputs be made available to farmers in remote areas, and what policy interventions are needed to facilitate and promote a widespread adoption of PA?

Therefore, this volume specifically focuses on PA technologies and application of modern innovations to enhance use efficiency of inputs through targeted management of soils and crops. The 15-chapter volume discusses (1) historical evolution, (2) soil variability at different scales, (3) soil fertility and nutrient management, (4) water quality, (5) land leveling techniques, and (6) special ecosystems involving small landholders and coastal regions.

This book provides the technological basis of adopting and promoting PA for addressing the issues of resource scarcity, environmental pollution, and climate change. Specific attention is given to scale-related issues and concerns of small landholders.

The editors thank all the authors for their outstanding contributions and for sharing their knowledge and experiences. Despite busy schedules and numerous commitments, all authors managed to produce their chapters in a timely manner; we greatly appreciate that. The editors also thank the editorial staff of Taylor & Francis Group for their help and support in publishing this book. The office staff of the Carbon Management and Sequestration Center provided support with the flow of manuscripts between authors and editors and made valuable contributions, and their help and support is greatly appreciated. In this context, special thanks are due to Laura Hughes who formatted the text and prepared the final submission. Help from Jennifer Donovan in the early phases of the project is thankfully acknowledged. It is a challenging task to thank by listing names of all those who contributed in one way or another to bringing this book to fruition. Thus, it is important to build upon the outstanding contributions of numerous soil scientists, agricultural engineers, and technologists whose research is cited throughout the book.

Rattan Lal
B.A. Stewart

Editors

Rattan Lal, PhD, is a distinguished university professor of soil science and director of the Carbon Management and Sequestration Center, The Ohio State University, and an adjunct professor of the University of Iceland. His current research focus is on climate-resilient agriculture, soil carbon sequestration, sustainable intensification, enhancing use efficiency of agroecosystems, and sustainable management of soil resources of the tropics. He received honorary degrees of Doctor of Science from Punjab Agricultural University (2001); the Norwegian University of Life Sciences, Aas (2005); Alecu Russo Balti State University, Moldova (2010); and the Technical University of Dresden (2015). He was president of the World Association of the Soil and Water Conservation (1987–1990); the International Soil Tillage Research Organization (1988–1991); the Soil Science Society of America (2005–2007); and is president-elect of the International Union of Soil Science. He was a member of the Federal Advisory Committee on U.S. National Assessment of Climate Change-NCADAC (2010–2013) and is a member of the SERDP Scientific Advisory Board of the US-DOE (2011–present); senior science advisor to the Global Soil Forum of the Institute for Advanced Sustainability Studies, Potsdam, Germany (2010–present); member of the Advisory Board of the Joint Program Initiative of Agriculture, Food Security and Climate Change (FACCE-JPI) of the European Union (2013–present); and chair of the Advisory Board of the Institute for Integrated Management of Material Fluxes and Resources of the United Nations University (UNU-FLORES), Dresden, Germany (2014–2017). Professor Lal was a lead author of IPCC (1998–2000). He has mentored 102 graduate students and 54 postdoctoral researchers, and hosted 140 visiting scholars. He has authored/coauthored 730 refereed journal articles and has written 12 and edited/coedited 58 books. In 2014, Reuter Thomson listed him among the world's most influential scientific minds.

B.A. Stewart, PhD, is director of the Dryland Agriculture Institute and a distinguished professor of soil science at West Texas A&M University, Canyon, TX. He is a former director of the USDA Conservation and Production Laboratory at Bushland, TX, past president of the Soil Science Society of America, and member of the 1990–1993 Committee on Long-Range Soil and Water Policy, National Research Council, National Academy of Sciences. He is a fellow of the Soil Science Society of America, American Society of Agronomy, Soil and Water Conservation Society, a recipient of the USDA Superior Service Award, a recipient of the Hugh Hammond Bennett Award of the Soil and Water Conservation Society, and was an honorary member of the International Union of Soil Sciences in 2008. In 2009, Dr. Stewart was inducted into the USDA Agricultural Research Service Science Hall of Fame. Dr. Stewart is very supportive of education and research on dryland agriculture. The B.A. and Jane Ann Stewart Dryland Agriculture Scholarship Fund was established in West Texas A&M University in 1994 to provide scholarships for undergraduate and graduate students with a demonstrated interest in dryland agriculture.

Contributors

Tahirou Abdoulaye
International Institute of Tropical Agriculture
  (IITA)
Ibadan, Nigeria

S.K. Ambast
Central Soil Salinity Research Institute
Karnal, India

Jeetendra P. Aryal
International Maize and Wheat Improvement
  Centre
New Delhi, India

Ezra D. Berkhout
Netherlands Environmental Assessment
  Agency (PBL)
Bilthoven, the Netherlands

Johan Bouma
Wageningen University
Wageningen, the Netherlands

Mark Brindal
School of Agriculture, Food and Wine
University of Adelaide
Glen Osmond, Australia

Du Changwen
Chinese Academy of Sciences
Nanjing, China

Song Cui
Middle Tennessee State University
Murfreesboro, Tennessee

J.A. Delgado
Soil Plant Nutrient Research Unit
USDA-ARS
Fort Collins, Colorado

G. Dercon
Department of Nuclear Sciences and
  Applications
International Atomic Energy Agency
Vienna, Austria

Ma Fei
Chinese Academy of Sciences
Nanjing, China

Ademir de Oliveira Ferreira
Laboratório de Matéria Orgânica do Solo
State University of Ponta Grossa
Ponta Grossa, Brazil

Lucimara Aparecida Ferreira
State University of Ponta Grossa
Ponta Grossa, Brazil

Allison José Fornari
Paiquerê Farm Manager
Piraí do Sul, Brazil

Angelinus C. Franke
University of Free State
Bloemfontein, South Africa

Flávia Juliana Ferreira Furlan
State University of Ponta Grossa
Ponta Grossa, Brazil

Bruno Gerard
International Maize and Wheat Improvement
  Centre
El-Batan, Texcoco, Mexico

Gerard Gill
Independent Monitoring and Evaluation
  Specialist
Glasgow, United Kingdom

Daniel Ruiz Potma Gonçalves
State University of Ponta Grossa
Ponta Grossa, Brazil

Wenxuan Guo
Monsanto Company
Chesterfield, Missouri

L.K. Heng
Department of Nuclear Sciences and
    Applications
International Atomic Energy Agency
Vienna, Austria

M.L. Jat
International Maize and Wheat Improvement
    Centre
New Delhi, India

Zhou Jianmin
Chinese Academy of Sciences
Nanjing, China

Sami Khanal
School of Environment and Natural Resources
Ohio State University
Wooster, Ohio

Raj Khosla
Precision Agriculture
Colorado State University
Fort Collins, Colorado

Lammert Kooistra
Laboratory of Geo-Information Science and
    Remote Sensing
Wageningen University
Wageningen, the Netherlands

Jianbin Lai
Institute of Geographical Science and Natural
    Resources Research, CAS
Beijing, China

Rattan Lal
School of Environment and Natural Resources
Carbon Management and Sequestration Center
The Ohio State University
Columbus, Ohio

Chin Ding Lim
Futures and Options Department
RHB Investment Bank Berhad
Negeri Sembilan, Malaysia

Henry Lin
Pennsylvania State University
University Park, Pennsylvania

L. Mabit
Department of Nuclear Sciences and
    Applications
International Atomic Energy Agency
Vienna, Austria

David Mulla
Department of Soil, Water & Climate
University of Minnesota
St. Paul, Minnesota

M.L. Nguyen
Department of Nuclear Sciences and
    Applications
International Atomic Energy Agency
Vienna, Austria

Nithya Rajan
Texas A&M University
College Station, Texas

N. Ravisankar
Project Directorate for Farming System
    Research
Modipuram, India

João Carlos de Moraes Sá
Department of Soil Science and Agricultural
    Engineering
State University of Ponta Grossa
Ponta Grossa, Brazil

K. Sakadevan
Department of Nuclear Sciences and
    Applications
International Atomic Energy Agency
Vienna, Austria

K.D. Shepherd
World Agroforestry Centre (ICRAF)
Nairobi, Kenya

H.S. Sidhu
Borlaug Institute for South Asia
Ludhiana, India

Yadvinder Singh
Punjab Agricultural University
Ludhiana, Punjab, India

B.A. Stewart
West Texas A&M University
Canyon, Texas

Clare Stirling
International Maize and Wheat Improvement
    Centre
El-Batan, Texcoco, Mexico

Jetse J. Stoorvogel
Soil Geography and Landscape Group
Wageningen University
Wageningen, the Netherlands

T.P. Swarnam
Division of Natural Resource Management
Central Island Agricultural Research Institute
Port Blair, India

Yeong Sheng Tey
Institute of Agricultural and Food Policy
    Studies
Universiti Putra Malaysia
Selangor, Malaysia

P. Tittonell
Farming Systems Ecology
Wageningen University (WUR)
Wageningen, the Netherlands

Jessica Torrion
Northwestern Agricultural Research Center
Montana State University
Kalispell, Montana

R. van Dis
Farming Systems Ecology
Wageningen University
Wageningen, the Netherlands

B. Vanlauwe
International Institute for Tropical Agriculture
    (IITA)
Nairobi, Kenya

A. Velmurugan
Division of Natural Resource Management
Central Island Agricultural Research Institute
Port Blair, India

Hailong Yu
Ningxia University
Ningxia, China

and

Department of Ecosystem Science and
    Management
Pennsylvania State University
University Park, Pennsylvania

Lu Yuzhen
Chinese Academy of Sciences
Nanjing, China

M. Zaman
Department of Nuclear Sciences and
    Applications
International Atomic Energy Agency
Vienna, Austria

F. Zapata
Department of Nuclear Sciences and
    Applications
International Atomic Energy Agency
Vienna, Austria

1 Historical Evolution and Recent Advances in Precision Farming

David Mulla and Raj Khosla

CONTENTS

1.1 INTRODUCTION AND SCOPE OF CHAPTER

Precision farming is one of the top 10 innovations in modern agriculture (Crookston 2006). Precision farming is generally defined as doing the right practice at the right location and time at the right intensity. Since its inception in the early 1980s, precision farming has been adopted on millions of hectares of agricultural cropland around the world. The objective of this chapter is to review the history of precision farming and the factors that led to its widespread popularity. The specific focus is on the following aspects of precision farming: soil sampling, geostatistics and Geographic Information Systems (GIS), farming by soil, variable rate fertilizer, site-specific farming, management zones, Global Positioning System (GPS), yield mapping, variable rate herbicides, variable rate irrigation, remote sensing, automatic tractor navigation and robotics, proximal sensing of soils and crops, and profitability and adoption of precision farming. For each topic, reference to key groups of researchers and the breakthroughs that helped propel precision farming onward are identified. The chapter concludes with a vision for the future of precision farming.

1.2   SOIL SAMPLING

Spot applications of fertilizer were advocated as early as the 1920s (Linsley and Bauer 1929), but cheap fertilizer and labor combined with the increasing area of farms caused most farmers to shift to uniform applications (Franzen and Peck 1994) until the revolution in precision farming took place during the 1980s. Between the 1920s and 1970s, interest in the variability of soil fertility was primarily motivated by the need to accurately determine a field average soil fertilizer recommendation (Kunkel et al. 1971; Franzen 2007). Many scientists (e.g., Sig Melsted and Ted Peck from the University of Illinois) recognized that variability in soil fertility was large and that sparse soil sampling was likely to be a poor representation of average fertilizer requirements (Melsted 1967). Melsted and Peck designed an intensive grid sampling study (at spacings of 24.3 m) for the Mansfield field near Urbana, Illinois in 1961 with a view toward designing sampling strategies that minimized the cost of determining average soil fertility (Franzen 2007). The variability in soil test values prompted Melsted to suggest philosophically that customized fertilizer requirements were more efficient than a single uniform recommendation (Melsted 1967). Intensive grid sampling was continued in the same field at regular time intervals until about 1994 (Figure 1.1). However, there was little practical application of this concept until several decades later.

In Washington State, Irv Dow and colleagues conducted over 70 field trials on irrigated farms during the period from 1963–1970 in which variations in soil fertility were quantified using intensive soil sampling (Dow et al. 1973a,b). They concluded that "soil test variation is not random and may lend itself to mapping and differential fertilization." They opined that "fertilizing according to information from one composite sample results in erroneous fertility programs." Their solution

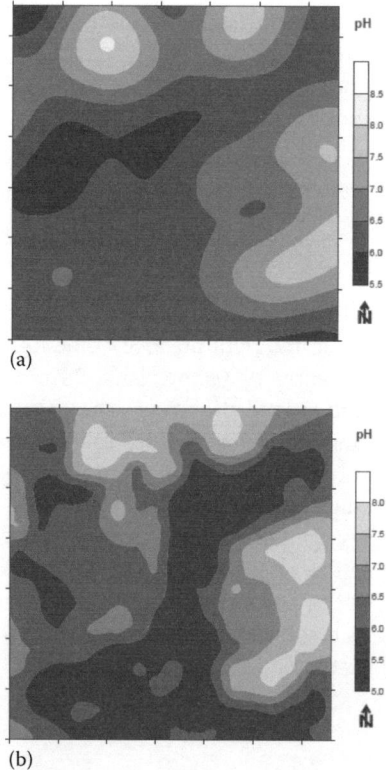

(a)

(b)

FIGURE 1.1   Interpolated soil pH values at Mansfield, IL from 1961 (a) to 1994 (b) based on intensive grid sampling by Melsted (1967) and his colleague Peck at 24.3 m intervals. (Courtesy of David Franzen.)

was to use "precision fertilization based on precision soil sampling." As with research conducted by Peck in Illinois, this idea languished for a decade because of the lack of technology to implement variable rate fertilization. Beginning in 1984, Mulla at Washington State University conducted intensive sampling for soil phosphorus across the eroded hilltops of the Palouse region (as reported in Veseth 1986). He found a good relationship between slope position and soil test phosphorus (P) values, modeled these relationships using geostatistics, and produced computerized contour and three-dimensional (3-D) maps of the relationships. He applied geostatistics and mapping techniques to this data as well as data collected previously by Dow from irrigated potato (*Solanum tuberosum*) farms in central Washington to show that applying variable rates of fertilizer was more efficient and cost-effective than applying uniform rates (Veseth 1986; Mulla and Hammond 1988; Hammond and Mulla 1989). Further, Mulla suggested that for accurate representation of spatial patterns in fertility, soil samples should be collected on a regular grid at spacings of between 30 and 60 m (Veseth 1986).

Wollenhaupt et al. (1994) compared traditional sampling strategies with those that involved estimating composite sample grid cell averages or using individual grid point estimates on some fields in Wisconsin and sampled at spacings of 32.3, 64.6, or 69.9 m. Grid-point sampling was the most accurate strategy for making variable rate P or potassium (K) fertilizer recommendations, followed by grid cell compositing. Traditional sampling for field average soil fertility was inadequate. They also compared different methods for interpolation of soil fertility data, including Delaunay triangulation, inverse distance weighting, and kriging. The most accurate sample spacing was 32.3 m, similar to results found by Mulla in Washington State. Soil fertility map accuracy was significantly degraded at sample spacings of 69.9 m. Several excellent summaries of soil sampling techniques are provided in the literature for those who wish to learn more about this topic (Wollenhaupt et al. 1997; Mulla and McBratney 2000).

1.3 GEOSTATISTICS AND GIS

The seeds for quantifying soil spatial variability were sown by soil scientists during the 1970s and 1980s. Soil physicists, led by Don Nielsen, studied the spatial variability of soil moisture and soil hydraulic properties (Nielsen et al. 1973). The Nielsen group was interested in quantifying the spatial variability of water and solute transport at the field scale, and promoted the use of geostatistics as a tool for doing so (Vieira et al. 1981). On the other hand, soil pedologists, led by Richard Webster, were interested in using geostatistics to quantify the spatial variability of soil properties that could be used to improve the precision of soil mapping (Burgess and Webster 1980). While both groups quantified soil spatial variability using geostatistics, neither group was particularly interested in studying practical issues such as variable rate fertilizer management. The Webster group studied soil sampling strategies for estimating soil properties that could be used for soil classification (McBratney et al. 1981; Webster and Burgess 1984), and later became interested in strategies for accurate estimation of the semivariogram (Webster and Oliver 1992) and interpolation by kriging (Oliver and Webster 1990).

Influenced by Nielsen's studies of field scale variability, during 1985 David Mulla became interested in the relationship between soil fertility and landscape position for rainfed wheat (*Triticum aestivum*) farms in eastern Washington state and irrigated potato farms in central Washington state (as reported by Veseth 1986). He used geostatistics to map soil test P levels, and showed that soil fertility varied significantly from bottom slope to hill crest positions in wheat farms, and that P fertilizer recommendations for a field could be mapped into different zones (Table 1.1). Parallel research on the spatial variability of soil P was also conducted by Assmus et al. (1985). Mulla's research caught the attention of Max Hammond, a crop consultant working for CENEX Land O'Lakes and Soil Teq, and in 1986 Soil Teq from Waconia, Minnesota hired Mulla as a consultant to write software that automatically reclassified and mapped soil fertility sampling data into fertilizer recommendation zones, which Mulla called "management zones." This was the first combined use of geostatistics and GIS for precision

TABLE 1.1

Early Advances in Research on Soil Sampling, Geostatistics, and GIS in Precision Farming

Research Area	Nature of Contribution	Key References
Soil sampling	Grid sampling recommendations	Melsted (1967), Dow et al. (1973b), Mulla and Hammond (1988), Wollenhaupt et al. (1994)
Geostatistics and GIS	Map interpolation and reclassification for soil fertility data	Mulla and Hammond (1988), Mulla (1989, 1991, 1993)

farming (Mulla 1988; Mulla and Hammond 1988; Hammond et al. 1988). Fertilizer recommendation maps were burned onto an E-Prom device by Soil Teq, fitted into a computer in the cab of a fertilizer spreader, and used to guide the delivery of variable rate fertilizer applications starting in the late 1980s. The combined use of geostatistics and GIS for precision farming was detailed in a series of papers by Mulla (1989, 1991, 1993). The use of geostatistics in precision agriculture is extensively documented by Oliver (2010).

1.4   FARMING BY SOIL

Pierre Robert is often regarded as the father of precision farming because of his active promotion of the idea and organization of the first workshop, "Soil Specific Crop Management," during the early 1990s. In 1982, Robert defended his PhD dissertation under the direction of Richard Rust in the University of Minnesota's Department of Soil Science. The dissertation was titled "Evaluation of Some Remote Sensing Techniques for Soil and Crop Management" (Robert 1982). Robert's research on 15 Minnesota commercial corn (*Zea mays*)-soybean (*Glycine max*) farms showed that color infrared (CIR) aerial photography could be used to detect "problems relating to drainage, erosion, germination, grass and weed control, crop stand and damage and machinery malfunction." Robert suggested that CIR data could be used to build a "farm information and management system containing precisely located natural and cultural data to improve cost efficiency of future cultural practices. Such improvement could come, for example, from adjusting seed density, herbicide control, or fertilization in response to detected field problems" (Robert 1982). In Robert's dissertation, he repeatedly notes that anomalous reflectance patterns from row-cropped fields were associated with soil series boundaries. He noted that "The important contribution of remote sensing in soil and crop management is not as a real-time tool but as an input to a geographic soil and crop management data base system" (Robert 1982), indicating that farm management for the following cropping season could be improved using CIR images from the previous year in association with soil series maps. Robert spent the next 3 years developing a computerized soil mapping database in close cooperation with Rust (Figure 1.2). The concept of farming by soil in Minnesota was formally introduced into scientific literature by Rust (1985), Larson and Robert (1991), and by Vetsch et al. (1993).

Carr et al. (1991) and Wibawa et al. (1993) conducted long transect trials to compare a farming by soil fertilizer management strategy with a uniform strategy in Montana and North Dakota, respectively. Results from several fields in Montana showed that rainfed wheat grain yields differed significantly across soil types. However, there were no significant differences in economic returns for the uniform versus the soil-based fertilizer management strategy in Montana. In North Dakota, the economically optimum strategy for growing rainfed barley (*Hordeum vulgare*) and wheat was either a uniform nitrogen (N) fertilizer application based on composite soil samples or a variable rate N strategy that involved compositing soil samples and yield goals by soil mapping unit. Variable rate fertilizer applications based on grid soil sample spacings of 15.2 to 30.4 m were generally able to increase crop yield in comparison with the uniform strategy, but they also incurred extra costs that made these strategies unprofitable.

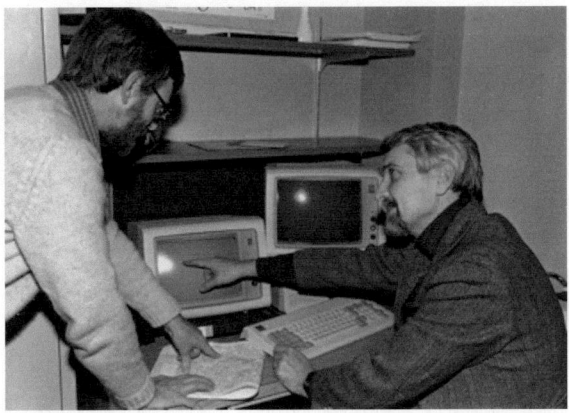

FIGURE 1.2 Pierre Robert explaining his computerized farming by soil map database (circa 1985) to Jim Anderson at the University of Minnesota.

1.5 VARIABLE RATE FERTILIZER

The idea of a variable rate fertilizer spreader was studied by several scientists during the early to mid-1980s, including John Hummel (Hummel 1985) working with the United States Department of Agriculture–Agricultural Research Service (USDA-ARS). Soil Teq of Waconia, Minnesota patented the first computer-controlled variable rate fertilizer spreading machine (Ortlip 1986; Schueller 1992). The system was apparently first tested in Minnesota using variable rate lime (Luellen 1985) or fertilizer applications (Schmitt et al. 1986). Guidance was possible using either dead reckoning or triangulation from radio beacons. Rates of fertilizer were varied according to digitized soil maps, hence the initial appellation "farming by soil" (Larson and Robert 1991). After 1987, using software written by Mulla from Washington State University, Soil Teq was able to vary rates of fertilizer application according to digital maps (Figure 1.3) that were based on soil fertility data obtained by grid sampling, hence the appellation "site-specific farming." This software evolved into Soil Geographic Information System (SGIS), which was marketed by SoilTeq and AgChem.

Gradually, the terms farming by soil and site-specific farming were replaced by variable rate technology (Sawyer 1994). Scientists at several U.S. universities started to investigate variable rate fertilizer applications in the late 1980s (Reichenberger and Russnogle 1989), including Mulla at

FIGURE 1.3 An early 1990s variable rate fertilizer applicator control system.

Washington State University (as reported by Miller 1988; Mulla et al. 1992), Kachanoski at the University of Guelph (Kachanoski et al. 1985) who had heard Mulla's reports on spatial variability of soil fertility at annual meetings of the W-188 Western Regional Committee on Spatial Variability, Searcy at Texas A&M (Borgelt et al. 1989, 1994) who used software written by Mulla, Robert at the University of Minnesota (Robert et al. 1990), and Jacobsen and Nielsen at Montana State University (Carr et al. 1991; Wibawa et al. 1993). Variable rate fertilizer applications were perceived as being both profitable and beneficial because they improved the efficiency of farm inputs, maintained or improved crop yield and quality, and protected water quality.

1.6   SITE-SPECIFIC FARMING AND MANAGEMENT ZONES

The philosophy of site-specific farming was distinct from farming by soil. Farming by soil posited that fertilizer requirements varied across soil series but were homogeneous within a given soil series. Site-specific farming posited that variability within soil series boundaries was significant, and the only way to identify fertilizer requirements was by grid or transect soil sampling across soil series boundaries. Field investigations involving these two contrasting philosophies were partially motivated by the need to document the profitability of variable rate fertilizer applications. To this end, Mulla and his colleagues from Cenex Land O'Lakes in Washington State established the first statistically rigorous field trials in 1987 comparing variable and uniform N and P fertilizer applications on commercial wheat farms (Mulla et al. 1992). Without GPS, they conducted long transect sampling at 15 m spacings across rolling landscapes, and used a manual controller to variably apply N and P according to a map developed by grouping and reclassification of soil fertility data from the transect sampling. While there were no statistically significant differences in crop yield between the uniformly and variably fertilized strips, the profitability of variable rate fertilizer was better than uniform management due to cost savings in fertilizer and improved protein content of wheat in the variably fertilized strips. Mulla introduced the concept of management zones into precision farming as a result of these and other early field trials on irrigated fields with variable rate fertilizer (Mulla 1991, 1993; Mulla et al. 1992). Management zones were relatively homogeneous regions within a larger field that differed from one another in fertilizer recommendations.

Site-specific farming was studied in irrigated agriculture beginning in 1986 where high-cash-value crops such as potatoes were grown (Mulla and Hammond 1988; Hammond et al. 1988, Hammond and Mulla 1989). The profitability of these irrigated crops justified intensive grid soil sampling that could be used to define management zones for variable P and K fertilizer applications. Through the use of geostatistics, it was determined that the optimum grid spacing for soil sampling was ~61 m (Mulla and Hammond 1988; Mulla 1991).

The terms site-specific and precision farming were introduced into scientific literature by John Schueller from the University of Florida (Schueller 1991, 1992). He helped organize an important symposium on this topic at the 1991 Annual Meeting of the American Society of Agricultural Engineers (ASAE) in Chicago. According to Schueller (1991), "the continuing advances in automation hardware and software technology have made possible what is variously known as spatially-variable, precision, prescription, or site-specific crop production."

The use of management zones in precision farming has persisted to the present day, but the concept and definition has shifted (Table 1.2). Mulla (1991) offered the first definition: "Each management zone should ideally represent portions of the field that are relatively similar and homogeneous in soil fertility status so that a different uniform fertilizer recommendation can be made for each zone." Doerge (1999) broadened the concept by saying that management zones "are sub-regions of a field that express a homogeneous combination of yield limiting factors for which a single crop input is appropriate." Fraisse et al. (2001) introduced the k-means clustering approach for delineation of management zones. Khosla et al. (2010) summarized an extensive body of scientific literature for delineating management zones for precision farming.

TABLE 1.2

Early Advances in the Concept and Testing of Farming by Soil, Variable Rate Fertilizer, Management Zones, and Precision Farming

Research Area	Nature of Contribution	Key References
Farming by soil	Proposed concept	Robert (1982), Rust (1985)
Variable rate fertilizer	Machinery development and field testing	Hummel (1985), Luellen (1985), Ortlip (1986), Schmitt et al. (1986), Borgelt et al. (1989, 1994), Carr et al. (1991), Mulla et al. (1992)
Management zones	Proposed and tested concept	Mulla (1991, 1993), Mulla et al. (1992), Doerge (1999), Fraisse et al. (2001)
Precision farming	Proposed concept	Schueller (1991, 1992)

They found that the most common approaches for delineating management zones were based on soil properties such as soil texture and soil organic matter (SOM) content, followed by sensing technologies such as electrical conductivity mapping and remote sensing. Other less common approaches for delineating management zones included yield mapping followed by elevation differences across a field.

1.7 GPS

Precise determination of location is essential for precision farming, especially for mapping the variability in soil fertility or crop yield, and in locating farm machinery that can spread variable rates of fertilizer relative to the information in these maps. When interest in precision farming first developed, there were two distinct philosophies about how to determine location within a field. The first was to use radio-based triangulation with strategically placed beacons (Palmer 1991). The main advantage of this approach was the ability to determine machinery position in real time at submeter accuracy without postprocessing of data. The main disadvantages were loss of signal in rolling topography and time and effort to establish beacon positions (Auernhammer and Muhr 1991). The alternative was the GPS, which was established for military purposes in the late 1970s (Larsen et al. 1988; Tyler 1993). GPS accuracies of 3 m could be achieved by the military during the early 1990s based on differential postprocessing of the P code, or 5 m accuracies based on processing of the C/A code (Tyler 1993). A GPS receiver in a fixed position was required for differential correction. Civilian users without recourse to differential correction could only achieve accuracies of between 10–100m with GPS receivers before military selective availability (spoofing) was turned off in 2000. Differential correction became very popular with agricultural users during the 1990s as a way to obtain acceptable accuracies before selective availability was turned off (Tyler et al. 1997). Real-time differential correction became possible when the Coast Guard and several companies such as Omnistar began to establish networks of GPS base stations whose real-time positions could be broadcast to roving machines with an FM receiver (Tyler et al. 1997). This real-time differential correction approach was preferred to real-time kinematic positioning, where the phase of signals from at least four satellites were counted continuously at both the base and roving receivers. Loss of signal often occurred when the roving receiver passed behind a tree, building, or hill, causing failure of the real-time kinematic approach.

The primary interest in GPS for precision farming was initially as a method for identifying the location of a combine that was collecting real-time data on spatial variability in crop grain yield (Auernhammer and Muhr 1991; Schueller and Wang 1994; Auernhammer et al. 1994). More information about yield mapping is provided in Section 1.7 below. Later, interest in GPS shifted to its use in navigating agricultural machinery (Zhang et al. 1999) and autosteering (Keller et al. 2001). More information about these applications is described in Section 1.8 below.

1.8 AUTOMATED TRACTOR NAVIGATION AND ROBOTS

Precise automated navigation has been one of the most intense areas of research and implementation over the last three decades. The advantages of this approach include reduced operator fatigue, elimination of machinery overlaps and skips, and improved efficiency in fuel usage and product application. Navigation of agricultural machinery has been studied for at least 75 years since Andrew (1941) patented a method for automated plowing of circular fields based on the distance to center using a cable spool system. Reid and Searcy (1987) used near-infrared computer vision to distinguish straight rows of crop from bare soil that could be used for straight-line navigation. Dust and vibration of the camera were the main limitations of computer vision (Reid et al. 2000; Wilson 2000). Triangulation of agricultural machinery positions using radio beacons (Palmer 1991, 1995) or microwave signals (Searcy et al. 1989b) required good line of sight and significant setup of equipment in the field.

The most frequent approach to precise automated navigation involves GPS, first proposed by Larsen et al. (1988). O'Connor et al. (1996) pioneered the use of real-time kinematic (RTK) GPS for automatic steering of a tractor along straight lines. This system initially involved four separate GPS receivers mounted on a tractor as well as a nearby GPS base station. An electrohydraulic steering unit on the tractor was automatically guided by the GPS. Accuracy of the RTK GPS method was better than a 2.5-cm standard deviation, which is better than accuracy (15–33 cm std. dev.) of the U.S. Federal Aviation Agency's Wide Area Augmentation System (WAAS), or the accuracy (5–12.5 cm std. dev.) of the commercial OmniStar system. A series of patents were subsequently issued for GPS-based navigation (Greatline and Greatline 1999; Keller et al. 2001; McClure 2005; Collins et al. 2006; McKay and Anderson 2007), leading to rapid commercialization and adoption of autosteer technology in agriculture, including the Beeline Navigator, which first appeared in Australia. Thuilot et al. (2002) used RTK GPS to guide a tractor along curved paths, which is more difficult than navigating along straight lines. Their accuracy was generally better than 2m deviations from prescribed pathways. Recent attempts to improve accuracy of navigation often involve multiple sensors, including GPS, geomagnetic direction sensors, and machine vision (Zhang et al. 1999). Japan has been a leader in adoption of all types of autosteer navigation technology, particularly on the small agricultural fields of Hokkaido Island (Torii 2000). Automated navigation of tractors can take various forms, including manual guidance with a lightbar, assisted steering, or autosteer (Berglund and Buick 2005), depending on the monetary investment.

1.9 YIELD MAPPING

The concept of detailed spatial mapping of crop yield was initially developed and tested by Schueller and Bae (1987) before the availability of GPS. They used variations in engine speed as a surrogate for grain flow to the combine, which was operated under constant throttle and cutting head height. Position of the combine was determined using real-time microwave ranging, with between 0.5 and 6 m accuracy. Variations in crop yield were aggregated to blocks of 10 × 10 m to reduce variability in measurements. Because measuring variations in engine speed could be inaccurate due to slippage of wheels and other factors, Bae et al. (1987) and Searcy et al. (1989a) also studied yield monitors that were based on measurements of the volume of grain leaving the grain auger. Grain volume flow in the combine could be estimated based on the rate of revolution of a rotating paddle in the auger. Each paddle had a fixed volume capacity for grain.

More sophisticated indirect techniques for measuring grain flow in the combine were developed by Vanischen and Baerdemaeker (1991), who used measurements of deflection in a curved plate caused by the impact of grain flow mass. Accuracy of this approach required sensitive measurements of swath width and combine speed, as well as filtering of the raw data signal from the deflection of the curved plate to overcome vibration and distortion effects. Most commercial yield monitoring devices are based on deflection of plates or fingers or on impact of grain on impact plates

(Pierce et al. 1997). Stafford et al. (1991, 1996) evaluated two other methods for yield mapping (Figure 1.4); namely; gamma ray detectors whose signal is attenuated based on the amount of grain flow, and a capacitive sensor whose response varies with the dielectric constant of the air/grain mixture. Neither technique was entirely satisfactory.

Errors in measuring grain yield with yield monitors are generally less than 5% (Pierce et al. 1997). The most common errors arise from inaccurate estimates of cutting swath width or combine travel distance; the latter is estimated from combine speed. Cutting swath width can be in error when overlap occurs between successive passes of the combine so that the effective cutting swath is less than the width of the cutting head. Use of centimeter-accuracy GPS systems on the combine can significantly reduce errors in estimating travel distance and can prevent cutting head overlap on successive passes of the combine. The primary error that persists in yield monitor data occurs when the combine turns around at the edge of a field. Because of a time lag between the measurement of grain yield relative to the location where the crop was harvested, the first few data points for crop yield after a combine turns will generally be too low.

Attempts to explain spatial variability of crop yield were initially based on the hypothesis that patterns in yield were determined by relationships with soil series mapping units. To test this hypothesis, Karlen et al. (1990) produced a detailed soil series map in an 8-ha field by intensive soil coring and profile descriptions in the Coastal Plains region of South Carolina (Karlen et al. 1990). They then studied spatial variations in corn, wheat, and sorghum (*Sorghum bicolor*) yield for this field from 1985–1988. Results showed that spatial variations in crop yield were extensive, and were significantly different across soil mapping units (Table 1.3). The best indicator of spatial patterns in crop yield was depth to argillic horizon, but spatial variation within soil map units was nearly as large as the variation in yield among map units. Sudduth et al. (1996) studied spatial relationships between crop yields and soil or topographic factors on two fields in central Missouri from 1993 to 1995. Using data from over 300 sampling locations, they found that advanced regression techniques could explain from 51%–77% of the variability in crop yield. The most important factor affecting crop yield was elevation, followed by topsoil depth, organic matter content, and soil test phosphorus. Both Karlen et al. (1990) and Sudduth et al. (1996) noted that there was significant variation in crop yield from one year to the next in response to variations in precipitation, but the primary focus of their analysis was on spatial variation.

Lamb et al. (1997) were the first to quantitatively compare spatial and temporal variations in continuous corn crop yield over a 5-year period for a 1.8-ha field in Minnesota. The field was divided into 60 grid cells of area 279 m² each. The highest and lowest crop yields for a given year differed between 2762 and 4519 kg/ha from one grid cell to another depending on the year, but spatial

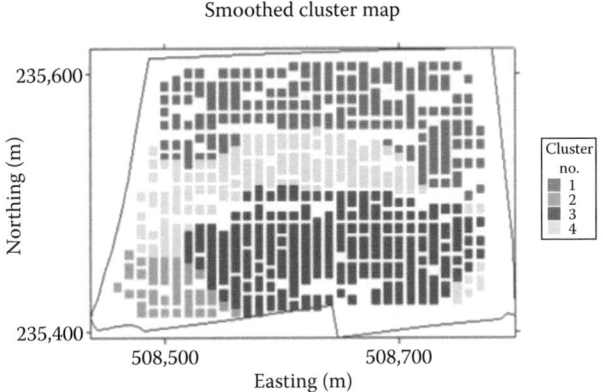

FIGURE 1.4 Clustered yield map for winter barley based on 3 years of data (1993–1995) in Cashmore field, UK. (Courtesy of John Stafford.)

TABLE 1.3

Early Advances in Research on GPS, Machinery Navigation, and Yield Mapping for Precision Farming

Research Area	Nature of Contribution	Key References
GPS	Technology adaptation to farming and testing	Larsen et al. (1988), Auernhammer and Muhr (1991), Tyler (1993)
Machinery navigation and autosteer	Technology development and testing	Reid and Searcy (1987), Palmer (1991), O'Connor et al. (1996), Greatline and Greatline (1999), Keller et al. (2001)
Yield mapping	Technology development and field testing	Bae et al. (1987), Schueller and Bae (1987), Searcy et al. (1989a), Karlen et al. (1990), Stafford et al. (1991)

patterns in crop yield were not temporally stable. Yield maps for the first 4 years of the study could only explain half of the spatial variability in grain yield for the last year of the study. These results showed that the magnitudes of both spatial and temporal variability were important, but temporal variability was not predictable from one year to the next.

McBratney and Whelan (1999) studied spatial and temporal variability in wheat crop yield in four fields located in Australia from 1995–1996. They noted that temporal variations across years were larger in magnitude than spatial variations within a year. This led them to propose the concept of using uniform field management techniques when yield stability across years was poor and variable management when yield was stable across years. Their null hypothesis was stated as: "given the large temporal variation evident in crop scale relative to the scale of a single field, then the optimal risk aversion strategy is uniform management." Blackmore et al. (2003) found that wheat yield variability across four fields in England for 6 years was unpredictable from one year to another, despite significant spatial variability in yield during any single year. They suggested that managing for the spatial variability that exists in a given year is better than trying to predict management needs from the previous year(s) yield map(s).

The focus on spatial and temporal variation in crop *yield* tends to overlook an important issue. For variable rate management, what is really critical is the crop *response* rather than the crop yield. Mamo et al. (2003) studied spatial and temporal variations in the response of a corn crop to variable rates of N fertilizer from 1995 to 1999 in Minnesota. Half of the field responded to N fertilizer in a given year, while 60% of the responsive areas were stable across years. A variable rate fertilizer strategy would have reduced N rates overall by 69 to 75 kg/ha in comparison to a uniform application of fertilizer. This study showed that crop yield by itself is a poor indicator of the crop response to N fertilizer. The implication is that defining management zones based on differences in crop yield is not generally an efficient approach for developing a variable rate fertilizer application strategy. Similar conclusions were reached by Scharf et al. (2006) in Missouri, who showed that economically optimum N rates were more strongly controlled by soil N supply than yield-controlled N uptake patterns.

1.10 VARIABLE RATE HERBICIDE APPLICATION

Weeds tend to occur in patches rather than in uniform coverage across fields, and if these patches exceed threshold populations, they can reduce crop yield and vigor (Coble and Mortensen 1992). Before the advent of Roundup Ready crops, there was significant interest in variable rate applications of preemergent or postemergent herbicides (Wiles et al. 1992).

Haggar et al. (1983) fitted an optical sensor to a handheld weed sprayer to test the concept of optically activated spot spraying. The optical sensor estimated the ratio of red to near-infrared reflectance of weeds on a background of bare soil. Guyer et al. (1986) and Thompson et al. (1990, 1991) suggested the concept of mapping weed locations using either machine vision, remote sensing, video cameras on tractors, or manual counts, and then spraying the field at a uniformly low rate with a higher dosage of herbicide in areas with weed patches. Thompson et al. (1990, 1991) believed that this approach would work well in fields where weed patches tend to occur in the same locations from one year to another. Thompson et al. (1991) discussed the potential for real-time mapping of weed populations in a growing crop, but this approach was rejected due to the difficulty of discriminating weeds from crop and low spatial resolution of aerial imagery. Felton and McCloy (1992) proposed a spot herbicide sprayer that was based on detection of weeds using red and near-infrared reflectance. This research led to the development of a commercial spot sprayer in Australia known as DetectSpray. Stafford and Miller (1993) built a variable rate herbicide sprayer that applied a low uniform rate of herbicide throughout the field and a higher rate where weed patches had been previously mapped (Figure 1.5). Sprayer position relative to the weed map was determined using differential GPS techniques. Weed patches were mapped using a model airplane equipped with a 35-mm color photography camera; this was the first use of an unmanned aerial vehicle for precision farming. Further development of the map-based approach to variable herbicide spraying in Europe was subsequently restricted as a result of legal rulings relating to an infringement of the SoilTeq patent for map-based variable rate applications.

Johnson et al. (1995a) mapped weed density in 12 corn or soybean fields in Nebraska for a single year. This research showed that weeds tended to occur in patches, and many areas of the field tended to be weed-free. They suggested that variable rate herbicide spray could be targeted to weed patches if the density of weeds in those patches exceeded an economic threshold. Research by Johnson et al. (1995b) then mapped weed patches for 2 years in 18 Nebraska corn and soybean fields. Results of this research showed that locations of weed patches were not stable from one year to the next.

As a result of research by Johnson et al. (1995b), interest in weed mapping soon turned to real-time mapping of weeds using photodetectors and variable herbicide spraying (Beck and Kinter 1998a,b) with what became known as WeedSeeker. Hanks and Beck (1998) evaluated two commercial sensors, DetectSpray and WeedSeeker, for their ability to identify and spray weeds; the former used passive reflectance, while the latter had an active sensor consisting of gallium-based light-emitting diodes. DetectSpray and WeedSeeker were both initially designed to sense green vegetation on a background of bare soil and were not appropriate for use in fields where crops and weeds were mixed together. Weeds were sprayed whenever detected by either system. When used in fields

FIGURE 1.5   Variable rate herbicide applicator developed by Stafford and Miller (1993).

where crops had already germinated, a spray hood was installed over the weed sensors and sprayer nozzles to prevent the application of glyphosate herbicide to growing crops. WeedSeeker performed better than DetectSpray in variable ambient lighting conditions because of its active sensor system. Sensor-based weed control reduced the volume of herbicide applied by 63%–85% relative to a uniform spray application (Hanks and Beck 1998). Using machine vision, Giles and Slaughter (1997) found that variable herbicide spray reduced application rates by 66%–80% in vegetable crops relative to a uniform spray application. Tian et al. (2000) developed a variable rate herbicide sprayer that used a low-resolution color video camera to identify clusters of weeds growing between rows of corn or soybean. Herbicide spray could be varied nozzle by nozzle depending on the weed density. For low-density weed cover, tests of this system showed that herbicide application amounts could be reduced by 71% relative to a uniform application rate (Tian 2002).

1.11 VARIABLE RATE IRRIGATION

Water conservation is a pressing issue in the face of drought and competition for water resources by agricultural, municipal, and industrial users. Overirrigation wastes water and leads to leaching and runoff losses that carry soluble pollutants such as nitrate-N and pesticides to ground or surface waters. McCann and Stark (1993) patented a method for variable application of irrigation water and chemicals applied through center pivot irrigation systems (Table 1.4). Aerial photography or soil sampling was used to identify management zones requiring different amounts of irrigation. Each nozzle on the irrigation spray boom could be independently controlled using a solenoid valve. A microprocessor was used to determine the location of each nozzle relative to mapped irrigation zones, and a control program then turned the nozzle on or off in order to deliver the required amount of water for that location. Variable rate irrigation was field-tested by King et al. (1996), who found that this method was able to accurately deliver recommended variations in irrigation water depth and associated nitrogen fertilizer requirements. Variable rate irrigation through linear move systems was developed by Fraisse (1994) at Colorado State University in parallel with the Washington State research by McCann and Stark (1993). In South Carolina, Omary et al. (1997) and Camp et al. (1998) placed multiple manifolds on a center pivot in order to have the flexibility of delivering one of eight possible rates of irrigation to any portion of the field.

Evans et al. (1996) collaborated with the Nelson Irrigation company in Walla Walla, Washington to install variable rate irrigation controllers on a center pivot system in a commercial farm. Thirty zones having two to four nozzles were cycled on and off by a master controller on an RS485 bus to achieve desired rates of water application. Water demands could be calculated using a potato growth simulation model based on landscape, soil, and climatic information. They found that the performance of the variable rate irrigation system was excellent, and that the main limitation in implementing the system "lies in the ability to interpret spatially variable data and develop rational and

TABLE 1.4
Early Research Advances in Variable Rate Herbicide for Weed Control and Variable Rate Irrigation

Research Area	Nature of Contribution	Key References
Variable rate herbicide	Technology development and testing	Haggar et al. (1983), Guyer et al. (1986), Beck and Kinter (1988a,b), Thompson et al. (1990, 1991), McCloy and Felton (1992), Stafford et al. (1993)
Variable rate irrigation	Technology development and testing	McCann and Stark (1993), Fraisse (1994), Evans et al. (1996)

coherent site-specific crop management prescriptions" (Evans et al. 1996). Distortion of irrigation spray patterns by wind was also a particularly vexing problem.

1.12 REMOTE SENSING

Remote sensing applications in precision agriculture are primarily based on reflectance of the sun's visible and near-infrared light by soils or crops. Remote sensing does not require contact between the sensor and the soil or crop and is usually achieved using cameras mounted on satellites, airplanes, towers, or unmanned aerial vehicles. Proximal sensing, discussed in Section 1.13 below, differs from the traditional definition of remote sensing in that proximal sensing involves sensors placed on ground vehicles rather than aerial platforms.

The earliest applications of remote sensing in agriculture were primarily focused on estimating crop yield (Pinter et al. 1981; Wiegand et al. 1991), although Al-Abbas et al. (1974) conducted laboratory studies of the spectral properties of corn leaves with various levels of nutrient stress. Robert (1982) used color infrared aerial photography in Minnesota for diagnosis of "problems related to drainage, erosion, germination, grass and weed control, crop stand and damage, and machinery malfunction."

Landsat imagery was investigated for diagnosis of agricultural problems by Robert (1982), but difficulties in processing satellite remote sensing data at that time prevented meaningful results. Zheng and Schreier (1988) and Bhatti et al. (1991) were the first to use aerial and satellite imagery, respectively, for the specific purpose of estimating spatial patterns in soil fertility that could be used to guide variable rate fertilizer applications. Zheng and Schreier (1988) found that potassium fertilizer recommendations for a bare field in British Columbia could be reduced relative to uniform applications if rates were varied according to spatial patterns in soil organic matter content identified using color aerial photographs. Bhatti et al. (1991) found that spatial patterns in soil organic matter from Landsat satellite imagery for bare soil on a commercial farm in Washington State were strongly related to patterns in soil phosphorus and wheat yield. They proposed that areas with low organic matter content and low crop productivity "could be managed with customized fertilizer and tillage practices" for environmental protection. During this early period in precision farming, satellite remote sensing imagery with Landsat was limited to 30 m spatial resolution with a return frequency of no better than 15 days. These factors, coupled with the problem of acquiring satellite imagery during cloudy days, limited the application of satellite imagery to precision farming during the 1990s.

Attention soon turned to using remote sensing to detect nitrogen deficiency in corn and other crops. Blackmer et al. (1995) used canopy reflectance measurements of single leaves with a spectroradiometer to confirm previous research by Walburg et al. (1982) that showed an increase in reflectance in the green spectrum (550 nm) with nitrogen stress. Blackmer et al. (1995, 1996a) then used black and white aerial photography of stressed and unstressed corn plants in Nebraska with a camera that filtered all light except green. Results showed an excellent relationship between canopy reflectance and grain yield across a large range of crop nitrogen stress. In a companion paper, Blackmer and Schepers (1996b) showed that crop nitrogen stress was accurately detected using the brightness of red in a digitized aerial color photograph of a cornfield in Nebraska. Bausch and Duke (1996) developed a green vegetation index (GVI) defined by the ratio of near-infrared to green (NIR/G) reflectance that accurately predicted differences in nitrogen stress for irrigated corn in Colorado. Research results showing that remote sensing could accurately identify areas of nitrogen stress in crops led directly to the development of proximal sensors (described in Section 1.13) for precision management of crop nutrient deficiencies.

Until launch of the commercial Ikonos satellite in 1999, there were few instances where satellite remote sensing was used for precision farming applications (Mulla 2013). Ikonos collected reflectance data using blue, green, red, and near-infrared bands at 1–4 m spatial resolution with a return frequency of 3 days, leading to immediate applications in precision farming such as diagnosis of crop nitrogen stress, fungal infestations, and soil drainage problems (Seelan et al. 2003). A second commercial satellite, Quickbird, launched in 2001, collected reflectance in the blue, green, red, and near-infrared bands at 0.6–2.4 m spatial resolution with a return frequency of 1–4 days. Quickbird normalized green normalized difference vegetation index (NGNDVI) data were used by Bausch and Khosla (2010) to identify locations experiencing crop N stress.

High-resolution, high-return-frequency commercial satellites launched from 2008–2009 included RapidEye, GeoEye1, and WorldView2. These satellites have spatial resolutions ranging from 6.5 to 0.5 m (Mulla 2013) and return frequencies ranging from 5.5 days to 1.1 days. Thus, they are highly suitable for applications in precision farming. More interestingly, these satellites offer additional spectral bands in comparison with earlier satellites. RapidEye collects reflectance data in the blue, green, red, red-edge, and near-infrared bands. Red-edge reflectance is highly sensitive to the chlorophyll status of growing crops. GeoEye1 collects reflectance data in the blue, green, red, and two near-infrared bands. The images in Google Earth are commonly obtained using GeoEye1 imagery. WorldView2 collects imagery in purple, blue, green, yellow, red, red-edge, and near-infrared bands. Despite the improvement in return frequencies for commercial satellites, difficulties persist in acquiring satellite imagery when needed due to cloud cover and competition for imagery among civilian and military users.

Remote sensing has been used in precision farming for a variety of purposes, including estimating spatial variability in soil organic matter (Bhatti et al. 1991; Mulla 1997; Fleming et al. 2004), in crop yield (Yang et al. 2000; Boydell and McBratney 2002; Garcia Torres et al. 2008), in crop water stress (Barnes et al. 1996; Meron et al. 2010; Rud et al. 2014), in insect infestations (Franke and Menz 2007; Prabhakar et al. 2011), in crop disease (Muhammad 2005; Huang et al. 2007; Mirik et al. 2011), and in weed infestations (Zwiggelaar 1998; Lamb and Brown 2001; Thorp and Tian 2004; Lopez-Granados 2011). By far the most common application of remote sensing in precision farming, however, is for detection of spatial and temporal patterns in crop nutrient deficiencies (Bausch and Duke 1996; Haboudane et al. 2002; Miao et al. 2007, 2009; Tremblay et al. 2012; Nigon et al. 2014).

In many of these applications, various combinations of spectral bands known as spectral indices are used to detect the property of interest (Haboudane et al. 2002, 2004; Thenkabail 2003; Mulla 2013). The spectral index most commonly used when a growing crop is present is the normalized difference vegetative index (NDVI) (Rouse et al. 1973), which is based on the sharp contrast in reflectance between the red and near-infrared portions of the spectrum. Plant pigments absorb radiation in narrow wavelength bands centered around 430 nm (blue or B) and 650 nm (red or R) for chlorophyll a and 450 nm (B) and 650 nm (R) for chlorophyll b. Wavelengths with low absorption characteristics conversely have high reflectance, particularly in the green (550 nm) wavelength. Remote sensing of crops in the near-infrared spectrum (particularly at 780, 800, and 880 nm) responds to crop canopy biomass and leaf area index (LAI), leaf orientation, and leaf size and geometry. NDVI has been used to detect crop nutrient deficiencies, patterns in crop yield, insect and weed infestations, and crop diseases (Mulla and Miao 2015). NDVI values are often not a good indicator of crop status due to either interference from bare soil reflectance or to insensitivity to changes in leaf chlorophyll in closed canopy crops when leaf area index values exceed 2 or 3 (Thenkabail et al. 2000).

As a result, there has been significant research effort devoted to finding broadband multispectral indices that can be used as an alternative to NDVI (Sripada et al. 2006, 2008; Miao et al. 2009). In general, there are three classes of broadband multispectral indices used in precision farming. These include soil-adjusted vegetation indices, ratios of green and near-infrared reflectance bands, and ratios of red and near-infrared reflectance bands (Thenkabail 2003; Mulla 2013). Soil-adjusted vegetation indices reduce reflectance from bare soil that interferes with the interpretation of reflectance from a growing crop before canopy closure. Red ratio indices typically are sensitive to absorption of radiation by leaf chlorophyll, while green ratio indices are sensitive to leaf pigments other than chlorophyll. In commonly used red and green ratio indices, either the red or green or the near-infrared reflectance can appear in the numerator of the ratio.

Hyperspectral remote sensing data involves the collection of reflectance data over the entire visible and near-infrared spectra in narrowbands typically of 10 nm or narrower width. In contrast, multispectral data typically involves reflectance in broadbands, 50 nm or wider, centered in the blue, green, red, and lower near-infrared portion of the spectrum. All of the broadband spectral indices calculated with multispectral data can be calculated as narrowband spectral indices using hyperspectral imaging. The advantage of doing this is that specific plant or soil responses that would be obscured by other plant or soil reflectance characteristics using broadband multispectral imagery become clear with narrowband hyperspectral imagery. For example, plants contain different pigments whose light absorption peaks at specific narrow wavelengths. Plant pigments such as chlorophyll strongly absorb radiation, particularly at wavelengths such as 430 (blue or B) and 660 (red or R) nm for chlorophyll a and 450 (B) and 650 (R) nm for chlorophyll b (Pinter et al. 2003). Other plant pigments such as anthocyanins and carotenoids absorb strongly at different wavelengths (Blackburn 2007). Crop reflectance also responds to changes in crop biomass, LAI, canopy structure, and leaf density in the red and near-infrared wavelengths. A narrow red band centered at 687 nm is sensitive to crop LAI and biomass, while a narrow near-infrared band centered at 970 nm is sensitive to crop moisture status (Thenkabail et al. 2010). Further examples of linking specific soil and crop characteristics with narrowband reflectance are given by Thenkabail et al. (2010). Hyperspectral imaging can be used to estimate narrowband spectral indices that have no analog in multispectral imagery. For example, several researchers have used red-edge reflectance in the spectral region between 700 and 740 nm to construct spectral indices that are sensitive to crop nitrogen status (Guyot et al. 1988; Datt 1999; Clarke et al. 2001; Haboudane et al. 2002; Gitelson et al. 2005; Fitzgerald et al. 2010; Shiratsuchi et al. 2011).

Interest is growing in the use of low-altitude unmanned aerial vehicles (UAVs) as a platform for remote sensing in precision farming (Zhang and Kovacs 2012). Imagery collected with UAVs can be high enough in resolution to view individual plants and leaves, although images at such high resolution have to be mosaicked to obtain complete coverage of a field. UAVs have been used to assess crop LAI, biomass, plant height, nitrogen status, water stress, weed infestation, and yield and grain protein content (Berni et al. 2009; Swain et al. 2010; Samseemoung et al. 2012; Bendig et al. 2013). Current limitations to using UAVs include governmental restrictions on their usage, light payloads, low power, and limited flight times.

1.13   PROXIMAL SENSING OF SOILS AND CROPS

Proximal sensing has been widely used in precision farming to map spatial patterns in soil or crop properties. Early advances in proximal soil sensing were initially based on geophysical prospecting techniques that were used to discover mineral reserves buried deep in the earth (Parasnis 1973). Two categories of geophysical prospecting techniques have been adapted for proximal sensing of soil in precision farming: electrical resistivity/conductivity methods and electromagnetic induction methods. Halvorson and Rhoades (1974) adapted electrical resistivity mapping methods to the problem of mapping soil salinity in agricultural fields based on the four-probe Wenner array developed in the mining industry. Wenner array probes were simply metal spikes inserted in soil along a straight line at fixed spacing. A battery supplied current to the soil through two of the spikes, while the other two served as voltage probes. The depth of measurement could be controlled by varying the spacing between metal electrodes. Carter et al. (1993) built on the research by Halvorson and Rhoades (1974) to pioneer continuous mobile electrical conductivity measuring equipment for soil salinity mapping. The mobile apparatus consisted of a battery attached to four equally spaced chisel blades mounted on a tractor. This apparatus was the inspiration for the Veris electrical conductivity mapping system (Christy and Lund 1998) based on equally spaced electrode disks that is widely used in precision farming. Colburn (1991) patented a device for a resistivity-based sensor mounted behind a moving fertilizer spreader that was claimed to accurately vary fertilizer rate in response to differences in soil nitrate-N concentrations, soil cation exchange capacity, organic matter content, and soil moisture. This device, called Soil Doctor, was widely marketed for applications in precision farming, although many scientists were skeptical of its accuracy in the absence of rigorous scientific testing.

There were several drawbacks of the Wenner array of electrodes for soil salinity mapping. First was the difficulty of ensuring good contact between electrodes and dry soil, and second was the need to eliminate site-specific calibration of electrical resistivity measurements with soil samples that were analyzed in the laboratory. To overcome these limitations, Rhoades and Corwin (1981) began working with Geonics Ltd. of Canada, who were commercial suppliers of electromagnetic induction probes for geophysical prospecting in the mining industry. Rhoades and Corwin (1981) suggested the development of a noncontacting electromagnetic induction probe that could be used specifically for shallow sensing of soil materials, and this suggestion led to the development of the EM-38 electromagnetic induction probe (Figure 1.6) that is widely used in precision farming (Lesch et al. 1992; Doolittle et al. 1994; Sudduth et al. 1995; Kitchen et al. 1999, 2003). Electromagnetic induction is a process where an electrical field generated above ground induces current loops in the soil that are proportional in magnitude to the soil's electrical conductivity. The primary current loops in the soil induce a secondary electromagnetic field whose strength is proportional to the current flowing in the loops. This secondary field is then measured by a receiver above ground to estimate soil electrical conductivity.

The commercially available EM-38 electromagnetic induction unit from Geonics Ltd. of Canada became a preferred tool to map soil salinity (Corwin and Lesch 2003). Lesch et al. (1992) reported the advantage of using an EM-38 unit that enhanced their ability to accurately predict spatial soil salinity patterns with 60% to 90% fewer soil samples. They concluded that EM-38 readings were a more practical and cost-effective tool for accurate mapping of spatial salinity patterns at the field scale than soil sampling. Likewise, Doolittle et al. (1994) in Missouri were able to quantify and map variations in the depth to claypan soils that restrict infiltration, influence the lateral movement of soil water and agrichemicals, and limit crop production. They found that EM techniques were noninvasive, less labor-intensive, more economical, and could produce large quantities of data in a relatively short period of time. Sudduth et al. (1995) continued the work of Doolittle et al. (1994) and found that by automating the process of collecting EM-38 data, they could map variations in soil properties over large areas for site-specific nutrient management. Kitchen et al. (1999) studied the relationships in spatial maps from EM-38 data and grain yield monitors. They found a significant relationship between yield maps and apparent electrical conductivity maps in nine out of 13 site years of data. Later, Kitchen et al. (2003) measured apparent soil electrical conductivity using a Veris electrical conductivity unit developed by

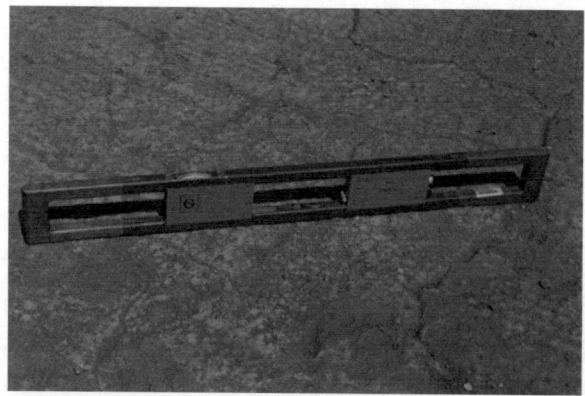

FIGURE 1.6 First commercial unit (circa 1980) of the Geonics EM-38 single dipole electromagnetic induction conductivity meter. (Courtesy of Dennis Corwin.)

Christy and Lund (1998) across three states (Missouri, Kansas, and Colorado), under four different crops (maize, wheat, soybean, and sorghum), and indicated that sensor-based soil information can greatly assist farmers in understanding yield variations for planning management decisions in precision farming.

Early proximal sensing in precision farming was also focused on the use of *in situ* reflectance methods to assess spatial patterns in soil organic matter content (Shonk and Gaultney 1988; Sudduth et al. 1991). Soil organic matter content is often correlated with other soil properties such as moisture content and nitrogen mineralization, each of which can affect potential crop yield. In addition, certain classes of crop protection chemicals are adsorbed by soil organic matter content. In either case, varying the rate of fertilizer or pesticide according to levels of soil organic matter content was the motivation for developing proximal sensing techniques for soil organic matter content. Gaultney et al. (1991) patented a device to measure soil organic matter content in real time that was based on laboratory calibration curves that depended on soil texture. The device consisted of an array of red-light-emitting diodes surrounding a photodiode sensor, both of which could be mounted on a chisel blade dragged by a tractor. McGrath et al. (1990) used the organic matter sensor to vary rates of herbicide applied in Midwestern fields, with good success at controlling weeds. Sudduth et al. (1991) developed an organic matter sensor that was based on near-infrared reflectance (Figure 1.7). When tested in the laboratory, the sensor was able to accurately detect differences in SOM content even when soil moisture content varied. However, field trials were initially unsatisfactory because reflectance values were sensitive to soil roughness (Hummel et al. 1996).

Electrochemical sensing of soil chemical properties was an important emphasis in early precision farming research (Colburn 1991; Adsett and Zoerb 1991; Birrell and Hummel 1993). Adsett and Zoerb (1991) adapted an ion-selective electrode that measured nitrate-N concentrations in soil solution to a real-time proximal sensor. A rather cumbersome apparatus was developed that sampled soil, mixed and stirred it with water, and then used the ion-specific electrode to measure nitrate concentrations on the go. The sample container was then dumped in the field and rinsed out automatically before taking another soil sample. Birrell and Hummel (1993) tested an ion-sensitive field effect transistor (ISFET) for nitrate-N concentration measurements in the laboratory. They found that samples could be analyzed every 1.5 seconds through a cycle that involved sample injection and washing of the sample container. ISFET sensors are smaller and have a faster response and higher signal-to-noise ratio than ion-specific electrodes. However, they also have greater electronic drift, requiring frequent recalibration. Flat surface ion-selective probes were developed by Adamchuk et al. (1999) and patented (Adamchuk et al. 2002) for real-time automated mapping of soil pH,

FIGURE 1.7 Soil organic matter sensor based on NIR reflectance. (From Sudduth, K. A. et al., Soil organic matter sensing: A developing science. In: G. A. Kranzler (ed.), *Automated Agriculture for the 21st Century.* ASAE Publication 11–91, St. Joseph, MI, pp. 307–316, 1991.)

while Adamchuk et al. (2003) developed ion-selective probes for real-time automated mapping of soil nitrate and potassium levels.

Interest in proximal sensing of crops is not new. Scientists and farmers alike will visually inspect crops to nondestructively estimate crop health status. Their visual inspection may often lead to a management decision such as fertilizer, irrigation, or pesticide application. Human eyes are in fact a pair of reflectance-based optical sensors. However, human eyes require experience to discern subtle differences in crop appearance due to various biotic and abiotic stresses. Machine-based optical sensors, on the other hand, can be used repeatedly without bias or need for experience. In the early 1970s, Leamer and his coworkers at the USDA-ARS unit in Weslaco, Texas, and his colleague Silva from the electrical engineering department at Purdue University, recognized the need for an instrument that would scan a specific field target through the visible and thermal infrared spectrum (Leamer et al. 1973). They designed and developed an instrument and provided specifications and plans to Exotech Inc. to build a sensor that could be used in fields for crop sensing. The instrument was built by Exotech, who also contributed many engineering concepts and the electronic circuits required to measure the reflected or emitted energy and to produce an electrical signal proportional to the energy detected by the instrument (Leamer et al. 1973).

The instrument (Exotech model 20-B Spectroradiometer) consisted of two systems, each made up of an optical unit and a control unit. One system covered the spectral range 0.37–2.52 μ; the other system covered the spectral range 2.76–13.88 μ. The optical units of the two systems were mounted side by side on a tiltable base mounted on an aerial lift truck. Separation between the objective lenses was about 30 cm to minimize parallax. The system was configured such that it can be operated separately or in the tandem, boresighted mode (Leamer et al. 1973). The control units were mounted in a camper-type equipment van and were connected to the optical units by 60 m of armored cable. Preamplifiers and auxiliary electronics in the optical units were designed to operate without picking up interference over this length of cable. The entire system was designed to operate in an outdoor environment.

The Laboratory for Applications of Remote Sensing (LARS) at Purdue University was among the early pioneers and leaders in advancing the science of remote sensing and its applications in agriculture. Much of the early work in the area of crop sensing came from LARS under the leadership of Bauer, Baumgardner, and many of their graduate students (Al-Abbas et al. 1974; Bauer 1975; Ahlrichs and Bauer 1978; Kumar et al. 1979; Daughtry et al. 1980; Walburg et al. 1982).

The 1980s witnessed the introduction of optical sensing into agriculture as a complementary monitoring tool to estimate crop health, growth, abiotic stresses, and crop yield (Bell and Xiong 2007). Daughtry et al. (1980), using an advanced version of Exotech Spectroradiometer (Exotech 100A), investigated the effects of various management practices (soil moisture, planting date, nitrogen fertilizer, and cultivar) on spring wheat crop canopies. They suggested that LAI, biomass, and percent soil cover can potentially be monitored by crop canopy sensing. Likewise, Walburg et al. (1982) studied the effects of N rates on growth, yield, and reflectance characteristics of corn canopies and reported that spectral reflectance differed significantly with N rate. Wanjura and Hatfield (1987) reported that vegetation indices had greater sensitivity to plant vegetation reflectance than did the reflectance of a single wavelength. Early work that utilized crop canopy sensing was primarily done to understand cause and effect relationships and to identify particular wavebands, their simple ratios, or various vegetative indices that were most effective in distinguishing the treatments of interest (Chappelle et al. 1992; Blackmer et al. 1994; Filella et al. 1995). Reflectance near the 550-nm wavelength was found to be the best at distinguishing nitrogen treatments in soybean (Glycine max [L.] Merr; Chappelle et al. 1992). Blackmer et al. (1994) reported similar findings for nitrogen in corn canopies. Around the same time, Filella et al. (1995) reported that reflectance at 550 and 680 nm was significantly correlated with canopy chlorophyll content across five N treatments in wheat canopies.

Around the same time, engineers Solie and Stone and agronomist Raun at Oklahoma State University were working on developing rapid ways of estimating leaf nitrogen concentrations and crop biomass using sensing technology (Solie et al. 1996). Stone developed a sensor with two detectors working in synchrony; one measured the incoming radiance and the other faced the plant canopy and measured the reflected radiance (Table 1.5). The irradiance reflected was divided by the incoming irradiance to determine reflectance (Bell and Xiong 2007). One of the limitations among the crop sensing devices of the 1980s and 1990s was that they were passive sensors (i.e., without their own sources of light energy). Hence, they were limited by the time of day when sensor readings were acquired in the field with and without cloud cover, and the need for constant white plate calibration (Rutto and Arnall 2009). To address this limitation, in 1998, the Oklahoma team joined hands with Mayfields (future founders of N-Tech Industries and manufacturers of the GreenSeeker™ sensor), who owned much of the relevant intellectual properties for active optical sensors, which they purchased from John Deere & Company, Moline, IL (Rutto and Arnall 2009). The team felt that the development of active sensors (sensors that have their own source of light energy) was necessary for such sensing devices to be incorporated into in-field and in-season precision nutrient recommendations.

The year 2002 marked the commercial release of the GreenSeeker active sensing device (Raun et al. 2002) by N-Tech Industries (Ukiah, CA). Since then other proximal crop-sensing devices such as Crop Circle active sensor (Holland Scientific, NE), the Yara N-Sensor ALS (active light sensor) by Yara International, Norway; and Isaria Crop Sensor by CLAAS, Germany have become commercially available. These reflectance-based active sensors are being widely used to guide variable rate nitrogen fertilizer applications around the world (Shanahan et al. 2008; Samborski et al. 2009; Kitchen et al. 2010).

Initially, the majority of the commercially available active sensors were reflectance-based sensors and were providing either NDVI or a modification of the NDVI readings. However, more recently there are fluorescence-based active sensors that are commercially available such as the Multiplex Fluorescence sensor by Force-A, France, and the MiniVeg laser-induced chlorophyll fluorescence sensor by Fritzmeier, Germany. Fluorescence sensors utilize either ultraviolet or laser light sources, or both, to stimulate a plant's chlorophyll to emit fluorescent light. Green plants emit fluorescence in the blue-green (440–520 nm) and in the red to far-red (690–740 nm) regions of the light spectrum when excited with a potent light source (Buschmann et al. 2000). Different combinations of red and far-red fluorescence ratios obtained with different excitation wavebands can be used to calculate a multitude of indices for the plant status (Tremblay et al. 2011). A fluorescence-based index called

---

TABLE 1.5
Early Research Advances in Remote Sensing, Proximal Sensing of Soils, and Proximal Sensing of Crops for Precision Farming

Research Area	Nature of Contribution	Key References
Aerial remote sensing	Ability to identify spatial patterns, identification of sensitive wavelengths	Robert (1982), Zheng and Schreier (1988), Bhatti et al. (1991), Blackmer et al. (1995), Barnes et al. (1996), Bausch and Duke (1996), Blackmer and Schepers (1996b), Haboudane et al. (2002), Thenkabail (2003)
Proximal sensing of soils	Development and testing of technology	Rhoades and Corwin (1981), Shonk and Gaultney (1988), Colburn (1991), Gaultney et al. (1991), Sudduth et al. (1991), Lesch et al. (1992), Carter et al. (1993), Doolittle et al. (1994), Christy and Lund (1998), Adamchuk et al. (1999)
Proximal sensing of crops	Development and testing of technology	Leamer et al. (1973), Daughtry et al. (1980), Walburg et al. (1982), Chappelle et al. (1992), Solie et al. (1996), Raun et al. (2002)

the nitrogen balance index (NBI) was developed to detect N variability by the ratio of chlorophyll content to the flavonoid content (Cartelat et al. 2005). In a recent study, Longchamps and Khosla (2014) reported early detection of nitrogen variability at the five-leaf growth stage of maize using the Multiplex 3 fluorescence sensor. They also reported that the fluorescence readings were not influenced by soil background noise when used at the recommended height of measurement. Such findings are significant for precision farming and are leading the way to enhance our ability to manage nutrients more efficiently. Interestingly, fluoresensing is not new. Lorenzen (1966) demonstrated that chlorophyll fluorescence can be used to assess plant chlorophyll content and photosynthetic activity through the state of photosystem II. However, such measurements were made in a dark chamber on a potted plant by inducing fluorescence with a laser beam. Today, similar measurements can be acquired using commercially available fluorescence sensors in broad daylight and in motion above the crop canopy. One can therefore envision that in the near future as technology and science continue to progress, scientists and farmers will be working with many innovative proximal crop sensors that are yet to be developed and commercialized to further the goals of precision farming.

1.14   ENVIRONMENTAL BENEFITS OF PRECISION FARMING

Precision farming allows for variation in the rate of applied fertilizer, manure, and pesticides to better match spatial patterns in soil fertility and pesticide adsorption, and to respond to changing temporal patterns in crop nutrient stress and infestations of weeds, insects, and disease. In addition, with autosteer technology, precision farming reduces overapplication of chemical inputs due to overlap between successive passes of chemical applicator machinery. All of these factors lead, conceptually, to improved environmental quality and sustainability (Larson et al. 1997; Bongiovanni and Lowenberg-DeBoer 2004).

The use of precision farming to reduce the risks of pesticide leaching to groundwater in sandy soils were first studied by Mulla et al. (1996) at a field site in Washington State. Measured concentrations in carbofuran applied at 8.1 kg ha^{-1} a.i. were measured to a depth of 1.8 m at 57 locations throughout the field and this data wad used to calibrate the convective-dispersive equation for pore water velocity, dispersion coefficient, and retardation factor. Results showed that there was significant spatial variability in pesticide leaching risks. Over half of the field was at high risk for leaching losses, and these losses could be significantly reduced by applying low rates of carbofuran in high-risk leaching areas. Spatial patterns in leaching risk were controlled primarily by variations in water movement rather than variations in soil organic matter content, making the *a priori* identification of high-risk leaching areas challenging.

Khakural et al. (1994, 1999) and Mulla et al. (2002) measured the impact of variable rate herbicide applications on herbicide losses in runoff, tile drainage, and eroded sediment for a fine textured soil in Minnesota. These studies showed that variable rate herbicide applications based on spatial patterns in soil organic matter content, soil pH, or weed populations resulted in significant reductions of herbicide applied to soil and subsequent reductions in the amount lost to surface waters relative to uniform herbicide applications. A number of other researchers showed that spatial variability in soil properties and weed populations could be used as the basis for variable rate herbicide applications that could lead to significant reductions in the amount of herbicide applied to fields (Mortensen et al. 1995; Johnson et al. 1997).

Mixed results have been found by a number of researchers concerning the environmental benefits of variable rate nitrogen fertilizer applications using either assessments of residual soil N and nitrogen use efficiency (NUE) or simulation modeling (Bongiovanni and Lowenberg-Deboer 2004). Hergert et al. (1996) compared uniform versus variable rate applications of N fertilizer in irrigated corn in Nebraska. They found that residual N concentrations in soil were significantly lower for variable rate N applications than uniform applications, but there were no differences in NUE for either strategy. On the other hand, Redulla et al. (1996) found no significant differences in residual soil N or NUE between uniform and variable rate N applications in irrigated Kansas corn fields.

Whitley et al. (2000) studied N leaching in irrigated potatoes grown in Washington State using either uniform or variable rate N applications. They found that N leaching was reduced in lower landscape positions with variable rate N application relative to uniform application.

A variety of simulation models have been used to assess the benefits of variable rate N applications on water quality. Larson et al. (1997) used Leaching Estimation and Chemistry Model (LEACHM) in rainfed corn produced in Minnesota to show that N leaching was reduced with variable rate N applications relative to uniform applications on a loamy sand, but not on a loamy soil. The Environmental Policy Integrated Climate (EPIC) model was used by Rejesus and Hornbaker (1999) and by English et al. (1999) to show that N losses to surface waters were reduced using variable rate N application in Illinois and Tennessee, respectively. Delgado et al. (2005) used the Nitrogen Loss and Environmental Assessment Package (NLEAP) model in Colorado to show that N leaching in irrigated corn could be reduced using variable rate N applications. Variable rate N applications were particularly effective at reducing N leaching losses in management zones with low crop productivity where crop uptake of N was limited.

1.15 PROFITABILITY OF PRECISION FARMING

Since the inception of precision farming, scientists, farmers, and practitioners alike have questioned the economic feasibility of precision farming. The perceptions among the early users of precision farming were (1) that it is technologically and resource-intensive and time consuming and hence it would be cost-prohibitive, and (2) that it primarily increases system efficiencies and not necessarily output (Napier et al. 2000). These perceptions were further strengthened by the limited number of early studies evaluating agronomic and economic gains of precision farming (Table 1.6). Hammond (1993) reported that variable management is unlikely to cause increases in crop yield or quality in high or intermediate fertility areas of the field. Furthermore, a majority of those early studies reported that precision farming was not profitable or that profitability was mixed at best (Lowenberg-DeBoer and Boehlje 1996). Table 1.6, adopted from Lowenberg-DeBoer and Boehlje (1996), summarizes the findings of 11 early studies evaluating the economics of precision farming. Five of the 11 studies reported precision farming as not being profitable, four studies reported mixed or inconclusive results, and only two studies showed potential profitability. They attributed the

---

TABLE 1.6
Profitability Findings from Eleven Precision Farming Studies

Study and Year	Crop(s)	Conclusion: Precision Farming Profitable or Not?
Beuerlein and Schmidt (1993)	Corn, soybean	No
Carr et al. (1991)	Wheat, barley	Mixed
Fiez et al. (1994)	Wheat	Yes, potentially
Hammond (1993)	Potato	Inconclusive
Hayes et al. (1994)	Corn	Yes, potentially
Hertz and Hibbard (1993)	Corn	No
Lowenberg-DeBoer et al. (1994)	Corn	No
Mahaman (1993)	Corn	Mixed
Wibawa et al. (1993)	Wheat	No
Wollenhaupt and Buchholz (1993)	Corn	Mixed
Wollenhaupt and Wolkowski (1994)	Corn	Mixed

Source: Modified and adapted from Lowenberg-DeBoer, J., and M. Boehlje, Revolution, evolution or dead-end: Economic perspectives on precision agriculture. *Proc. 3rd Intl. Conf. Precision Agriculture*, pp. 923–944, 1996.

failure to demonstrate economic gains to two primary reasons: (1) these studies compared whole-field gains to precision farming treatments, and (2) the findings depended heavily on the cost of sampling and variable rate applications that exceeded any gains made in yields. Lowenberg-DeBoer and Boehlje (1996) argued that whole-field profit maximizing conditions differ from the site-specific conditions and that there has to be a more economical, less expensive way of preparing variable rate prescription maps for precision farming. In a follow-up study, Lowenberg-DeBoer and Swinton (1997) suggested that many previous studies failed to indicate whether problems with the technology or with its management were behind the apparent low profitability findings.

The limitations identified by Lowenberg-DeBoer and Boehlje (1996) and others (Khosla and Alley 1999) highlighted the need to develop techniques to replace expensive grid soil sampling for the purpose of variable rate management of crop and soil inputs. Khosla et al. (2002) reported significant agronomic gains from using site-specific management zones to variably apply nitrogen in comparison with conventional uniform applications of nitrogen across the entire field. A more detailed, in-depth study on economic feasibility by Koch et al. (2004) clearly documented for the first time significant economic gains from using precision farming. An increase in net return was reported ranging from $18.21 to $29.57 ha^{-1} higher for variable-rate N application than for uniform N management (Koch et al. 2004). Interestingly, most studies in the late 1990s and early 2000s focused on the precision nutrient management aspects of precision farming, logically so, because this was the first precision farming technology that was technically feasible (Lowenberg-DeBoer and Swinton 1997).

In the following years, novel technologies and improved management techniques continued to accelerate the adoption of precision farming, including, but not limited to, precision autoguidance, lightbar, autopilot systems, active crop sensing, precision irrigation systems, and smart sampling (Jones 2004; McClure 2005; Inman et al. 2007; Shaner et al. 2008). Griffin et al. (2005) investigated the economics of using a lightbar and autoguidance GPS navigation technologies and found that a high-precision real-time kinematic (RTK) GPS system was less profitable to operate compared with no GPS use on a farm. However, when a farmer begins using GPS, its profitability increases with the level of GPS precision (Griffin et al. 2005). They found that the lightbar was the most profitable navigation technology; however, as farm size increases, RTK-GPS becomes profitable and it outranks lightbar-GPS-based precision farming operations (Griffin et al. 2005). More recently, Shockley et al. (2011, 2012) demonstrated that autosteer technologies can influence machinery selection and replacement decisions, increase net returns, and reduce production risk. This is even more significant because their work incorporated over 500 real-world cropland fields from farms in Colorado, Kansas, and Nebraska.

The economics of precision farming continue to improve as technologies and management techniques improve. However, a major factor that previously has and will continue to greatly influence the economics of precision farming is commodity prices. During the last decade we have witnessed substantial fluctuations in commodity prices. For example, corn prices have gone from approximately $80 Mg^{-1} in early 2000 to over $200 Mg^{-1} recently (USDA-ERS 2014). Such increases in commodity prices are a welcome relief for farmers and have resulted in higher economic gains and investment of such gains back into innovative technologies and techniques, including precision farming. However, recent declines in commodity prices may start to reverse the growing adoption of precision farming.

1.16 ADOPTION OF PRECISION FARMING

Precision farming has been termed the most significant innovation introduced to U.S. agriculture in the mid- to late 1980s (Napier et al. 2000). Yet its adoption rate, at least for the first 10 to 15 years, did not meet expectations. Yapa and Mayfield (1978) indicated a number of reasons that could lead to nonadoption of an innovation. These include, but are not limited to, lack of sufficient information, lack of favorable attitude, lack of economic means to acquire technology, and lack of physical availability of technology. Nowak (1992) narrowed the reasons down to two major factors: willingness and ability to adopt. Rogers (1995) argued that adoption of technological innovations will not occur unless potential adopters become aware that a problem exists that can be resolved in a more efficient manner by new technologies. The early practitioners of precision farming highlighted the problem they were trying to address. However, the perception that it amounts to mere improvement in efficiency of farming was not enough to attract adopters. Nowak (1992) emphasized that the willingness to adopt depends on access to information about the technology, confidence in that information, and belief that the information provides a basis for favorable outcomes. Fairchild and Duffy (1993) suggested that profitability is one favorable outcome desired by virtually all producers in a market economy and it is a necessary condition for extensive, voluntary technology adoption. It is therefore no surprise that the pace of adoption of precision farming techniques and technologies was slow at best until the mid- to late 1990s because there was no clear evidence of economic feasibility or profitability associated with precision farming (Lowenberg-DeBoer and Boehlje 1996). By early 2000, it was widely recognized among the precision farming community that for adoption of precision farming to occur, the technology needed to be relevant in addressing problems, be easy to use, and above all be profitable to farming (Napier et al. 2000; Kitchen et al. 2002). Batte and Arnholt (2003) examined case studies of six leading-edge adopters and found that profitability was the biggest motivating factor in using precision agriculture tools.

The first decade of the twenty-first century saw accelerated adoption of precision farming. Over the past decade, the adoption and use of different precision agricultural technologies among producers and commercial agricultural businesses increased steadily (Whipker and Akridge 2006). There are a number of reasons that can be attributed to the change in pace of adoption of precision farming techniques and technologies over the past decade: (1) certain components of precision farming such as precision nutrient management have been thoroughly tested and evaluated and were being reported to enhance efficiency, productivity, and profitability in an environmentally responsible manner (Franzen et al. 2000; Khosla et al. 2002; Koch et al. 2004), (2) positive change in commodity prices that increased multifold over the last 10 years (USDA-ERS 2014), which led to higher net returns to farmers and their respective investment into advanced technologies, (3) introduction of new technologies such as autoguidance systems that in addition to improving efficiencies and profitability, reduced farmers' fatigue and allowed them to work longer hours (Griffin et al. 2005), and (4) enhanced capacity and availability of a skilled labor force, trainers, and practitioners who understood both technology and agriculture. While these may only serve as a partial list of factors responsible for the enhanced level of adoption over the last decade, it does reflect on bridging the gap in many areas as pointed out by Nowak (1992), Fairchild and Duffy (1993), and Kitchen et al. (2002).

Unique partnerships among industries, the media, and academic institutions have enabled monitoring and documentation of change, progress, and adoption of precision farming among U.S. farmers. For the past 15 years, Purdue University in partnership with CropLife Media and recently Trimble Inc., has employed a survey tool to quantify changes in precision technologies, their respective adoption rates, and predicted future trends by agricultural retailers, practitioners, and farmers. Figure 1.8 reflects a summary of the most recent survey conducted in 2011 (Whipker and Erickson 2013). It lists the top 11 precision technologies that have gained popularity and adoption among farmers, agricultural retailers, and practitioners. The top two are both related to precision guidance that allow farmers to increase speed of field operations, work longer days, provide greater flexibility in hiring labor, have a more appropriate placement of inputs, and reduce overlap or chemical use (Griffin et al. 2005). It is interesting to note that some of the latest innovations such as crop and soil sensing devices that are still undergoing extensive research across academic institutions are being adopted by very few farmers and practitioners. Overall it is encouraging that about 60% of the agricultural retailers surveyed are offering precision farming services to their clientele. Review of the survey reports from previous years (data not presented) indicates that precision farming technologies and its adoption continue to grow among farmers.

While some may term the adoption of precision farming as slow, it is still remarkable given that precision farming was a new concept only about 25 years ago. In contrast, machine harvesters and

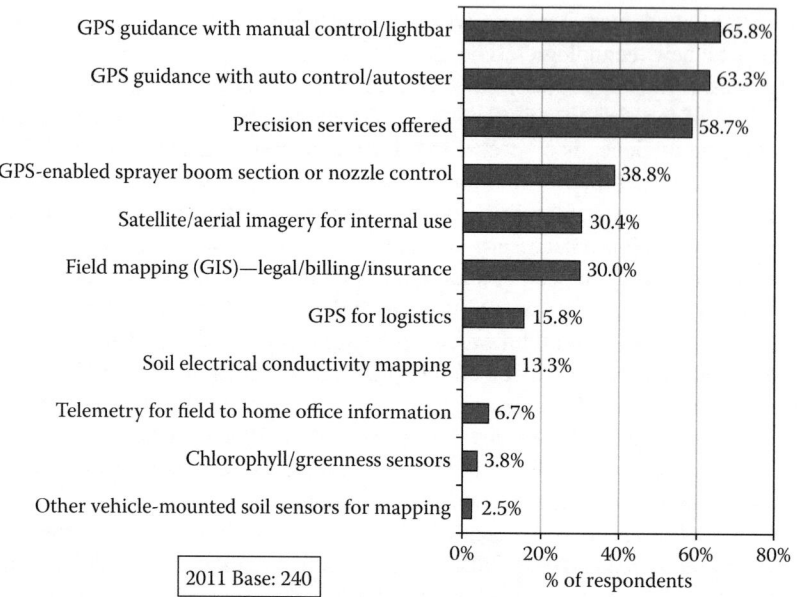

FIGURE 1.8 Summary of a nationwide survey reflecting the use of precision technologies across agricultural states in the United States. (Modified and adapted from Whipker, L. D., and B. Erickson, 2011 Precision agriculture services dealership survey results. Staff paper, Department of Agricultural Economics, Purdue University. W. Lafayette, IN, 2013.)

combines that are used today by most farmers in North America and elsewhere were invented in the early 1800s, commercialized in the late 1800s, and it took over 50 years or more since commercialization before they were used in large numbers on agricultural farms (Wikipedia 2014). Nevertheless, precision farming technologies continue to evolve and change and its impacts are reflected in overall production, efficiency, and environmental footprints of farming operations.

1.17 SUMMARY AND FUTURE TRENDS

The history of precision agriculture has shown that it is more strongly influenced by technological innovations rather than innovations in information analysis and decision support. For example, when first introduced, GPS and yield monitors were viewed as technological advances that could be added to existing farm equipment to add value. Later, agribusiness began embedding both GPS and yield monitors onto farm combines as part of the standard sales package. This combination of technology is now widely adopted by farmers so that it is used by practitioners of precision farming as much as by practitioners of conventional farming. The addition of GPS to farm equipment enabled many other technological breakthroughs in precision farming, such as autosteer, and furthermore, machine location was essential for variable rate fertilizer application technology.

In contrast, information analysis and decision support systems for deriving management zones or making variable rate recommendations have not largely been embedded in routine farm operations. In many cases these functions are performed by crop retailers, consultants, and agribusiness service providers for a fee. There seems to be a trend toward more focus on information analysis and decision support systems in precision agriculture (McBratney et al. 2005). In particular, large corporations and researchers are beginning to focus their attention on big data issues, involving combinations of spatially and temporally varying yield monitor, soil fertility, crop stress, and climate data. This data is overlain from many separate farming operations with a view toward identifying and modeling relationships with soil or landscape features that could be used to create knowledge that informs precision farming decisions. In general, the volume, variety, and value of large databases is increasing, while the scale at which management decisions are being visualized and implemented is becoming finer. Increasingly, there is likely to be a trend toward stronger reliance on forecasting precision farming operations based on short-term weather forecasts and expert system simulation models and delivering recommendations to farmers via the Internet and smartphones.

Within the technology realm, there is an increasing convergence between proximal sensing and robotics. Sensors mounted on aerial and ground robots are increasingly being used to scout for and mitigate damages caused by crop stress. Significant research efforts are being directed toward improved software algorithms that are devoted to improved navigation and coordination between swarms of aerial and ground robots deployed in large agricultural fields. However, this convergence between robotics and proximal sensing will not be successful without increased emphasis on information analysis and decision support systems that allow massive amounts of data collected with these technologies to be quickly and accurately turned into useful recommendations and management strategies. A wide range of analytic tools are increasingly being used for this purpose, including partial least squares analysis, neural network analysis, and computer vision.

The spatial and temporal resolution of remote sensing information has dramatically improved since the inception of precision farming. In the early years of precision farming, spatial resolutions of satellite data were on the order of 30 m, while temporal resolutions were on the order of weeks to months. Today, spatial resolutions are as good as several centimeters, while temporal resolutions are as good as a few days. With this level of spatial and temporal resolution, it is likely that precision farming practitioners will be able in the near future to develop customized management recommendations on a weekly basis for every single plant growing in their field.

REFERENCES

Adamchuk, V. I., M. T. Morgan, and D. R. Ess. 1999. An automated sampling system for measuring soil pH. *Trans. ASAE* 42:885–891.

Adamchuk, V. I., M. T. Morgan, and D. R. Ess. 2002. System and method for automated measurement of soil pH. U.S. Patent No. 6,356,830, issued March 12, 2002.

Adamchuk, V. I., E. Lund, A. Dobermann, and M. T. Morgan. 2003. On-the-go mapping of soil properties using ion-selective electrodes, pp. 27–33. In: J. Stafford and A. Werner (eds.), *Precision Agriculture*. Wageningen Academic Publishers, Wageningen, the Netherlands.

Adsett, J. F., and G. C. Zoerb. 1991. Automated field monitoring of soil nitrate levels, pp. 326–335. In: G. A. Kranzler (ed.), *Automated Agriculture for the 21st Century*. ASAE Publ. 11-91, St. Joseph, MI.

Ahlrichs, J. S., and M. E. Bauer. 1978. Relation of crop canopy variable to the multi-spectral reflectance of spring wheat. Tech. Report 072479, Laboratory for Applications in Remote Sensing, Purdue University, W. Lafayette, IN.

Al-Abbas, A. H., R. Barr, J. D. Hall, F. L. Crane, and M. F. Baumgardner. 1974. Spectra of normal and nutrient deficient maize leaves. *Agron. J.* 66:16–20.

Andrew, F. W. 1941. Automatic tractor control. U.S. Patent No. 2,259,193, issued October 14, 1941.

Assmus, R. A., P. E. Fixen, and P. D. Evenson. 1985. Detection of soil-phosphorus spatial variability through the use of semivariograms and strip sampling. *J. Fert. Issues* 2:136–143.

Auernhammer, H., and T. Muhr. 1991. GPS in a basic rule for environment protection in agriculture, pp. 394–402. In: G. Kranzler (ed.), *Automated Agriculture for the 21st Century*. ASAE Publ. 11-91, St. Joseph, MI.

Auernhammer, H., M. Demmel, T. Muhr, J. Rottmeier, and K. Wild. 1994. GPS for yield mapping on combines. *Comput. Electron. Agri.* 11:53–68.

Bae, Y. H., S. C. Borgelt, S. W. Searcy, J. K. Schueller, and B. A. Stout. 1987. Determination of spatially variable yield maps. *Am. Soc. Agric. Eng.* (microfiche collection).

Barnes, E. M., M. S. Moran, P. J. Pinter, and T. R. Clarke. 1996. Multispectral remote sensing and site-specific agriculture: Examples of current technology and future possibilities. *Precis. Agric.* 3:845–854.

Batte, M. T., and M. W. Arnholt. 2003. Precision farming adoption and use in Ohio: Case studies of six leading-edge adopters. *Comput. Electron. Agri.* 38:125–139.

Bauer, M. E. 1975. The role of remote sensing in determining the distribution and yield of crops. *Adv. Agron.* 27:271–304.

Bausch, W. C., and H. R. Duke. 1996. Remote sensing of plant nitrogen status in corn. *Trans. ASAE* 39:1869–1875.

Bausch, W. C., and R. Khosla. 2010. QuickBird satellite versus ground-based multi-spectral data for estimating nitrogen status of irrigated maize. *Precis. Agric.* 11:274–290.

Beck, J. L., and M. L. Kinter. 1998a. Photodetector circuit for an electronic sprayer. U.S. Patent No. 5,763,873, issued June 9, 1998.

Beck, J. L., and M. L. Kinter. 1998b. Agricultural implement having multiple agents for mapping fields. U.S. Patent No. 5,809,440, issued September 15, 1998.

Bell, G, E., and X. Xiong. 2007. The history, role and potential of optical sensing for practical turf management, pp. 641–660. In: M. Pessarakli (ed.), *Handbook of Turfgrass Management and Physiology.* CRC Press, Boca Raton, FL.

Bendig, J., A. Bolten, and G. Bareth. 2013. UAV-based imaging for multi-temporal, very high resolution crop surface models to monitor crop growth variability. *Photogrammetrie-Fernerkundung-Geoinformation* 6:551–562.

Berglund, S., and R. Buick. 2005. Guidance and automated steering drive resurgence in precision farming. *Precis. Agric.* 5:39–45.

Berni, J. A. J., P. J. Zarco-Tejada, L. Suárez, and E. Fereres. 2009. Thermal and narrowband multispectral remote sensing for vegetation monitoring from an unmanned aerial vehicle. *IEEE Trans. Geosci. Remote Sens.* 47:722–738.

Beuerlein, J., and W. Schmidt. 1993. Grid soil sampling and fertilization. Ohio State University Extension, Agronomy Technical Report 9302.

Bhatti, A. U., D. J. Mulla, and B. E. Frazier. 1991. Estimation of soil properties and wheat yields on complex eroded hills using geostatistics and Thematic Mapper images. *Remote Sens. Environ.* 37:181–191.

Birrell, S. J., and J. W. Hummel. 1993. Multi-ISFET sensors for soil nitrate analysis, p. 349. In: P. C. Robert, R. H. Rust, and W. E. Larsen (eds.), Soil Specific Crop Management. American Society of Agronomy, Crop Science Society of America, Soil Science Society of America, Madison, WI.

Blackburn, G. A. 2007. Hyperspectral remote sensing of plant pigments. *J. Exp. Bot.* 58:855–867.

Blackmer, T. M., J. S. Schepers, and G. E. Varvel. 1994. Light reflectance compared with other nitrogen stress measurements in corn leaves. *Agron. J.* 86:934–938.

Blackmer, T. M., J. S. Schepers, and G. E. Meyer. 1995. Remote sensing to detect nitrogen deficiency in corn, pp. 505–512. In: P. C. Robert, R. H. Rust, and W. E. Larson (eds.), *Site-Specific Management for Agricultural Systems.* ASA-CSSA-SSSA. Madison, WI.

Blackmer, T. M., J. S. Schepers, G. E. Varvel, and G. E. Meyer. 1996a. Analysis of aerial photography for nitrogen stress within corn fields. *Agron. J.* 88:729–733.

Blackmer, T. M., and J. S. Schepers. 1996b. Aerial photography to detect nitrogen stress in corn. *J. Plant Physiol.* 148:440–444.

Blackmore, S., R. J. Godwin, and S. Fountas. 2003. The analysis of spatial and temporal trends in yield map data over six years. *Biosyst. Eng.* 84:455–466.

Bongiovanni, R., and J. Lowenberg-DeBoer. 2004. Precision agriculture and sustainability. *Precis. Agric.* 5:359–387.

Borgelt, S. C., S. W. Searcy, B. A. Stout, and D. J. Mulla. 1989. A method for determining spatially variable liming rates. Paper No. 89-1034, International Summer Meeting, American Society of Agricultural Engineers, Quebec City, Quebec, Canada, June 25–28, 1989.

Borgelt, S. C., S. W. Searcy, B. A. Stout, and D. J. Mulla. 1994. Spatially variable liming rates: A method for determination. *Trans. ASAE* 37(5):1499–1507.

Boydell, B., and A. McBratney. 2002. Identifying potential within-field management zones from cotton-yield estimates. *Precis. Agric.* 3:9–23.

Burgess, T. M., and R. Webster. 1980. Optimal interpolation and isarithmic mapping of soil properties. *J. Soil Sci.* 31:315–331.

Buschmann, C., G. Langsdorf, and H. K. Lichtenthaler. 2000. Imaging of the blue, green, and red fluorescence emission of plants: An overview. *Photosynthetica* 38:483–491.

Camp, C. R., E. J. Sadler, D. E. Evans, L. J. Usrey, and M. Omary. 1998. Modified center pivot system for precision management of water and nutrients. *Applied Eng. Agric.* 14:23–32.

Carr, P. M., G. R. Carlson, J. S. Jacobsen, G. A. Nielsen, and E. O. Skogley. 1991. Farming soils, not fields: A strategy for increasing fertilizer profitability. *J. Prod. Agric.* 4:57–61.

Cartelat, A., Z. G. Cerovic, Y. Goulas, S. Meyer, C. Lelarge, J. L. Prioul, A. Barbottin, M. H. Jeuffroy, P. Gate, and G. Agati. 2005. Optically assessed contents of leaf polyphenolics and chlorophyll as indicators of nitrogen deficiency in wheat (*Triticum aestivum* L.). *Field Crops Res.* 91:35–49.

Carter, L. M., J. D. Rhoades, and J. H. Chesson. 1993. Mechanization of soil salinity assessment for mapping. ASAE Paper 931557.

Chappelle, E. W., M. S. Kim, and F. E. McMurtrey. 1992. Ratio analysis of reflectance spectra (RARS): An algorithm for the remote estimation of the concentrations of chlorophyll a, chlorophyll b, and carotenoids in soybean leaves. *Remote Sens. Environ.* 39:239–247.

Christy, C., and E. Lund. 1998. Device for measuring soil conductivity. U.S. Patent No. 5,841,282, issued November 24, 1998.

Clarke, T. R., M. S. Moran, E. M. Barnes, P. J. Pinter, and J. Qi. 2001. Planar domain indices: A method for measuring a quality of a single component in two-component pixels. In: *Proc. IEEE International Geosci. Remote Sens. Sympos.* (CD-ROM), Sydney, Australia, July 9–13, 2001.

Coble, H. D., and D. A. Mortensen. 1992. The threshold concept and its application to weed science. *Weed Technol.* 6:191–195.

Colburn, J. W., Jr. 1991. Soil chemical sensor and precision agricultural chemical delivery system and method. U.S. Patent No. 5,033,397, issued July 23, 1991.

Collins, D. M., K. D. Funk, R. W. Heiniger, J. A. McClure, and J. T. E. Timm. 2006. Automatic steering system and method. U.S. Patent No. 7,142,956, issued November 28, 2006.

Corwin, D. L., and S. M. Lesch. 2003. Application of soil electrical conductivity to precision agriculture: Theory, principles, and guidelines. *Agron. J.* 95:455–471.

Crookston, K. 2006. A top 10 list of developments and issues impacting crop management and ecology during the past 50 years. *Crop Sci.* 46:2253–2262.

Datt, B. 1999. A new reflectance index for remote sensing of chlorophyll content in higher plants: Tests using eucalyptus leaves. *J. Plant Physiol.* 154:30–36.

Daughtry, C. S. T., M. E. Bauer, D. W. Crecelius, and M. M. Hixson. 1980. Effects of management practices on reflectance of spring wheat canopies. *Agron. J.* 72:1055–1066.

Delgado, J. A., R. Khosla, W. C. Bausch, D. G. Westfall, and D. J. Inman. 2005. Nitrogen fertilizer management based on site-specific management zones reduces potential for nitrate leaching. *J. Soil Water Conserv.* 60:402–410.

Doerge, T. 1999. Defining management zones for precision farming. *Crop Insights* 8:1–5.

Doolittle, J. A., K. A. Sudduth, N. R. Kitchen, and S. J. Indorante. 1994. Estimating depths to claypans using electromagnetic induction methods. *J. Soil Water Conserv.* 49:572–575.

Dow, A. I., and D. W. James. 1973a. *Intensive Soil Sampling: A Principle of Soil Fertility Management in Intensive Irrigation Agriculture.* Washington Agricultural Experiment Station Bulletin 781. Washington State University, Pullman, WA.

Dow, A. I., D. W. James, and T. S. Russell. 1973b. *Soil Variability in Central Washington and Sampling for Soil Fertility Tests.* Washington Agricultural Experiment Station Bulletin 788, Washington State University, Pullman, WA.

English, C. B., S. B. Mahajanashetti, and R. K. Roberts. 1999. Economic and environmental benefits of variable rate application of nitrogen to corn fields: Role of variability and weather, pp. 8–11. Selected paper presented at the American Agricultural Economics Association Meeting, Nashville, TN.

Evans, R. G., S. Han, M. W. Kroeger, and S. M. Schneider. 1996. Precision center pivot irrigation for efficient use of water and nitrogen, pp. 75–84. In: P. C. Robert, R. H. Rust, and W. E. Larson (eds.), *Precision Agriculture.* Proc. 3rd Intl. Conf. ASA-CSSA-SSSA, Madison, WI.

Fairchild, D., and M. Duffy. 1993. Working Group Report. In: P. C. Robert, R. H. Rust, and W. E. Larson (eds.), Proc. Soil Specific Crop Management Workshop. ASA-CSSA-SSSA, Madison, WI.

Felton, W., and K. McCloy. 1992. Controller for agricultural sprays. U.S. Patent No. 5,144,767, issued September 8, 1992.

Fiez, T. E., B. C. Miller, and W. L. Pan. 1994. Assessment of spatially variable nitrogen fertilizer management in winter wheat. *J. Prod. Agric.* 7:86–93.

Filella, I., L. Serrano, J. Serra, and J. Peñuelas. 1995. Evaluating wheat nitrogen status with canopy reflectance indices and discriminant analysis. *Crop Sci.* 35:1400–1405.

Fitzgerald, G., D. Rodriguez, and G. O'Leary. 2010. Measuring and predicting canopy nitrogen nutrition in wheat using a spectral index—The canopy chlorophyll content index (CCCI). *Field Crops Res.* 116:318–324.

Fleming, K. L., D. F. Heermann, and D. G. Westfall. 2004. Evaluating soil color with farmer input and apparent soil electrical conductivity for management zone delineation. *Agron. J.* 96:1581–1587.

Fraisse, C. W. 1994. Variable water application with moving irrigation systems. PhD dissertation. Colorado State University, Ft. Collins, CO.

Fraisse, C. W., K. A. Sudduth, and N. R. Kitchen. 2001. Delineation of site-specific management zones by unsupervised classification of topographic attributes and soil electrical conductivity. *Trans. ASAE* 44: 155–166.

Franke, J., and G. Menz. 2007. Multi-temporal wheat disease detection by multi-spectral remote sensing. *Precis. Agric.* 8:161–172.

Franzen, D. W., and T. R. Peck. 1994. Sampling for site specific management. pp. 535–551. In: P. C. Robert (ed.), *Proceedings of Site-Specific Management for Agricultural Systems: Second International Conference*, March 27–30, 1994.

Franzen, D. W., A. D. Halvorson, and V. L. Hofman. 2000. Management zones for soil N and P levels in the Northern Great Plains. In: P. C. Robert, R. H. Rust, and W. E. Larson (eds.), *Precision Agriculture*. ASA-CSSA-SSSA, Madison, WI.

Franzen, D. W. 2007. Lessons learned from 40 years of grid-sampling in Illinois. 2007 Indiana CCA Conference Proceedings, pp. 1–11.

García Torres, L., J. M. Pena-Barragan, F. Lopez-Granados, M. Jurado-Exposito, and R. Fernandez-Escobar. 2008. Automatic assessment of agro-environmental indicators from remotely sensed images of tree orchards and its evaluation using olive plantations. *Comput. Electron. Agr.* 61:179–191.

Gaultney, L. D., D. G. Schulze, J. L. Shonk, and G. E. Van Scoyoc. 1991. Real-time soil organic matter sensor. U.S. Patent No. 5,044,756, issued September 3, 1991.

Giles, D. K., and D. C. Slaughter. 1997. Precision band spraying with machine-vision guidance and adjustable yaw nozzles. *Trans. ASAE* 40:29–36.

Gitelson, A. A., A. Viña, V. Ciganda, D. C. Rundquist, and T. J. Arkebauer. 2005. Remote estimation of canopy chlorophyll content in crops. *Geophys. Res. Lett.* 32:L08403.1–L08403.4.

Greatline, M. W., and S. E. Greatline. 1999. Method and apparatus for prescription application of products to an agricultural field. U.S. Patent No. 5,919,242, issued July 6, 1999.

Griffin, T., D. Lambert, and J. Lowenberg-DeBoer. 2005. Economics of lightbar and auto-guidance GPS navigation technologies, pp. 581–587. In: J. V. Stafford (ed.), *Precision Agriculture '05*. Wageningen Academic Publishers, Wageningen, the Netherlands.

Guyer, D., G. Miles, M. Schreiber, O. Mitchell, and V. Vanderbilt. 1986. Machine vision and image processing for plant identification. *Trans. ASAE* 29:1500–1506.

Guyot, G., F. Baret, and D. J. Major. 1988. High spectral resolution: Determination of spectral shifts between the red and infrared. *Int. Arch. Photogram. Remote Sens.* 11:750–760.

Haboudane, D., J. R. Miller, N. Tremblay, P. J. Zarco-Tejada, and L. Dextraze. 2002. Integrated narrow-band vegetation indices for prediction of crop chlorophyll content for application to precision agriculture. *Remote Sens. Environ.* 81:416–426.

Haboudane, D., J. R. Miller, Elizabeth Pattey, P. J. Zarco-Tejada, and I. B. Strachan. 2004. Hyperspectral vegetation indices and novel algorithms for predicting green LAI of crop canopies: Modeling and validation in the context of precision agriculture. *Remote Sens. Environ.* 90:337–352.

Haggar, R., C. Stent, and S. Isaac. 1983. A prototype hand-held patch sprayer for killing weeds, activated by spectral differences in crop/weed canopies. *J. Agr. Eng. Res.* 28:349–358.

Hammond, M. W., Mulla, D. J., and Fairchild, D. S. 1988. Development of management maps for soil variability, pp. 67–76. In: Proc. 39th Annual Far West Regional Fertilizer Conference, Bozeman, MT, July 11–13, 1988.

Hammond, M. W., and D. J. Mulla. 1989. Field variation in soil fertility: Its assessment and management for potato production. In: Proc. 28th Annual Washington State Potato Conference, Moses Lake, WA, February 2, 1989.

Hammond, M. W. 1993. Cost analysis of variable fertility management of phosphorus and potassium for potato production in central Washington, pp. 213–228. In: P. C. Robert, R. H. Rust, and W. E. Larson (eds.), *Soil Specific Crop Management*, American Society of Agronomy, Crop Science Society of America, Soil Science Society of America, Madison, WI.

Halvorson, A. D., and J. D. Rhoades. 1974. Assessing soil salinity and identifying potential saline-seep areas with field soil resistance measurements. *Soil Sci. Soc. Am. J.* 38:576–581.

Hanks, J. E., and J. L. Beck. 1998. Sensor-controlled hooded sprayer for row crops. *Weed Tech.* 12:308–314.

Hayes, J. E., A. Overton, and J. W. Price. 1994. Feasibility of site-specific nutrient and pesticide applications, pp. 62–68. In: K. L. Campbell, W. D. Graham, and A. B. Bottcher (eds.), *Environmentally Sound Agriculture*. Proc. Second Conference, April 20–22, 1994, Orlando, FL. American Society of Agricultural Engineers, St. Joseph, MI.

Hergert, G. W., R. B. Ferguson, C. A. Gotway, and T. A. Peterson. 1996. The impact of variable rate N application on N use efficiency of furrow irrigated corn, pp. 389–397. In: P. C. Robert, R. H. Rust, and W. E. Larson (eds.), *Precision Agriculture*. Proc. 3rd Intl. Conference. ASA-CSSA-SSSA. Madison, WI.

Hertz, E. A., and J. D. Hibbard. 1993. A preliminary assessment of the economics of variable rate technology for applying phosphorus and potassium in corn production. *Farm Economics Facts & Opinions* 93-14, Department of Agricultural and Consumer Economics, College of Agricultural, Consumer and Environmental Sciences, University of Illinois, Champaign-Urbana, IL.

Huang, W., D. W. Lamb, Z. Niu, L. Liu, and J. Wang. 2007. Identification of yellow rust in wheat by in situ and airborne spectrum data. *Precis. Agric.* 8:187–197.

Hummel, J. W. 1985. Monitoring and control of field machines. *Ill. Res.* 27(4):8–10.

Hummel, J. W., L. D. Gaultney, and K. A. Sudduth. 1996. Soil property sensing for site-specific crop management. *Comput. Electron. Agr.* 14:121–136.

Inman, D., R. Khosla, R. Reich, and D. Westfall. 2007. Active remote sensing and grain yield in irrigated agriculture. *Precis. Agric.* 8:241–252.

Johnson, G. A., D. A. Mortensen, and A. R. Martin. 1995a. A simulation of herbicide use based on weed spatial distribution. *Weed Res.* 35:197–205.

Johnson, G. A., D. A. Mortensen, L. J. Young, and A. R. Martin. 1995b. The stability of weed seedling population models and parameters in eastern Nebraska corn (Zea mays) and soybean (Glycine max) fields. *Weed Sci.* 43:604–611.

Johnson, G., J. Cardina, and D. Mortensen. 1997. Site-specific weed management: Current and future directions, pp. 131–147. In: F. Pierce and E. Sadler (eds.), *The State of Site-Specific Management for Agriculture*, ASA-CSSA-SSSA, Madison, WI.

Jones, H. G. 2004. Irrigation scheduling: Advantages and pitfalls of plant-based methods. *J. Exp. Biol. Water-Saving Agriculture Special Issue* 55:2427–2436.

Kachanoski, R. G., R. P. Voroney, E. De Jong, and D. A. Rennie. 1985. The effect of variable and uniform N-fertilizer application rates on grain yields. In: *Proc. Soil and Crop Workshop*, pp. 123–132.

Karlen, D. L., E. J. Sadler, and W. J. Busscher. 1990. Crop yield variation associated with Coastal Plain soil map units. *Soil Sci. Soc. Am. J.* 54:859–865.

Keller, R. J., M. E. Nichols, and A. F. Lange. 2001. Methods and apparatus for precision agriculture operations utilizing real time kinematic global positioning system systems. U.S. Patent No. 6,199,000, issued March 6, 2001.

Khakural, B., P. Robert, and W. Koskinen. 1994. Runoff and leaching of alachlor under conventional and soil-specific management. *Soil Use Manage.* 10:158–164.

Khakural, B. R., G. A. Johnson, P. C. Robert, D. J. Mulla, R. Oliveira, and W. C. Koskinen. 1999. Site-specific herbicide management for preserving water quality, pp. 1719–1732. In: P. C. Robert, R. H. Rust, and W. E. Larson (eds.), *Precision Agriculture*, Proc. 4th Intl. Conf. ASA-CSSA-SSSA, Madison, WI.

Khosla, R., and M. M. Alley. 1999. Soil-specific nitrogen management on mid-Atlantic coastal plain soils. *Better Crops Plant Food* 83:6–7.

Khosla, R., K. Fleming, J. Delgado, T. Shaver, and D. Westfall. 2002. Use of site specific management zones to improve nitrogen management for precision agriculture. *J. Soil Water Conserv.* 57:513–518.

Khosla, R., D. G. Westfall, R. M. Reich, J. S. Mahal, and W. J. Gangloff. 2010. Spatial variation and site-specific management zones. In: M. A. Oliver (ed.), *Geostatistical Applications for Precision Agriculture.* Springer, Dordrecht, Netherlands.

King, B. A., J. C. Stark, I. R. McCann, and D. T. Westermann. 1996. Spatially varied nitrogen application through a center pivot irrigation system, pp. 85–94. In: P. C. Robert, R. H. Rust, and W. E. Larson (eds.), Proc. 3rd Intl Conf. Prec. Agric. ASA-CSSA-SSSA, Madison, WI.

Kitchen, N. R., K. A. Sudduth, and S. T. Drummond. 1999. Soil electrical conductivity as a crop productivity measure for claypan soils. *J. Prod. Agric.* 12:607–617.

Kitchen, N. R., C. J. Snyder, D. W. Franzen, and W. J. Weibold. 2002. Educational needs of precision agriculture. *Precis. Agric.* 3:341–351.

Kitchen, N. R., S. T. Drummond, E. D. Lund, K. A. Sudduth, and G. W. Buchleiter. 2003. Soil electrical conductivity and topography related to yield for three contrasting soil–crop systems. *Agron. J.* 95:483–495.

Kitchen, N. R., K. A. Sudduth, S. T. Drummond, P. C. Scharf, H. L. Palm, D. F. Roberts, and E. D. Vories. 2010. Ground-based canopy reflectance sensing for variable-rate nitrogen corn fertilization. *Agron. J.* 102:71–84.

Koch, B., R. Khosla, W. M. Frasier, D. G. Westfall, and D. Inman. 2004. Economic feasibility of variable-rate nitrogen application utilizing site-specific management zones. *Agron. J.* 96:1572–1580.

Kumar, R., B. Robinson, and L. Silva. 1979. Calibration of long wavelength Exotech model 20C spectroradiometer. *Appl. Opt.* 18:2334–2341.

Kunkel, R., C. D. Moodie, T. S. Russell, and N. Holstad. 1971. Soil heterogeneity and potato fertilizer recommendations. *Am. Potato J.* 48:163–173.

Lamb, J. A., R. H. Dowdy, J. L. Anderson, and G. W. Rehm. 1997. Spatial and temporal stability of corn grain yields. *J. Prod. Agric.* 10:410–414.

Lamb, D. W., and R. B. Brown. 2001. Remote-sensing and mapping of weeds in crops. *J. Agr. Eng. Res.* 78:117–125.

Larsen, W. E., D. A. Tyler, and G. A. Nielsen. 1988. Field navigation using the global positioning system (GPS). *Am. Soc. Agric. Eng.* ASAE Paper 88-1604. St. Joseph, MI.

Larson, W. E., and P. C. Robert. 1991. Farming by soil, pp. 103–112. In: R. Lal and F. J. Pierce (eds.), *Soil Management for Sustainability.* Soil and Water Conservation Society, Ankeny, IA.

Larson, W., J. Lamb, B. Khakural, R. Ferguson, and G. Rehm. 1997. Potential of site-specific management for nonpoint environmental protection, pp. 337–367. In: F. Pierce and E. Sadler (eds.), *The State of Site-Specific Management for Agriculture.* ASA-CSSA-SSSA, Madison, WI.

Leamer, R. W., V. I. Myers, and L. F. Silva. 1973. A spectroradiometer for field use. *Rev. Sci. Instrum.* 44:611–614.

Lesch, S. M., J. D. Rhoades, L. J. Lund, and D. L. Corwin. 1992. Mapping soil salinity using calibrated electromagnetic measurements. *Soil Sci. Soc. Am. J.* 56:540–548.

Linsley, C. M. and F. C. Bauer. 1929. Test your soil for acidity. Univ. IL College Agric. Agric. Exp. Station Circular 346.

Longchamps, L., and R. Khosla. 2014. Early detection of nitrogen variability in maize using fluorescence. *Agron. J.* 106:511–518.

López-Granados, F. 2011. Weed detection for site-specific weed management: Mapping and real time approaches. *Weed Res.* 51:1–11.

Lorenzen, C. J. 1966. A method for the continuous measurement of in vivo chlorophyll concentration. *Deep-Sea Res. Oceanogr. Abstr.* 13:223–227.

Lowenberg-DeBoer, J., R. Nielsen, and S. Hawkins. 1994. Management of intrafield variability in large scale agriculture: A farming systems perspective. *Proc. Intl. Symposium on Systems Research in Agriculture and Rural Development*, Montpelier, France, pp. 551–555.

Lowenberg-DeBoer, J., and M. Boehlje. 1996. Revolution, evolution or dead-end: Economic perspectives on precision agriculture. *Proc. 3rd Intl. Conf. Precision Agriculture*, pp. 923–944.

Lowenberg-DeBoer, J., and S. M. Swinton. 1997. Economics of site-specific management in agronomic crops, pp. 369–396. In: F. Pierce and E. Sadler (eds.), *The State of Site-Specific Management for Agriculture.* ASA-CSSA-SSSA, Madison, WI.

Luellen, W. R. 1985. Fine-tuned fertility: Tomorrow's technology here today. *Crops Soils* 38:18–22.

Mahaman, M. I. 1993. An evaluation of soil chemical properties variation in northern and southern Indiana. PhD thesis, Department of Agronomy, Purdue University, W. Lafayette, IN.

Mamo, M., G. L. Malzer, D. J. Mulla, D. J. Huggins, and J. Strock. 2003. Spatial and temporal variation in economically optimum N rate for corn. *Agron. J.* 95:958–964.

McBratney, A. B., R. Webster, and T. M. Burgess. 1981. The design of optimal sampling schemes for local estimation and mapping of regionalized variables—I: Theory and method. *Comput. Geosci.* 7:331–334.

McBratney, A. B., and B. M. Whelan. 1999. The null hypothesis of precision agriculture, pp. 947–956. In: J. V. Stafford (ed.), *2nd European Conference on Precision Agriculture.* Sheffield Academic Press, Sheffield, UK.

McBratney, A., B. Whelan, T. Ancev, and J. Bouma. 2005. Future directions of precision agriculture. *Precis. Agric.* 6: 7–23.

McCann, I. R., and J. C. Stark. 1993. Method and apparatus for variable application of irrigation water and chemicals. U.S. Patent No. 5,246,164, issued September 21, 1993.

McClure, J. A. 2005. Method and system for implement steering for agricultural vehicles. U.S. Patent No. 6,865,465, issued March 8, 2005.

McGrath, D. E., J. P. Ellingson, and A. O. Leedahl. 1990. Variable application rates based on soil organic matter. ASAE Paper No. 901598, St. Joseph, MI.

McKay, M. D., and M. O. Anderson. 2007. Auto-steering apparatus and method. U.S. Patent No. 7,191,061, issued March 13, 2007.

Melsted, S. W. 1967. The philosophy of soil testing, pp. 13–23. In: *Soil Testing and Plant Analysis Part I.* SSSA Special Publication Series No. 2. Soil Science Society of America, Inc., Madison, WI.

Meron, M., J. Tsipris, V. Orlov, V. Alchanatis, and Y. Cohen. 2010. Crop water stress mapping for site-specific irrigation by thermal imagery and artificial reference surfaces. *Precis. Agric.* 11:148–162.

Miao, Y., D. J. Mulla, G. W. Randall, J. A. Vetsch, and R. Vintila. 2007. Predicting chlorophyll meter readings with aerial hyperspectral remote sensing for in-season site-specific nitrogen management of corn, pp. 635–641. In: J. V. Stafford (ed.), *Precision Agriculture '07.* Wageningen Academic Publishers, Wageningen, the Netherlands.

Miao, Y., D. J. Mulla, G. Randall, J. Vetsch, and R. Vintila. 2009. Combining chlorophyll meter readings and high spatial resolution remote sensing images for in-season site-specific nitrogen management of corn. *Precis. Agric.*10:45–62.

Miller, P. 1988. Applying space-age technology to age-old hills. *Cooperative Partners Magazine* November/ December, pp. 12–13.

Mirik, M., Y. Aysan, and F. Sahin. 2011. Characterization of *Pseudomonas cichorii* isolated from different hosts in Turkey. *Int. J. Agric. Bio.* 13:203–209.

Mortensen, D. A., G. A. Johnson, D. Y. Wyse, and A. R. Martin. 1995. Managing spatially variable weed populations, pp. 397–415. In: P. C. Robert, R. H. Rust, and W. E. Larson (eds.), *Site Specific Crop Management.* ASA-CSSA-SSSA. Madison, WI.

Muhammad, H. H. 2005. Hyperspectral crop reflectance data for characterising and estimating fungal disease severity in wheat. *Biosys. Eng.* 91:9–20.

Mulla, D. J. 1988. Using geostatistics and spectral analysis to study spatial patterns in the topography of south-eastern Washington State, U.S.A. *Earth Surf. Processes* 13:389–405.

Mulla, D. J. and Hammond, M. W. 1988. Mapping of soil test results from large irrigation circles. pp. 169–176. In: *Proc. 39th Annual Far West Regional Fertilizer Conference,* Bozeman, MT, July 11–13.

Mulla, D. J. 1989. Soil spatial variability and methods of analysis: I. Kriging soil fertility patterns, pp. 185–204. In: C. E. Whitman, J. F. Parr, R. I. Papendick, and R. E. Meyer (eds.), *Rainfed Agriculture in the Near East Region: Soil, Water and Crop/Livestock Management Systems.* USDA/USAID, Washington, D.C.

Mulla, D. J. 1991. Using geostatistics and GIS to manage spatial patterns in soil fertility, pp. 336–345. In: G. Kranzler (ed.), *Automated Agriculture for the 21st Century.* American Society of Agricultural Engineers, St. Joseph, MI.

Mulla, D. J., A. U. Bhatti, M. W. Hammond, and J. A. Benson. 1992. A comparison of winter wheat yield and quality under uniform versus spatially variable fertilizer management. *Agric. Ecosyst. Environ.* 38:301–311.

Mulla, D. J. 1993. Mapping and managing spatial patterns in soil fertility and crop yield, pp. 15–26. In: P. Robert, W. Larson, and R. Rust (eds.), *Soil Specific Crop Management.* American Society of Agronomy, Madison, WI.

Mulla, D. J., C. A. Perillo, and C. G. Cogger. 1996. A site-specific farm-scale GIS approach for reducing groundwater contamination by pesticides. *J. Environ. Qual.* 25:419–425.

Mulla, D. J. 1997. Geostatistics, remote sensing and precision farming, pp. 100–119. In: A. Stein and J. Bouma (eds.), *Precision Agriculture: Spatial and Temporal Variability of Environmental Quality.* Ciba Foundation Symposium 210. Wiley, Chichester, UK.

Mulla, D. J., and A. B. McBratney. 2000. Soil spatial variability, pp. A321–A352. In: M. E. Sumner (ed.), *Handbook of Soil Science.* CRC Press, Boca Raton, FL.

Mulla, D. J., P. Gowda, W. C. Koskinen, B. R. Khakural, G. Johnson, and P. C. Robert. 2002. Modeling the effect of precision agriculture: Pesticide losses to surface waters, pp. 304–317. In: E. Arthur, A. Barefoot, and V. Clay (eds.), *Terrestrial Field Dissipation Studies.* ACS Symposium Series No. 842, ACS, Washington, DC.

Mulla, D. J. 2013. Twenty five years of remote sensing in precision agriculture: Key advances and remaining knowledge gaps. *Biosys. Eng.* 114:358–371.

Mulla, D. J., and Y. Miao. 2015. Remote sensing for precision farming. In: P. Thenkabail (ed.), *Advances in Land Remote Sensing: Last 50 Years.* Accepted, Taylor & Francis.

Napier, T. L., J. Robinson, and M. Tucker. 2000. Adoption of precision farming within three Midwest water-sheds. *J. Soil Water Conserv.* 55:135–141.

Nielsen, D. R., J. W. Biggar, and K. T. Erh. 1973. Spatial variability of field-measured soil-water properties. Hilgardia 42:215–259.

Nigon, T. J., D. J. Mulla, C. J. Rosen, Y. Cohen, V. Alchanatis, and R. Rud. 2014. Evaluation of the nitrogen sufficiency index for use with high resolution, broadband aerial imagery in a commercial potato field. *Precis. Agric.* 15:202–226.

Nowak, P. 1992. Why farmers adopt production technology. *Soil Water Conserv.* 47:14–16.

O'Connor, M., T. Bell, G. Elkaim, and B. Parkinson. 1996. Automatic steering of farm vehicles using GPS, pp. 767–777. In: P. C. Robert, R. H. Rust, and W. E. Larson (eds.). *Precision Agriculture.* Proc. 3rd Intl. Conf. ASA-CSSA-SSSA, Madison, WI.

Oliver, M. A., and R. Webster. 1990. Kriging: A method of interpolation for geographical information systems. *Int. J. Geogr. Inf. Syst.* 4:313–332.

Oliver, M. A. 2010. *Geostatistical Applications for Precision Agriculture.* Springer, Dordrecht, Netherlands.

Omary, M., C. R. Camp, and E. J. Sadler. 1997. Center pivot irrigation system modification to provide variable water application depths. *Appl. Eng. Agric.* 13(2):235–239.

Ortlip, E. W. 1986. Method and apparatus for spreading fertilizer. U.S. Patent No. 4,630,773, issued December 23, 1976, Assignee: Soil Teq.

Palmer, R. J. 1991. Progress report of a local positioning system, pp. 403–408. In: G. Kranzler (ed.), *Automated Agriculture for the 21st Century*. American Society of Agricultural Engineers, St. Joseph, MI.

Palmer, R. J. 1995. Positioning aspects of site-specific applications, pp. 613–618. In: *Site-Specific Management for Agricultural Systems*. American Society of Agronomy, Crop Science Society of America, Soil Science Society of America.

Parasnis, D. S. 1973. *Mining Geophysics*, 2nd Edition. Elsevier Scientific Publishing Company, Amsterdam.

Pierce, F. J., N. W. Anderson, T. S. Colvin, J. K. Schueller, D. S. Humburg, and N. B. McLaughlin. 1997. Yield mapping, pp. 211–243. In: *The State of Site-Specific Management for Agriculture*. ASA-CSSA-SSSA, Madison, WI.

Pinter, P. J., Jr., R. D. Jackson, S. B. Idso, and R. J. Reginato. 1981. Multidate spectral reflectance as predictors of yield in water stressed wheat and barley. *Int. J. Remote Sens.* 2:43–48.

Pinter, P. J., Jr., J. L. Hatfield, J. S. Schepers, E. M. Barnes, M. S. Moran, C. S. T. Daughtry, and D. R. Upchurch. 2003. Remote sensing for crop management. *Photogr. Engin. Remote Sens.* 69:647–664.

Prabhakar, M., Y. G. Prasad, M. Thirupathi, G. Sreedevi, B. Dharajothi, and B. Venkateswarlu. 2011. Use of ground based hyperspectral remote sensing for detection of stress in cotton caused by leafhopper (Hemiptera: Cicadellidae). *Comput. Electron. Agric.* 79:189–198.

Raun, W. R., J. B. Solie, G. V. Johnson, M. L. Stone, R. W. Mullen, K. W. Freeman, W. E. Thomason, and E. V. Lukina. 2002. Improving nitrogen use efficiency in cereal grain production with optical sensing and variable rate application. *Agron. J.* 94:815–820.

Redulla, C., J. Havlin, G. Kluitenberg, N. Zhang, and M. Schrock. 1996. Variable N management for improving groundwater quality, pp. 1101–1110. In: P. C. Robert, R. H. Rust, and W. E. Larson (eds.), *Precision Agriculture*. Proc. 3rd Intl. Conference. ASA-CSSA-SSSA, Madison, WI.

Reichenberger, L., and J. Russnogle. 1989. Farm by the foot. *Farm J.* 11–15.

Reid, J. F., and S. W. Searcy. 1987. Vision-based guidance of an agriculture tractor. *IEEE Control Syst. Mag. N Y* 7:39–43.

Reid, J. F., Q. Zhang, N. Noguchi, and M. Dickson. 2000. Agricultural automatic guidance research in North America. *Comput. Electron. Agric.* 25:155–167.

Rejesus, R. M., and R. H. Hornbaker. 1999. Economic and environmental evaluation of alternative pollution-reducing nitrogen management practices in central Illinois. *Agric. Ecosyst. Environ.* 75:41–53.

Rhoades, J. D., and D. L. Corwin. 1981. Determining soil electrical conductivity-depth relations using an inductive electromagnetic soil conductivity meter. *Soil Sci. Soc. Am. J.* 45:255–260.

Robert, P. C. 1982. Evaluation of some remote sensing techniques for soil and crop management. PhD dissertation, University of Minnesota.

Robert P., S. Smith, W. Thompson, W. Nelson, D. Fuchs, and D. Fairchild. 1990. Soil specific management, pp. 54–55. In: A Research Report on Field Research in Soils. Minnesota Ag Exp Sta Misc Pub 62. Univ. Minnesota, St. Paul, MN.

Rogers, M. E. 1995. *Diffusion of Innovations*. Free Press, Division of Simon and Schuster, Inc., New York.

Rouse, J. W., Jr., R. H. Hass, J. A. Schell, and D. W. Deering, 1973. Monitoring vegetation systems in the Great Plains with ERTS, pp. 309–317. In: *Proceedings 3rd Earth Resources Technology Satellite (ERTS) Symposium*, Vol. 1. Washington, DC, NASA SP-351, NASA.

Rud, R., Y. Cohen, V. Alchanatis, A. Cohen, A. Levi, R. Brikman, C. Shenderey, B. Heuer, T. Markovits, Z. Dar, C. Rosen, D. Mulla, and T. Nigon. 2014. Crop water stress index derived from multi-year ground and aerial thermal images as an indicator of potato water status. *Precis. Agric.* 15:273–289.

Rust, R. H. 1985. Computerized soil maps benefit crop management. *Farming with Pride Magazine,* 4–10.

Rutto, E., and D. B. Arnall. 2009. The history of the GreenSeeker™ sensor. Oklahoma Cooperative Extension Service PSS-2260. Oklahoma State University, Stillwater, OK.

Samborski, S. M., N. Tremblay, and E. Fallon. 2009. Strategies to make use of plant sensors-based diagnostic information for nitrogen recommendations. *Agron. J.* 101:800–816.

Samseemoung, G., P. Soni, H. P. W. Jayasuriya, and V. M. Salokhe. 2012. Application of low altitude remote sensing (LARS) platform for monitoring crop growth and weed infestation in a soybean plantation. *Precis. Agric.* 13:611–627.

Sawyer, J. E. 1994. Concepts of variable rate technology with considerations for fertilizer application. *J. Prod. Agric.* 7:195–201.

Scharf, P. C., N. R. Kitchen, K. A. Sudduth, and J. G. Davis. 2006. Spatially variable corn yield is a weak predictor of optimal nitrogen rate. *Soil Sci. Soc. Am. J.* 70:2154–2160.

Schmitt, M. A., W. G. Walker, and D. Fairchild. 1986. Computerized fertilizer application by soil type, pp. 1–5. In: Proc. Great Plains Soil Fertility Workshop, Denver, CO, March 4–5, 1986. Kansas State Univ., Manhattan, KS.

Schueller, J. K., and Y. H. Bae. 1987. Spatially attributed automatic combine data acquisition. *Comput. Electron. Agr.* 2:119–127.

Schueller, J. K. 1991. In-field site-specific crop production, pp. 291–292. In: G. Kranzler (ed.), *Automated Agriculture for the 21st Century*. American Society of Agricultural Engineers, St. Joseph, MI.

Schueller, J. K. 1992. A review and integrating analysis of spatially-variable control of crop production. *Fert. Res.* 33:1–34.

Schueller, J. K., and M.-W. Wang. 1994. Spatially-variable fertilizer and pesticide application with GPS and DGPS. *Comput. Electron. Agr.* 11:69–83.

Searcy, S. W., J. K. Schueller, Y. H. Bae, S. C. Borgelt, and B. A. Stout. 1989a. Mapping of spatially-variable yield during grain combining. *Trans. ASAE* 32:826–829.

Searcy, S. W., J. K. Schueller, H. B. Yeong, and B. A. Stout. 1989b. Measurement of agricultural field location using microwave frequency triangulation. *Comput. Electron. Agr.* 4:209–223.

Seelan, S. K., S. Laguette, G. M. Casady, and G. A. Seielstad. 2003. Remote sensing applications for precision agriculture: A learning community approach. *Remote Sens. Environ.* 88:157–169.

Shanahan, J. F., N. R. Kitchen, W. R. Raun, and J. S. Schepers. 2008. Responsive in-season nitrogen management for cereals. *Comput. Electron. Agr.* 61:51–62.

Shaner, D., R. Khosla, M. Brodhal, G. Buchleiter, and H. Farahani. 2008. How well does zone sampling based on soil electrical conductivity maps represent soil variability? *Agron. J.* 100:1472–1480.

Shiratsuchi, L., R. Ferguson, J. Shanahan, V. Adamchuk, D. Rundquist, D. Marx, and G. Slater. 2011. Water and nitrogen effects on active canopy sensor vegetation indices. *Agron. J.* 103:1815–1826.

Shockley, J. M., C. R. Dillon, and T. S. Stombaugh. 2011. A whole farm analysis of the influence of auto-steer navigation on net returns, risk, and production practices. *J. Agric. Appl. Econ.* 43:57–75.

Shockley, J. M., C. R. Dillon, and T. S. Stombaugh. 2012. The influence of auto-steer on machinery selection and land acquisition. *J. Am. Soc. Farm Managers Rural Appraisers* June, pp. 1–7.

Shonk, J. L., and L. D. Gaultney. 1988. Spectroscopic sensing for the determination of organic matter content. *Am. Soc. Agric. Engin.* (Microfiche collection) Paper No. 88-2142. St. Joseph, MI.

Solie, J. B., W. R. Raun, R. W. Whitney, M. L. Stone, and J. D. Ringer. 1996. Optical sensor based field element size and sensing strategy for nitrogen. *Trans. ASAE* 39:1983–1992.

Sripada, R. P., R. W. Heiniger, J. G. White, and R. Weisz. 2006. Aerial color infrared photography for determining late-season nitrogen requirements in corn. *Agron. J.* 97:1443–1451.

Sripada, R. P., J. P. Schmidt, A. E. Dellinger, and D. B. Beegle. 2008. Evaluating multiple indices from a canopy reflectance sensor to estimate corn N requirements. *Agron. J.* 100:1553–1561.

Stafford, J. V, B. Ambler, and M. P. Smith. 1991. Sensing and mapping grain yield variation, pp. 356–365. In: G. Kranzler (ed.), *Automated Agriculture for the 21st Century*. American Society of Agricultural Engineers, St. Joseph, MI.

Stafford, J. V., and P. C. H. Miller. 1993. Spatially selective application of herbicide to cereal crops. *Comput. Electron. Agr.* 9:217–229.

Stafford, J. V., B. Ambler, R. M. Lark, and J. Catt. 1996. Mapping and interpreting the yield variation in cereal crops. *Comput. Electron. Agr.* 14:101–119.

Sudduth, K. A., J. W. Hummel, and M. D. Cahn. 1991. Soil organic matter sensing: A developing science, pp. 307–316. In: G. A. Kranzler (ed.), *Automated Agriculture for the 21st Century*. ASAE Publication 11–91, St. Joseph, MI.

Sudduth, K. A., N. R. Kitchen, D. F. Hughes, and S. T. Drummond. 1995. Electromagnetic induction sensing as an indicator of productivity on claypan soils, pp. 671–681. In: *Site-Specific Management for Agricultural Systems*. American Society of Agronomy, Crop Science Society of America, Soil Science Society of America.

Sudduth, K. A., S. T. Drummond, S. J. Birrell, and N. R. Kitchen. 1996. Analysis of spatial factors influencing crop yield, pp. 129–139. In: P. C. Robert, R. H. Rust, and W. E. Larson (eds.), *Precision Agriculture*. ASA-CSSA-SSSA, Madison, WI.

Swain, K. C., S. J. Thomson, and H. P. W. Jayasuriya. 2010. Adoption of an unmanned helicopter for low-altitude remote sensing to estimate yield and total biomass of a rice crop. *Trans. ASABE* 53:21–27.

Thenkabail, P. S., R. B. Smith, and E. De Pauw. 2000. Hyperspectral vegetation indices and their relationships with agricultural crop characteristics. *Remote Sens. Environ.* 71:158–182.

Thenkabail, P. S. 2003. Biophysical and yield information for precision farming from near-real-time and historical Landsat TM images. *Int. J. Remote Sens.* 24:2879–2904.

Thenkabail, P. S., J. G. Lyon, and A. Huete. 2010. Hyperspectral remote sensing of vegetation and agricultural crops: Knowledge gain and knowledge gap after 40 years of research. Ch. 28, pp. 705. Boca Raton, FL: CRC Press.

Thompson, J. F., J. V. Stafford, and P. C. H. Miller. 1990. Selective application of herbicides to UK cereal crops. Paper, American Society of Agricultural Engineers 90-1629.

Thompson, J. F., J. V. Stafford, and P. C. H. Miller. 1991. Potential for automatic weed detection and selective herbicide application. *Crop Prot.* 10:254–259.

Thorp, K. R., and L. Tian. 2004. A review on remote sensing of weeds in agriculture. *Precis. Agric.* 5:477–508.

Thuilot, B., C. Cariou, P. Martinet, and M. Berducat. 2002. Autonomous guidance of a farm tractor relying on a single CP-DGPS. *Auton. Robots* 13:53–71.

Tian, L., J. F. Reid, and J. W. Hummel. 2000. Development of a precision sprayer for site-specific weed management. *Trans. ASAE* 42:893–902.

Tian, L. 2002. Development of a sensor-based precision herbicide application system. *Comput. Electron. Agr.* 36:133–149.

Torii, T. 2000. Research in autonomous agriculture vehicles in Japan. *Comput. Electron. Agr.* 25:133–153.

Tremblay, N., E. Fallon, and N. Ziadi. 2011. Sensing of crop nitrogen status: Opportunities, tools, limitations, and supporting information requirements. *Horttechnology* 21:274–281.

Tremblay, N., Z. Wang, and Z. G. Cerovic. 2012. Sensing crop nitrogen status with fluorescence indicators. A review. *Agron. Sustain. Dev.* 32:451–464.

Tyler, D. A. 1993. Positioning technology (GPS), pp. 159–166. In: P. C. Robert, R. H. Rust, and W. E. Larsen (eds.), *Proc. Soil Specific Crop Management.* American Society of Agronomy, Madison, WI.

Tyler, D. A., D. W. Roberts, and G. A. Nielsen. 1997. Location and guidance for site-specific management. In: F. J. Pierce and E. J. Sadler (eds.), *The State of Site Specific Management for Agriculture.* ASA/CSSA/SSSA, Madison, WI. pp. 161–180.

USDA-ERS. 2014. Report on recent cost on returns: Corn for USA 1996–2000; 2001–04; 2005–09; and 2010–13. http://www.ers.usda.gov/data-products/commodity-costs-and-returns.aspx, accessed August 23, 2014.

Vanischen, R., and J. De Baerdemaeker. 1991. Continuous wheat yield measurement on a combine, pp. 346–355. In: G. Kranzler (ed.), *Automated Agriculture for the 21st Century.* American Society of Agriculture Engineers, St. Joseph, MI.

Veseth, R. 1986. Managing variable soils. *STEEP Extension Farming Update*, Fall issue, 29–33.

Vetsch, J. A., G. L. Malzer, P. C. Robert, and W. W. Nelson. 1993. Nitrogen specific management by soil condition, p. 377. In: P. C. Robert, R. H. Rust and W. E. Larson (eds.), *Soil Specific Crop Management.* American Society of Agronomy, Crop Science Society of America, Soil Science Society of America, Madison, WI.

Vieira, S. R., D. R. Nielsen, and J. W. Biggar. 1981. Spatial variability of field-measured infiltration rate. *Soil Sci. Soc. Am. J.* 45:1040–1048.

Walburg, G., M. Bauer, and C. S. T. Daughtry. 1982. Effects of nitrogen nutrition on the growth, yield and reflectance characteristics of corn canopies. *Agron. J.* 74:677–683.

Wanjura, D. F., and J. L. Hatfield. 1987. Sensitivity of spectral vegetative indices to crop biomass. *Trans. ASAE* 30:810–816.

Webster, R., and T. M. Burgess. 1984. Sampling and bulking strategies for estimating soil properties in small regions. *J. Soil Sci.* 35:127–140.

Webster, R., and M. A. Oliver. 1992. Sample adequately to estimate variograms of soil properties. *J. Soil Sci.* 43:177–192.

Whipker, L. D., and J. D. Akridge. 2006. Precision agricultural services dealership survey results. Staff paper, Department of Agricultural Economics, Purdue University, W. Lafayette, IN.

Whipker, L. D., and B. Erickson. 2013. 2011 Precision agriculture services dealership survey results. Staff paper, Department of Agricultural Economics, Purdue University, W. Lafayette, IN.

Whitley, K. M., J. R. Davenport, and S. R. Manley. 2000. Differences in nitrate leaching under variable and conventional nitrogen fertilizer management in irrigated potato systems, pp. 1–9. In: P. C. Robert, R. H. Rust, and W. E. Larson (eds.), *Precision Agriculture.* Proc. 5th Intl. Conference. ASA-CSSA-SSSA, Madison, WI.

Wibawa, W. D., D. L. Dludlu, L. J. Swenson, D. G. Hopkins, and W. C. Dahnke. 1993. Variable fertilizer application based on yield goal, soil fertility, and soil map unit. *J. Prod. Agric.* 6:255–261.

Wiegand, C. L., A. J. Richardson, and D. E. Escobar. 1991. Vegetation indices in crop assessment. *Remote Sens. Environ.* 35:105–119.

Wikipedia. 2014. Combine harvester. http://en.wikipedia.org/wiki/Combine_harvester, accessed August 24, 2014.

Wiles, L. J., G. G. Wilkerson, H. J. Gold, and H. D. Coble. 1992. Modeling weed distribution for improved postemergence control decisions. *Weed Sci.* 40:546–553.

Wilson, J. N. 2000. Guidance of agricultural vehicles—A historical perspective. *Comput. Electron. Agric.* 25:3–9.

Wollenhaupt, N. C., and D. D. Buchholz. 1993. Profitability of farming by soils, pp. 199–211. In: P. C. Robert, R. H. Rust, and W. E. Larson (eds.), *Proc. Soil Specific Crop Management: A Workshop on Research and Development Issues*, April 14–16, 1992, Minneapolis, MN. ASA-CSSA-SSSA. Madison, WI.

Wollenhaupt, N. C., and R. P. Wolkowski. 1994. Grid soil sampling for precision and profit. Unpublished manuscript, Department of Soil Science, University of Wisconsin, Madison, WI. Modified from a paper prepared for 24th North Central Extension-Industry Soil Fertility Workshop, St. Louis, MO, October 26–27, 1994.

Wollenhaupt, N. C., R. P. Wolkowski, and M. K. Clayton. 1994. Mapping soil test phosphorus and potassium for variable-rate fertilizer application. *J. Prod. Agric.* 7:441–448.

Wollenhaupt, N. C., D. J. Mulla, and C. A. Gotway. 1997. Soil sampling and interpolation techniques for mapping spatial variability of soil properties, pp. 19–54. In: F. J. Pierce and E. J. Sadler (eds.), *The State of Site Specific Management for Agriculture.* ASA/CSSA/SSSA, Madison, WI.

Yang, C., J. H. Everitt, J. M. Bradford, and D. E. Escobar. 2000. Mapping grain sorghum growth and yield variations using airborne multispectral digital imagery. *Trans. ASAE* 43:1927–1938.

Yapa, L. S., and R. C. Mayfield. 1978. Non-adoption of innovations: Evidence from discriminant analysis. *Econ. Geogr.* 54:145–156.

Zhang, Q., J. F. Reid, and N. Noguchi. 1999. Agricultural vehicle navigation using multiple guidance sensors, pp. 293–298. In: *Proc. Intl. Conf. Field Service Robotics.*

Zhang, C., and J. M. Kovacs. 2012. The application of small unmanned aerial systems for precision agriculture: A review. *Precis. Agric.* 13:693–712.

Zheng, F., and H. Schreier. 1988. Quantification of soil patterns and field soil fertility using spectral reflection and digital processing of aerial photographs. *Fert. Res.* 16:15–30.

Zwiggelaar, R. 1998. A review of spectral properties of plants and their potential use for crop/weed discrimination in row-crops. *Crop Prot.* 17:189–206.

2 Managing Soil Variability at Different Spatial Scales as a Basis for Precision Agriculture

Jetse J. Stoorvogel, Lammert Kooistra, and Johan Bouma

CONTENTS

2.1   INTRODUCTION

A wide variety of farming systems occur in different agroecological zones of the world and focus on the production of dairy and crops. The advance of information and communication technology (ICT) is causing fundamental and rapid change to farming enterprises all over the world, although most developments have occurred thus far in industrialized, developed countries. The introduction of ICT is most clearly evident in dairy farming (for a good overview of recent developments see e.g., Endres 2013). Sensors for individual cows allow fine-tuning of feeding patterns as a function of milk production and also allow early detection of infections. Milk robots focus on production patterns of individual cows and allow milking at different times during the day. For each individual animal, variation as a function of time governs the monitoring process. ICT is fundamentally changing dairy farming and new technology is being rapidly accepted by dairy farmers despite significant initial investments because feeding costs are significantly reduced, cows are healthier, and the drudgery of having to milk cows twice a day becomes part of history. Most important, however, is that farmers' net incomes tend to increase substantially.

Conditions in crop production are fundamentally different. Attention cannot be focused on a well-defined entity, such as an animal, but deals with fields as the elementary production unit. Fields are usually quite heterogeneous and contain an enormous quantity of individual plants that together produce a yield. But looking at plants alone cannot, of course, explain the dynamics of production systems under natural conditions because soils play a key role in the transfer of water and agrochemicals to plants through their root systems. Therefore, linking plant and soil is essential when analyzing plant production systems. There is, however, an encouraging unifying principle: every plant, genetically modified or not, needs water and agrochemicals to grow, along with energy by the sun and carbon dioxide from the air. Just as with the economic return of a given animal, the difference between inputs and outputs determines the economic return of a given field. But while an animal is a discrete entity with a reasonably well known biological system, where inputs and outputs can be rather well defined, fields are much larger and are spatially heterogeneous in terms of soils and topography. Crop development is also unpredictable because weather conditions vary considerably during a growing season and cannot be predicted beyond a period of a few days. To further complicate matters, yields may be strongly affected by weeds, pests, and diseases. As a result, the input/output analysis of the cropping system is much more difficult than the analysis of a dairy system, but it is not impossible, as we will be demonstrate in this chapter, when we recognize that soils, just like animals, are also living and dynamic entities to which the laws of nature apply. Whereas variation in time governs the monitoring process for individual animals, variation in space and time has to be considered when dealing with heterogeneous fields, the exception being crops or vegetables grown in greenhouses where homogeneous substrates are used and where only the variation in time has to be considered. This chapter focuses on crop growth under field conditions.

Soils are an essential element of the production system to make it possible for crops to grow and to allow a variety of management measures, such as soil tillage and soil traffic. Soils also play a crucial role in protecting the environmental quality of air and water. Crops disappear after every harvest, but the soil remains, affecting growing conditions in the following season. Poor management may result in soil compaction or other forms of irreversible degradation. Aside from a focus on the role of soils in a given growing season, attention should be focused on the long-term effects of the selected management schemes on soil quality. Despite more complicated conditions compared to dairy farming, arable farming can still benefit from the application of ICT, as evidenced by the rapid development of precision agriculture (PA) that in principle allows tillage and soil traffic at moments when no structural damage is done and application of the right quantity of water, agrochemicals, and biocides to the crop at the right time and place during the growing season within any given field, avoiding losses to the environment. PA can therefore be defined as observing, measuring, and responding to intra-field spatial and temporal variability in crop development with the objective to develop an operational decision support system (DSS) for farm management, aimed at optimizing financial returns while preserving natural resources. Most fields are heterogeneous in terms of soil conditions, and a key activity of PA is therefore the determination of management units within a field that have such a degree of internal homogeneity that they can be subjected to a specific form of management that is significantly different from that applied to adjacent management units. Different spatial scales can be distinguished: differences between farms, between fields in a given farm, and within separate fields. Successful application of PA practices increases the efficiency of the production process as resources are preserved, thereby reducing costs, while environmental pollution of water and air is either reduced or, ideally, omitted.

Fortunately, PA receives much attention in international research (e.g., Stafford 2012 and 2013; Oliver et al. 2013). Originating in the United States in the 1980s, where the late Dr. Pierre Robert of the University of Minnesota was a driving force, active programs now are proceeding in many other countries such as Australia, United Kingdom, Argentina, and Brazil.

This study focuses on the role of farmers considering the possibility of embracing the PA concept and managing their soil variability. This involves strategic, tactical, and operational decisions. First, the strategic decision has to be made to either adopt PA practices or not. Once a positive decision has been made, a selection of procedures is required in a tactical decision process in which many options exist to delineate management units and to select operational equipment. Next, operational decisions have to be made during the growing season. The PA program involves a continuous learning process as problems are likely to arise along the way or new procedures may become available in this rapidly developing field of activity.

This chapter begins with an overview of different soil- and crop-oriented methods to describe the spatial variability within fields. Then, the main question that remains is how to translate this information into practical management. Management decisions will be discussed on the basis of four operational low- and high-tech PA systems. General conclusions will be drawn from the case studies in terms of the scale of PA management, the technological implications and requirements of PA, and the general policy implications.

2.2 METHODOLOGIES TO ESTABLISH MANAGEMENT UNITS OR IDENTIFY CONTINUOUS VARIATION

2.2.1 INTRODUCTION

Production of high-quality crops with high economic returns is the primary objective of any arable farmer. Plants reflect growth conditions, integrating various yield-limiting factors such as nutrient and/or water deficiency, poor weather, water logging, soil compaction, inappropriate crop species, excessive weed growth, and the occurrence of pests and diseases. Many reasons for suboptimal growth are, directly or indirectly, soil-related. Farmers have been aware of the fact that their fields did not offer homogeneous growing conditions for their crops (Heijting et al. 2011). Many have used their experience to somewhat adjust their management practices taking into account major

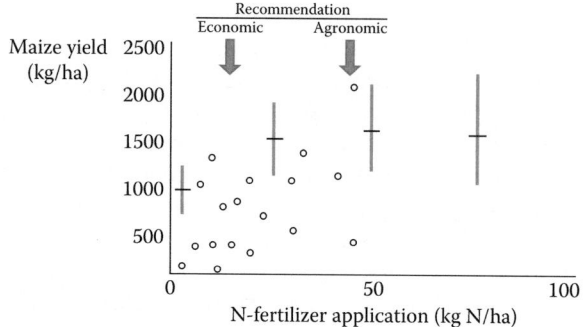

FIGURE 2.1 Analysis of fertilizer use recommendation for Machakos (Kenya) with trial data (bars representing the mean ± s.d.) from the Fertilizer Use Recommendation Program (Schnier, H. F. et al., *Nutri. Cycl. Agroecosys.* 47:213–26, 1997) and dots from the Leinuts farm survey (De Jager, A. et al., *Agric. Syst.* 79:205–23, 2004).

observed differences. This becomes more important as farm mechanization and rationalization leads to increasingly larger fields that are bound to be more heterogeneous. Soil fertility researchers and extension services have not recognized the heterogeneity problem because they often have defined fertilizer recommendations based on "average" soil fertility values for any given field based on elaborate sampling procedures. Averaging out soil fertility differences implies that locally either too much or too little fertilizer is applied, reducing the efficiency of the production process and increasing the risk for environmental pollution. In addition, fertilizer application recommendations are based on multiyear data from experimental plots on representative soils. The effects of different weather and soil conditions are reflected when recommended fertilization rates are related to measured yields by single curves, ignoring the enormous variability and scatter of individual plot data. An example of such a derivation of a fertilizer recommendation for maize production in the Machakos district by the Kenyan Fertilizer Use Recommendation Program (FURP 1987 and 1994; Schnier et al. 1997) is provided in Figure 2.1. The experiment was made on a representative soil type for the region (ferralo-orthic Acrisol/oxic Paleustult*). The results show a large variation in the experiment and an even larger variation if compared to farm survey data from the region (de Jager et al. 2004) but led to a single fertilizer use recommendation for the region.

Two fundamentally different approaches can be followed when tackling the spatial heterogeneity problem. First, attempts can be made to define stable, relatively homogeneous subareas in a field that can be managed in a particular manner. These management units are defined as "sub-regions in a field that express a homogeneous combination of yield limiting factors for which a single crop input is appropriate to attain maximum efficiency of farm inputs" (Doerge 1999). Second, the variation can be described continuously. An increasing number of techniques are available that can describe soil and crop variation in a continuous way. Sensing techniques can be applied in combination with variable rate applicators to manage the variability in a continuous way. In fact, sensing techniques also represent recognition of management units but they are small and dynamic. The size of the management units is then determined by the speed and accuracy of the observation techniques in relation to the speed of the agricultural machinery, the technical capacity of the equipment to generate quick responses and the decision rules transforming the measured signal into a variable application rate. This section discusses the state-of-the art techniques to define management units or identify continuous variation. Three procedures can be followed to define stable management units. The first one is soil-oriented (Section 2.2) and can follow six procedures: farmers' experience (Section 2.2.1), traditional soil survey (Section 2.2.2), digital soil mapping (Section 2.2.3), soil

---

* All soils in this chapter are classified according to the FAO-UNESCO Soil Map of the World (FAO 1990) and Soil Taxonomy (Soil Survey Staff 1994).

sensing *in situ* networks (Section 2.2.4), proximal soil mapping (Section 2.2.5), and remote-sensing based soil mapping (Section 2.2.6). The second procedure is crop-based (Section 2.3) and can follow four procedures: visual crop mapping (Section 2.3.1), proximal crop sensing (Section 2.3.2), remote-sensing-based crop mapping (Section 2.3.3), and yield mapping (Section 2.3.4). Finally, the third procedure integrates soil and crop aspects by functional soil characterization (Section 2.4).

2.2.2 Soil-Oriented Management Units

2.2.2.1 Expert/Farmer-Knowledge-Based Units

2.2.2.1.1 Description

Farmers everywhere experience differences in crop growth within their fields but they often are not aware of the underlying mechanisms, nor is it reasonable to expect that they will be. This makes it difficult for them to select proper management procedures to improve conditions and increase yields in areas with lower yields. For example, local poor drainage or fertility call for different measures than locally compacted soil. Pronounced heterogeneity can, of course, be well interpreted by a farmer. For example, a farmer will recognize sandy spots within an otherwise clayey field or poorly drained spots in lower parts of a field, both calling for different types of management.

2.2.2.1.2 Implications for PA

When innovative PA technology is implemented it is important to consider farmers' experience (Heijting et al. 2011). The description of variability does not only cover many years and a wide range of conditions, it is also the current perception of the farmer. An example is provided in Figure 2.2 where a Kenyan farmer identified the main soil differences on his farm that were subsequently confirmed by chemical and physical soil analysis. In practice, a farmer may modify the application rate of chemical fertilizers by manipulating the handle of his or her fertilizer applicator as he or she drives over the field. The farmer may also apply more organic manure on problem spots. Researchers can define the underlying soil mechanisms contributing to the observed heterogeneity, and in turn, propose possible management procedures that can overcome local problems being experienced. Only after this analysis has been made should attention be focused on the type of technology that might be applied successfully under the circumstances that are not only defined by soil factors but also by operational aspects that are valid for the particular farmer. This can, obviously, be improved by better understanding of the underlying soil processes and PA can make a significant contribution here.

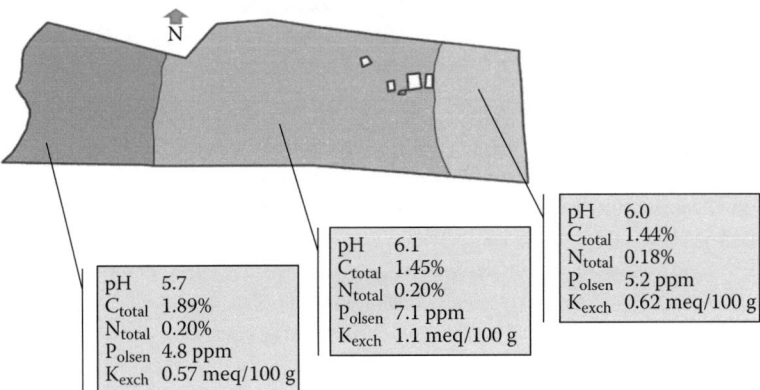

FIGURE 2.2   Soil variability on a 3-ha farm on the slopes of Mt. Kenya (Embu province, Kenya) as identified by the farmer, which were subsequently confirmed by physical and chemical analysis. (From Stoorvogel, J.J., Bonzi, M., and Gicheru, P., 2000. Spatial variation in soil nutrient stocks in sub-Saharan farming systems. Wageningen University. Wageningen, The Netherlands.)

2.2.2.2   Traditional Soil Survey

2.2.2.2.1   Description

Different soils occur at different landscape positions and observing and interpreting landscape patterns has been one of the major methods applied by soil surveyors (Soil Survey Division Staff 1993). The landscape scale (1:50,000 and smaller) has therefore been most effective when preparing soil maps. Soils by themselves are invisible and can only be observed by augering or by digging pits, a time-consuming activity. Therefore, the number of observations has to be limited and surveyors must use their interpretation of landscapes as a basis to select observation points, a procedure that is far from random. Soil surveys are made at different spatial scales and the rule of thumb suggests that for every square centimeter map area, one to two observations are needed (Soil Survey Division Staff 1993). Only then can different map units correctly be delineated. At the farm or field level less landscape features can be identified and often grid or random sampling is suggested. As well, legends of soil maps are based on soil classification that, in turn, is based on soil genesis, and therefore focusing on permanent soil properties as classifications should not change as a result of different types of soil management. Units on the soil map are named after the classification unit that is considered to be representative for that particular unit, ignoring likely internal variability.

2.2.2.2.2   Implications for PA

To characterize within field variation, detailed soil maps of at least a scale of 1:5000 are needed. Smaller-scale maps do not allow for the expression of details on field scale, where relations between soils and landscape features are often less obvious. The 1:5000 scale requires many observations (4–8 observations/km^2) and results in a costly procedure that yields a map with pedological units that, as such, have little direct significance for plant growth. Figure 2.3 shows a traditional 1:5000

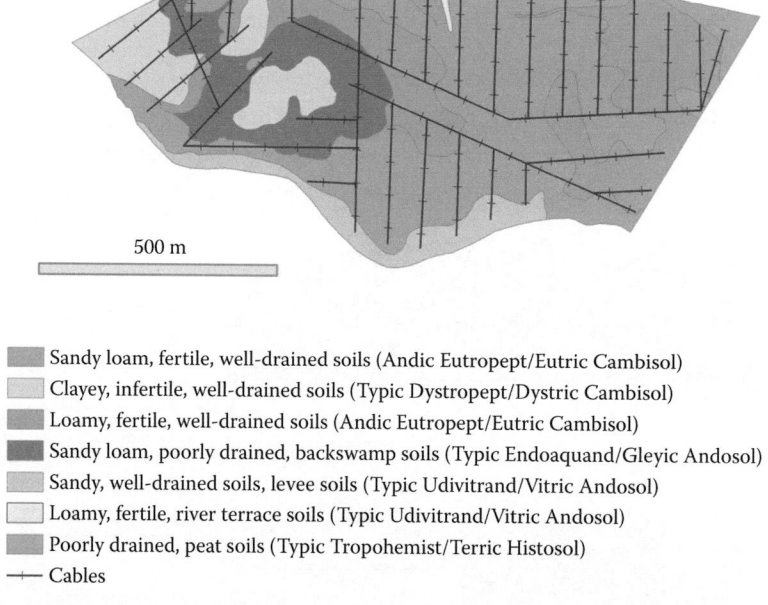

500 m

    Sandy loam, fertile, well-drained soils (Andic Eutropept/Eutric Cambisol)

    Clayey, infertile, well-drained soils (Typic Dystropept/Dystric Cambisol)

    Loamy, fertile, well-drained soils (Andic Eutropept/Eutric Cambisol)

    Sandy loam, poorly drained, backswamp soils (Typic Endoaquand/Gleyic Andosol)

    Sandy, well-drained soils, levee soils (Typic Udivitrand/Vitric Andosol)

    Loamy, fertile, river terrace soils (Typic Udivitrand/Vitric Andosol)

    Poorly drained, peat soils (Typic Tropohemist/Terric Histosol)

    Cables

FIGURE 2.3   A traditional soil survey for the La Rebusca banana plantation in the northern Atlantic zone of Costa Rica (Kooistra et al. 1997); soil classification according to FAO (1990) and Soil Taxonomy (Soil Survey Staff 1994) between parentheses. The black lines represent the cable structure that is the core of all management operations on the plantation. A further description is provided in Section 2.3.4.

soil survey of a Costa Rican banana plantation (for further details see Section 2.3.4). To translate the soil units into site-specific management, specific sampling, another interpretation of the representative soil profiles of the various mapping units, or *in situ* monitoring networks (see Section 2.2.2.4) are required to translate the soil types into relevant soil qualities that can be matched to crop requirements. Differences on the soil map can, but do not necessarily, result in differences in management recommendations for the mapping units. Soil mapping units are implicitly assumed to be homogeneous, theoretically allowing an internal variation of a maximum of 15%. This, most likely, does not necessarily apply to most soil mapping units even at the 1:5000 scale, representing a key problem when applying traditional soil maps for PA. Soil survey focuses on establishing different pedological units that are assumed to be homogeneous, while PA focuses on functional differences between subareas within a field. These functional differences do not necessarily coincide with pedological boundaries on the soil map.

2.2.2.3 Digital Soil Mapping

2.2.2.3.1 Description

Soil surveys have been made since the early 1900s and are based on relationships between soil types and soil forming factors such as parent material, climate, vegetation and fauna, topography, time, and humans (Jenny 1941; Turk et al. 2012). In contrast to the traditional soil survey described in Section 2.2.2.2, modern soil survey increasingly applies additional auxiliary data and ICT technology to quantify the relationships between soil properties and auxiliary data. Under the basic assumption that auxiliary information such as digital elevation models, satellite imagery, and weather data provide insight into the continuous variation of basic soil-forming factors, the new surveys make use of new techniques such as geostatistics and geographical information technologies to derive quantitative relationships between auxiliary data and important soil properties. These new survey techniques, called digital soil mapping (DSM), have a number of key differences compared to traditional soil survey (e.g., McBratney et al. 2000 and 2003; Van Zijl et al. 2014). First, DSM techniques are much cheaper due to the limited number of field observations that are required by making extensive use of auxiliary data. Second, digital soil mapping provides quantitative maps of soil properties. The focus was initially on topsoil and the techniques are being further developed to provide three-dimensional (3-D) descriptions of the soil properties (e.g., Kempen et al. 2011; Meersmans et al. 2009). Alternatively, digital soil mapping is also being applied on qualitative soil information such as soil types (e.g., Kempen et al. 2009). Third, the maps resulting from digital soil mapping describe variation in soil properties continuously. The resolution of the maps often corresponds to the resolution of the auxiliary information that is being used, which can be highly variable.

2.2.2.3.2 Implications for PA

The advantage of digital soil mapping techniques is that the variability is described continuously for the entire field or farm. If so desired the information can be generalized into a number of discrete units. Typical problems in the implementation of digital soil mapping techniques is the focus on a limited number of soil properties that are evaluated independently whereas the crop-soil interface relies on the entire pedon. For practical applications in PA it is required that one has *a priori* knowledge of which soil properties are most relevant to crop growth and possible management variation, but such knowledge is not necessarily available. Digital soil mapping has proven to be a very efficient technique in comparison to traditional soil survey. However, it strongly relies on the use of auxiliary data, which needs to have a degree of detail corresponding with variability encountered when managing the soil. Given the spatial scale of precision management, digital soil mapping for PA originally relied strongly on digital elevation models as the key explanatory factor. However, increasingly proximal and remote sensing images (see also Sections 2.2.2.5 and 2.2.2.6) are used in digital soil mapping exercises. Lagacherie et al. (2006) and the *Journal of Precision Agriculture* provide a range of different applications where digital soil mapping is used to support PA.

2.2.2.4 Soil Sensing *In Situ* Networks

2.2.2.4.1 Description

Over the past decade *in situ* networks of soil sensors are increasingly being used for high-frequent monitoring of soil conditions (Zerger et al. 2010). Individual sensor nodes of the network located at different positions and depths within an agricultural parcel provide spatiotemporal insight into soil variation. The main development of *in situ* sensor observation systems is focused on the characterization of soil moisture content and soil temperature at multiple depths in the soil profile (López Riquelme et al. 2009). For example, soil moisture sensors have been combined with data from weather stations and remote sensing data on crop status for irrigation scheduling (Vellidis et al. 2008). More research-oriented systems have been evaluating networks consisting of ion-selective electrodes to monitor ammonium, calcium, carbonate, chloride, pH, reduction-oxidation, and nitrate (Ramanathan et al. 2006). However, further development is required to reduce the effect of sensor response variability. Wireless data transfer is a critical requirement to use soil-sensing networks for real-time support of PA practices such as precision irrigation. With systems communicating aboveground, robust results have been achieved. While good performance has been shown for underground sensor networks (Ritsema et al. 2009) as well, communication through soils can be limiting, especially when distances between nodes are longer.

2.2.2.4.2 Implications for PA

Soil sensor networks provide high-frequent observations without complete spatial coverage of the parcel. To optimize the representativeness of sampling within a parcel, the location of the sensor nodes could be based on delineated management zones within a parcel. Off-the-shelf wireless soil-sensing systems are available ranging from advanced systems that combine soil moisture sensors with meteorological measurements (e.g., rainfall) to low-cost miniature sensors capable of measuring one specific soil variable (e.g., Ibutton for soil temperature). Wireless communication allows real-time data acquisition and direct communication from the sensor node with farm management systems. By combining the high-frequent sensor observations with spatial-explicit soil hydrological models, the spatial-temporal variability of the soil water balance within a parcel can be characterized. This would allow the support of precision irrigation practices taking into account spatial soil variation within the parcel.

2.2.2.5 Proximal Soil Mapping

2.2.2.5.1 Description

Proximal soil sensing techniques can be defined as field-based techniques that can be used to measure soil chemical, physical, biological, and mineralogical properties from a distance of approximately less than 2 m above or within the soil surface (Bartholomeus et al. 2011). Over the past decade a broad variety of proximal sensing techniques has been developed ranging from diffuse reflectance spectroscopy using visible (VIS), near-infrared (NIR), or mid-infrared (MIR) wavelengths (Viscarra Rossel et al. 2006), electrical conductivity (EC) (Mertens et al. 2008), gamma-ray radiometry (GRR) (Robinson et al. 2009), and ground-penetrating radar (GPR) (Lambot et al. 2006). VIS, NIR, and MIR techniques can provide information about several soil properties of topsoil (e.g., soil organic carbon, texture, nitrogen content, and pH) (Lee et al. 2010). Viscarra Rossel et al. (2006) compared the performance of different wavelength ranges using VIS, NIR, and MIR. Prediction accuracy varied greatly depending on the soil property. GRR and EC provide subsurface textural and soil-structure information. GPR also quantifies the soil properties of the subsurface and is capable of providing information about soil moisture. In general, the techniques are based on local statistical calibration of an indirect relation between measured signal and quantitative soil properties derived from lab-analyzed soil samples. As such the underlying soil processes are not well described, which hampers the development to more general applicable prediction relations or models. Next to the indirect soil property maps the variation of the measured signal is also used to characterize soil variation within a parcel or as a basis for soil stratification and sampling.

2.2.2.5.2 Implications for PA

In general, proximal soil sensing techniques provide detailed spatial sampling over the parcel and as such can provide a good continuous description of soil variability. Depending on the technique, it covers the upper 0.5 cm (VIS, NIR, and MIR) of the soil or acquires an averaged signal over the soil profile up to 1m (GRR, EC, GPR). Earlier studies have shown that based on local calibrations proximal soil sensing can provide good prediction and characterization for texture (EC), organic matter (VIS, NIR, and MIR), soil water (EC), soil compaction (mechanical), and pH (electrochemical) (Viscarra Rossel et al. 2010). While in some cases individual soil properties or proxies are used, a combination may be more appropriate given the complex nature of soil and crop interface (e.g., Scudiero et al. 2013). Several proximal soil-sensing approaches are past the initial commercialization stage and are ready for operational use (see Section 2.3.6), often through specialized companies that thus could provide a soil mapping service to farmers. Figure 2.4 shows the spatial patterns of EC for an arable field 1 month before the planting of potato. For operational application in PA, site-specific calibration and validation of the proximal soil-sensing technique remains an important step in the mapping process. In the long term, systematic sampling of sensor measurements and associated laboratory analysis for a large number of soils and storage of the results in a large-scale calibration database could reduce this effort and potentially reduce the costs. Every soil-sensing technology has strengths and weaknesses. However, the integration of multiple proximal soil sensors provides the ability to improve the prediction accuracy of individual soil properties. Kweon (2012) showed, for example, that combining EC, VIS, and NIR measurements acquired all at once improved the prediction of organic matter and the cation exchange capacity at the soil surface.

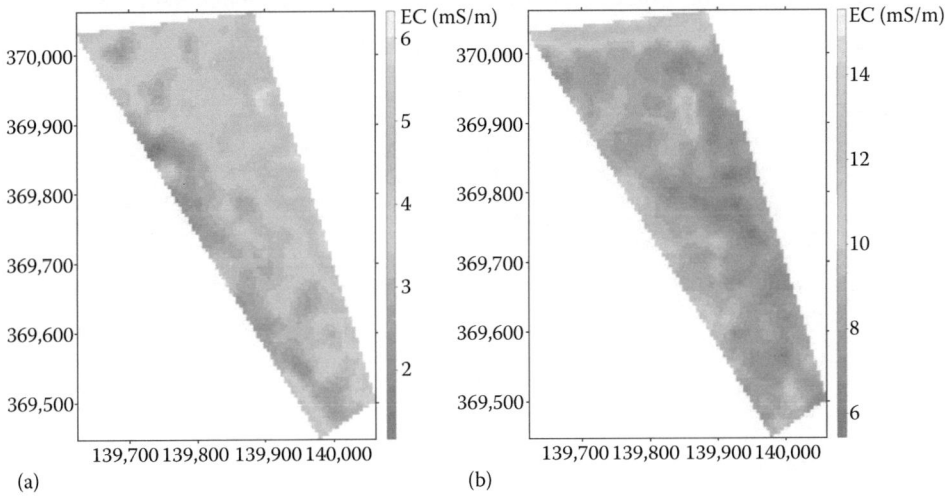

FIGURE 2.4 Spatial patterns of EC (mS/m) measured with an EM38-MK2 sensor (Geonics Ltd., Canada) for experimental field at van den Borne Aardappelen in the Netherlands (Section 2.3.6). Panel (a) shows the signal in horizontal mode until 0.38 m representing soil organic matter content and panel (b) until 0.50 m in vertical mode representing the depth of the A horizon. The *X*- and *Y*-axes represent coordinates in meters. EC = electrical conductivity.

2.2.2.6 Remote-Sensing-Based Soil Mapping

2.2.2.6.1 Description

In this context, remote sensing refers to observation of the earth from aerial or satellite-based platforms measuring the amount of reflected electromagnetic energy to determine soil properties and associated soil variation within a parcel. Two recent review papers by Yufeng et al. (2011) and Mulder et al. (2011) show the broad range of sensing platforms, sensor types, and data analysis techniques that quantify a large array of agriculturally important soil properties (including texture, organic and inorganic carbon content, macro- and micronutrients, moisture content, cation exchange capacity, electrical conductivity, pH, and iron) using remote sensing with varying degree of success. Most studies thus far have been performed on a local scale and to date no coherent remote sensing methodology has been established for complete area soil mapping (Mulder et al. 2011). This is for the most part explained by the complexity and indirect relations between soil components that makes the remotely sensed soil spectrum a difficult signal to translate into soil properties (Ben-Dor et al. 1999). With regard to satellite-based platforms, multispectral sensors like Landsat, ASTER, MODIS, and MERIS provide images with a spatial resolution of 30–300 m. Typically, these images support the segmentation of the agricultural landscape into homogeneous soil-landscape units and the spatial interpolation of sparsely sampled soil property data based on digital soil mapping approaches (Mulder et al. 2011). The availability of remote sensing systems with a high number of spectral bands, so-called imaging spectrometers, has improved the retrieval of soil properties because these systems detect the subtle differences between and within soil spectral signatures (Stevens et al. 2008). Using multivariate regression techniques, soil properties can be estimated from spectroscopic remote sensing data with increased accuracy compared to multispectral remote sensing data. Currently, imaging spectrometers are mainly available from airborne platforms, but in the coming decade several satellite-based imaging spectrometers will be launched. Another application from airborne platforms has shown that high-resolution aerial photographs can be adopted for quantitative analysis of topsoil organic matter (Figure 2.5) and automated delineation of management zones (Bartholomeus and Kooistra 2012). Variation at the level of organic matter within the field is an especially important driver for potential potato yield. This principle was also evaluated for a parcel on a sandy soil system using very high-resolution red-green-blue (RGB)

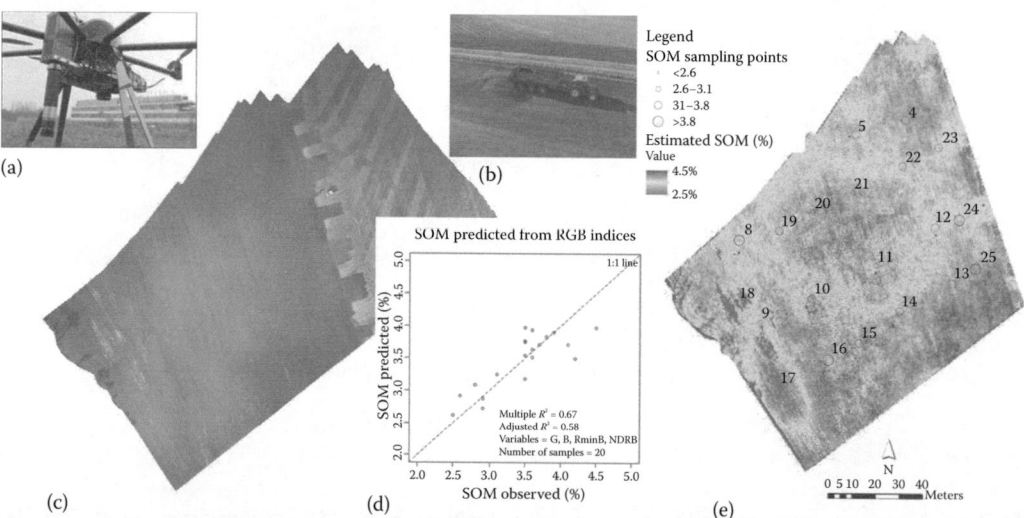

FIGURE 2.5 Variable rate application of compost for an experimental field at van den Borne Aardappelen in 2013 (Section 2.3.6). Through the use of an unmanned multirotor system (a) compost was applied at variable rates (b) using a high-resolution RGB image acquired before planting of the potato (c) that was interpreted (d) as a soil organic matter map (e).

imagery acquired with an unmanned aerial multirotor system (Figure 2.5). Using a multilinear regression model ($R^2 = 0.58$) and a set of calibration samples, a soil organic matter map was prepared. Due to the darkening effect of soils with higher organic matter content, spatial variation of organic matter based on local calibration is quantified. An important recent development is the increased availability of unmanned aerial systems (UASs) that can provide imagery with a high spatial resolution, flexible timing, and image acquisition programming and relatively low cost of operation for agricultural monitoring (Zhang and Kovacs 2012).

2.2.2.6.2   Implication for PA

Remote-sensing-based soil mapping allows characterization of soil variability over relatively large surface areas; however, compared to proximal soil sensing the spatial detail is often coarser. This means that for remote sensing systems with a spatial resolution >100 m, characterization of within-parcel soil variability will be limited. In addition, remote-sensing-based characterization only represents the upper few millimeters of the soil and does not provide information on the variability within the soil profile. An important limitation for remote-sensing-based soil mapping is the timing of bare soils within the cropping system since for a large part of the year the soil is covered with crops or crop residues. In temperate regions this bare soil coincides with wintertime, which is also the period with low incoming solar radiance, resulting in lower signal quality of the remote sensing images. This limits image acquisition for bare soil studies in the cropping system to only a few weeks (Bartholomeus et al. 2011). Although several research projects have shown the capabilities of remote-sensing-based soil mapping (Mulder et al. 2011; Yufeng et al. 2011), there is no standardized soil remote sensing product yet available. For example, the current MODIS product list offers over 70 remote-sensing-derived products but no soil-related products. However, satellite-based remote sensing images are an important source as covariate for prediction of soil classes or soil properties in digital soil mapping (Section 2.2.2.3). In addition, the increasing availability of imaging spectroscopy-based sensors even in combination with UASs will also allow increased capability for quantitative assessment of soil variability with high spatial detail within the parcel (Bartholomeus et al. 2014).

2.2.3   Crop-Based Management Units

2.2.3.1   Visual Crop Characterization

2.2.3.1.1   Description

As long as farmers have been growing crops on their fields, they have been observing variability in crop development and yield differences within the parcel. Often together with agronomic advisors, they try to diagnose the reasons for the visually detected variation and identify management interventions to improve crop performance. The management interventions are based both on factors related to crop and soil management. Characterization of crop variation is mainly related to visual observation and the agronomic experience of farmers. In addition, specialized agronomic laboratories provide services for high-quality soil or crop analysis to translate the visual observation into appropriate management. Recently, an increasing amount of field kits, sensors (Samborski et al. 2009), and features and apps on smartphones (Ferster and Coops 2013) allow farmers to prepare for a more quantitative characterization of the crop status.

2.2.3.1.2   Implication for PA

Visual crop characterization links closely to the farmers' experience on crops grown and parcels under management. Earlier studies (Heijting et al. 2011) have shown that farmers have considerable spatial knowledge of their fields and they apply this knowledge intuitively during various field management activities such as fertilizer application, soil tillage, and herbicide application. As a first phase in PA this spatial knowledge would need to be recorded, for example, by delineating

| May 31 | June 11 | June 23 | July 5 | July 14 | Aug 19 | Sept 6 |

FIGURE 2.6 Example of visual crop observation using oblique photographs showing the development of a potato crop in one of the experimental plots as part of the experimental field in 2010 at the farm of van den Borne Aardappelen, the Netherlands (Section 2.3.6).

main homogeneous units and their associated agricultural management history within a parcel. Over the growing season, farmers' observations of crop development or anomalies can be localized and stored as photographs using the Global Positioning System (GPS) and camera features of a smartphone (see Figure 2.6). This information together with other localized sensor-based crop observations can be directly communicated to agronomic experts for evaluation or during the after-season evaluation of PA practices within the parcel to support the diagnosis of soil- or crop-related limitations.

2.2.3.2 Proximal Crop Sensing

2.2.3.2.1 Description

Considerable progress has been made in the use of proximal crop-sensing techniques with handheld and tractor-mounted sensors for crop monitoring in PA (Goffart et al. 2008; Samborski et al. 2009). For these so-called near-sensing systems, different commercial devices (e.g., Yara, Greenseeker, Cropcircle, and Isaria) are currently on the market that measure reflectance in a small number of relatively broad spectral bands using their own active light source. In an operational setting, the sensors are mounted on the tractor or the spraying boom and measurements are acquired when agricultural activities on the field are carried out (Tremblay et al. 2009). This results in a regular point sampling of the field depending on the number of sensors and distance between them and the velocity of the mobile platform during acquisition. The output of near-sensing instruments consists either of more green biomass-related vegetation indices such as the normalized difference vegetation index (NDVI) and red-edge position (REP) or system-specific indices that represent the relative difference in crop conditions. Currently, proximal crop sensing is primarily used for optimization of nutrient management within a parcel mainly related to nitrogen deficiency. This also requires a set of decision rules next to the measurement system that translates the measured vegetation index value into a fertilizer recommendation. These are then transmitted to a spreader equipped with a rate controller to apply appropriate amounts of fertilizer on the go. For example, Berntsen et al. (2006) adopted the Yara sensor to target nitrogen fertilizer in fields of winter wheat. Their results showed that better relationships between sensor measurements and grain yield could be achieved when improved sensors would be able to describe additional crop features (e.g., leaf area index (LAI) and canopy characteristics) or soil properties (e.g., soil organic matter, water content). An intercomparison study by Tremblay et al. (2009) showed that although two commercial sensors (Greenseeker and Yara) were both capable of characterizing differences in crop growth resulting from differences in nitrogen status, marked differences in NDVI development over the growing season between sensors were observed. Both studies indicate that integration between near-sensing instruments or combined use of near-sensing with other sensors still requires further improvement.

2.2.3.2.2 Implication for PA

Proximal crop-sensing techniques provide detailed measurements in space and depending on the observation frequency also over the growing season. Figure 2.7 shows the variation of the weighted

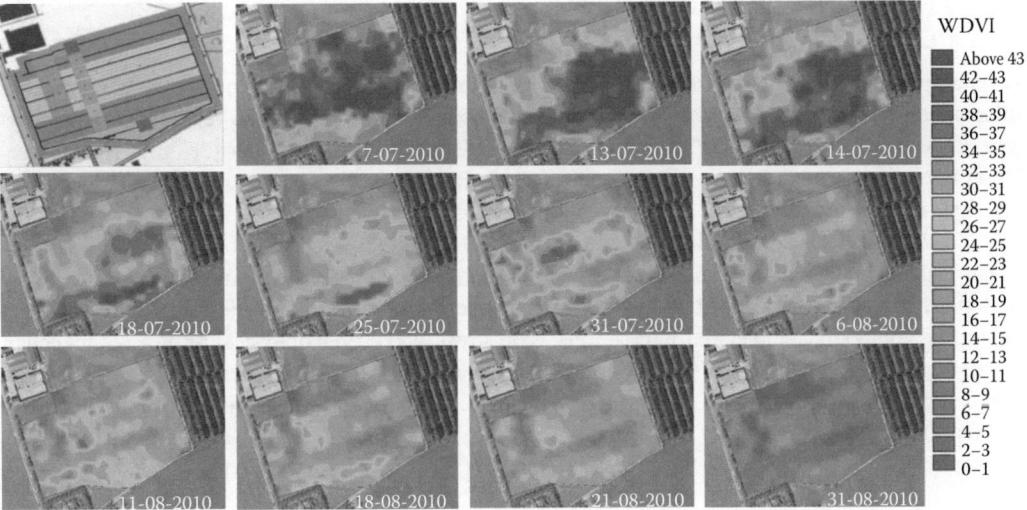

FIGURE 2.7 Development of potato crop for experimental field at van den Borne Aardappelen, the Netherlands (Section 2.3.6) in 2010 based on measurements of weighted difference vegetation index from proximal crop sensing. The upper left map shows the location of experimental plots and the layout of the fertilization experiment that was used to establish a relationship between sensor measurements and the crop variables aboveground biomass and nitrogen content.

difference vegetation index (WDVI) as an indicator of aboveground biomass for an experimental field of van den Borne Aardappelen in the Netherlands in 2010 (see also Section 2.3.6). Clear differences in biomass development for initial fertilization levels can be observed. The sensor observations are translated into spectral vegetation indices; that is, WDVI to characterize aboveground biomass (AGB) (kg/ha) and REP to monitor total aboveground nitrogen (AGN) content (kg/ha). Dedicated fertilization experiments were set up between 2010 and 2013 to establish the relation between the vegetation indices, crop variables, and nitrogen rates. Based on these relations and additional agronomic crop modeling, decision rules were established translating the sensor measurements into fertilization rates that could be used for variable rate application (VRA) of fertilization of nitrogen application over the growing season. The frequency of observation depends on the management intensity of the crop under cultivation but compared to remote-sensing-based systems it is independent of atmospheric limitations (e.g., clouds). As crop growth is closely linked to available water and nutrient resources in the soil system, which means that although proximal crop-sensing techniques give a direct measurement of crop status, they could also be adopted as an indirect method to characterize soil variability within a parcel. This will not allow direct determination of soil properties when the driving factors underlying the differences are also unknown. By analyzing the spatial-temporal variation in the proximal crop-sensing signal over the parcel in relation to additional soil data sources, potential problem areas can be identified where crop development was reduced or delayed (Thessler et al. 2011). This allows segmenting the agricultural parcel into discrete management units for which further diagnosis is required, including potential soil-related limitations. This segmentation could be used as a basis for a further functional soil characterization. As a next step, predictions of crop nitrogen availability derived from proximal crop-sensing techniques could be integrated with plant production models to monitor fertilization status over the growing season (Van Alphen and Stoorvogel 2000). Recommendations for site-specific in-season fertilization are then based on balanced nutrient availability in both the soil and plant system.

2.2.3.3 Remote-Sensing-Based Crop Mapping

2.2.3.3.1 Description

A range of satellite- and airborne-based remote sensing techniques providing spatially continuous information at different scales have already proven to be relevant for many requirements of crop inventory and monitoring (Atzberger 2013; Haboudane et al. 2002). In the past decade, high spatial resolution operational satellites (e.g., Ikonos, QuickBird, RapidEye, GeoEye, and WorldView-2) with a small number of broad spectral bands (usually blue, green, red, and NIR) have aimed at providing solutions for PA. Earlier studies have shown the application of these satellites for anomaly detection in wheat (Eitel et al. 2007) and maize (Bausch and Koshla 2010) for specific moments over the growing season. The increasing spectral bandwidth of hyperspectral remote sensing allows improved analysis of specific compounds, molecular interactions, crop stress, and crop biophysical or biochemical characteristics (Mulla 2013). Recently, simulation studies have shown that the increased spectral resolution of upcoming superspectral sensors like Sentinel-2 and VENµS and hyperspectral systems like EnMap provide the opportunity to estimate biophysical properties like LAI (Herrmann et al. 2011), chlorophyll (Verrelst et al. 2012), and leaf water content (Clevers et al. 2008) with improved accuracy.

2.2.3.3.2 Implications for PA

Variation in remotely sensed spectral vegetation indices within a parcel can be explained by the interaction between the soil and plant system assuming that the same crop variety has been planted or sown over the complete parcel. Remote-sensing-based crop mapping provides continuous spatial coverage over the parcel with up to a pixel size of 1 m depending on the remote sensing system. An example is provided in Figure 2.8 for plant density and yields derived for a Costa Rican banana plantation based on Landsat and Ikonos images (see also Section 2.3.4). The combined spatial-temporal detail of high-frequent remote sensing observations allows the characterization of soil variability within the parcel in an indirect way (e.g., Song et al. 2009). The identified variability could directly optimize soil sampling procedures within the parcel. Although optical remote sensing is a promising technique for crop detection and diagnosis in PA, there are still some important limitations related to continuity, near-real-time availability, and usability of data streams that need to be tackled. More specifically, remote-sensing-based crop monitoring requires higher-resolution systems (<30 m) that can provide the required spatial detail. Currently, the revisit periods of these satellites systems (e.g., Landsat-TM, SPOT-HRV) are too long relative to the active vegetative growing period; however, new satellite systems have considerably better revisit capabilities (e.g., Worldview-2, RapidEye, future Sentinel-2). In the case of cloud cover during critical moments over the growing season, alternative technologies like radar and UAS-based sensing are able to acquire data by sensing through or under the clouds, respectively.

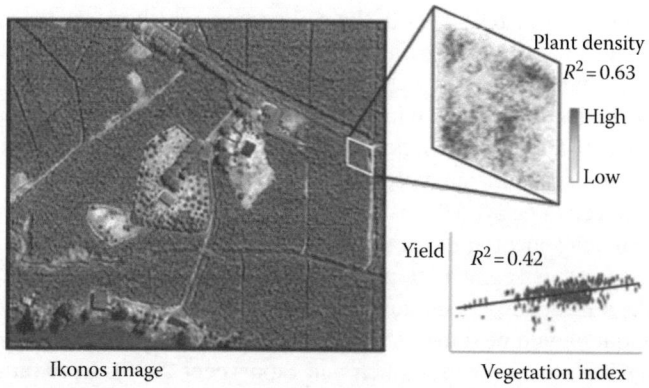

FIGURE 2.8 Derivation of crop characteristics such as plant density and yield from satellite imagery (Ikonos and Landsat7) for the Rebusca banana plantation in the Atlantic zone of Costa Rica.

2.2.3.4   Yield Mapping

2.2.3.4.1   Description

A critical component in PA that is becoming a more widely adopted practice is yield sensing and mapping during harvesting. Yield mapping systems usually incorporate data from a yield sensor, a moisture sensor, a ground speed sensor, an elevator speed sensor, and a Differential Global Positioning System (DGPS) receiver to relate sensed crop yield to a location (Arslan and Colvin 2002). Most major agricultural equipment companies now provide optional yield-mapping systems for their combine harvesters. As a result, yield mapping is most common for grain-based crops but commercial yield mapping systems are also available for tuber-based crops like potato and sugar beet. In literature, several studies have been conducted to assess the accuracy of yield measurement and yield mapping systems (see Thessler et al. 2011 and references therein). As has been indicated by Robinson and Metternich (2005), several factors can influence the accuracy of the yield mapping system, which include yield map smoothing errors, unknown crop width entering the header during harvest, missing data, and positional GPS errors due to stopping or turning of system at the end of the parcel. Figure 2.9 shows an example of a yield map for a potato crop (see Section 2.3.6). The driving paths are clearly visible and between these clear variations in yield are observed. The map also shows some areas along the driving lines with missing data and increased uncertainty in parts where the yield sensor needed to turn (Figure 2.9b). Next to sensor-based yield mapping systems, examples of whole-farm-based mapping systems have been developed for bananas that use specific procedures for weighing spatially localized bunches using the existing cable infrastructure as a localization grid (Stoorvogel et al. 2004). Crop response can vary between years.

2.2.3.4.2   Implications for PA

Yield maps are a critical information source to evaluate the efficiency of PA. On the one hand, yield maps provide an integrated signal of influencing soil and crop factors on crop yield measured at one point at the end of the growing season. On the other hand, for the farmer it provides the direct link to economic efficiency of the agricultural management that has been carried out over the growing season. In the case of PA, often these measures have been carried out spatially in an explicit manner and by comparing the adopted rates of variation with the spatial variation in yield. Thus, the efficiency of the measures can be evaluated and a cost-benefit analysis can be prepared. The absolute yield level within and between parcels is highly dependent on year-to-year variation of weather conditions as well as the effects of tara (i.e., soil and crop remains included in weight of harvested

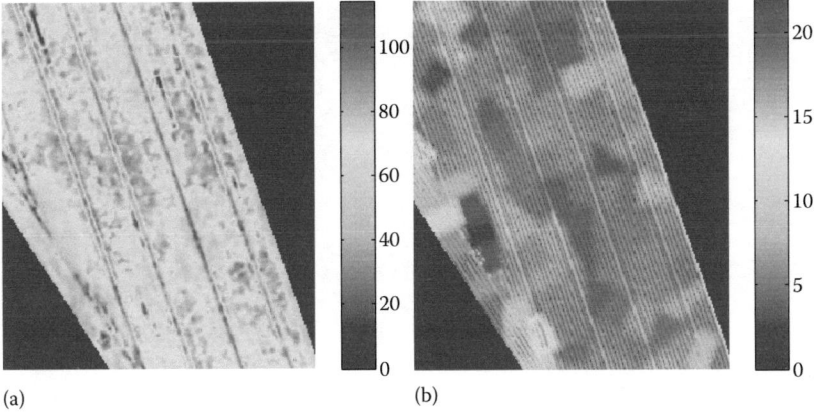

(a)   (b)

FIGURE 2.9   Yield map for potato crop (a) in ton/ha derived from interpolation of original point observations of harvesting equipment. Panel (b) shows the associated prediction errors. Yield map for a 2013 parcel from the farm management system of van den Born Aardappelen, the Netherlands (Section 2.3.6).

crop), and yield quality has to be taken into account. Therefore, several years of yield mapping data are required to characterize the spatial variation of potential yield within a parcel. Although the yield maps provide an *ex post* evaluation of the crop soil interface and therefore are unsuitable for operational decisions, multiple yield maps are an excellent source from which to derive stable management units (e.g., Brock et al. 2005).

2.2.4 Soil and Crop-Based Functional Management Units

2.2.4.1 Description

Traditional soil surveys make use of one of the available soil classification systems that focus on pedogenesis and the soil-forming factors rather than on soil properties. A functional soil characterization is an alternative approach where the individual soil observations are interpreted in a functional way and that their behavior forms the basis for the classification and interpolation. An example is provided by Van Alphen and Stoorvogel (2000). The procedure is illustrated in Figure 2.10. The soil data is collected in a similar fashion as for traditional soil survey. The data is supplemented by specific soil chemical and physical properties using pedotransfer functions (e.g., Wösten et al. 2001) since actual measurements for all the observations are practically impossible. Soil characterization focused on functional properties describing soil-specific characteristics in terms of water regimes and nutrient dynamics: water stress in a dry year, nitrogen (N)-stress in a wet year, N-leaching from the root zone in a wet year, and residual N-content at harvest in a wet year. These functional properties were quantified for individual soil profiles using the mechanistic–deterministic simulation model Water and Agrochemical in Soil and Vadose Environment (WAVE) by Vanclooster et al. 1994. A fuzzy c-means classification (FCM) was applied to identify classes of

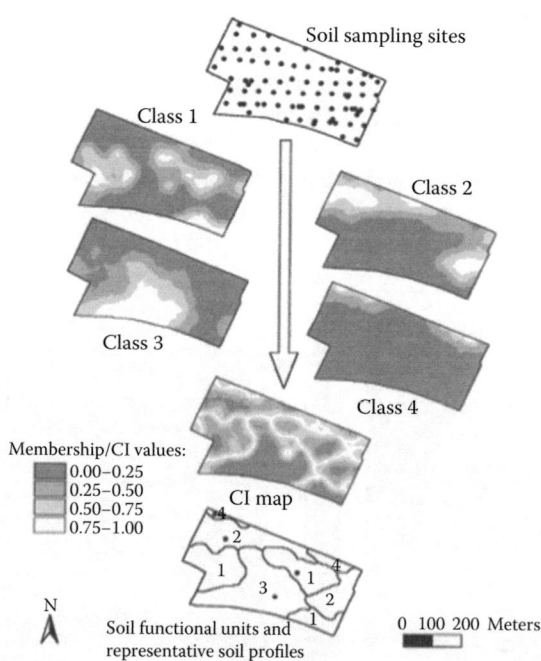

FIGURE 2.10 Derivation of the soil functional management units for a field at a Dutch arable farm. After an initial survey with six soil sampling points per ha, water stress, N-stress and N-leaching, and residual N are simulated. Sampling points are classified into different soil classes on the basis of the functional properties, after which the membership values are interpolated leading to a final definition of the management units. (From Van Alphen, B. J., and J. J. Stoorvogel. *Soil Sci. Soc. Am. J.* 64:1706–13, 2000.)

functional similar soil profiles. Standard interpolation techniques yielded maps of the membership values. A confusion index (CI) was calculated identifying the location where the classification does not provide a clear membership to one of the classes. The CI is the basis for a boundary detection algorithm that identified soil functional units. The management units described more than 65% of the soil variation.

2.2.4.2 Implications for PA

The study by Van Alphen and Stoorvogel is a good example where the soil observations are interpreted in a functional way for crop production. The functional properties will obviously depend on the specific conditions for each case. The functional soil characterization has a number of advantages. The management units may have a strong relation to crop behavior and automatically focus on the relevant soil properties. As such, the units are strongly related to the core of precision agriculture to optimize production with minimal environmental impacts. A major drawback for the procedure is that it requires serious data collection and data analysis. It should also be taken into consideration that the mapping units will vary for different crops. This basically means that if farmers are using a crop rotation the units may vary over the rotation. These units, can, however, be based on the point data originally obtained. Functional mapping, as described, results in a permanent database for the farmer—a modern soil map. The one-time costs of preparing such a map were quite acceptable to the farmer involved. Model calculations needed for the functional characterization can be realized ever more quickly and cheaply as computing capacities increase. This is well illustrated by a side study on application of various biocides, which varied strongly in terms of costs. Point data of the farm database were used to derive maps by simulation of biocide transport showing the leaching potential of different biocides (Van Alphen and Stoorvogel 2002).

2.2.5 DISCUSSION

The proper scale of the management units is not only determined by the inherent variability within fields but also by the spatial resolution that the farmer can manage considering his or her attention span and the possibilities of his or her technical equipment. This issue, to be further discussed in the case studies of Section 2.3, should be considered when designing operational PA systems. The introduction of new measurement and monitoring techniques has revolutionized both soil and crop science. Traditional methods of soil and crop analysis were time-consuming in terms of sampling and required transportation of samples to the laboratory followed by costly analyses with specialized equipment. Days or weeks were needed to obtain experimental data. Introduction of a variety of sensors has resulted in a scientific paradigm shift: now a variety of soil and crop data are instantly available, allowing, in principle, direct application when managing soils and crops. Limitations do, however, apply.

Taking the only realistic perspective of a farmer facing yet another growing season, he or she has to manage a highly complex soil-plant-atmosphere system with a high degree of uncertainty because weather conditions can only be predicted for, perhaps, a 1-week period. Management therefore has a highly adaptive character with a need to not only respond to unpredictable weather conditions but also to sudden, unexpected outbreaks of pests and diseases. The central challenge is to stay ahead of unfavorable developments rather than react to them. Soil traffic, either representing tillage or movement of equipment, should not occur at moisture contents that are too high as this may result in compaction or puddling of soil. Here, remote and proximal soil sensors are very useful even though they only reflect moisture conditions at the very surface that may not be representative for moisture contents in the upper soil horizons. Moisture sensors placed *in situ* are, however, also available and could play an important role here (Cosh et al. 2012). Crop sensors are very useful to signal critical crop conditions, such as water stress, nutrient deficiency, or effects of pests and diseases. The real challenge is, of course, to avoid such conditions because damage has already occurred when such symptoms are observed, and, obviously, sensing can only represent present conditions, while the farmer has to look ahead as well. As sensors cannot be helpful here, simulation modeling of the soil-crop-atmosphere system presents a clear alternative. Models are by now well tested (e.g., Michalczyk et al. 2014). For example, when a sensor observes N deficiency in plant leaves, yields are already suffering. When the farmer would know how much N was present at any time in the root zone of the plant during the growing season, he or she could keep track of daily plant consumption and only replenish the N by fertilization when N contents reached a threshold level of a remaining supply of, say, 3 days. Then the plants would not suffer, while excessive N application (and leaching) is avoided, with positive environmental results. Operational sensors for the *in situ* monitoring of soil nitrogen are not yet available. Modeling is therefore a suitable alternative (Michalczyk et al. 2014; Van Alphen and Stoorvogel 2002) but the availability of sensing data of both soils and crops is crucial to calibrate and validate simulation models. Soil sensing of only the surface presents a problem because crop growth is governed by the interaction between roots and soils in the root zone that may extend to more than 1m below the soil surface.

2.3 MANAGING THE MANAGEMENT UNITS: CASE STUDIES OF PA

2.3.1 GENERAL CONSIDERATIONS: THE COMPLEX NATURE OF MANAGING HETEROGENEITY IN SPACE AND TIME

The various methods described in Section 2.2 have to fit into an overall management plan that all farmers have to devise regardless of where they operate. Restricting attention to cropping would still include soil tillage, selection of a crop variety, sowing or planting, possible choice of a cover crop, timing of fertilization practices including the choice of fertilizers, amount and timing of irrigation if applicable, weed control and control of possible other pests and diseases, and harvest, followed by either a second crop or by tillage. The primary driver for PA is the heterogenous spatial development of the crop. This spatial variation can have several underlying causes that are associated to elements of the overall management plan. For example:

1. Tillage can result in local subsoil compaction at relatively wet locations in a field early in the growing season, and this in turn can restrict rooting patterns of the crop and reduce growth later in the season when wetness has disappeared. In wet climates subsoil compaction may also result in temporarily perched water tables also restricting root growth.
2. A selected crop variety may be more suitable for certain areas of a field than others when soil conditions vary considerably. As a result, crop performance will vary within the field.
3. Sowing or planting involves, again, pressure by agricultural equipment on the soil, possibly leading to local compaction that shows up in growth patterns of the crop, and at the same time small differences in plant densities may occur. Good examples of this are shown in Figures 2.6, 2.7, and 2.9, where the patterns are management-related.
4. Differences in soil fertility may lead to local nutrient limitations resulting in local differences in crop response.
5. The development of weeds and pests may have specific spatial patterns starting at certain point infestations or from weed development starting at field boundaries.

There are a number of reasons for observing patterns of heterogeneity and defining management units using the methods of Section 2.2, many of them soil-related. Optimal soil management can only be achieved when the farmer not only recognizes crop patterns but when he or she also understands the underlying processes. Only then is he or she able to select proper corrective management measures such as variable fertilization, application of biocides to combat pests and diseases, irrigation, tillage to remove compacted layers, and surface mulching to present crust formation. The case studies in this section will analyze which decisions the farmer has made during the growing season to make his or her management as effective as possible.

2.3.2   THREE TYPES OF DECISIONS TO BE MADE BY THE FARMER

Any farmer considering application of PA techniques has to make three types of decisions: strategic, tactical, and operational.

The *strategic* decision has a go/no-go character to adopt PA or not. The decision depends on quite a number of considerations in the context of a multiyear time frame. When differences in yield or crop quality within the farmer's fields have been relatively small over the years, there is little need to consider PA. But how does one define a level of heterogeneity where PA becomes relevant? What is the tipping-point? This definition will not only depend on variability as such, but also on the entrepreneurial character of the farmer and his or her willingness to take risks. As well, the local availability of equipment and expertise will play an important role. So far, research on variability of soils and of plant growth has been focused on variability as such, but not on defining its practical implications in terms of distinguishing tipping points. This is very much needed because farmers are confronted with many competing commercial firms that are primarily focused on selling their own equipment. The scientific community has a major responsibility to provide an objective feasibility test allowing farmers to make a correct strategic decision on PA that could invoke major capital investments such as variable rate applicators or site-specific yield monitoring.

Once a decision has been made to apply PA, *tactical* decisions, in the time frame of a growing season, have to be made as to which procedures are to be followed. This depends again on estimated costs versus benefits and availability of local support and supporting data. The decision to manage fields continuously or in discrete management units is an essential tactical decision, just as is how to define the management units. Section 2.2 gave an indication of the plethora of different ways to define management units that a farmer has to choose from. Some of these methods are linked to operational procedures that allow one to reach the ultimate objective of PA, which is to apply the right type of management to the right place at the right time. In practice, precision management will most often consist of application of agrochemicals but the concept also applies to irrigation and tillage.

After the tactical decision the farmer will be faced by numerous *operational* decisions during the growing season, including the entire cropping calendar. For each management operation, the farmer has to decide when and how (including how much for agricultural inputs) for each location (in the case of continuous management) or management unit (in the case of discrete units). The decisions are increasingly complex as we are moving away from a fixed cropping calendar with single management for the entire field to a dynamic calendar that may vary through the field.

Choosing a set of PA procedures that appears to the farmer to best suit his or her particular management objectives represents only the start of a learning process because PA techniques are new, every farm and farmer is different, and field applications are bound to present many problems and challenges. In fact, only field testing of procedures and on-the-go modifications of what appeared to be perfect techniques on paper can produce truly operational systems in the end. Here, close cooperation between farmers and researchers over an extended period of time is of crucial importance for both, as only a joint learning process will produce truly operational PA systems. This stage of development has not been reached as yet with PA. This joint learning process associated with *practical* PA will be further illustrated in the case studies in the following sections.

2.3.3   LOW-TECH PA IN SMALLHOLDER AGRICULTURE IN DEVELOPING COUNTRIES

2.3.3.1   General Context

Four fifths of the developing world's food is produced on about half a billion small farms, supporting more than at least 1 billion people (IFAD 2012). Smallholder farmers live and earn their livelihoods in the world's most ecologically and climatically vulnerable landscapes such as hillsides, drylands, and floodplains. Many belong to the poorest group of people and they have to rely on weather-dependent natural resources with little or no access to modern technology. The term "smallholder agriculture" covers a wide variety of farming systems in different parts of the world. In many developing countries small farms are often less than a few hectares in size supporting only the farming family. Recent reviews (e.g., IFAD 2012) emphasize the enormous diversity of small farming systems in developing countries, requiring a focus on local conditions when formulating desirable future developments as generalizations may be meaningless. A common feature of smallholder farms is the heterogeneous character of their land that results in highly variable yields within single fields. Near-the-farm household yields are usually relatively high because manure and waste products are deposited there. Yields decrease with increasing distance from the household. Sometimes, local spots of higher production are associated with locations where trees were growing in the past (e.g., Brouwer and Bouma 1997). As heterogeneity is the prime driver for PA, application in smallholder agriculture would appear to be of utmost relevance. However, smallholder farmers face significant problems, as will be discussed.

2.3.3.2   Strategic Decisions

Smallholder farmers in developing countries are certainly aware of not only their low yields but also of the heterogeneity of yields within their fields. A positive strategic decision as to whether or not to adapt their management, taking this heterogeneity into account, is easily made. But how to go about it? A specific element for the subsistence farmer is that within-field variation can play an important role in minimizing risks. Low parts of the field may give an acceptable yield in dry years due to increased moisture availability whereas the higher parts of the field may provide an acceptable yield in wet years.

2.3.3.3   Tactical Decisions

Typically, smallholders do not have the funds or the expertise to select the costly PA techniques discussed above. In any case, they are not even available in the areas concerned. However, larger farms can do so as is shown, for example, on wheat farms in India (e.g., Khurana et al. 2008). Giller et al. (2011), realizing that farmers have only a limited amount of chemical fertilizers available, advise them to apply it near their farms rather than in areas farther away where soils are often so depleted that a limited application of chemical fertilizer is ineffective and therefore wasteful. When discussing nutrient management in resource-poor areas, the Sustainable Development Solutions Network (SDSN 2013) advocates application of integrated soil fertility management (ISFM) strategies that make use of mineral fertilizers and locally available organic amendments but also promote other good management practices that are considered to be a key to increasing agricultural productivity and improving poor soils in Sub-Saharan Africa. This approach appears to ignore the key feature of this type of production system, which is its high internal heterogeneity. The reality is that most farmers and extension offices still rely on the typical blanket fertilizer use recommendations established through large-scale traditional field experiments (e.g., FURP 1994). Smallholders are left to use their common sense to distribute their limited quantities of organic and chemical fertilizer in the most efficient way. The definition of the management units often relies on farmer experience (as illustrated in Figure 2.2) or observed variation in crop performance in previous years.

2.3.3.4   Operational Decisions

Smallholder farmers will need assistance from the research community to fine-tune their empirical trial and error procedures, implementing, in fact, the most elementary form of PA. Traditional soil fertility research does not address the issue of how a limited amount of fertilizer can be applied in the most effective manner in fields with a pronounced fertility gradient. This needs attention in soil fertility research. Also, Bouma et al. (2014) pointed out that current studies often ignore the type of soil that occurs, thereby failing to discover that crop reactions to fertilizers are significantly different in inherently richer soils as compared with poorer ones. As is, smallholder farmers are left to their own creativity to receive maximum output of their, by necessity, very limited input. So far, the research community offers little assistance. Science tries to support the farmer using crop growth simulation models but even those efforts of two decades ago (e.g., Smaling and Janssen 1993) are not widely adopted because detailed farm-level recommendations require site-specific data that is lacking. New programs focusing on proximal sensing or digital soil mapping (See Section 2.2) try to fill these gaps but it is too early to judge their success.

2.3.4 LOW-TECH PRECISION AGRICULTURE FOR LARGE-SCALE COMMERCIAL PLANTATIONS

2.3.4.1 General Context

Globally, Costa Rica is the fourth largest banana exporting country. Currently, bananas are produced at large plantations that are either owned by independent producers (50%) or multinational companies (50%). All producers are organized in the Costa Rican Banana Corporation (Corbana). The banana producers have to operate in a complex set of political conditions. Costa Rica provides over 25% of the bananas consumed in Europe. The European Union supported banana production in Africa, the Caribbean, and the Pacific (ACP) through a complex tariff quota system restricting imports from, among others, Costa Rica. The use of biocides is constrained by the regulatory agencies in the United States and Europe. They determine which pesticides are allowed and how they can be applied. A plethora of different labels have been developed in Europe and the United States focusing on chemical use (organic bananas), social production conditions (fair trade), and the environment (ISO 14001, the environmental management standard). In many cases producers have to follow the guidelines of these labels to be able to export. Large buyers of bananas (e.g., supermarket chains) force the banana plantations to produce under a detailed tracing and tracking system. Plantation agriculture in the tropics is in many aspects similar to Western agriculture. It is therefore an agroecosystem that is suitable for the implementation of PA. A more detailed look reveals a number of operational problems. Many crops (coffee, tea, bananas, rubber) in tropical plantations cannot be characterized by mechanistic simulation models and the processes are less well understood than in Western agriculture. Second, we find that farm management relies to a large extent on manual labor. This makes the technological innovations that triggered the introduction of PA more difficult to implement. A medium-tech approach for PA was therefore developed on the *La Rebusca* banana plantation in the northern Atlantic zone of Costa Rica (Stoorvogel et al. 2004). The plantation measures 100 ha and is located on dominantly fertile, well-drained soils. It uses a perennial production system with intensive use of agrochemicals to replenish soil nutrients and to control pests and diseases. The need for changes in the management system is based on a reduction of production costs and the pressure of environmental organizations as well as consumers.

2.3.4.2 Management Decisions

2.3.4.2.1 Strategic Decisions

As a commercial enterprise the first objectives of management are economical and can be described as profit maximization. Nevertheless, the owner of the plantation is also looking at the environment and tries to minimize the use and indirectly the emissions of agrochemicals to the environment. In contrast to other plantations the owner is seeking innovations in the production system. This can be explained in part by a vision toward the future where he expects environmental regulations to tighten. After confronting the owner with the concepts of PA, the owner was highly interested. He was well aware of the large differences in soil conditions and crop performance. As a result, he saw opportunities in terms of chemical use efficiency and the registration of production activities. The implementation of PA required the development of a system for yield mapping and a soil survey to describe the spatial variability on the plantation.

2.3.4.2.2 Tactical Decisions

In contrast to common practices in PA, standard yield mapping techniques as described in Section 2.3.4 could not be used. In addition, the technological focus of PA was not easily adoptable in a continuous cropping system with almost no mechanization and management by manual labor. Banana plantations in Costa Rica are frequently (two–three times per week) screened for bunches ready for export. Bunches are harvested and transported by a dense cable system (every 100 m) to the packing plant. For the yield monitoring, groups of bunches are coded based on their origin (cable number and the location within the cable). A weighing balance is installed in the main cable just before the packing plant. Codes are registered at the balance and bunches are weighed. A software package called BanMan was developed for data processing and to create and analyze the yield maps. In addition to the yield maps a detailed soil survey (Section 2.2.2.2) has been carried out for the Rebusca plantation (Kooistra et al. 1997) resulting in the soil map of Figure 2.3. Traditionally, banana plantations operated the plantations using the cable infrastructure (Figure 2.3) as the basic management units for fertilization, harvesting, and so forth. The yield maps and soil survey indicated that the accumulative yields of cables described very poorly (e.g., only 23% of the yield variability is explained by the cables) the variation in the plantation as they are located in the direction of the slope (note that slopes are minimal <1%) to facilitate transport. To maximize the description of the variability but at the same time have management units that are easy for management operations, management units were defined by dividing the existing cable into different compartments on the basis of soil and yield maps. There are many different factors that may cause yield differences in the banana crop ranging from drainage and soil fertility to pest and diseases. It was therefore extremely difficult to interpret the yield maps in combination with the soil maps. The yield maps were evaluated by calculating relative yields compared to the average for the different soil types as illustrated in Figure 2.11. Subsequently, critical sites (i.e., sites with low yields compared to the average of the soil type) were visited and evaluated through field observations and additional sampling. If possible, the problems were corrected through site-specific management. If no problems could be identified and the problem was persistent, it was concluded that it may be a problem of the deterioration of the plant material and in several cases it was decided to renew that part of the plantation.

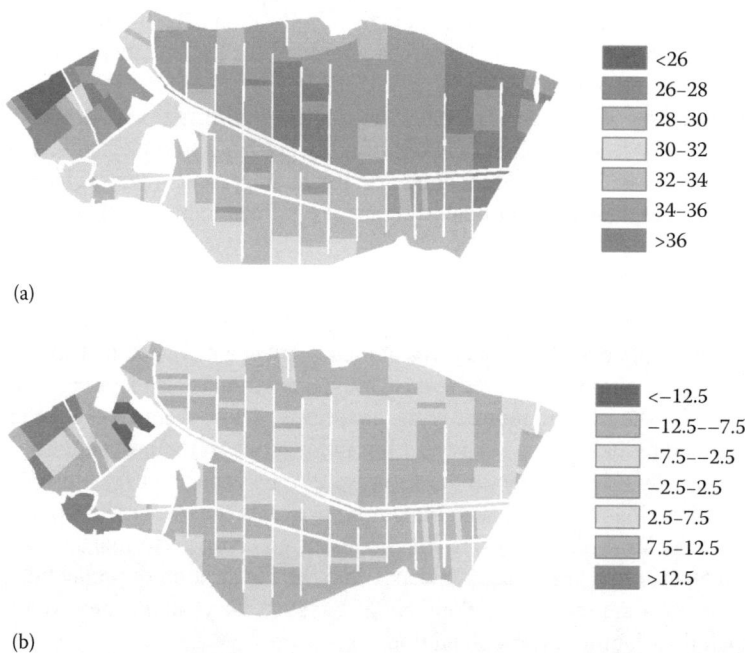

FIGURE 2.11 Selection of critical areas in a Costa Rican banana plantation based on the yield map in terms of bunch size (kg) (a). Bunch sizes are averaged per soil type (see Figure 2.3) and expressed in terms of deviations to the average bunch size for the soil type (%) (b).

2.3.4.2.3   Operational Decisions

A key element in an effective system for PA is the translation of site-specific soil and production data into management recommendations. There is an increasing call for the use of crop growth simulation models. However, we have to realize that for many tropical crops these simulation models do not exist. In this case we analyze the yield maps in relation to the soil conditions and determine the location of so-called problem areas. These areas are characterized by relatively low production compared to the average production for areas with similar soil conditions. This provides an important signaling function after which more detailed studies have to pinpoint the exact causes of the low production so that management can intervene. The analysis initially resulted in specific fertilization or pesticide application but in a later stage also resulted in the renewal of entire sections of the plantation. This was done if no specific reason for reduced yields could be found and deterioration of planting was suspected.

Fertilization programs in Costa Rica are established based on general knowledge of fertilizer response for two main soil types: the volcanic ash soils west of the Reventazón river and the nonvolcanic soils located east of the Reventazón river. Experts from Corbana or the fertilizer companies advise the farmers based on these general insights and soil testing. For PA more detailed insights in fertilizer response are required. Therefore the plantation owner asked Corbana and Wageningen University to design specific fertilizer experiments that could be carried out in a plantation without intervening too much with standard farm management. Three main soil types in the plantation where identified during a period of a year 1-ha blocks received 75%, 100%, and 150% of conventional fertilization. In these blocks crop performance as well as nutrient leaching was monitored. Depending on the soil type, changes in fertilization resulted in changes in production and/or nutrient leaching. Results for the three main soil types are presented in Table 2.1. In the fertile soils almost all additional nitrogen is taken up by the plant if we move from 75% to 100% fertilization. However, further increasing fertilization resulted in increased leaching. In the infertile soil, however, leaching already increased, moving from 75% to 100% (with a stable production), indicating that fertilization could be reduced. The low production was caused by the heavy soil textures, poor internal drainage, and low pH. These results formed the basis for site-specific fertilizer recommendations. The recommendations resulted in a decrease in fertilizer use of 12%.

Due to the perennial character of the banana plantation with weekly harvests of banana, monthly fertilizer applications, and weekly fungicide applications, there are many additional operational decisions to be taken. The operational decisions can be subdivided in two large groups. A number of decisions are taken in an *ex ante* manner. Although the crop is still in good condition, problems are anticipated and management is already adapted. Good examples include the change toward less soluble, slow-release fertilizers and the intensification of fungicide applications in periods with excessive rainfall. In addition there are the *ex post* management changes in the case of yield reduction (observed by the BanMan yield monitoring system) or if signs of nutrient deficiency are observed.

TABLE 2.1

Effect of Fertilization Rates on a Costa Rican Banana Plantation on the Nitrogen Concentration (mg/l) in Soil Water under Different Soil Types[a]

	Fertilization		
Soil Type	**75%**	**100%**	**150%**
Sandy loam, fertile, well-drained soil	3.61	3.81	4.96
Loamy, fertile, well-drained soil	4.12	4.08	6.36
Clayey, infertile, well-drained soil	1.91	2.34	3.29

[a]   See soil map in Figure 2.3.

The crop basically requires continuous monitoring and adaptation of the tactical management strategies for a given year. Due to the continuous cropping system, yield maps can also be created on a weekly basis, providing a continuous monitoring of the plantation.

2.3.4.3 Evaluation

The experiences with PA in the banana crop have been very favorable. Farm management has been able to improve the productivity of the plantation and it has provided them with an improved insight into the performance of the plantation. In addition, they are equipped with an excellent system for the registration of farm management and productivity to answer questions in terms of tracking and tracing. Since the concepts of PA have been introduced into the Costa Rican banana sector many companies and plantations have used particular elements of PA. A good example is the site-specific application of nematocides in those areas where nematode concentrations exceed the threshold of 16,000 per 100 g of roots. However, the spatial variation in nematode populations makes sampling and detection an awkward practice. The success of PA at the Rebusca plantation can be attributed to the integrated approach in which all the elements of the cropping system are included and the intensive involvement of the owner of the plantation in the development and implementation phase. Corbana recognizes the value of PA and has started to link their extension work to PA. Currently, integrated management systems have become common practice in the Atlantic zone of Costa Rica. The strategy of Corbana is that the sector needs to take a proactive approach. Tighter regulations with respect to the environment, the social conditions of workers, and track and trace are to be expected. New global threats to the banana sector such as the recent outbreak of Panama disease (García-Bastidas et al. 2014) make intensive monitoring even more relevant.

2.3.5 High-Tech PA for Dutch Wheat Farming

2.3.5.1 General Context

The role of agriculture in a densely populated country like the Netherlands has surprised many people. An intensive system for spatial planning has reserved large parts of the Netherlands for agriculture. Nevertheless, only those farms that produce highly efficient can survive under current market conditions. This has resulted in intensively managed farms with increasing farm sizes. The Netherlands has been struggling with the environmental impacts of its agricultural sector. Intensive management and high groundwater tables are two important causes of these problems. Driven by European legislation, the Dutch government introduced the Mineral Accounting System (MINAS) system for soil nutrient management. MINAS defined limits for budgetary N surpluses (e.g., 100 kg N ha^{-1} for arable land in 2003) and imposes levies when these surpluses are exceeded (Oenema 2004). These surpluses correspond with the allowed (gaseous and leaching) losses as specified in environmental regulations. After the introduction of MINAS, fertilizer use has been reduced by almost 30% while maintaining productivity. However, the system does not meet the European criteria and the European Court has disapproved the system, requiring a focus on the original requirements of the EU Nitrate Guideline, defining a limit of 170 kg N from manure/ha. This puts increased emphasis on N transformations in soil.

The concepts of PA are often related to very large farms as they occur in, for example, the United States and Canada. Although farm size in the Netherlands is significantly smaller, farmers experience significant soil variability in the highly heterogeneous landscape with complex alluvial deposits and cover sands. Farms are managed intensively and with production levels close to the optimum. Nevertheless, there is still room for improvement. These improvements can be found in terms of environmental impacts through increased nutrient use efficiency, the quality of the agricultural produce, and the control of pests and diseases. As a result there is quite some potential for PA. N emissions to air, ground, and surface waters have become a major concern in many regions. In reaction, policy makers are tightening environmental constraints on agriculture, resulting in a call for more efficient management systems. An operational system for PA was developed and tested for the commercial arable farm of Van Bergeijk in the central-western part of the Netherlands (51°17′N, 4°32′E; Van Alphen and Stoorvogel 2001). The farm covers an area of approximately 100 ha and applies a crop rotation of winter wheat, consumption potato, and sugar beet. Soils originate from marine deposits and are generally calcareous and have textures ranging from sandy loam to clay (Calcaric Fluvisol/Typic Fluvaquents).

2.3.5.2 Management Decisions

2.3.5.2.1 Strategic Decisions

The decision of the farmer to adopt and develop PA management in the early 1990s has to be seen in the context of a number of specific conditions:

- The political environment of further constraining fertilizer management, increased legislation, and the increasing attention to water quality
- The farmer did not only manage his own farm but rented out services to a large number of farms
- The increasing availability of new technology (GPS, variable rate technology, sensors) on new equipment (the farmer frequently purchased new equipment due to the size of his operation)
- The close ties with the scientific community with his son studying farm technology at Wageningen University
- The farmer was very much interested in the new technological developments

Given the above, the farmer made the decision to not only adopt, but also develop PA. A methodology for precision N fertilization in high-input farming systems was developed in close collaboration with the farmer by applying split fertilizer strategies.

2.3.5.2.2 Tactical Decisions

PA was based on an integrated soil and crop oriented approach using mechanistic simulation models. Given the fact that the simulation model could only be run for a limited number of management units, the identification of these management units to properly describe the spatial variability was a key challenge. The functional approach described in Section 2.4 combined the traditional field survey with a functional interpretation of soil differences.

Once the management units had been established, precision N management was applied in two consecutive years (Van Alphen and Stoorvogel 2002). The proposed methodology was tested in a winter wheat (*Triticum aestivum L.*) field during the 1998 growing season. Six experimental strips were delineated receiving either precise or traditional fertilization. Precision management used real-time simulation to monitor soil mineral N levels in each management unit.

2.3.5.2.3 Operational Decisions

Early warning was provided when mineral N concentrations dropped below a critical threshold of kg/ha (Figure 2.12). Used as a *threshold*, this information served to optimize the timing of three consecutive N fertilizations (Figure 2.12). Thresholds are not static, but defined in relation to actual uptake rates that are estimated with real-time modeling. Spatial variation is incorporated through the concept of management units (i.e., stable units with relatively homogeneous characteristics in terms of water regimes and nutrient dynamics). Separate simulations are conducted for each management unit based on selected representative soil profiles. If N concentrations dropped under the threshold in one of the management units all the management units were fertilized but with varying fertilization rates. Fertilizer rates were determined through exploratory simulations, which calculated the amount of mineral N required under "normal" conditions. Compared to conventional

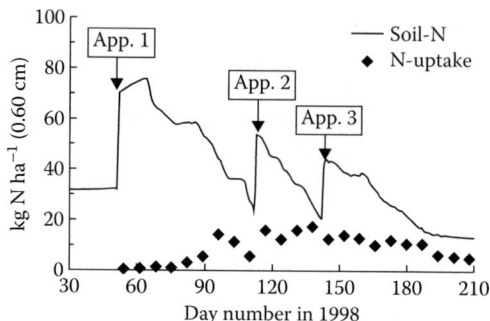

FIGURE 2.12   Fertilizer applications (App. 1–3) in winter wheat on a Dutch arable farm using a real-time monitoring system and simulation modeling. (From Van Alphen, B. J., and J. J. Stoorvogel. *Precis. Agric.* 2:319–32, 2001.)

management, fertilizer input was reduced by 15%–27% without affecting grain yield and quality. Precision fertilization proved efficient in reducing fertilizer inputs −23%, while slightly improving grain yields +3% and hectoliter weights +4%. Results clearly illustrate the significance of precision management in the process of increasing fertilizer use efficiency.

2.3.5.3   Evaluation

The study that was carried out showed a great potential to increase the efficiency of fertilizer use through managing the spatial and temporal variation on the farm on a number of experimental strips in the farm. A logical step would have been to expand the tested management practices toward the entire farm. However, at the moment that the experimental phase ended, it became very attractive for the farmer to use the freely available organic manure that is produced in excess in the livestock industry in the Netherlands. With ever-restricting environmental laws in the Netherlands, it became increasingly difficult to get rid of the manure. As a result it became a readily available source of nutrients for many arable farmers. It took away the major incentive of the farmer to continue with the PA procedures described above. However, it may become attractive again under different legislation.

2.3.6   High-Tech PA in Potato Cropping on Sandy Soils

2.3.6.1   General Context

Potato is the fifth most important stable food crop worldwide and is being processed for the convenience food and snack industries. Compared to grain crops, PA practices for potato have only recently been developed. As potato has strict fertilization and water requirements for optimal yield in terms of quantity and quality, PA measures have great potential for potato growth. In addition, potato is grown in a 4-year rotation and previous crops have an important effect on the soil quality status for potato growing after every fourth year. For the Netherlands, with a cultivated area of 155.000 ha in 2009 (16% of total area of arable land), potato are a major staple and export crop (e.g., seed potato) (CBS 2010).

   This case study describes the PA cycle of a large-scale potato farm (51° 19′ 04.55″ N and 5° 10′ 11.29″ E), van den Borne Aardappels, in the south of the Netherlands close to the village of Reusel. On the farm, a yearly area of 400 ha of potato is grown but with 1600 ha of land within the rotation cycle. With only 30% of owned land this means that over 1100 ha of land is rented from other farms. As a result the knowledge on soil conditions in the year of potato cultivation and variation within parcels is limited or uncertain. In general, the soils are sandy with a thick black A horizon on top of shallow nondeveloped B horizon followed by densely packed sandy C horizon (Humic Gleysol/ Typic Haplaquod).

2.3.6.2   Management Decisions

2.3.6.2.1   Strategic Decisions

Field management activities are highly mechanized with GPS guidance of tractors for preparation of the soil before planting, site-specific fertilization and irrigation during the growing season, and yield mapping during harvest of the potato. All field activities are registered and spatially referenced, which means that maps of planting densities of individual tubers, fertilization rates, and site-specific yield estimates can be visualized in maps. The availability of spatial base layers allows ICT integration of (sensor) data acquisition, data analysis, and transformation to management decisions. To achieve long-term viability of the potato farm, van den Borne Aardappels would like to increase average yearly potato yield by 2%. In addition, they would like to improve the quality of the yield measured according to the size distribution of the potato and the so-called underwater weight, which is a measure of the starch content. From these objectives and in order to deal with unknowns and uncertainty on variation in soil conditions in sandy soils at the start of growing season, it was decided that the adoption of PA would be the most efficient way to achieve the targeted objectives.

2.3.6.2.2   Tactical Decisions

The PA management cycle (Figure 2.13) was used as a starting point to decide on tactical decisions on PA practices that could be developed and implemented. As a first step a combination of proximal crop sensing and remote-sensing-based crop sensing (Sections 2.2.3.2 and 2.3.3.3) was adopted to optimize nutrient availability over the growing season (Kooistra et al. 2012). At the start of the growing season, detailed information on soil nutrient availability for the more than 100 fields under cultivation is unknown. For all fields a base application of organic manure is therefore applied before planting of the potato. During the growing season the crop is monitored and crop-based sensor measurements are available on a weekly basis. The sensor observations are translated into AGB (kg/ha) and AGN (kg/ha). Dedicated fertilization experiments were set up between 2010 and 2013 to establish the relationship between the vegetation indices and crop variables. Based on these relationships and additional agronomic crop modeling, decision rules were established translating the sensor measurements into fertilization rates, which could be used for VRA of fertilization of N application over the growing season.

Starting in 2013, the opportunities of proximal soil mapping techniques have been investigated as a basis for soil property mapping. Insight into the spatial variability of soil properties would allow management practices to improve the soil system before the start of the growing season (e.g., application of compost) and support VRA practices for fertilization and irrigation during the growing season. Two sensing techniques were evaluated: (1) EC using the EM38 sensor (Geonics, United States) and (2) very-high-resolution reflectance measurement in the visible part of the spectrum derived from aerial photographs taken from (un)manned platforms (Bartholomeus and Kooistra 2012). The main objective was to map variation in organic matter within the field before planting of the potato. Within the sandy soils, organic matter is the main source of both water and nutrient availability. Thus variation in the level of organic matter within the field is an important driver for potential potato yield. Earlier research showed that spectral reflectance based indices could be used for soil organic matter quantification (Bartholomeus et al. 2008). This principle was also evaluated for this sandy soil system using very-high-resolution RGB imagery acquired with an unmanned aerial multirotor system (Figure 2.5). Using a multilinear regression model ($R^2 = 0.58$) and a set of calibration samples, a soil organic matter map was prepared.

2.3.6.2.3   Operational Decisions

From 2011 until 2013 the established VRA rules were applied for N fertilization of all fields of the farm using Isaria crop sensors (Fritzmeier, Germany) measuring both the required spectral indices for AGB and AGN. The newly adopted fertilization practices resulted in improved tuber quality based on measurements of mean underwater weight, which increased for the experimental fields

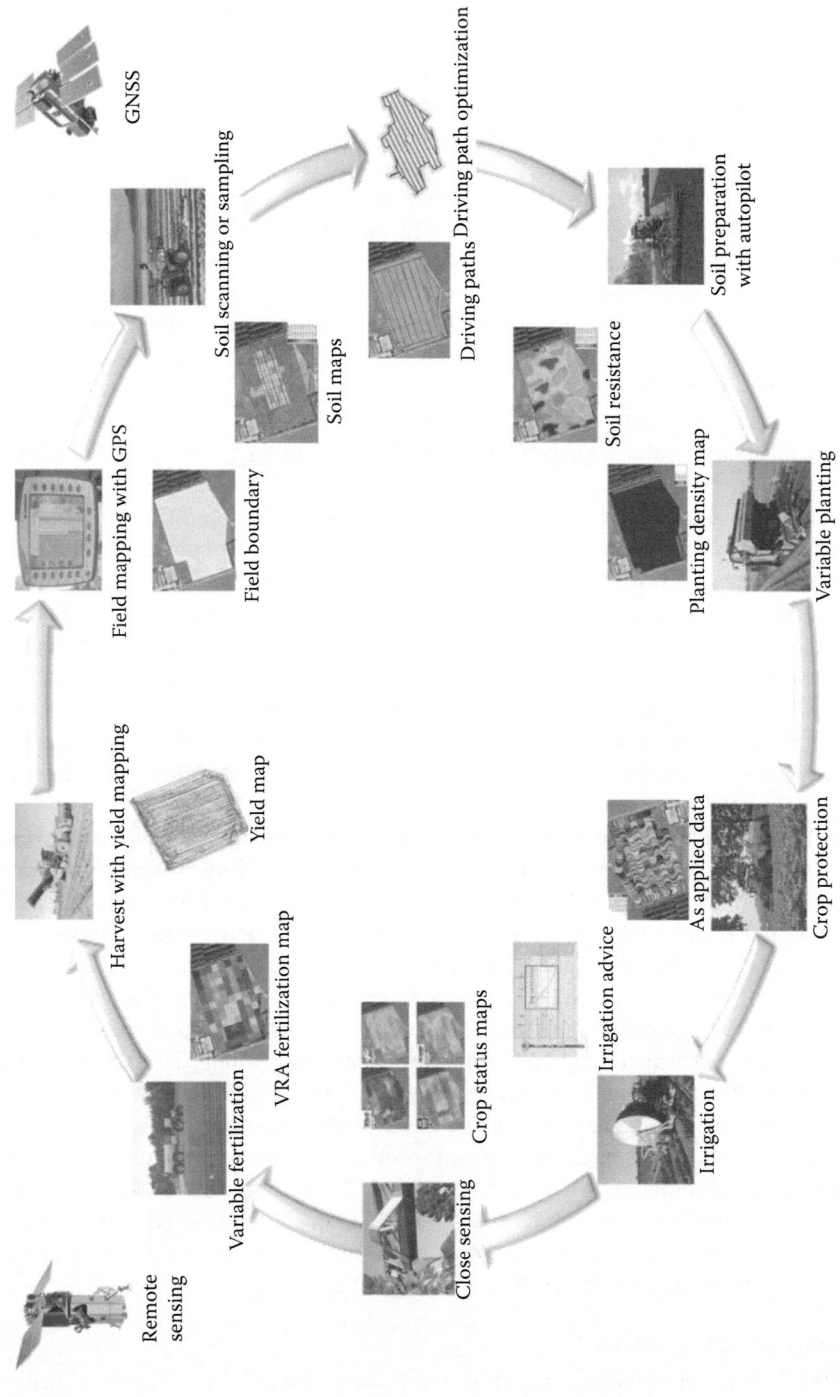

FIGURE 2.13   Precision agriculture cycle for the large-scale potato farm of van den Borne Aardappelen in the Netherlands.

from 342 in 2010, 360 in 2011, 404 in 2012, and 411 in 2013. The average potato yield at farm level showed a yearly increase of 1.5% over this time period, which was mainly achieved by improving the yield levels for lower-yielding fields.

In 2014, EC maps acquired using EM38 were adopted as a base for VRA of organic compost (see the inset in Figure 2.5) to increase the soil organic level in the soil for a selection of 10 fields. Additional research is required to investigate the effect on organic matter levels. The selection of the so-called key soil variables for the delineation of potential management classes (Van Meirvenne et al. 2013) will also be investigated, combining observations from different soil sensing systems and making a comparison with the available crop yield maps.

Additional PA practices implemented at the farm of van den Borne include site-specific planting of potato tubers with larger planting distances for areas that get lower incoming radiation levels due to the shadowing of trees along the boundaries of the field. For this practice, all field boundaries have been combined with a 3-D model of tree lines and a radiation model to calculate radiation levels within the field. Another practice is focused on a more efficient procedure for haulm killing of the AGB of the potato crop at the end of the growing season (Van Evert et al. 2012). Based on biomass mapping using WDVI, VRA maps are prepared for site-specific application resulting in reduced use of herbicides.

2.3.6.3 Evaluation

A complete cost-benefit analysis to assess the efficiency of the applied PA practices has not yet been made. Certainly, as van den Borne Aardappels can be considered an early adopter, the acquisition of new sensor technology can be relatively expensive and the operational implementation also requires additional effort since not all parts of the processing chain are available off the shelf. In time, as PA is more widely adopted, costs are bound to decrease. With increased adoption of PA technology the different components of this chain and the required adaptation to specific requirements of an agricultural system (e.g., soil- and crop-specific) will also be available at reduced cost. As an early adopter, the farmer is also facing many operational problems while using proximal sensors and translating sensor information into management operations. Finally, the knowledge and best practices derived from the field experiments are shared with other farmers through presentations, a detailed description on the website of the company, and by setting up cooperation networks between early adopting farmers from different regions in the Netherlands.

2.3.7 DISCUSSION

2.3.7.1 Taking Stock: What Is the Current Practical Status of PA?

PA has now been studied and implemented for several decades and at this moment in time it would appear advisable to take stock and analyze what has been accomplished and what lies ahead. PA has inspired much excellent research, but studying the literature and the proceedings of many international conferences leaves the impression that the field suffers from a degree of atomization as certain aspects and specific PA methodologies have received ample attention while analyses of the entire soil-water-plant production system, ranging from the selection process of methodologies to operational application on farm level, have not received the same degree of attention. It should be recognized that many different approaches to the research are being implemented (Le Gal et al. 2011), but much research is technology-driven and this is quite valuable but may not always be helpful to farmers attempting to apply PA: nobody shows them the forest, so they only see lots of individual trees. This is particularly true because producers of certain types of equipment are primarily interested in selling their products. There is a need to develop and report comprehensive studies covering the entire range of strategic, tactical, and operational decisions that farmers have to make, including their experiences as they apply the selected procedures. This very much includes negative experiences, which are important for the overall learning process. The premise has to be that there will be no standard recipe, that all farms and farmers are different and that only tailor-made procedures represent best practices for any particular farmer, offering attractive opportunities for knowledge brokers with a soil science background. Overall, the dynamic character of PA needs to be stressed: new techniques are being developed all the time and tested in the field. The scientific community would be well advised to keep and periodically update a record of techniques, as presented in Section 2.2, while also presenting complete case studies based on a systems analysis including a colearning approach involving farmers.

2.3.7.2   Central Role of the Farmer

In this chapter primary emphasis has been placed on the farmer who is confronted with strategic, tactical, and operational questions when dealing with PA.

Strategic decisions are not only a matter of a cost/benefit analysis but also include implications of the farming style that the farmer is comfortable with. On one end of the scale, some farmers are technology buffs and are intrigued by introducing high-tech methods and tools in their farming operations even though costs may far exceed benefits. On the other end, some farmers like to stay as close as possible to natural processes, as in biological agriculture. The examples of the previous sections show that the principles of PA apply to the entire ranges of farming styles. As is, PA has a high-tech image that is experienced as quite negative by farmers, citizens, and action groups who advocate—in their eyes—a nature-friendly approach to agriculture. Reframing of the PA approach is therefore necessary and the best way to do so is by presenting specific case studies, including, as in this chapter, ones on subsistence farming in Africa and banana farms in Costa Rica.

However, this does not address the basic problem of the strategic decision that farmers have to make whether to apply PA in whatever form it may be. For many farmers consideration of costs versus benefits will be of crucial importance and this value is difficult to define, more so since technological developments in the field and the associated costs are rapidly changing. As well, a procedure that may produce an attractive economic spinoff in, for example, a dry year may have no effect in an average year and may even be negative in a wet year. It is impossible for an individual farmer to see all implications of a decision on PA, whether positive or negative. Here, extension services have a clear role to play by presenting the services of independent counselors with not only solid technical know-how but also with a basic understanding of farming operations and the wide range of farming styles. Such counselors can only effectively operate if they go beyond a one-shot round of advice. A major program on sustainable agriculture in the Netherlands showed the crucial role that so-called knowledge brokers played when realizing plans for innovative and sustainable agricultural systems. They focused on inserting the right type of knowledge at the right time and place to the right person in the right way. This is different from traditional extension and was therefore called Extension 2.0 (Bouma et al. 2011). Traditional extension communicated results of research to farmers in a rather linear manner that mainly focused on production: this is the problem, this is what you should do. Extension 2.0 is about colearning as conditions which farmers (and researchers!) face go beyond production alone and also include environmental and societal aspects. Knowledge brokers can help farmers to discover which factors are key in determining spatial heterogeneity of crop yields, as discussed in Section 2.3.1. A key factor for PA should be its focus on individual farmers. There is no standard recipe for PA. A tailor-made solution is needed for any farm in terms of defining a best practice that is to be fine-tuned not only to the particular conditions on any given farm but also on the wishes and perspectives of any given farmer. In this context, the Extension 2.0 concept has major potential for the future and soil scientists with a high social intelligence would seem to be particularly equipped to play the role of knowledge brokers for PA in the future.

2.3.7.3   PA in the Context of Sustainable Development

Successful introduction of PA represents a significant contribution to sustainable development. The Rio+20 conference in 2012 produced an agreement to develop a set of Sustainable Development Goals (SDGs), building on the earlier and partly successful Millennium Development Goals (MDGs), aiming for a final plan of action in 2015 and ranging to 2030. Initial proposals by the

TABLE 2.2
UN "Sustainable Development Goals" for the Period 2015–2030 (Status July 2014)

Goal 1	End poverty in all its forms everywhere.
Goal 2	End hunger, achieve food security and improved nutrition, and promote sustainable agriculture.
Goal 3	Ensure healthy lives and promote well-being for all at all ages.
Goal 4	Ensure inclusive and equitable quality education and promote lifelong learning opportunities for all.
Goal 5	Achieve gender equality and empower all women and girls.
Goal 6	Ensure availability and sustainable management of water and sanitation for all.
Goal 7	Ensure access to affordable, reliable, sustainable, and modern energy for all.
Goal 8	Promote sustained, inclusive, and sustainable economic growth, full and productive employment, and decent work for all.
Goal 9	Build resilient infrastructure, promote inclusive and sustainable industrialization, and foster innovation.
Goal 10	Reduce inequality within and among countries.
Goal 11	Make cities and human settlements inclusive, safe, resilient, and sustainable.
Goal 12	Ensure sustainable consumption and production patterns.
Goal 13	Take urgent action to combat climate change and its impacts.
Goal 14	Conserve and sustainably use the oceans, seas, and marine resources for sustainable development.
Goal 15	Protect, restore, and promote sustainable use of terrestrial ecosystems, sustainably manage forests, combat desertification, halt and reverse land degradation, and halt biodiversity loss.
Goal 16	Promote peaceful and inclusive societies for sustainable development, provide access to justice for all, and build effective, accountable, and inclusive institutions at all levels.
Goal 17	Strengthen the means of implementation and revitalize the global partnership for sustainable development.

Source: Available at http://sustainabledevelopment.un.org/focussdgs.html.

SDSN suggest that SDGs will have an *ecological ceiling* based on criteria for planetary well-being and a *social floor* considering basic human well-being. The 17 goals are listed in Table 2.2. Note that goals 2, 7, 12, 13, and 15 have direct relevance for agriculture and the environment, and implicitly, for soil science. This presents an excellent and unique window of opportunity for the soil science profession (see also http://sustainabledevelopmentsolutionsnetwork.org and SDSN 2013). The significance of PA for agricultural development, which is widely considered to be a top priority in view of the need to feed 9 billion people by 2050 while protecting the environment, can be communicated to a wide audience by framing PA activities in the context of the SDGs.

2.4   CONCLUSIONS

1. PA presents a crucial technology to improve soil management in terms of tillage and the use-efficiency of agrochemicals, thereby reducing or omitting the negative impact of many more traditional agricultural practices on the environmental quality of soil, water, and air. The concept is not restricted to high-tech applications but applies to all types of production systems, including subsistence farming. PA can make a significant contribution toward achieving the SDGs defined by RIO+20 that emphasize food production and environmental quality in a broad societal context.

2. When discussing PA, primary emphasis should be on farmers who are faced with strategic decisions (to apply or not to apply PA), tactical decisions (which techniques to select), and operational decisions (how to proceed in practice). So far, PA research appears to be rather technology-driven, emphasizing development of new sensing and measuring techniques. This has been quite successful and has transformed the research arena. However, farmers are faced with decisions that cover a complex, weather-dependent soil-water-atmosphere production system requiring adaptive management. More attention should be paid to advising

individual farmers. There are no general recipes for PA. Every farm management system is unique and only tailor-made approaches can be successful and they evolve as available technologies evolve. Here, knowledge brokers can play an important role in linking research with farmers' practice, and soil scientists are well equipped to perform this role based on colearning provided they have familiarized themselves with all the technologies involved.

3. Different new sensing techniques for soils and crops provide excellent possibilities to characterize their status in space and time. However, many soil sensors only focus on the soil surface and cannot express conditions in the root zone, which determines plant growth. Proximal and remote crop sensors express crop conditions as expressed by plant leaves but cannot explain why certain unfavorable conditions prevail, let alone what can be done to correct them. However valuable, sensors can only express actual conditions while farmers face an uncertain future and their main challenge is to define proactive management measures that can avoid future problems. Here, real-time computer simulations of crop production, considering the soil-water-atmosphere system, can play a crucial role as was demonstrated in the case study in Section 2.3.6. Sensing data is very important to calibrate and validate such models.

4. The complexity of soil-water-atmosphere plant production systems and the possible role of PA can best be illustrated by farmer-oriented narratives, as presented in Section 2.3, based on specific case studies covering strategic, tactical, and operational decisions that have been made by certain farmers in cooperation with soil and crop researchers and knowledge brokers. Narratives should include negative experiences and results as part of a joint learning process for farmers and researchers. This way, technical aspects are considered in a broader societal context that is crucial from an operational point of view.

REFERENCES

Arslan, S., and T. S. Colvin 2002. Grain yield mapping: Yield sensing, yield reconstruction, and errors. *Precis. Agric.* 3:135–54.

Atzberger, C. 2013. Advances in remote sensing of agriculture: Context description, existing operational monitoring systems and major information needs. *Remote Sens.* 5:949–81.

Bartholomeus, H. M., M. E. Schaepman, L. Kooistra, A. Stevens, W. B. Hoogmoed, and O. S. P. Spaargaren 2008. Spectral reflectance based indices for soil organic carbon quantification. *Geoderma* 145:28–36.

Bartholomeus, H. M., L. Kooistra, A. Stevens, M. van Leeuwen, B. Van Wesemael, E. Ben-Dor, and B. Tychon 2011. Soil organic carbon mapping of partially vegetated agricultural fields with imaging spectroscopy. *Int. J. Appl. Earth Obs. and Geoinformation* 13:81–88.

Bartholomeus, H. M., and L. Kooistra 2012. Use of aerial photographs for assessment of soil organic carbon and delineation of agricultural management zones. In *Proceedings Geophysical Union (EGU) Conference*, p. 212, April 22–27, 2012, Vienna, Austria.

Bartholomeus, H. M., J. Suomalainen, and L. Kooistra 2014. Estimation of within field variation of SOM using UAV based RGB and elevation data. EGU General Assembly 2014. *Geophysical Research Abstracts* 16, EGU2014-5660.

Bausch, W. C., and R. Khosla 2010. QuickBird satellite versus ground-based multi-spectral data for estimating nitrogen status of irrigated maize. *Precis. Agric.* 11:274–90.

Ben-Dor, E., J. R. Irons, and G. F. Epema 1999. Soil Reflectance. In *Remote Sensing for the Earth Sciences: Manual of Remote Sensing*, A. N. Rencz (ed.), pp. 111–88. New York: John Wiley & Sons, Inc.

Berntsen, J., A. Thomsen, K. Schelde, O. M. Hansen, L. Knudsen, N. Broge, H. Hougaard, and R. Hørfarter 2006. Algorithms for sensor-based redistribution of nitrogen fertilizer in winter wheat. *Precis. Agric.* 7:65–83.

Bouma, J., A. C. van Altvorst, R. Eweg, P. J. A. M. Smeets, and H. C. Van Latesteijn 2011. The role of knowledge when studying innovation and the associated wicked sustainability problems in agriculture. *Adv. Agron.* 113:283–312.

Bouma, J., N. Batjes, M. P. W. Sonneveld, and P. Bindraban 2014. Enhancing soil security for smallholder agriculture. In *Soil Management of Smallholder Agriculture*. R. Lal and B. A. Stewart (eds). Adv. Soil Sci. pp. 17–37. Baco Raton, FL; CRC Press.

Brock, A., S. M. Brouder, G. Blumhoff, and B. S. Hofmann 2005. Defining yield-based management zones for corn–soybean rotations. *Agron. J.* 97:1115–28.

Brouwer, J., and J. Bouma 1997. Soil and crop growth variability in the Sahel: Highlights of research (1990–1995) at ICRISAT Sahelian Center. Inform. Bulletin 49. Patancheru, India: ICRISAT, and Netherlands: Wageningen University.

CBS 2010. *Statistical Yearbook of the Netherlands, 2010.* Voorburg, Netherlands: Statistics Netherlands.

Clevers, J. G. P. W., L. Kooistra, and M. E. Schaepman 2008. Using spectral information from the NIR water absorption features for the retrieval of canopy water content. *Int. J. Appl. Earth Observ. Geoinform.* 10:388–97.

Cosh, M. H., S. R. Evett, and L. McKee 2012. Surface soil water content spatial organization within irrigated and non-irrigated agricultural fields. *Adv. Water Res.* 50:55–61.

De Jager, A., D. Onduru, and C. Walaga 2004. Facilitated learning in soil fertility management: Assessing potentials of low-external-input technologies in east African farming systems. *Agric. Syst.* 79:205–23.

Doerge, T. 1999. Management Zone Concepts. Retrieved from http://www.ipni.net/publication/ssmg.nsf/0 /C0D052F04A53E0BF852579E500761AE3/$FILE/SSMG-02.pdf (Accessed on December 21, 2014)

Eitel, J. U. H., D. S. Long, P. E. Gessler, and A. M. S. Smith 2007. Using in-situ measurements to evaluate the new RapidEye™ satellite series for prediction of wheat nitrogen status. *Int. J. Remote Sens.* 28:4183–90.

Endres, M., ed. 2013. *Proc. Precision Dairy Conference.* Rochester MN, June 26–27, 2013. University of Minnesota Extension and College of Food, Agriculture and Natural Resource Sciences, University of Minnesota, MN.

FAO 1990. *FAO-UNESCO Soil Map of the World. Revised Legend.* World Resources Report 60. Rome: FAO.

Ferster, C. J., and N. C. Coops 2013. A review of earth observation using mobile personal communication devices. *Comput. Geosci.* 51:339–49.

FURP 1987. *The Fertilizer Use Recommendation Project (FURP).* Final Report. Annex I: Fertilizer trial documentation (FERDOC). Nairobi: Ministry of Agriculture.

FURP 1994. *Fertilizer Use Recommendations*, Volumes 1 to 24. Nairobi: Kenya Agricultural Research Laboratory.

García-Bastidas, F., N. Ordóñez, J. Konkol, M. Al-Qasim, Z. Naser, M. Abdelwali, N. Salem, C. Waalwijk, R. C. Ploetz, and G. H. J. Kema. 2014. First report of Fusarium oxysporum f. sp. cubense Tropical Race 4 associated with Panama disease of banana outside Southeast Asia. Disease notes. *Plant Dis.* 98:694.

Giller, K. E., P. Tittonell, M. C. Rufino, M. T. van Wijk, S. Zingore, P. Mapfumo, S. Adjei-Nsiah, M. Herrero, R. Chikowo, M. Corbeels, E. C. Rowe, F. Baijukya, A. Mwijage, J. Smith, E. Yeboah, W. J. van der Burg, O. M. Sanogo, M. Misiko, N. de Ridder, S. Karanja, C. Kaizzi, J. K'ungu, J. Mwale, D. Nwaga, C. Pacini, and B. Vanlauwe 2011. Communicating complexity: Integrated assessment of trade-offs concerning soil fertility management within African farming systems to support innovation and development. *Agric. Syst.* 104:191–203.

Goffart, J. P., M. Olivier, and M. Frankinet 2008. Potato crop nitrogen status assessment to improve N fertilization management and efficiency: Past-present-future. *Potato Res.* 51:355–83.

Haboudane, D., J. R. Miller, N. Tremblay, P. J. Zarco-Tejada, and L. Dextraze 2002. Integrated narrow-band vegetation indices for prediction of crop chlorophyll content for application to precision agriculture. *Remote Sens. Environ.* 81:416–26.

Heijting, S., D. De Bruin, and A. K. Bregt 2011. The arable farmer as the assessor of within-field soil variation. *Precis. Agric.* 12:488–507.

Herrmann, I., A. Pimstein, A. Karnieli, Y. Cohen, V. Alchanatis, and D. J. Bonfil 2011. LAI assessment of wheat and potato crops by VENμS and Sentinel-2 bands. *Remote Sens. Environ.* 115:2141–51.

IFAD 2012. *Sustainable Smallholder Agriculture: Feeding the World, Protecting the Planet.* Proc. of Gov. Council. Rome: IFAD.

Jenny, H. 1941. *Factors of Soil Formation.* New York: McGraw Hill Book Co.

Kempen, B., D. J. Brus, G. B. M. Heuvelink, and J. J. Stoorvogel 2009. Updating the 1:50,000 Dutch soil map using legacy soil data: A multinomial logistic regression approach. *Geoderma* 151:311–26.

Kempen, B., D. J. Brus, and J. J. Stoorvogel 2011. Three-dimensional mapping of soil organic matter content using soil type–specific depth functions. *Geoderma* 162:107–23.

Khurana, H. S., S. B. Phillips, B. Singh, M. M. Alley, A. Dobermann, A. S. Sidhu, Y. Singh, and S. Peng 2008. Agronomic and economic evaluation of site-specific nutrient management for irrigated wheat in northwest India. *Nutr. Cycling Agroecosyst.* 82:15–31.

Kooistra, L., J. J. Stoorvogel, and R. Schuiling 1997. *Soil Survey of the Rebusca Plantation.* Wageningen, Netherlands: Laboratory of Soil Science and Geology, Wageningen University.

Kooistra, L., E. A. Beza, J. Verbesselt, J. van den Borne, and W. van der Velde 2012. Integrating remote-, close range- and in-situ sensing for high-frequency observation of crop status to support precision agriculture. In *Proceedings Sensing a Changing World*, pp. 15–20. May 9–11, 2012. Wageningen, Netherlands: Wageningen University.

Kweon, G. 2012. Towards the ultimate soil survey: Sensing multiple soil and landscape properties in one pass. *Agron. J.* 104:1547–57.

Lagacherie, P., A. B. McBratney, and M. Voltz 2006. *Digital Soil Mapping. An Introductory Perspective.* Developments in Soil Science, Vol. 31. Amsterdam: Elsevier.

Lambot, S., E. C. Slob, M. Vanclooster, and H. Vereecken 2006. Closed loop GPR data inversion for soil hydraulic and electric property determination. *Geophys. Res. Lett.* 33:L21405.

Le Gal, P.-Y., P. Dugué, G. Faure, and S. Novak 2011. How does research address the design of innovative agricultural production systems at the farm level? A review. *Agric. Syst.* 104:714–28.

Lee, W. S., V. Alchanatis, C. Yang, M. Hirafuji, D. Moshou, and C. Li 2010. Sensing technologies for precision specialty crop production. *Comput. Electron. Agr.* 74:2–33.

López Riquelme, J. A., F. Soto, J. Suardíaz, P. Sánchez, A. Iborra, and J. A. Vera 2009. Wireless sensor networks for precision horticulture in southern Spain. *Comput. Electron. Agr.* 68:25–35.

McBratney, A. B., I. O. A. Odeh, T. F. A. Bishop, M. S. Dunbar, and T. M. Shatar 2000. An overview of pedometric techniques for use in soil survey. *Geoderma* 97:293–327.

McBratney, A. B., M. L. Mendonca Santos, and B. Minasny 2003. On digital soil mapping. *Geoderma* 117:3–52.

Meersmans, J., B. van Wesemael, F. De Ridder, and M. Van Molle 2009. Determining soil organic carbon for agricultural soils: A comparison between the Walkley & Black and the dry combustion methods (north Belgium). *Soil Use Manage.* 25:346–53.

Mertens, F. M., S. Pätzold, and G. Welp 2008. Spatial heterogeneity of soil properties and its mapping with apparent electrical conductivity. *J. Plant Nutri. Soil Sci.* 171:146–54.

Michalczyk, A., K. C. Kersebaum, M. Roelcke, T. Hartmann, S. C. Yue, X. P. Chen, and F. S. Zhang 2014. Model-based optimisation of nitrogen and water management for wheat–maize systems in the North China Plain. *Nutr. Cycl. Agroecosyst.* 98:203–22.

Mulder, V. L., S. de Bruin, M. E. Schaepman, and T. R. Mayr 2011. The use of remote sensing in soil and terrain mapping—A review. *Geoderma* 162:1–19.

Mulla, D. J. 2013. Twenty five years of remote sensing in precision agriculture: Key advances and remaining knowledge gaps. *Biosyst. Eng.* 114:358–71.

Oenema, O. 2004. Governmental policies and measures regulating nitrogen and phosphorus from animal manure in European agriculture. *J. Anim. Sci.* 82:E196–206.

Oliver, M., T. Bishop, and B. Marchant, eds. 2013. *Precision Agriculture for Sustainability and Environmental Protection.* London: Routledge.

Ramanathan, N., T. Schoellhammer, D. Estrin, M. Hansen, T. Harmon, E. Kohler, and M. Srivastava 2006. The final frontier: Embedding networked sensors in the soil. CENS Technical Report #68, Los Angeles: UCLA.

Ritsema, C. J., H. Kuipers, L. Kleiboer, E. Van den Elsen, K. Oostindie, J. G. Wesseling, J. Wolthuis, and P. Havinga 2009. A new wireless underground network system for continuous monitoring of soil water contents. *Water Resour. Res.* 45:W00D36.

Robinson, T. P., and G. Metternich 2005. Comparing the performance of techniques to improve the quality of yield maps. *Agri. Syst.* 85:19–41.

Robinson, N. J., P. C. Rampant, A. P. L. Callinan, M. A. Rab, and P. D. Fisher 2009. Advances in precision agriculture in south-eastern Australia. II. Spatio-temporal prediction of crop yield using terrain derivatives and proximally sensed data. *Crop Past. Sci.* 60:859–69.

Samborski, S. M., N. Tremblay, and E. Fallon 2009. Strategies to make use of plant sensors-based diagnostic information for nitrogen recommendations. *Agron J.* 101:800–16.

Schnier, H. F., H. Recke, F. N. Muchena, and A. W. Muriuki 1997. Towards a practical approach to fertilizer recommendations for food crop production in smallholder farms in Kenya. *Nutri. Cycl. Agroecosys.* 47:213–26.

Scudiero, E., P. Teatini, D. L. Corwin, R. Deiana, A. Berti, and F. Morari 2013. Delineation of site-specific management units in a saline region at the Venice Lagoon margin, Italy, using soil reflectance and apparent electrical conductivity. *Comput. Electron. Agr.* 99:54–64.

Smaling, E. M. A., and B. H. Janssen 1993. Calibration of QUEFTS, a model predicting nutrient uptake and yields from chemical soil fertility indices. *Geoderma* 59:21–44.

Soil Survey Division Staff 1993. *Soil Survey Manual.* Agriculture Handbook 18. Washington, DC: NRCS.

Soil Survey Staff 1994. *Keys to Soil Taxonomy*, 6th Edition. Washington, DC: United States Department of Agriculture, Soil Conservation Service.

Song, X., J. Wang, W. Huang, L. Liu, G. Yan, and R. Pu 2009. The delineation of agricultural management zones with high resolution remotely sensed data. *Precis. Agric.* 10:471–87.

Stafford, J., ed. 2012. *Proceedings 11th International ISPA Conference on Precision Agriculture,* July 15–18, 2012, Indianapolis, IN. Berlin: Springer Verlag.

Stafford, J., ed. 2013. *Proceedings 9th European Conference on Precision Agriculture*, July 7–11, 2013, Lleida, Spain. Berlin: Springer Verlag.

Stevens, A., B. van Wesemael, H. Bartholomeus, D. Rosillon, B. Tychon, and E. Ben-Dor 2008. Laboratory, field and airborne spectroscopy for monitoring organic carbon content in agricultural soils. *Geoderma* 144:395–404.

Stoorvogel, J. J., J. Bouma, and R. A. Orlich 2004. Participatory research for systems analysis: Prototyping for a Costa Rican banana plantation. *Agron. J.* 96:323–36.

SDSN 2013. *Solutions for Sustainable Agriculture and Food Systems.* Technical Report for the post-2015 Development Agenda by the Thematic Group on Sustainable Agriculture and Food Systems. New York: Sustainable Development Solutions Network.

Thessler, S., L. Kooistra, F. Teye, H. Huitu, and A. Bregt 2011. Geosensors to support crop production: Current applications and user requirements. *Sensors* 11:6656–84.

Tremblay, N., Z. Wang, B. L. Ma, C. Belec, and P. Vigneault 2009. A comparison of crop data measured by two commercial sensors for variable-rate nitrogen application. *Precis. Agric.* 10:145–61.

Turk, J. K., O. A. Chadwick, and R. C. Graham 2012. Pedogenic processes. In *Handbook of Soil Sciences*, Second Edition. Part V. Pedology. P. M. Huang, Y. Li, and M. E. Summer (eds.), pp. 30-1–30-29. New York: CRC Press.

Van Alphen, B. J., and J. J. Stoorvogel 2000. A functional approach to soil characterization in support of precision agriculture. *Soil Sci. Soc. Am. J.* 64:1706–13.

Van Alphen, B. J., and J. J. Stoorvogel 2001. A methodology for precision nitrogen fertilization in high-input farming systems. *Precis. Agric.* 2:319–32.

Van Alphen, B. J., and J. J. Stoorvogel 2002. Effects of soil variability and weather conditions on pesticide leaching—A farm-level evaluation. *J. Env. Qual.* 31:797–805.

Van Evert, F. K., P. van der Voet, E. van Valkengoed, L. Kooistra, and C. Kempenaar 2012. Satellite-based herbicide rate recommendation for potato haulm killing. *Eur. J. Agron.* 43:49–57.

Van Meirvenne, M., M. M. Islam, P. De Smedt, E. Meerschman, E. Van de Vijver, and T. Saey 2013. Key variables for the identification of soil management classes in the aeolian landscapes of north-west Europe. *Geoderma* 199:99–105.

Van Zijl, G. M., D. Bouwer, J. J. Van Tol, and P. A. L. leRoux 2014. Functional digital soil mapping: A case study from Namarroi, Mozambique. *Geoderma* 219–220:155–61.

Vanclooster, M., P. Viane, J. Diels, and K. Christaens 1994. *WAVE, A Mathematical Model for Simulating Water and Agrochemicals in the Soil and Vadose Environment.* Reference and Users' Manual, Leuven, Belgium: Institute for Land and Water Management.

Vellidis, G., M. Tucker, C. Perry, C. Kvien, and C. Bednarz 2008. A real-time wireless smart sensor array for scheduling irrigation. *Comput. Electron. Agr.* 61:44–50.

Verrelst, J., J. Muñoz, L. Alonso, J. Delegido, J. P. Rivera, G. Camps-Valls, and J. Moreno 2012. Machine learning regression algorithms for biophysical parameter retrieval: Opportunities for Sentinel-2 and -3. *Remote Sens. Environ.* 118:127–39.

Viscarra Rossel, R. A., D. J. J. Walvoort, A. B. McBratney, L. J. Janik, and J. O. Skjemstad 2006. Visible, near infrared, mid infrared or combined diffuse reflectance spectroscopy for simultaneous assessment of various soil properties. *Geoderma* 131:59–75.

Viscarra Rossel, R. A., V. I. Adamchuk, K. A. Sudduth, N. J. McKenzie, and C. Lobsey 2010. Proximal soil sensing: An effective approach for soil measurements in space and time. *Adv. Agronomy* 113:237–82.

Wösten, J. H. M., Ya. A. Pachepsky, and W. J. Rawls 2001. Pedotransfer functions: Bridging the gap between available basic soil data and missing soil hydraulic characteristics. *J. Hydro.* 251:123–50.

Yufeng, G., A. Thomasson, and R. Sui 2011. Remote sensing of soil properties in precision agriculture: A review. *Front. Earth Sci.* 5:229–38.

Zerger, A., R. A. Viscarra Rossel, D. L. Swain, T. Wark, R. N. Handcock, V. A. J. Doerr, G. J. Bishop-Hurley, E. D. Doerr, P. G. Gibbons, and C. Lobsey 2010. Environmental sensor networks for vegetation, animal and soil sciences. *Int. J. Appl. Earth. Obs.* 12:303–16.

Zhang, C., and J. M. Kovacs 2012. The application of small unmanned aerial systems for precision agriculture: A review. *Precis. Agric.* 13:693–712.

3 Precision Spacing and Fertilizing Plants for Maximizing Use of Limited Resources

B.A. Stewart

CONTENTS

3.1 INTRODUCTION

The efficient use of resources for crop production is challenging. Water is in many cases the first limiting factor, particularly in semiarid regions, followed by nutrients. When water is limiting, it is extremely difficult to determine the most suitable crop and the best combination of row spacing, plant population, length of maturity, and date of seeding to make the most efficient use of the limited water. While the most important goal is perhaps to deplete as much of the plant-available water as possible from the soil profile before the end of the growing season, depleting it too early will result in a lot of biomass production early in the season and a critical shortage of water during the reproductive stage late in the season. A shortage of water late in the season generally results in low harvest index (HI) values for crops like maize (*Zea mays*), sorghum (*Sorghum bicolor*), wheat (*Triticum aestivium*), and cotton (*Gossypium hirsutum*). The HI is the weight of the harvested product as a percentage of the total aboveground biomass.

Maize and other crops that are commonly grown in rows were historically planted in rows 100 cm apart. This was done because the crops were often planted and cultivated with horses or bullocks which required that amount of space to move through the fields. The spacing of plants within the rows varied with the crop species and the desired plant population. With the transition from animal power to tractors, row spacing remained similar in the beginning but has become increasingly narrower in recent years. The tendency has been to move toward uniform spacing to achieve greater efficiency of light interception. Such changes may improve water use efficiency in some cases but may decrease efficiency in other cases. Loomis (1983) advocated that crops could be manipulated to increase the efficient use of water by stating "A useful generalization emerges: where soil resources are limiting (in this case, water), non-uniform treatment of the land or the crop can be an advantage. Where soil resources are non-limiting, however, uniform cropping will provide the greatest efficiency in light interception and photosynthesis."

For a given plant species, the amount of water transpired during photosynthesis is primarily dependent on the difference in the vapor pressure inside and outside the leaf surface. The primary factors are temperature, radiation, wind speed, and relative humidity. In simple terms, it depends on (1) how hot it is, (2) how sunny it is, (3), how windy it is, and (4) how dry the air is at the underside of the leaf where the stomata are located that open to allow carbon dioxide to enter for photosynthesis. While the stomata are open for entry of carbon dioxide, water is lost from the plant to the atmosphere. These factors are commonly measured at a weather station and the values reported can be vastly different from those at the underside of the leaves of a growing crop. The values for the crop leaves are affected by how close the plants are to one another, size of plants, overlapping of leaves that provide partial shading, wetness of the soil, plant population that affects wind speed, and numerous other factors. For example, the microclimatic conditions for a fully irrigated maize crop with 100,000 plants ha^{-1} would be vastly different than for a dryland crop with 30,000 plants ha^{-1} growing in an adjacent field than would be estimated from data measured at a weather station. Loomis (1983) presented an overview of how crops could possibly be manipulated to increase the efficiency of water use.

In recent years, technological advances have made it feasible to control the seeding and fertilizing of crops with high precision. Therefore, opportunities may exist to manipulate crops to improve the microclimate so that the vapor pressure deficit at the leaf surface would be reduced. The objective of this chapter is to review some of the past research and present some possible ways that the use of precision technology tools can utilize limited water supplies and fertilizer more efficiently. Seeding and fertilizing technologies that can place materials at specific depths and spaces are becoming increasingly available and have the potential to improve the management of spatial and temporal variability, which may in turn improve crop performance and environmental quality.

3.2   RELATIONSHIPS BETWEEN WATER USE AND CROP PRODUCTION

Generalized relationships between water use and production of crops are well established; see Figure 3.1 for some examples, which are similar to those presented by Loomis (1983) and Howell (1990). The amount of biomass produced is directly related to the amount of water transpired, which is shown in Figure 3.1 with a straight line (A) moving through the origin, indicating that no biomass

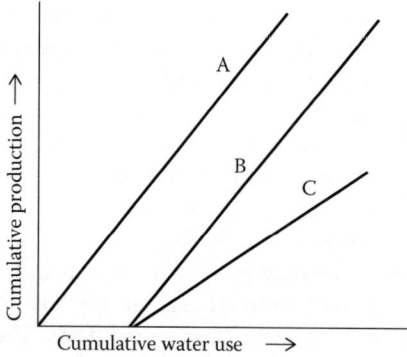

FIGURE 3.1   Generalized relationships between (A) transpiration and aboveground biomass, (B) evapotranspiration and aboveground biomass, and (C) evapotranspiration and grain yield of a grain crop using maize as an example.

is produced without transpiration and that each additional unit of biomass produced requires a constant number of water units. This, of course, assumes that the transpiration environment remains constant and the slope of the line is greater for a favorable transpiration environment than for an unfavorable environment. Sinclair and Weiss (2010) state that C_4 crops such as maize and sorghum growing in a somewhat average transpiration environment of 2 kPa have a transpiration rate of approximately 220 g of water per day for each gram of plant growth. In an arid environment with a vapor pressure deficit of 2.5 kPa, the rate increases to about 330 and decreases to about 160 in a humid environment where the vapor pressure deficit is about 1.5 kPa. In contrast, a C_3 species growing in an average transpiration environment will transpire approximately 330 g of water per day for each gram of plant growth. Sinclair and Weiss (2010) further stated, "Despite claims that crop yields will be substantially increased under water-stressed conditions by the application of technology, the physical linkage between growth and transpiration imposes a limit that is not amenable to genetic alteration." As already discussed, the transpiration rate is controlled by the vapor pressure deficit at the leaf surface and there has been little or no evidence that this has been changed by genetics. Line B in Figure 3.1 is also a straight line in parallel to line A but differs in that it intersects the x-axis rather than passing through the origin.

1 is also a straight line parallel to line A but differs in that it intersects the x-axis rather than passing through the origin. Line B is the relationship between evapotranspiration (ET), which is the sum of the water transpired and the water lost during the growing season by evaporation from the soil surface. The amount of water represented between the origin and the point of intersecting the x-axis represents the amount lost by evaporation. The portion of ET that is used for transpiration (T) is of extreme importance in utilizing water efficiently and will be discussed in more detail later. Line C in Figure 3.1 represents the yield of grain that is also a straight-line relationship with ET, as was the aboveground biomass represented by line B, but the slope of line C is less. The weight of grain as a ratio of the weight of aboveground biomass is the HI, and the HI value increases with increasing yield. The maximum HI values for various crops are genetically controlled. Prihar and Stewart (1990) estimated maximum HI values by estimating upper bounds of grain yield versus dry matter yield taken from literature reports. They estimated maximum HI values of about 0.60, 0.53, and 0.47 for wheat. Similar values have been reported in numerous publications and it is clearly understood that HI values are generally low at low yield levels and increase with increasing grain yield toward the genetic potential value. In the generalized relationships shown in Figure 3.1 that use maize as an example, the HI would range from near zero at low levels of biomass production to more than 0.5 at the higher levels of biomass production.

Based on the relationship between the transpiration environment and the crop species that results in a transpiration rate (Sinclair and Weiss 2010) and the relationships between water use and crop production illustrated in Figure 3.1, Stewart and Peterson (2014) developed an equation to express the yield of grain crops or other crops where the harvested product is expressed as a percentage of the aboveground biomass. The equation is

$$GY = ET \times T/ET \times 1/TR \times HI \qquad (3.1)$$

where GY is kg ha^{-1} of dry grain yield, ET is kg ha^{-1} of ET (water use by evaporation from the soil surface and transpiration by the crop between seeding and harvest), T/ET is the fraction of ET transpired by the crop, TR is the transpiration ratio (number of kilograms of water transpired to produce 1 kg of aboveground biomass), and HI is the harvest index (kg dry grain/kg aboveground dry biomass). While this equation applies to all situations where grain crops are produced, the ranges of values for each of the components become considerably greater and more variable in dryland farming areas. In general, a change in any one of the components will result in a change in the other three. Before addressing how precision technologies can possibly improve grain yield, each of the components will be briefly discussed.

3.2.1 ET

When water is added to a field by precipitation or irrigation, it can be stored in the soil, transpired by the crop, or lost as runoff, evaporation, or drainage. Water transpired by the crop plus the water lost by evaporation from the soil surface between seeding and harvesting is ET. It is sometimes referred to as crop water use or consumptive use. When measured, it includes the growing season precipitation plus the change in plant-available soil water in the root zone between time of seeding and harvesting plus any water added as irrigation during the growing season. The amount of plant-available water stored in the soil profile at the time of seeding plays a significant role in determining the amount of ET, particularly in dryland systems where growing season precipitation is limited. During the past several decades, major emphasis in dryland areas has been on (1) maximization of precipitation capture in the soil, (2) minimization of stored soil water evaporation, and (3) maximization of water-use efficiency (Peterson et al. 2012). Mulch cover is the most effective method of reducing evaporation from the soil surface, and crop residues are the most practical mulch source although they are often in short supply in dryland areas and are used for fuel and fodder in many areas, particularly in developing countries. However, reduced tillage, no-till, and conservation agriculture systems have been very successful in capturing and storing more precipitation for use by subsequent crops in areas of low and variable precipitation (Peterson et al. 2012; Stewart et al. 2006, 2010; Unger et al. 2006, 2012). Using the U.S. Great Plains as an example, the storage of 40% or more of the precipitation occurring during fallow periods is now possible, which can result in an increase of 70 to 100 mm ET for the succeeding crop. Without question, managing crop residues on the soil surface by reduced and no-till systems has increased the amount of water available for ET, reduced wind and water erosion, and improved the overall environment.

Equation 3.1 clearly indicates that GY increases linearly with increasing ET so it is important in water-deficient areas to maximize ET by exploiting as much of the stored plant available soil water as feasible. ET is the most important factor in Equation 3.1 because it tends to increase all the other factors in positive ways, as will be discussed. Because Equation 3.1 is a linear equation, increasing any factor will increase GY, but changing all the factors in a positive direction will result in much larger increases.

3.2.2 TRANSPIRATION-EVAPOTRANSPIRATION RATIO

The portion of ET that is used as T is affected by many variables and varies greatly within short periods as well as over the entire growing season. Stewart and Peterson (2014) stated that growing season transpiration-evapotranspiration (T/ET) ratios are likely to be in the range of 0.5 for dryland crops in semiarid regions to perhaps as high as 0.8 under highly irrigated conditions. Stirzaker (2010) stated that as a general guide, it would be considered excellent if 65% of ET was transpired and just 35% evaporated. Ritchie and Burnett (1971) quantified the effect of leaf area index (LAI) on T/ET when ET was not constrained by available water in the root zone (Figure 3.2). They concluded that T/ET when the soil surface was dry ranged from 0 for an LAI of 0, to about 0.5 for an LAI of 1, and to approximately 1 for an LAI of 3 and greater. When the soil is wetted by irrigation or precipitation, the T/ET ratio decreases and considerable water can be lost as evaporation. Surface mulch can reduce the evaporation but the amount and position of mulch are important. Smika (1983) measured soil water losses that occurred during a 35-day period without precipitation. Soil water losses were 23 mm from bare soil, 20 mm with flat wheat residues, 19 mm with 75% flat and 25% standing residues, and 15 mm with 50% flat and 50% standing residues on the surface. This indicates that standing residue is more effective than flat residue, which was likely due to slowing the wind speed. Steiner (1986) found in general that the higher the plant population and the narrower the row spacing, the higher the ratio of T/ET and the greater the exploitation of the stored plant-available soil water.

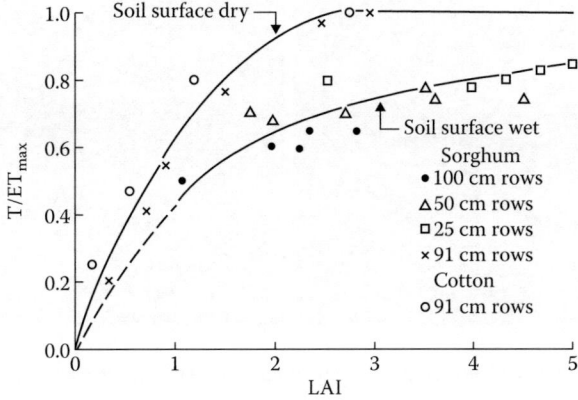

FIGURE 3.2 Transpiration (T) relative to evapotranspiration (ET) as influenced by leaf area index (LAI) where water in the root zone is nonlimiting for dry and wet soil surfaces. (From Ritchie, J.T. and E. Burnett, *Agron. J.* 63:56–62, 1971.)

3.2.3 TRANSPIRATION RATE

Aboveground plant biomass accumulation is directly related to availability of water (Sinclair 2009a). He states that the difference in vapor pressure inside and outside the leaf (VPD) controls the water loss through the stomata. The VPD in arid regions is large because the vapor pressure of the atmosphere is very low relative to humid areas. For a given environment, the VPD cannot be controlled—it is what it is. Sinclair (2009b) stated "Despite claims that crop yields will be substantially increased by the application of biotechnology, the physical linkage between growth and transpiration imposes a barrier that is not amenable to genetic alteration." While many plant scientists disagree, there is little evidence to date to repute it (Gurian-Sherman 2012). The transpiration environment of a growing crop is controlled by many factors, but primarily by temperature, radiation, wind speed, and humidity. Although Sinclair (2009b) stated that VPD cannot be controlled, it is important to note that the VPD is at the underside of the leaf and not in the environment at large. In that sense, VPD can be controlled to some degree and that is the main premise of this chapter. Plants can be manipulated to change the VPD, and precision technologies are becoming available that can possibly make such changes practical. The maize crops shown in Figure 3.3 could very well be growing in adjacent fields where the temperature, radiation, wind speed, and relative humidity measurements at the weather station were used to characterize the environment. However, the

(a)   (b)

FIGURE 3.3 The transpiration environment (vapor pressure deficit) of plant leaves is affected by plant density (80,000 ha^{-1} [a] and 30,000 ha^{-1} [b]), ground cover, and factors other than temperature, radiation, humidity, and wind speed measured at a weather station.

FIGURE 3.4   Indigenous maize growing on Hopi Indian Reservation in Arizona in 2006 under extremely arid conditions.

transpiration environments of the two crops are very different. The irrigated maize plants shown in Figure 3.3a have a much lower VPD than the dryland maize plants in Figure 3.3b. Therefore, the plant density of the irrigated crop is about three times higher than the dryland crop and this changes all four of the factors that affect VPD. The Hopi and Papago American Indians have manipulated plants for thousands of years to grow maize grain under desert conditions and continue this practice today (Figure 3.4). As a graduate student, William Brown, who later became president of Pioneer Hi-Bred International, Inc. (Brown 1985), investigated the origin, evolution, and culture of some of the important crops that furnish the major sources of food for certain Native Americans of the desert southwestern United States. A typical Hopi farm lies at the base of one of many mesas from which seep limited amounts of moisture. Precipitation during the growing season is a bonus that cannot be counted on. Ten to 12 maize seed are planted as much as 25 cm deep in hills about 2 m apart. Modern maize seed would not emerge from these depths but natural selection over centuries under these conditions has resulted in the evolution of genotypes whose hypocotyls are of sufficient length to fully emerge from that depth. The Hopi learned that clumping plants within a relatively small space reduces the desiccation of the foliage, anthers, and silk, thereby allowing normal fertilization to occur in extremely arid conditions. Although this may be an extreme example, it clearly shows that plants can be manipulated to change the microenvironment, which can result in an improvement in the transpiration environment by reducing the VPD.

Sinclair and Weiss (2010) state that the transpiration ratio of C_4 crops, like maize and grain sorghum, is about 220 g water day^{-1} for each gram of biomass produced when growing in a transpiration environment of 2 kPa, which they classify as somewhat average. For an arid region with a transpiration environment of 2.5 kPa, they state the transpiration ratio for C_4 crops increases to about 280 g for each gram of biomass, but decreases to about 160 g when the crop is growing in a humid transpiration environment of 1.5 kPa. For C_3 crops such as wheat, they indicate the transpiration ratio is about 1.5 times greater than for C_4 crops.

3.2.4   HI

Donald (1962) proposed the use of HI, defined as the ratio of grain yield to aboveground dry matter, to permit better analytical interpretation of genotypic and environmental effects than possible from yield alone. Snyder and Carlson (1984) stated "synthesis, translocation, partitioning and accumulation of photosynthetic products within the plant are controlled genetically and influenced by environment." Important environmental factors affecting HI include planting density, water and nutrient availability, and temperature, radiation, wind speed, and humidity that determine the VPD of plant leaves. Environmental factors differ widely, making it difficult to compare cultivars. Prihar and Stewart (1990) proposed that the slope of an upper-bound in the grain yield versus dry matter plot passing through the origin approximates the genetic HI because the highest grain yields against given dry matter represent the least-stressed and/or stress-adapted plants and passage through the origin is necessary to satisfy the definition of HI. Points lying below the upper bound manifest the degree of stress experienced by the crop. Using data from the literature, they concluded that the genetic HI values of maize, grain sorghum, and wheat ranged between narrow limits of 0.58 to 0.60, 0.48 to 0.53, and 0.38 to 0.47, respectively. Their analysis showed that the HI for a given species or cultivar is not determined by the size of the plant but by the stress. Water stress is in most cases the main cause of reduced yields, and it is well known that HI values are generally low at low grain yield levels and increase with increasing grain yields toward the genetic potential value (Howell 1990; Steiner et al. 1994). Although maize and grain sorghum are both C_4 crops, maize generally has higher HI values at high grain yield levels but lower values when yields are low. Under severe water stress, the HI can even be zero. In contrast, grain sorghum will generally produce some grain even under severe water stress. For all crops, however, the HI factor in Equation 3.1 is of great importance and can be significantly driven by management.

3.3 USING PRECISION AGRICULTURE TECHNOLOGIES TO IMPROVE CROP PRODUCTION IN LIMITED WATER AREAS

The preceding sections have shown that grain yield is determined by four factors that are closely linked and can be expressed by a simple mathematical equation as shown in Equation 3.1. This is a linear equation so increasing any one of the factors by 5% will increase the grain yield (GY) by 5%. However, because the factors are so closely linked, changing one factor almost invariably changes the other factors, but not always in the same direction. Large increases in GY only result when all the factors become more positive simultaneously. This is the reason that irrigation is so highly beneficial in low precipitation areas and why GYs increase sharply with increasing precipitation. Using maize growing across the U.S. Corn Belt as an example, Stewart and Peterson (2014) proposed some hypothetical values for Equation 3.1 based on published literature as annual precipitation increased from 500 to 1000 mm (Table 3.1). In general, increasing the amount of ET allows the use of a higher plant population that provides more plant canopy to shade the soil surface, which increases the T/ET ratio, improves the microclimate to decrease the VPD that will in turn decrease the units of water required to produce a unit of dry matter, and increases the HI. Thus, if each of the factors were increased by 5%, the GY would be increased by 20%. The data in Table 3.1 shows that doubling the annual precipitation increased GY almost fourfold because all four of the factors in Equation 3.1 were increased. Unfortunately, under limited water conditions, increasing one factor can lead to a decrease in another factor, which can negate a benefit or even cause a decrease in GY. For example,

---

TABLE 3.1

Hypothetical Values of Components in the Equation GY = ET × T/ET × 1/TR × HI for Corn Production in Areas of Increasing Annual Precipitation

Precipitation (mm)	500	600	700	800	900	1000
ET (mm)	320	375	435	500	570	650
T/ET ratio	0.55	0.58	0.61	0.64	0.67	0.70
TR (kg water/kg biomass)	270	258	246	234	222	210
Harvest index	0.44	0.46	0.48	0.5	0.53	0.55
Grain yield (kg ha^{-1} dry wt.)	2860	3865	5160	6815	9090	10,575

Source: Stewart, B.A. and G. Peterson, *Agron J.* 1006:1–10, 2014.

Note: ET = evapotranspiration, T/ET = transpiration/ET, TR = transpiration ratio.

in water-deficient areas, plant population must be decreased significantly and this reduces the T/ET ratio, which may also increase the transpiration ratio (TR). Once a good understanding of these factors and how they affect one another is achieved, it may be feasible to use precision agriculture technologies that can address these factors. At the same time, Equation 3.1 clearly shows that GY is inextricably dependent on water so that increasing GY in areas of low precipitation will be challenging and limited.

3.3.1 Plant Spacing

Studies to determine the optimum number of plants and the ideal way to space them have been conducted since the beginning of crop production. The number of past studies is countless but literally hundreds of new studies are conducted each succeeding year. The common belief of many researchers is "The ideal plant spacing for any crop is one where each plant is the same distance from each of its neighbors (equidistant spacing)" (Beuerlein 2014). While it is easy to understand this thinking, there are many reasons to challenge this belief. As already discussed in Section 3.1, Loomis (1983) made the case that while this belief may be valid for cases where resources are not limited, it was not valid under conditions of limited resources such as water. Under conditions of limited resources, he advocated that plants could be manipulated in nonuniform ways to an advantage. The belief that equidistant spacing is best seems to be mainly rooted in the idea that it will provide the greatest efficiency in light interception and photosynthesis. It is well understood that the early dominant space between rows was 1m because this allowed sufficient space for draft animals. Later, tractors were manufactured to work with these rows. Over time, width between rows became narrower and in recent years, 0.75 m rows have become the most common. However, with the rapidly expanding interest and availability of precision technologies, there are numerous machines being developed with the objective of improving germination and spacing. While the objective is often to move closer and closer to equidistant spacing, there are exceptions.

The objective of this chapter is to encourage readers to use Equation 3.1 to evaluate how changing a crop production factor will likely change one or more of the four factors that determine yield. A careful examination of Equation 3.1 suggests that equidistant spacing of plants is likely not the best geometry when resources are limited. Precision agriculture technologies make it much more feasible to strive for equidistant spacing, but precision technologies also make it more feasible to achieve spacing patterns to manipulate plants to increase the efficient use of limited resources as suggested by Loomis (1983).

Using water as an example, a study of Equation 3.1 strongly suggests that equidistant spacing of plants would be desirable when water is not limiting. This is because ET would increase as long as there was sufficient light, temperature, and fertility, and that pests were controlled. As ET increased, there would be more biomass produced and with equidistant spacing the soil surface would be shaded more quickly and more completely, which would increase the T/ET factor. Hence, the plant density would be high enough to maintain a favorable microenvironment that would result in a favorable vapor pressure deficit. With all of these factors being favorable, the HI would also be favorable. In contrast, under conditions where water is limited, and particularly so where water is severely limited, equidistant spacing is likely not the best option. Loomis (1983) discusses the many challenges for using water efficiently when it is limited. The soil texture and frequency of precipitation are very important. The evaporation (E) from moist soil is energy-dependent and approaches potential ET. However, as the soil dries, E declines rapidly since movement of water from lower depths in the profile to the soil surface becomes slow. The rate is also affected by the soil-air interface and crop residue mulch can be effective, but many soils are somewhat self-mulching on drying. Loomis (1983) stresses the important point is that after the soil surface dries, E is less than T, so by reducing the plant cover below a LAI of 1, ET per land area can be reduced. Although a greater proportion of the water may be lost as E, more water is available for T per unit cover and crop success is improved, particularly for phasic crops or long-season crops. This exemplifies how complex using limited water efficiently is, and when this is coupled with how unpredictable and variable precipitation is in limited water areas, developing optimum cropping systems is difficult. In principle, plants can be spaced in a way that influences the time when stored moisture is used. With equidistant spacing, roots can reach all of the soil mass in a uniform manner. While this is good for exploiting stored plant-available soil water, which is desirable for water-limited conditions, if most of the stored water is used for vegetative growth, the plants become entirely dependent on growing season precipitation during their reproductive growth stages. Where plants are closely spaced within rows, hills, or clumps, roots may reach the interrow soil mass much later in the season and benefit from having water available during their critical reproductive period. An example of growing grain sorghum under water-deficient conditions in the Texas High Plains is presented in Table 3.2. During the early growth stages when the plants are small and transpiration use is relatively small, the growing season precipitation is generally adequate in years of average or above-average precipitation. However, the variation in precipitation between years is high, as shown by the coefficient of variation (CV) values. In contrast, the average growing season precipitation during the later growth stages are less than one-third of the potential ET when the plants are large and transpiration needs are roughly equal to potential ET values. This is when the crop needs the plant-available water that was present in the soil profile at the time of seeding the crop. However, equally spaced plants would have already used most if not all of this water for vegetative growth and the amount of water available for grain filling would be woefully inadequate, which drastically reduces the HI and results in low yields. Bandaru et al. (2006) compared growing grain sorghum in clumps to equally spaced plants keeping the plant population the same. The schematic in Figure 3.5 shows the plant geometries used in their study. The grain yields were higher for the clump treatments and the authors attributed this to fewer tillers formed in the clumps and an improved microclimate. When grain sorghum plants are several centimeters apart within rows and early growing conditions are favorable, one to three tillers are often formed that use water and nutrients but produce little or no grain under dryland conditions because water becomes limited and prevents the tillers from producing grain. Although the Texas High Plains example differs from other areas where water is limited, it is useful for illustrating how Equation 3.1 can be used for assessing how different technologies might affect yields. As discussed earlier in Section 3.3, increasing ET provides the greatest benefit because this generally also increases the other factors. In the absence of irrigation, however, ET for a given year is limited to how much stored plant-available water at time of seeding can be extracted plus growing season precipitation. This can be estimated by measuring the plant-available water in the soil profile and assuming growing season precipitation by using probability amounts or long-term averages. In many dryland areas, the actual ET amounts will be less than 50% of the requirement. Therefore, producers must alter some of their practices. The first change is generally to significantly reduce the plant density. While this is an easy decision to make, the extent that plant density should be decreased is

TABLE 3.2

Long-Term Average Precipitation during Various Growth Stages of Grain Sorghum Seeded on June 1 in Bushland, TX

Crop Stage	Days	PET	Precipitation	Pct./PET[a]
		mm		%
Day 1 to 3-leaf	23 (9)	64 (11)	50 (87)	78
3-leaf to flag leaf	30 (7)	151 (9)	64 (65)	42
Flag leaf to flowering	21 (10)	131 (11)	37 (69)	28
Flowering to black layer	37 (11)	191 (8)	57 (71)	30
Total	111 (6)	537 (7)	208 (44)	39

Source: Bandaru, V. et al., *Agron. J.* 98:1109–1120, 2006.

Note: Numbers in parentheses are coefficients of variation values.

[a] Percentage of potential evapotranspiration supplied by precipitation for the various growth stages.

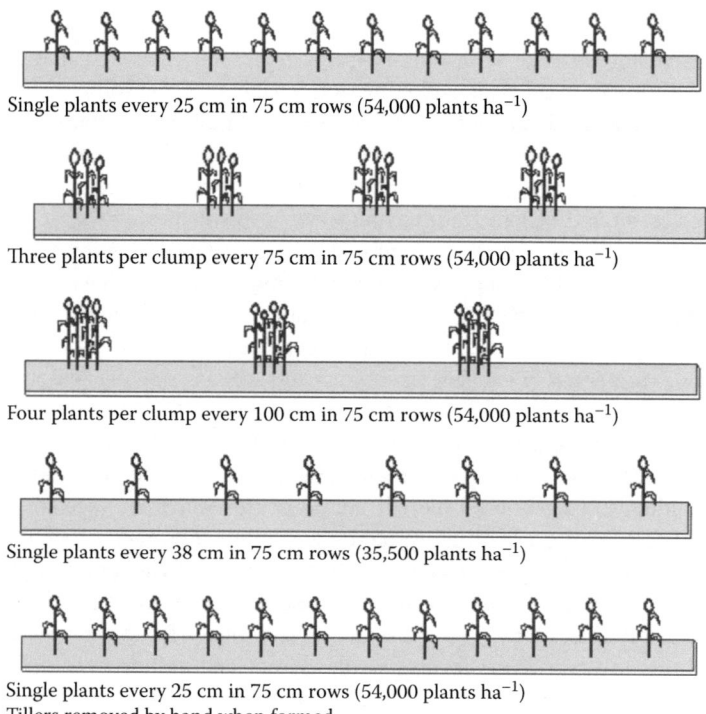

Single plants every 25 cm in 75 cm rows (54,000 plants ha^{-1})

Three plants per clump every 75 cm in 75 cm rows (54,000 plants ha^{-1})

Four plants per clump every 100 cm in 75 cm rows (54,000 plants ha^{-1})

Single plants every 38 cm in 75 cm rows (35,500 plants ha^{-1})

Single plants every 25 cm in 75 cm rows (54,000 plants ha^{-1})
Tillers removed by hand when formed

FIGURE 3.5 Schematic showing different geometries used by Bandaru et al. (2006) for manipulating grain sorghum plants to reduce formation of tillers and enhance microclimate.

difficult to determine. The density should remain high enough to fully exploit the soil mass to remove as much as possible of the stored plant-available soil water. Even if most of the water is used for ET, reducing plant density results in less LAI so there is more evaporation from the soil surface that negatively affects the T/ET factor. The TR also negatively affected because the VPD becomes higher because the microclimate is less favorable when plant density is low. The HI will likely be increased, but even the best-case scenario for reducing plant density is that one of the factors (ET) will remain constant, two (T/ET and 1/TR) will be decreased, and only one (HI) will likely be increased. Another strategy applied in dryland areas is to use skip rows. The theory is that having rows of plants widely separated will delay extraction of much of the stored plant-available water until the plants reach the reproductive stage of growth. While this is a successful practice in some cases, an analysis of Equation 3.1 suggests that the amount of ET may be reduced and the T/ET factor will be less favorable. It is conceivable that if the number of plants is held constant, the plants in the planted rows will be closer together and this could possibly improve the microclimate and reduce the VPD. Even in this case, some of the factors are being negatively affected and the only factor that may be significantly affected is the HI. Therefore, skip-row technologies are more likely to reduce risk of a failed crop than to increase yield. Clark and Knight (1996) in the southwestern United States showed increased grain sorghum yields only when GYs were less than about 2 Mg ha^{-1} and Spackman et al. (2001) reported similar findings in Australia. More recent studies in Kansas and Nebraska in the United States (Lyon et al. 2009) recommended that risk-averse maize growers using 75-cm row spacing use a plant-2-skip-2 pattern where GYs were likely to be less than 4700 kg ha^{-1}. Where GYs were likely to be between 4700 and 6000 kg ha^{-1}, moderate risk-averse producers should use a plant-1-skip-1 pattern. Mesfin et al. (2014) found no advantage of skip-row planting in Ethiopia where maize GYs averaged 3100 kg ha^{-1}. They stated that there might be greater advantage with skip-row planting if some form of reduced tillage or no tillage is practiced coupled with maintaining crop residues on the surface to reduce evaporation. However, crop residues are highly valued and conservation agriculture practices are rarely used by smallholder farmers in Ethiopia (Mesfin et al. 2014). Wortmann et al. (2009) reported that 30% to 40% of the value of a sorghum crop in Ethiopia is for the crop residue. Mulch can positively affect all of the factors in Equation 3.1. Unfortunately, the lack of large amounts of mulch material and the demand of residues for competing uses such as animal feed and fuel prevent its use for increasing the T/ET factor significantly. In parts of China plastic film is widely used for maize production. In Gansu Province, there were 896,000 ha of maize in 2012 and 92% was produced using plastic mulch (personal communication, Dr. Fan Tinglu, Gansu Academy Agricultural Science, Langzhou). This practice greatly increases the T/ET factor and allows satisfactory maize production in relatively low precipitation areas. While this practice may not be practical for large commercial farms, it may be feasible for smallholder farmers in countries other than China.

Some recent studies indicate that a useful strategy in water-deficient areas is to grow plants in clumps to decrease tiller formation in widely spaced plants and to improve the microclimate. The common practice of reducing plant density of maize and particularly grain sorghum results in increased formation of tillers when early season growing conditions are favorable. However, when water becomes limited during later growth stages, many of the tillers result in little or no grain production but have used water and nutrients. Several studies have shown that the formation of tillers is minimized or eliminated when three or four plants are grown in clumps (Bandaru et al. 2006; Kapanigowda et al. 2010; Krishnareddy et al. 2010; Haag 2013) and yields are generally increased when water is limited during reproductive growth stages. Increases in HI values have been consistent in all of the studies, and spacing of the clumps less than 1m apart appears satisfactory for exploiting plant-available water from the soil mass by the end of the growing season. A common belief is that plants adjacent to one another result in decreased productivity but these studies have shown otherwise when total plant density of the clump geometry is the same as for plants evenly spaced within the rows. However, Doerge et al. (2002) found benefits from improved within-row plant spacing in an extensive study across four diverse maize-growing environments. Missing, misplaced, and extra plants were found to have different effects on yield. Poorly spaced and missing plants decreased yield while occasional extra plants tended to increase yields. They consistently found that occasional closely spaced plants were the highest yielding and there was no evidence of increased barrenness of ears on these plants.

3.3.2 Application of Nutrients

Perhaps second only to water, the lack of one or more of the essential nutrients limits crop production. Nitrogen (N) is generally the first essential nutrient that limits GY, followed by phosphorus (P), but there are exceptions. The timing and placement of these nutrients are important and countless studies have been conducted to develop optimum application technologies. In general, the proper placement of P is more critical than N because it is less soluble and more prone to being fixed in the soil in forms that are unavailable for uptake by plants. It is generally believed the best placement of fertilizer is approximately 5 cm below and to the side of where seed are placed. The placement of P is more critical than N because plants primarily take up N as nitrate, which is highly soluble and is absorbed mainly by mass flow. In contrast, P uptake is mostly by diffusion to the plant root so having a high concentration of P near the plant roots is highly desirable to increase P fertilizer use efficiency. Therefore, P is often applied in bands in rows parallel to the seed rows. Advances in precision application equipment has made it possible to not only apply the fertilizer more evenly and at a desirable distance from the seed, but even to vary the application rate for different parts of the field based on soil tests showing different levels of soil fertility or GY maps from previous crops. Plant density is also important for maximizing fertilizer use efficiency. Referring again to Figure 3.3a, the maize field shown has a much higher plant density requiring more fertilizer. Therefore, P fertilizer placed in a band will be at a substantially higher concentration, and because the plants are more closely spaced within rows, uptake efficiency will be higher than for the low-density plants shown in Figure 3.3b that are fertilized at a much lower application rate.

It is well understood that optimizing nutrient availability enhances water use efficiency (WUE). Numerous studies have shown that when fertilizers or organic amendments are added to low fertility soils, yields are increased, resulting in higher WUE values. Referring again to Equation 3.1, with a fixed amount of ET the increased yield from optimizing fertility occurs because the plants become larger with a more extensive root system to fully exploit water from the soil profile, which provides more canopy to increase the T/ET value. The higher canopy also enhances the microclimate that tends to reduce the VPD that in turn reduces the TR, and the HI is generally increased because of higher yields. Even though the amount of water available for plant use is not increased, optimizing fertility tends to increase all of the other factors shown in Equation 3.1, so yields can be increased significantly. Therefore, the interaction between soil fertility and water should be carefully considered in optimizing plant geometries.

3.3.3 DEVELOPING PRECISION TECHNOLOGIES

In recent years, technologies for planting seeds at specific depths and interrow spacing have advanced to unparalleled precisions. Fertilizers can also be applied with extreme precision even to the point of varying the rate with changes in the soil within fields. In contrast, 60% of the more than 34 Mha of maize in developing countries is planted by hand (Oklahoma State University 2014). In parts of Africa, a traditional method is to plant maize and sorghum in pits, as shown in Figure 3.6. Fertilizer or manure is often added at the time of planting the seeds. This method has been used for centuries and is similar to the method used by the Hopi in America for thousands of years

FIGURE 3.6 Traditional method of growing sorghum in parts of Africa. (From TerrAfrica, Sustainable Land Management in Practice: Guidelines and Best Practices for Sub-Saharan Africa. A TerrAfrica Partnership Publication. Available at http://knowledgebase.terrafrica.org/fileadmin/user_upload/terrafrica/docs/topic_page/SLM_in_Practice_english.pdf, 2011.)

(Figure 3.4). Many of the fields in Africa are being transitioned to more modern practices as shown in Figure 3.7. The focus is to move toward uniform spacing of the plants and banding fertilizers close to the plant rows. This will likely be successful in favorable areas because the technologies are being patterned after those used in developed countries where GYs are several times higher than those in many areas of Africa. In areas of limited water and low fertility soils, however, uniform spacing of low-density plants may not be advantageous because added fertilizer is more dispersed and the microclimate around widely spaced plants is poorer. Loomis (1983) stated "A useful generalization emerges: where soil resources are limiting, non-uniform treatment of the land can be an advantage. Where soil resources are non-limiting, however, uniform cropping will provide the greatest efficiency in light interception and photosynthesis." Modern technologies may provide new opportunities for evaluating and putting into practice precision agriculture methods that can improve the efficient use of water and fertilizers. An American farmer in a 425-mm annual average precipitation is presently building a prototype planter that can be used for seeding maize, sorghum, and wheat. The concept is illustrated in Figure 3.8. The planter will seed twin rows about 10 cm

FIGURE 3.7   Mechanized seeding practices being introduced in parts of Africa. (From TerrAfrica, Sustainable Land Management in Practice: Guidelines and Best Practices for Sub-Saharan Africa. A TerrAfrica Partnership Publication. Available at http://knowledgebase.terrafrica.org/fileadmin/user_upload/terrafrica /docs/topic_page/SLM_in_Practice_english.pdf, 2011.)

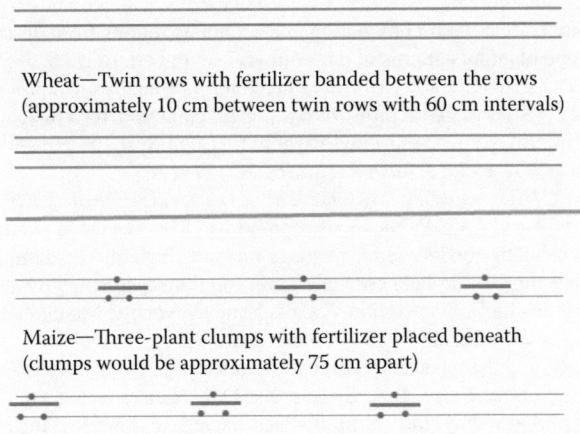

Wheat—Twin rows with fertilizer banded between the rows
(approximately 10 cm between twin rows with 60 cm intervals)

Maize—Three-plant clumps with fertilizer placed beneath
(clumps would be approximately 75 cm apart)

FIGURE 3.8   Hypothetical concept of a precision planter that would place maize or wheat seed in rows along with fertilizer strategically placed near the seed.

apart, and then have a skip of about 60 cm to the next set of twin rows. As an example, three maize seeds will be dropped about every 0.75 m and staggered in alternate rows. Fertilizer will be banded between the twin rows and a sensor will turn the fertilizer on and off so that fertilizer is placed only where the seeds are placed and it will be about 5 cm deeper than the seed. The concept is that this will improve fertilizer use efficiency, provide a better microclimate for the maize plants, and ration the use of stored plant-available water so that more of it is available for use during the reproductive growth stages. Referring back to Equation 3.1, this system should fully utilize the ET, decrease the TR, and increase the HI. In addition, the producer plans to use little or no tillage, which will increase the T/ET factor. For seeding wheat, the plan is to place the seed continuously in the rows and place one band of fertilizer between the rows. While this prototype seeder may not work satisfactorily, the concepts appear scientifically sound, supporting the views of Loomis (1983) that plants can be manipulated to take better advantage of limited resources.

3.4 CONCLUSION

The general perception is that the ideal plant spacing for any crop is one where each plant is of the same distance from each of its neighbors. While this belief may be valid for many situations, particularly where resources are not limited, it is not valid under conditions of limited resources such as water and fertilizers. Under conditions of limited resources, it is conceivable that plants can be manipulated in non-uniform manners to improve microclimate, increase fertilizer use efficiency, influence the time when stored soil water is used, and to improve other factors that affect crop yields. Agricultural technologies are rapidly emerging that allow the placement of seed, fertilizer, and irrigation water with unparalleled precision. Therefore, additional studies are needed that look at new ways of manipulating plants to utilize resources, particularly when those resources are limited.

REFERENCES

Bandaru, V., B.A. Stewart, R.L. Baumhardt, S. Ambati, C.A. Robinson, and A. Schlegel. 2006. Growing grain sorghum in clumps to reduce vegetative growth and increase yield. *Agron. J.* 98:1109–1120.

Beuerlein, J. 2014. Soybean plant spacing, the last frontier. Ohio State University Extension Fact Sheet, AGF-140-01. Columbus, OH. Available at http://ohioline.osu.edu/agf-fact/pdf/0140.pdf (verified April 22, 2014).

Brown, W.L. 1985. New technology related to water policy–plants. In *Water and Water Policy in World Food Supplies*, W.R. Jordan (ed.). Proceedings of the Conference May 26–30, 1985, 37–41. Texas A&M University, College Station, TX. Texas A&M University Press, College Station, TX.

Clark, L.E. and T.O. Knight. 1996. Grain production and economic returns from dryland sorghum in response to tillage systems and planting patterns in the semi-arid southwestern USA. *J. Prod. Agric.* 9:249–256.

Doerge, T., T. Hall, and D. Gardner. 2002. New research confirms benefits of improved plant spacing in corn. *Crop Insights* 12(2):1–5. Available at https://www.google.com/#q=Crop+Insights++New+research+confirms+benefits+of+improved+plant+spacing+in+corn (verified April 28, 2014).

Donald, C.M. 1962. In search of yield. *J. Aust. Agric. Sci.* 28:171–178.

Gurian-Sherman, D. 2012. *High and Dry: Why Genetic Engineering Is Not Solving Agriculture's Drought Problem in a Thirsty World.* UCS Publications, Cambridge, MA. Available at http://www.ucsusa.org/food_and_agriculture/our-failing-food-system/genetic-engineering/high-and-dry.html (verified April 15, 2014).

Haag, L.A. 2013. Ecophysiology of dryland corn and grain sorghum as affected by alternative planting geometries and seeding rates. Ph.D. dissertation, Kansas State University, Manhattan, KS. Available at http://krex.k-state.edu/dspace/handle/2097/16277 (verified April 28, 2014).

Howell, T.A. 1990. Relationships between crop production and transpiration, evapotranspiration, and irrigation. In *Irrigation of Agricultural Crops*, B.A. Stewart and D.R. Nielsen (eds.), 391–434. Agronomy No. 30. American Society of Agronomy, Inc., Crop Science Society of America, Inc., Soil Science Society of America, Inc., Madison, WI.

Kapanigowda, M, B.A. Stewart, T.A. Howell, H. Kadasrivenkata, and R.L. Baumhardt. 2010. Growing maize in clumps as a strategy for marginal climatic conditions. *Field Crops Res.* 118:115–125.

Krishnareddy, S., B.A. Stewart, W.A. Payne, and C.A. Robinson. 2010. Grain sorghum tiller production in clump and uniform planting geometries. *J. Crop Improv.* 24:1–11.

Loomis, R.S. 1983. Crop manipulations for efficient use of water: An overview. In *Limitations to Efficient Water Use in Crop Production*, H.M. Taylor, W.R. Jordan, and T.R. Sinclair (eds.), 345–380. American Society of Agronomy, Inc., Crop Science Society of America, Inc., Soil Science Society of America, Inc., Madison, WI.

Lyon, D.J., A.D. Pavlista, G.W. Hergert, R.N. Klein, C.A. Shapiro, S.C. Mason, L.A. Nelson, D.D. Baltensperger, R.W. Elmore, M.F. Vigil, A.J. Schlegel, B.L. Olson, and R.M. Aiken. 2009. Skip-row planting patterns stabilize corn grain yields in the Central Great Plains. Plant Management Network. Available at http://www .agronext.iastate.edu/corn/contact/roger_elmore/docs/skip.pdf (verified April 28, 2014).

Mesfin, T., J. Mohammed, A. Taklete, F. Merga, and C. Wortmann. 2014. Skip-row planting of maize and sorghum in semi-arid Ethiopia. *Afr. J. Plant Sci.* 8(3):140–146.

Oklahoma State University. 2014. OSU Hand Planter for the Developing World. Available at http://nue.okstate .edu/Hand_Planter.htm (verified April 29, 2014).

Peterson, G.A., D.G. Westfall, and N.C. Hansen. 2012. Enhancing precipitation-use efficiency in the world's dryland agroecosystems. In *Soil Water and Agronomic Productivity*, R. Lal and B.A. Stewart (eds.), 455–476. CRC Press, Boca Raton, FL.

Prihar, S.S. and B.A. Stewart. 1990. Using upper-bound slope through origin to estimate genetic harvest index. *Agron. J.* 82:1160–1165.

Ritchie, J.T. and E. Burnett. 1971. Dryland evaporative flux in a subhumid climate: I. Icrometeorological influences. *Agron. J.* 63:56–62.

Sinclair, T.R. 2009a. Taking measure of biofuel limits. *Am. Sci.* 97:400–407.

Sinclair, T.R. 2009b. Taking measure of biofuel limits. Available at http://climatesanity.wordpress.com /2009/09/24/taking-measure-of-biofuel-limits (verified April 15, 2014).

Sinclair, T.R. and A. Weiss. 2010. *Principles of Ecology in Plant Production*, 2nd edition. CAB International, Cambridge, MA.

Smika, D.E. 1983. Soil water changes as related to position of straw mulch on the soil surface. *Soil Sci. Soc. Am. J.* 47:988–991.

Snyder, F.W. and G.E. Carlson. 1984. Selecting for partitioning of photosynthetic products in crops. *Adv. Agron.* 37:47–73.

Spackman, G.B., K.J. McCoske, A.J. Farquharson, and M.J. Conway. 2001. Innovative management of grain sorghum in central Queensland. In *Science and Technology: Delivering Results for Agriculture*. Available at http://www.regional.org.au/au/asa/2001/1/a/spackman.htm (verified June 9, 2015).

Steiner, J.L. 1986. Dryland grain sorghum use, light interception, and growth responses to planting geometry. *Agron. J.* 78:720–726.

Steiner, J.L., H.H. Schomberg, and J.E. Morrison, Jr. 1994. Measuring surface residue and calculating losses from decomposition and redistribution. In *Crop Residue Management to Reduce Erosion and Improve Soil Quality–Southern Great Plains*, B.A. Stewart and W.C. Moldenhauer (eds.), 21–32. Conservation Research Report Number 37, Agricultural Research Service, USDA, Washington, DC.

Stewart, B.A., P. Koohafkan, and K. Ramamoorthy. 2006. Dryland agriculture defined and its importance in the world. In *Dryland Agriculture*, G.A. Peterson, P.W. Unger, and W.A. Payne (eds.), 1–26. Agronomy Monograph No. 23, 2nd edition. American Society of Agronomy, Inc., Crop Science Society of America, Inc., and Soil Science Society of America, Inc., Madison, WI.

Stewart, B.A., R.L. Baumhardt, and S.R. Evett. 2010. Major advances of soil and water conservation in the U.S. Southern Great plains. In *Soil and Water Conservation Advances in the United States*, T.M. Zobeck and W.F. Schillinger (eds.), 103–129. SSSA Special Publication 60. Soil Science Society of America, Inc., Madison, WI.

Stewart, B.A. and G. Peterson. 2014. Managing green water in dryland agriculture. *Agron. J.* 1006:1–10. doi: 10.2134/aronj14.0038.

Stirzaker, R. 2010. *Out of the Scientist's Garden: A Story of Water and Food*. Commonwealth Scientific and Industrial Research Organization. CSIRO Publishing, Collingwood VIC, Australia.

TerrAfrica. 2011. *Sustainable Land Management in Practice: Guidelines and Best Practices for Sub-Saharan Africa*. A TerrAfrica Partnership Publication. Available at http://knowledgebase.terrafrica.org/fileadmin /user_upload/terrafrica/docs/topic_page/SLM_in_Practice_english.pdf.

Unger, P.W., W.A. Payne, and G.A. Peterson. 2006. Water conservation and efficient use. In *Dryland Agriculture*, G.A. Peterson, P.W. Unger, and W.A. Payne (eds.), 39–85. Agronomy Monograph No. 23, 2nd edition. American Society of Agronomy, Inc., Crop Science Society of America, Inc., and Soil Science Society of America, Inc., Madison, WI.

Unger, P.W., R.L. Baumhardt, and F.J. Arriga. 2012. Mulch tillage for conserving soil water. In *Soil Water and Agronomic Productivity*, R. Lal and B.A. Stewart (eds.), 427–453. CRC Press, Boca Raton, FL.

Wortmann, C.S., M. Mamo, C. Mburu, E. Letayo, G. Abebe, K.C. Kayuki, M. Chisi, M. Mativavaria, S. Xerinda, and T. Nidacyayisenga. 2009. Atlas of sorghum (*Sorghum bicolor* (L.) Moench) production in eastern and southern Africa. Available at http://scholar.google.com/scholar?q=Atlas+of+sorghum+(Sorghum+bicolor +(L.)+Moench)+production+in+eastern+and+southern+Africa&hl=en&as_sdt=0&as_vis=1&oi=scholar t&sa=X&ei=XVl3VaTgN8PaoASX64HYBg&ved=0CBsQgQMwAA (verified June 9, 2015).

4 4 Rs Are Not Enough

We Need 7 Rs for Nutrient Management and Conservation to Increase Nutrient Use Efficiency and Reduce Off-Site Transport of Nutrients

J.A. Delgado

CONTENTS

4.1 INTRODUCTION

Agricultural production is affected by many factors and management decisions at a field level that can impact yields, nutrient use efficiency, and soil and water conservation. Management decisions to increase agricultural production at the site should also consider how site-specific landscape properties interact with local weather and management to potentially impact sustainability at the site and the environment off-site. Management decisions made for a given site-specific field affect the potential flow of nutrients, agrochemicals, and sediment, contributing to their total mass flow out of the watershed. Managing spatial and temporal variability will be important in minimizing the potential flows of nutrients, agrochemicals, and sediment transport, especially since there is a need to reduce the massive flows coming from hot spot areas of fields and watersheds.

It is important to consider how precision farming in combination with precision conservation can help us address today's challenges and how they will contribute to climate change mitigation and adaptation (Delgado et al. 2011; Lal et al. 2011). Some of today's important challenges are a changing climate, feeding a continuously growing world population, the depletion of water resources to use for irrigation, desertification, deforestation, and the need for sustainable soils in order to maintain and/or increase agricultural production. Conservation practices and management will be key to managing these challenges and for food security (Delgado et al. 2011; Lal et al. 2011). Among the new technologies that can help us increase food production, efficiency, and conservation are geospatial technologies for precision agriculture and precision conservation. Managing spatial and temporal variability with precision farming, precision conservation, and precision harvesting could contribute to increased resource use efficiency, sustainability, and reduced environmental impacts.

4.1.1 Climate Change and Population Growth

Climate change and population growth are tremendous challenges that are interconnected, with anthropogenic activities reported to increase the potential for negative impacts on natural systems and agriculture (IPCC 2007). Agriculture in the twenty-first century faces great challenges, and soil scientists, agronomists, and nutrient managers will need to cooperate with farmers and peers from other disciplines to find ways to significantly increase yields per unit area while reducing the environmental impacts of agricultural activities. Fortunately, we can use conservation practices and management decisions to adapt to a changing climate (Delgado et al. 2011; Lal et al. 2011; Walthall et al. 2012).

Several researchers have reported that climate change will potentially negatively impact agricultural production due to changes in weather patterns such as maximum and minimum temperatures, frost days, heat stress, droughts, floods, rapid snow melts, lower snowcaps, higher evapotranspiration rates, and other climatological events (IPCC 2007; Walthall et al. 2012). One of the major effects of climate change that can have a negative impact on productivity and sustainability is the increase in soil erosion, especially from increases in extreme events (Nearing 2001; Nearing et al. 2004; Pruski and Nearing 2002a,b; SWCS 2003, 2007). Climate change can significantly increase erosion potential and it has been reported that for every 1% increase in total precipitation, soil erosion will be increased by 1.7% (Nearing 2001; Nearing et al. 2004; Pruski and Nearing 2002a,b). Recent studies predict that climate change will contribute to increases in soil erosion in the United States (Segura et al. 2014; Zhang et al. 2012), so conservation practices to adapt to climate change will be critical (Delgado et al. 2011).

Potential increases in soil erosion in the United States due to climate change are of concern because for each 10 cm of soils that we lose, we reduce potential productivity by 4.3% to 29.6% (Bakker et al. 2004). It takes hundreds of years to form 1 inch (2.54 cm) of soil (SSSA 2013), and we could lose more than that in an extreme event. Fortunately, conservation practices can be used to adapt to climate change (Delgado et al. 2011, 2013; Lal et al. 2011; Walthall et al. 2012). Precision farming and precision conservation are tools that can be used to adapt to climate change and to help increase efficiency of conservation practices and efficiency of water resource use by accounting for temporal and spatial variability.

4.1.2 DEPLETION OF WATER RESOURCES IS OF GREAT CONCERN

One of the great challenges that the humans will face during the twenty-first century is depletion of water resources that are needed for irrigation practices. This is especially important because we know that irrigated systems have yields that are double those of nonirrigated dryland systems (Bucks et al. 1990; Rangely 1987; Tribe 1994). Several papers have reported on some areas where the irrigation resources are being depleted (Hu et al. 2005; Walthall et al. 2012). Some factors that have been reported to reduce the available water for irrigation are a smaller snowcap and/or a faster depletion of the snowcap early in the spring, contributing to greater early runoff, and to a smaller available supply of water for later in the growing season when the evapotranspiration water demands are higher (Knowles et al. 2006; Walthall et al. 2012). Water management will be more important during the twenty-first century due to higher demands for food and the challenges of water depletion in some regions. It will be important to increase water conservation and maintain water quality.

For rainfed systems, droughts have a significant negative impact on average yields (Al-Kaisi et al. 2013; Auffhammer 2011; Lal et al. 2012; Lobell et al. 2011; Peng et al. 2004), so management practices that can contribute to the conservation of soil and water can have an important role in climate change adaptation (Delgado et al. 2011, 2013; Lal et al. 2011; Walthall et al. 2012). Water management could also help reduce loss of soil particles, nutrients, and soil organic matter via erosion. Site-specific practices for precision agriculture and precision conservation could potentially contribute to increased water use efficiency, soil quality, and sustainability (Delgado et al. 2011, 2013; Lal et al. 2011; Walthall et al. 2012). The right product, at the right rate, at the right time, and at the right place (the 4 Rs) are not enough to reduce off-site transport of nutrients (Figures 4.1 and 4.2). We need 7 Rs (right product, right rate, right method, right practice, right place, right scale, and right time) for nutrient management and conservation to increase nutrient use efficiency and reduce off-site transport of nutrients. (See details about 4 Rs and 7 Rs in Section 4.6.5.)

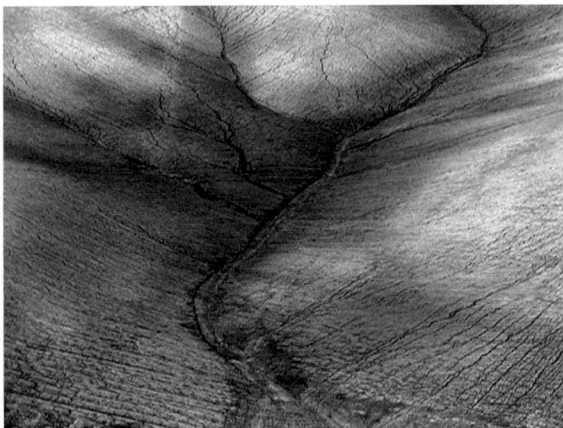

FIGURE 4.1   Location where the 4 Rs alone will not reduce the off-site transport of nutrients (see Section 4.6.2 regarding the development of ephemeral gullies). Precision conservation needs to be merged with precision farming. Nutrient management and conservation need to incorporate 7 Rs to increase nutrient use efficiency and reduce off-site transport of nutrients. (From NRCS, Development of Ephemeral Gullies.) Potential precision conservation practices that can be applied at this site are grass waterways, buffers, crop residue management, no-till, and others.

FIGURE 4.2 This site demonstrates the need for the 7 Rs in nutrient management and conservation to increase nutrient use efficiency and reduce off-site transport of nutrients and environmental impacts. (From NRCS, Development of Ephemeral Gullies.)

4.1.3 Environmental Impacts of Off-Site Transport of Nitrogen and Phosphorus

Crop yields respond significantly to nitrogen inputs, and this key element helps maximize agricultural production. However, when N management decisions result in application of more N than necessary, they contribute to increased losses of nitrogen from the cropping system via several different pathways to the environment, so management is an important factor in reducing nitrogen losses (Shaffer and Delgado 2002). The more information we have about spatial and temporal variability at a site, the more informed management decisions can be. Across the literature there are reports of environmental impacts from nitrogen losses, and average nitrogen use efficiency has been reported to be as low as 33% for grains (Raun and Johnson 1999) and as high as 50% (Baligar et al. 2001).

Fixen and West (2002) and Snyder and Bruulsema (2007) reported that nitrogen use efficiency has been increasing for corn systems across the United States during the last three decades and that during this period the average corn yields in the United States increased while the average fertilizer application rate was maintained at the same levels. In contrast, a report by Ribaudo et al. (2011) suggests there is a need to increase nitrogen use efficiency, since it found that there is still a need to improve nitrogen management practices. Ribaudo et al. (2011) found that only about a third of the cropland in the United States met three criteria for best nitrogen management practices (using the best nitrogen application rate, best method of nitrogen application, and best time of nitrogen application), suggesting that there is still a great need to improve nitrogen management practices to reduce nitrogen losses to the environment. Independent of these reports on nitrogen use efficiency from Fixen and West (2002), Ribaudo et al. (2011), Snyder and Bruulsema (2007), and Snyder (2012), environmental indicators for nitrogen (e.g., NO_3-N concentrations in surface and/or groundwater) strongly suggest that the increase in nitrogen use efficiency over the last three decades has not contributed to reduced total losses of nitrogen across the United States (USGAO 2013). They also show that nutrient losses and the off-site transport of reactive nitrogen to the environment is negatively impacting water bodies (Dubrovsky et al. 2010; Goolsby et al. 2001; Robertson et al. 2009; USGAO 2013).

A recent example of the impacts of nitrogen losses to the environment and how fast these flows of nitrogen move across the environment is the recent report that the highest-ever nitrate concentrations in the Des Moines and Raccoon Rivers were measured during March 2013 (Associated Press 2013). A hypothesis that could explain where the nitrogen that contributed to these high, record-breaking nitrate concentrations came from is that it was the result of lower corn yields due to the drought of 2012. The hypothesis is that not all of the fertilizer applied to the corn was taken up by the plants because the drought reduced the yields, thus reducing the nitrogen uptake demand of the crop and leaving a higher amount of residual soil nitrate in the profile that was susceptible to being leached. If this hypothesis is correct, it reflects how fast nitrogen moves across the environment. Nitrogen is highly dynamic and mobile, and there is a need to quantify nitrogen losses so that we can evaluate how management impacts the pathways for flows of nitrogen in the environment and then apply management decisions that minimize the losses of reactive nitrogen to the environment (Delgado 2002). This hypothesis is in sync with Al-Kaisi et al. (2013) paper that reported that the lower yields from the 2012 drought could potentially contribute to lower N uptake and higher residual nitrate with the potential to leach out and impact water quality.

If the above hypothesis about the relationship between residual soil nitrate and record-breaking nitrate levels in the Des Moines and Raccoon Rivers is true, then it suggests that there is a very fast pathway that connects field management and pathways for losses with surface water bodies of the USA, and it suggests there is a need for site-specific precision agriculture and precision conservation to reduce N losses to the environment and maintain water quality. Independent of whether this hypothesis is correct, the fact that record-breaking nitrate concentrations did occur, along with other water quality reports about negative impacts to water bodies (Dubrovsky et al. 2010; Goolsby et al. 2001; Robertson et al. 2009; USGAO 2013), suggest that there is a real need to improve nitrogen management and increase nitrogen use efficiency to protect water quality, and that doing so should be a key goal at the field and watershed levels. Climate change will create new challenges for nutrient management even with the 4 Rs; for example, if we get a drought that reduces yields, the potential to have a larger amount of residual soil nitrate available to leach may increase due to lower crop N uptake.

These studies from environmental indicators are also in agreement with a recent report from Randall et al. (2008) that evaluated a large number of ^{15}N isotopic studies and reported that about 33% of the fertilizer-applied N is leaving agricultural systems and is not recovered in the crop or in the soil. The ^{15}N-labeled fertilizer makes it possible to trace the fate of the fertilizer-nitrogen. These types of ^{15}N studies show the importance of management; for example, cover crops can contribute to the recovery of nitrogen by scavenging nitrogen from the soil profile and moving it to the plant compartment, minimizing its loss and reducing the losses of N from the system. Additionally, the N from the cover crop is recycled with a loss of 13%, which is lower than the average 33% loss from inorganic fertilizer (Delgado et al. 2010). Ribaudo et al. (2011) reported that removing nitrate from U.S. drinking water that came from agricultural systems costs about 1.7 billion dollars per year. Improving site-specific nitrogen management practices together with precision conservation could significantly reduce nitrogen losses to the environment and the negative economic impacts of these nitrogen losses.

It is well known that nitrogen losses to the environment via surface runoff (Bjorneberg et al. 2002), nitrate leaching (Davies and Sylvester-Bradley 1995; Madison and Brunett 1985; Milburn et al. 1990), ammonia volatilization (Fox et al. 1996; Freney et al. 1981; Peoples et al. 1995; Sharpe and Harper 1995), and emissions of nitrous oxide (N_2O) (IPCC 2007) are negatively affecting the environment. For example, surface losses of nitrogen and nitrate leaching are reported to be contributing to the transport of nitrogen to water bodies and to issues such as the hypoxia problem in the Gulf of Mexico (Goolsby et al. 2001; Robertson et al. 2009). The impacts of NO_3-N leaching on groundwater is another example of the effects of land use management practices that have been correlated with groundwater nitrate levels (De Paz et al. 2009; Fletcher 1991; Hallberg 1989; Juergens-Gschwind 1989). Irrigation and larger-than-necessary organic and/or inorganic nitrogen fertilizer applications have been found to be correlated with these higher nitrate levels (De Paz et al. 2009; Hall et al. 2001). Increasing nitrogen use efficiency and implementing conservation practices are important management tools to reduce the off-site transport of nitrogen and to reduce impacts to water bodies, especially from risky landscape-management combinations that have higher potential for leaching. Phosphorus losses from agricultural systems have also been reported to contribute to environmental problems across the United States, and there is the potential to use site-specific management of phosphorus to reduce these losses (Chang et al. 2003; Fu et al. 2010; Sawchik and Mallarino 2007; Secoges et al. 2013).

4.1.4 IMPACTS OF EROSION ON SOIL QUALITY

Erosion has been reported to contribute to lower yields and to affect soil quality properties such as soil fertility, soil organic matter, and water holding capacity, which are needed to maintain sustainability; erosion also contributes to the degradation of the soil system (Andraski and Lowery 1992; Bakker et al. 2004; Follett and Stewart 1985; Lal 1995, 1998; Quine and Zhang 2002). We need to minimize erosion to maintain sustainability of cropping systems since the degradation of worldwide soils threatens the potential to maintain and/or achieve food security during the twenty-first century (Lal 1995, 1998; Montgomery 2007). We will also need to increase yields to meet the increasing demands for food from the ever-growing world population.

It is well established that erosion in agricultural systems will generally have a negative impact on productivity. Even if some landscape positions that are receiving the eroded soil material may benefit from this process, since erosion in general threatens to degrade worldwide cropping/soil systems, it must be minimized (Lal 1995, 1998; Montgomery 2007). Erosion could reduce yields by 10%–20% or even more (Bakker et al. 2004; Quine and Zhang 2002). Several scientists have raised the issue of the potential negative impacts that removing crop residue for biofuel systems may have on soil quality due to the correlation between higher erosion rates and the removal of crop residue (Figure 4.3) (Cruse and Herndl 2009; Johnson et al. 2010; Karlen et al. 2009; Lal 2004). It is important that we consider site-specific factors and precision conservation when we manage soils and nutrients in order to maintain soil quality and minimize erosion and off-site transport of nutrients (Berry et al. 2003).

Bakker et al. (2004) evaluated the results from 24 soil erosion studies and found that the impact of a loss of a soil depth as small as 5 cm could reduce the yields by up to 95%. This is significant and shows that the site-specific impacts of erosion could be devastating. On average they found that for every 10 cm of soil lost, the productivity is reduced by 4.3%–29.6% depending on the methodology used to assess the effects of erosion. Additionally, they found that with the next 10 cm of soil lost, the loss in productivity will be much higher since it is not a linear relationship, and the more degraded the soil gets, the higher the impact on the initial loss of productivity. They concluded that a loss of productivity of 4% for every 10 cm of soil lost is a realistic loss. If we are to increase yields to feed a growing world population, we cannot afford to reduce yields at that rate. Precision farming and precision conservation will be important tools for maintaining worldwide sustainability and productivity.

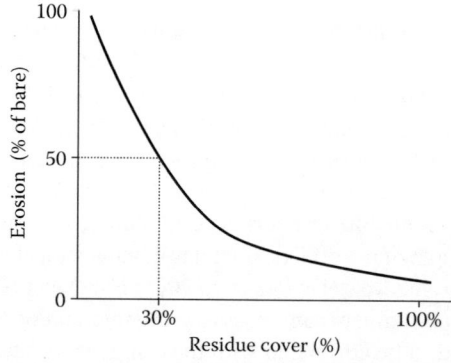

FIGURE 4.3 Effect of residue cover on soil erosion expressed as a percent of erosion occurring on a bare, residue-free surface. (From Cruse, R. M., and C. G. Herndl, *J. Soil Water Conserv.* 64:286–291, 2009.)

4.2   GEOSPATIAL TECHNOLOGIES AND TOOLS TO INCREASE EFFICIENCY WITH PRECISION FARMING AND PRECISION CONSERVATION

We have technologies and tools that are helping us collect detailed information across the landscape to understand and manage temporal and spatial variability. We can use a combination of the Global Positioning System (GPS), Geographic Information Systems (GIS), remote sensing (RS), and modeling tools to collect site-specific field information and analyze site-specific properties related to yields and productivity. These tools also make it possible to trace the fate of nutrients and soil across the landscape. We can use these tools to assess the effects of management practices and how they can impact the transport of nutrients such as nitrogen and phosphorus, as well as how predicted values for site-specific transport of these nutrients are correlated with observed values at different points in the landscape (Cho et al. 2010; Saleh et al. 2011).

Collecting a large amount of information, overlapping the information in layers, and then looking for correlations can help us identify factors that are related to nutrient management and soil and water conservation. Precise landscape topography using lidar technologies with other geospatial tools can help us assess the flows of water, the potential for soil erosion and surface transport of nutrients, and where to place conservation practices such as buffers to reduce off-site nutrient transport. This can also be joined with temporal variability data to improve site-specific nutrient management and conservation over time (Galzki et al. 2011).

4.3   NUTRIENT MANAGEMENT

4.3.1   SPATIAL VARIABILITY OF YIELDS AND PATHWAYS FOR NUTRIENT LOSSES

Research has shown that the properties that affect crop nitrogen response, nitrogen losses via different pathways, and nitrogen availability are spatially variable across the field (Basso et al. 2013; Cambardella et al. 1994; Delgado and Bausch 2005; Jaynes et al. 2011; Jokela and Randall 1989; Khosla et al. 2002; Legg and Meisinger 1982; Roberts et al. 2012; Shanahan et al. 2008). At some site-specific locations, these properties that are related to soil texture are correlated with nitrate leaching potential and residual soil nitrate across a field (Delgado 1999, 2001; Delgado and Bausch 2005; Delgado et al. 2001a,b, 2005). Some researchers have also reported that the position in a catena is also correlated with nitrogen cycling and nitrogen availability (Ortega et al. 1997; Schimel et al. 1985, 1986). Several scientists have reported that there is the potential to use this spatial variability to manage nitrogen and increase nitrogen use efficiency (Basso et al. 2013; Delgado and Bausch 2005; Ferguson et al. 1996; Gotway et al. 1996; Hergert et al. 1996; Jaynes et al. 2011; Raun et al. 2002; Redulla et al. 1996; Roberts et al. 2012; Scharf et al. 2002, 2011; Shanahan et al. 2008).

The coarser soils are more aerated than the fine-textured soils, have less soil organic matter content (Parton et al. 1987), and the denitrification potential is lower for these soils (Meisinger and Randall 1991) than for the finer-textured areas of the field. Other factors such as manure and the water table will also affect the denitrification potential (Meisinger and Randall 1991). Mosier et al. (1991, 1996) reported that the coarser back slopes had lower total N_2O emissions than the fine-textured swale soils.

We could use these relationships to try to use site-specific information for precision conservation to manage flows of nutrients across the landscape and reduce denitrification, N_2O emissions, and nitrate leaching (Delgado and Berry 2008; Pennock 2005). If we are to reduce nutrient losses via different pathways, we will have to implement nutrient management that incorporates soil and water conservation practices (e.g., use of cover crops). Delgado (1999, 2001) and Delgado et al. (2001a,b) found that the sandier, coarser areas of the field had much higher nitrate leaching than the fine-textured areas of the field. In these studies cover crops were used to protect soil water quality and increase nitrogen use efficiency in the system. The cover crops scavenged a large amount of nitrogen from the soil profile that had been left after a vegetable harvest (178 kg N ha^{-1}) in the finer-textured areas of the field (sandy loam), but only scavenged 95 kg N ha^{-1} in the coarser textured areas of the field (loamy sand) (Dabney et al. 2001; Delgado et al. 1999). This nitrogen was returned and cycled back to the following crop, increasing the nitrogen use efficiency of the system (Delgado et al. 1998).

4.3.2 VARIABLE RATES

There is the potential to apply nutrients at a variable rate to increase nitrogen use efficiency (Bausch and Delgado 2003; Biermacher et al. 2006; Jaynes et al. 2011; Khosla et al. 2002; Raun et al. 2002; Roberts et al. 2012; Scharf et al. 2002, 2011; Shanahan et al. 2008). The spatial variability requirement for a crop can be assessed by taking intensive spatial samples using a grid system to assess the soil status of the different nitrogen pools such as the residual soil nitrate and ammonium that is available to the crop, plus the amount of soil organic matter that can provide additional mineralization of the soil organic matter. We can integrate this spatial information and variability of inorganic and/or organic nitrogen pools with other soil spatial information to make management decisions that integrate information across space to develop a variable rate to increase nitrogen use efficiency (Khosla and Alley 1999). Two major disadvantages are that this intensive method has a higher cost and requires more time in order to reach a decision about the amount of nutrients required to meet the crop's needs (Masek et al. 2001). An alternative that requires much less time in collecting representative spatial soil samples (and at a much reduced cost due to a lower number of samples to collect and process) was developed by using a new, simpler management zone approach that incorporates spatial information provided by incorporating yield history, aerial photographs, and farmers' knowledge and experience with the field (Fleming et al. 1999; Khosla et al. 2002). There are also other viable approaches to determine site-specific management zones (e.g., terrain and soil attributes, electrical conductivity, soil fertility) (Davatgar et al. 2012; Fleming et al. 2004; Jaynes et al. 2011; Ortega and Santibanez 2007; Peralta and Costa 2013; Roberts et al. 2012; Xin-Zhong et al. 2009).

Recent advances in sensing technology have allowed us to develop sensing techniques that are capable of being used *in situ* to assess the crop nitrogen status (Raun et al. 2002; Scharf et al. 2002, 2011; Shanahan et al. 2008). We can use these RS techniques to integrate the variability of soils in supplying nitrogen to the crop by using the status of nitrogen in the crop as a proxy. Sensor prices are coming down and are becoming a more economical management tool that can integrate the nitrogen factors correlated with crop yields (Raun et al. 2002; Scharf et al. 2002, 2011; Schepers et al. 2004; Shanahan et al. 2001). Plant sensors can provide real-time information about the nitrogen status of a plant during the growing season (Bausch and Delgado 2003; Bausch and Diker 2001; Bausch and Duke 1996; Biermacher et al. 2006; Raun et al. 2002; Scharf et al. 2002, 2011; Shanahan et al. 2001, 2008). These crop sensors are capable of determining the areas of the field that are deficient in supplying nitrogen to the crop (Bausch and Delgado 2003; Raun et al. 2002; Scharf et al. 2002, 2011; Shanahan et al. 2001, 2008). One approach is to use a reference strip across the field that has been supplied with a large amount of nitrogen to avoid deficiency (Raun et al. 2002; Schepers et al. 1992a,b, 1998; Shanahan et al. 2001, 2008). By comparing the spatial variability between the crop nitrogen status across the reference strip and the rest of the field, the equipment integrates all this information using specialized algorithms to spray a variable rate of nitrogen across the field (Raun et al. 2002; Shanahan et al. 2001, 2008).

Management zones can significantly contribute to increased nitrogen use efficiency and reduced losses of nitrogen to the environment (Bausch and Delgado 2003; Delgado and Bausch 2005; Delgado et al. 2005; Khosla et al. 2002). The yields and nitrogen uptake potential can be accurately predicted by management zones (Bausch and Delgado 2003; Delgado and Bausch 2005; Delgado et al. 2005; Inman et al. 2005). Delgado (1999, 2001), Delgado et al. (2001a,b), and Delgado and Bausch (2005) described the factors that drive the higher nitrate leaching potential in the sandier, coarse areas of the field compared to the finer-textured areas of the field that were managed similarly with the same irrigation and nitrogen fertilizer rates.

The nitrogen uptake patterns can be accurately characterized by management zones, where there is a higher nitrogen uptake rate in the zones with the higher yields (Delgado et al. 2005; Delgado and Bausch 2005; Inman et al. 2005). Several scientists have shown that the higher yields are correlated with the management zones with the best productivity (Hornung et al. 2003; Inman et al. 2005; Khosla et al. 2002). The hot zones have greater leaching potential (Delgado 1999, 2001; Delgado and Bausch 2005). Management zones have the potential to increase nutrient use efficiency and economic returns for farmers (Fleming et al. 1999; Gotway et al. 1996; Khosla et al. 2002; Koch et al. 2003). The use of variable rates and management zones has been found to increase nutrient use efficiency and reduce nitrogen losses via leaching (Delgado and Bausch 2005; Delgado et al. 2005).

4.4   WATER MANAGEMENT: SPATIAL IRRIGATION

Spatial irrigation for water management has been reported to increase water use efficiency and to reduce the potential for nitrate leaching by integrating irrigation with the spatial properties of soils and spatial hydraulic properties (Sadler et al. 2005, 2010). By matching irrigation with the areas with high infiltration rates and the soil water content of these areas, water use efficiency was increased (Sadler et al. 2005, 2010). Sadler et al. (2005) reported that the spatial irrigation approach can potentially be used to increase water use efficiency (Sadler et al. 2005). Sadler et al. (2005) reported that precision irrigation can increase water use efficiency by 10% to 15%; in other words, it can reduce the water applied by 10% to 15%. For the cases that they studied, Sadler et al. (2005) reported that there were other added benefits such as lower occurrence of diseases and lower nitrate leaching.

Evans et al. (2013) reported on the status and adoption of site-specific variable rate sprinkler irrigation systems. They reported that although there are more than 20 years of research in this field, the transfer of technology to an applied level is minimal. One reason for this is that there are limited reports proving the advantages of this technology. Sadler et al. (2005, 2010) also reported on the limitations as far as adoption of the technology at a field level. Evans et al. (2013) reported on the need to develop dynamic management zones that could change depending on real-time scenarios that may change as the crop is growing. They listed several factors that will affect the spatial dynamic variability of management zones such as soil properties, topography, hydrological properties (e.g., percolation versus irrigation), and other factors that could affect plant growth (e.g., diseases). They reported that there were about 175,000 center pivots and linear systems in the United States but approximately less than 200 systems that have the capability to apply variable rates (USDA NASS 2009).

4.5   RS AND SENSORS FOR PRECISION FARMING AND MANAGEMENT ZONES

RS and sensors for precision farming and management zones are becoming more advanced and applicable at the field level. This geotechnology makes it possible to quickly obtain large data sets about the *in situ* crop nitrogen status and to make immediate management decisions to improve the nitrogen status of the crop (Biermacher et al. 2006; Raun et al. 2002; Scharf et al. 2002, 2011; Shanahan et al. 2001, 2008). The nitrogen status of a crop is a characteristic that affects the reflectance properties of a plant and can potentially be assessed with sensors (Al-Abbas et al. 1974; Stanhill et al. 1972). Once these correlations were established, scientists tried to develop techniques that could assess these reflectance properties of a plant as a function of the nitrogen status with satellite, aerial, or ground-based sensors (Bausch and Delgado 2003; Bausch and Diker 2001; Bausch et al. 1996; Biermacher et al. 2006; Blackmer et al. 1996; Franzen et al. 1999; McMurtrey et al. 1994; Raun et al. 2002; Scharf et al. 2002; Schepers et al. 1992a,b, 1998). Scientists developed different types of indices for the spectral characteristics of plants to assess their nitrogen status (Bausch and Duke 1996; Stone et al. 1996; Tucker 1979; Wood et al. 1999). Different decision algorithms have been developed to assess the nitrogen needed by the crop and calculate the application of nitrogen during the growing season (Bausch and Diker 2001; Raun et al. 2002; Scharf et al. 2002, 2011; Shanahan et al. 2008).

Recently, Scharf et al. (2011) showed that by using sensor technology to integrate spatial information across the field about the nitrogen status of the crop and the capability of the soil to provide nitrogen, the sensor outperformed the management decisions of farmers. The sensor technology was tested across a large number of commercial farms. The yields were slightly higher with the sensor and the nitrogen applications were lower, increasing nitrogen use efficiency. The concept of using sensors to make management decisions under commercial field operations was also supported earlier in studies by Bausch and Delgado (2003) and Delgado and Bausch (2005), which showed that the nitrogen applications can be cut to half the rate of traditional management practices, reducing nitrate leaching losses by half, and thereby increasing nitrogen use efficiency and significantly decreasing losses of reactive nitrogen to the environment without reducing yields. This sensor technology can therefore be used to increase nitrogen use efficiency, yields, and/or economic returns for farmers (Biermacher et al. 2006; Scharf et al. 2011; Shanahan et al. 2008).

4.6 PRECISION CONSERVATION FOR IMPROVING MANAGEMENT WITHIN AND OFF THE FIELD

4.6.1 SPATIAL EROSION

The data and modeling exercise from Quine and Zhang (2002) shows the correlation of spatial variability in erosion rates across a field and that those areas of the field with the higher erosion rates will have lower yields. This is in agreement with the spatial assessment of soil erosion conducted by Papiernik et al. (2005), who found similar responses. A spatial assessment of soil erosion conducted by Schumacher et al. (2005) using Cesium-137 techniques found that those areas of the field with the higher slope are the areas of the field with the higher erosion rate. Schumacher et al. (2005) also conducted a spatial modeling assessment with the water and tillage erosion model (Figure 4.4). Schumacher et al. (2005) suggested that this type of information could be used to develop site-specific applied conservation practices such as cover crops, organic matter additions, and no-till practices. There is a need for site-specific precision conservation to address these higher erosion

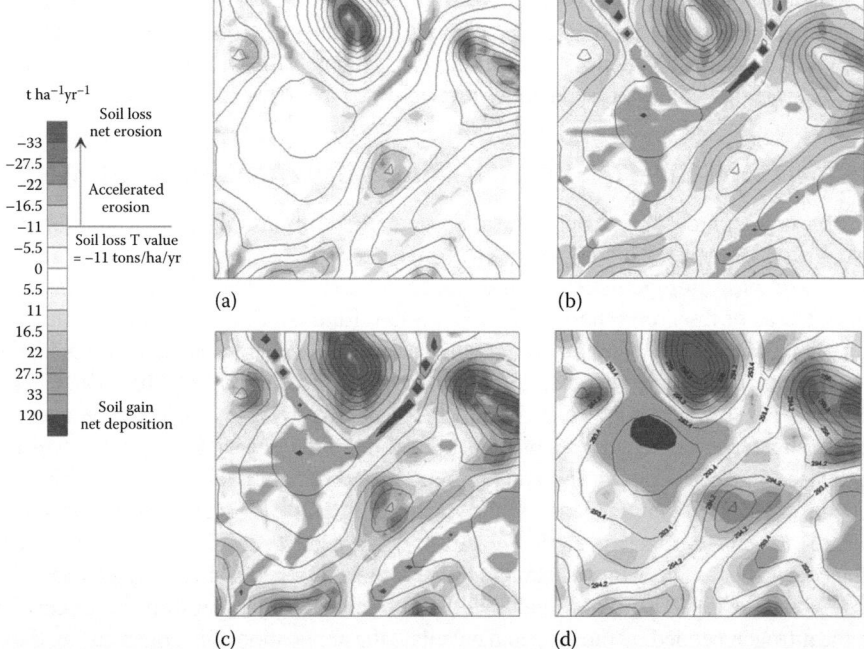

FIGURE 4.4 Erosion patterns developed from (a) tillage, (b) water, (c) tillage-water, and (d) total erosion (cesium-137 [137Cs]) modeling of the research field are displayed. Cesium-137 sampling sites are also displayed on a contour map of slope percent for the field. (From Schumacher, J. A. et al., *J. Soil Water Conserv.* 62:355–362, 2005.)

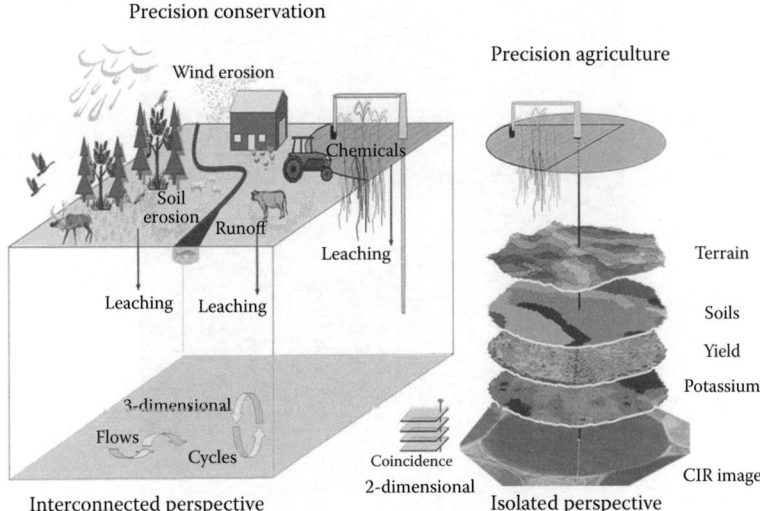

FIGURE 4.5 The site-specific approach can be expanded to a three-dimensional scale approach that assesses inflows and outflows from fields to watershed and region scales. (From Berry, J. K. et al., *J. Soil Water Conserv.* 58:332–339, 2003.)

rates that will contribute to significant degradation of soil quality and productivity in these areas (Figure 4.5) (Berry et al. 2003; Papiernik et al. 2005; Schumacher et al. 2005). This spatial erosion will also contribute to the redistribution of nutrients and soil organic matter across the field, lowering the nutrient status and soil organic matter content of the eroded areas (Delgado and Berry 2008).

4.6.2 GRASSED WATERWAYS

For some fields these higher spatial erosion rates will contribute to the development of ephemeral gullies across the field and significantly accelerate the erosion rates of a field, decreasing its productivity. We can use the site-specific terrain attributes/characteristics of a field to identify the areas of the field that are more fragile and susceptible to being impacted when precipitation generates flows that converge at these landscape positions, increasing the force needed to develop higher erosion rates and ephemeral gullies (Moore et al. 1988; Srivastava and Moore 1989; Thorne et al. 1986). The development of ephemeral gullies across the field will contribute to mass transfer of nutrients and soil organic matter out of the field, and to the acceleration of erosion, which will contribute to further degradation of the field and further movement of agrochemicals and nutrients off-site.

Mueller et al. (2005) developed a new approach of using a logistic regression model to predict the spatial occurrence of soil erosion considering site-specific information and to identify the areas of the field where accelerated erosion rates will be occurring. Mueller et al. (2005) found there is the potential to use this spatial approach to develop more effective and economical soil and water conservation, management, and planning scenarios that integrate spatial and temporal information for better management decisions. Grass waterways are an excellent management practice that has been reported to reduce the movement of soil, agrochemicals, and nutrients from the field (Briggs et al. 1999; Fiener and Auerswald 2003; Udawatta et al. 2004). This approach, using new geotechnologies such as precise real-time kinematic GPSs together with modeling assessments, can potentially help conservationists make decisions based on site-specific information to pinpoint with accuracy those areas of the field that are susceptible to erosion and where the grass waterways should be placed. Precision conservation can increase conservation effectiveness of grassed waterways (Luck et al. 2010; Mueller et al. 2005; Pike et al. 2009). Figures 4.6 and 4.7 shows how these new model approaches could contribute to the identification of the best positions for grass waterways.

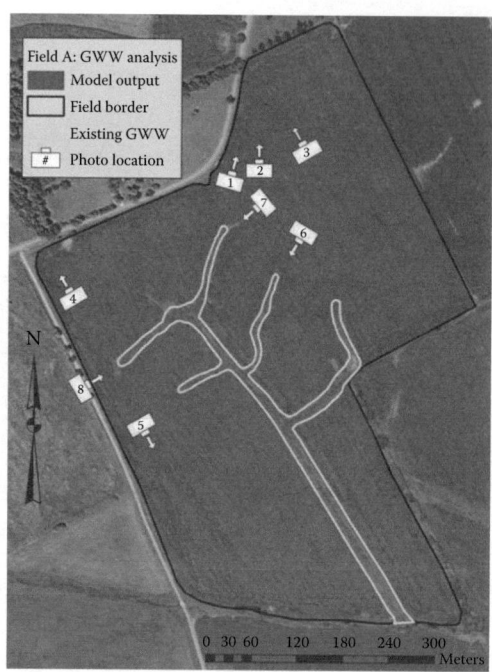

FIGURE 4.6 Aerial photograph of field A identifying probability of erosion model output (values >0.5 are shown in red), existing grassed waterways (GWWs), and locations of photographed eroded areas (1 m contour lines are shown in gray). (From Luck, J. D. et al., *J. Soil Water Conserv.* 65:280–288, 2010.)

4.6.3 Contour Planting with Deep Furrows to Collect Water

Another approach using new geotechnologies is to use real-time kinematic GPSs to develop precise surface models to develop precise contour planting accompanied by strategically placed contour furrows to capture and move precipitation water to reduce the erosion force and minimize erosion (Williams et al. 2011). Williams et al. (2011) used this precise digital elevation data collected with a real-time kinematic GPS to develop site-specific terrain models for contour planting. They used this site-specific, precise information to develop precisely aligned deep furrows at key positions of the field along the contour of the field so they could trap and intercept water runoff and minimize erosion potential. By doing this and developing furrows that could handle a 25-year storm at the site, they are using site-specific temporal and spatial data to minimize not only erosion potential, but also the off-site transport of soil, nutrients, and soil organic matter (Figure 4.8).

4.6.4 Management of Critical Areas

Critical areas of the field and the watershed are important because they contribute to higher rates of losses of nutrients and agrochemicals. They are hot spots of the fields and watersheds, and precision conservation approaches are a management tool to minimize the environmental impacts of these hot spot areas. Galzki et al. (2011) reported that critical areas of the field and the watershed are the dominant sources of contaminants, nutrients, and pesticides to surface water. If we are to reduce the transport of nutrients to water bodies we must address the hot spot areas of the landscape, which contribute a larger proportion of nutrients to water bodies (Figure 4.9). They were able to use lidar elevation geotechnology to precisely identify the critical areas that are pathways for higher flows of sediments and nutrients to ditches, which will enable the implementation of precision conservation practices to manage these spatial effects more accurately. This is another example of how to use

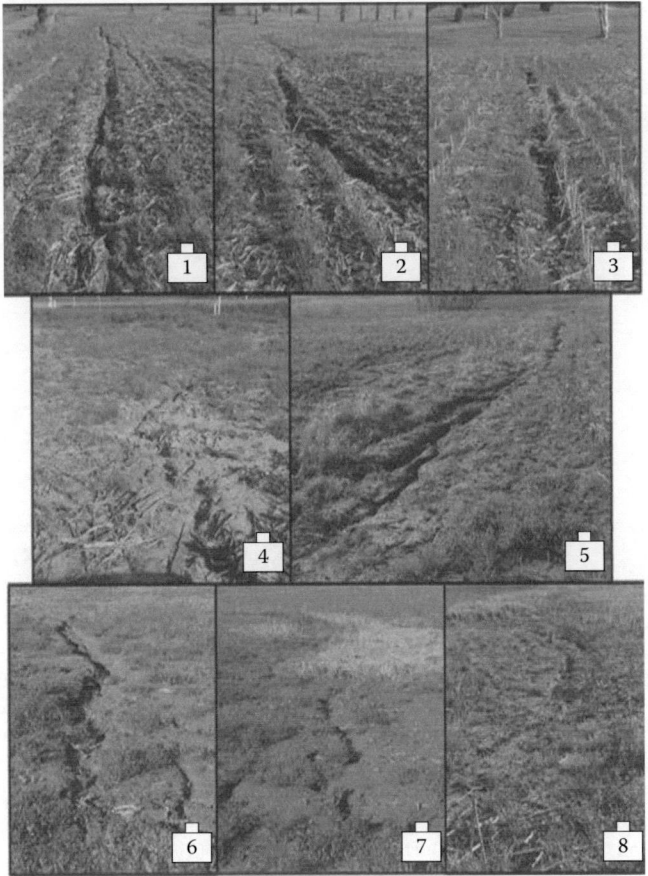

FIGURE 4.7 Photographs of eroded areas in field A taken in April 2009 (locations and viewpoint identified in Figure 4.6). (From Luck, J. D. et al., *J. Soil Water Conserv.* 65:280–288, 2010.)

precision conservation to reduce the off-site transport of nutrients from the field. Galszi et al. (2011) reported that this is a viable technology that is economical and can help identify critical areas and pathways for nutrient transport off-site. By quickly identifying these critical areas, we can apply precision conservation and improve the nutrient management at the site as far as recycling and maintaining nutrients within the field and reducing the off-site transport of these nutrients to water bodies.

4.6.5 Precision Harvesting of Crop Residue

It is well known that nitrogen is spatially variable, that some areas of the field have a higher soil organic matter content than others, and that there is more leaching of nitrate in some areas than in others. As far as maintaining a nutrient balance and reducing transport of nutrients from the field, we need to implement a holistic approach. Crop residue is an important part of nutrient management, and it is clear that removing crop residue from the field is also harvesting nutrients from the field. Not only does the removal of crop residue remove nutrients from the field, it can also increase the potential for soil erosion, contributing to further removal of nutrients off-site. Several scientists have reported on the potential negative impacts of removing crop residue, such as the degradation of soil quality and decreased soil fertility (Cruse and Herndl 2009; Delgado 2010; Johnson et al. 2010; Karlen et al. 2009; Lal 2004; Newman et al. 2010). Crop residue is key in returning a large

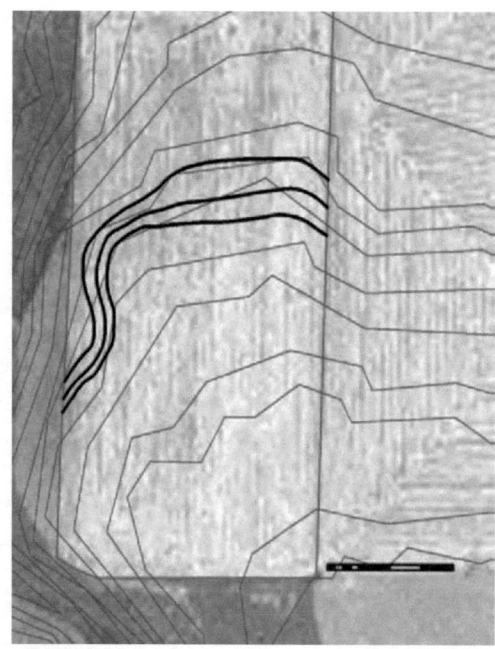

FIGURE 4.8 Schematic of contour furrow application to a winter wheat field. The field is bounded on the south and west by a perennial grass draw and on the east by a neighboring field, both of which are hydrologic boundaries resulting from tillage erosion. Light gray elevation contours (2 m) were developed from a 10 m digital elevation model. Precision contours shown in black were generated using a laser level. Arrows indicate direction of plateau seeding, which controls flow in the contributing area above the shoulder slope. Only one of the three precision contour locations shown here would be chosen for creating on-contour furrows to capture runoff from the plateau. (From Williams, J. D. et al., *J. Soil Water Conserv.* 66:355–361, 2011.)

percentage of the macro- and micronutrients absorbed by crops (Delgado and Follett 2002) and for conservation of the biosphere (Delgado 2010).

Schumacher et al. (2005) reported that they could target areas of the field that have high erosion rates with precision conservation using cover crops, minimum tillage, and other site-specific practices. Delgado and Berry (2008) reported that they could use precision harvesting to increase conservation of these areas of the field by spatially managing how the residue was harvested (e.g., keeping the residue in the areas of the fields that are more prone to soil erosion or have lower soil organic matter). Using modeling evaluations of cropping systems of the upper Mississippi River Basin, Meki et al. (2011) concluded that we could use a precision conservation approach where we use site-specific farm landscape variability and information about soil types, soil texture, and soil hydrology to make spatial management decisions that reduce erosion and increase conservation effectiveness.

Leaving crop residue on the field surface is a mechanism to reduce soil erosion, especially for the most risky soil types and areas of the landscape (Meki et al. 2011). Increasing the crop residue removal rate from 0% to 40%, 60%, or 80% significantly increased erosion for the highly erodible soils and nonhighly erodible soil types. Losses of nitrogen and phosphorus were higher from the highly erodible soil types than from the nonhighly erodible soil types (Meki et al. 2011). This is in agreement with previous studies from Laflen and Colvin (1981) that found a similar response to crop residue cover and erosion. This is an example showing that using the 4 Rs is not enough as far as controlling the rate of nutrient losses from a field and that we need to also consider merging them with the four Rs of precision conservation to minimize the off-site nutrient transport from a given field and to increase nutrient use efficiency.

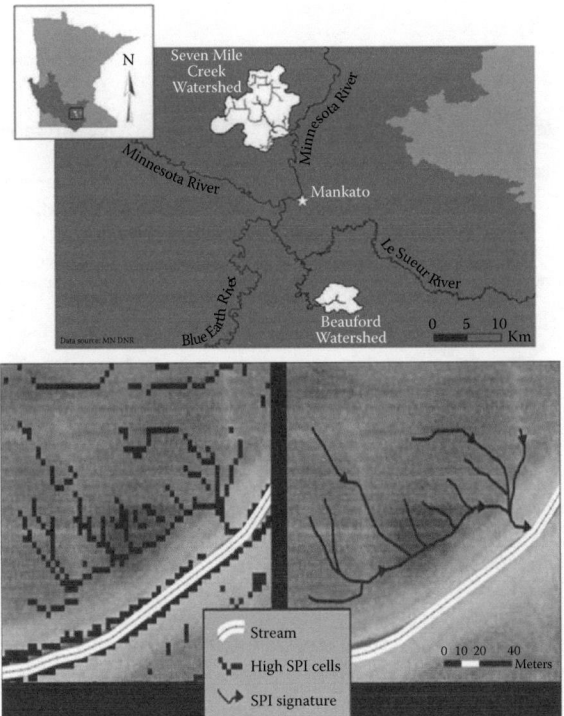

FIGURE 4.9 Location of Seven Mile Creek watershed and Beauford watershed within the Minnesota River Basin. Stream power index (SPI) signature represented by (a) 3 m raster data and (b) a visual interpretation of potential flow paths and their interface with the stream channel. (From Galzki, J. C. et al., *J. Soil Water Conserv.* 66:423–430, 2011.)

Meki et al. (2011) found that over time, removing the crop residue significantly decreased the soil organic matter and soil organic nitrogen. These findings from Meki et al. (2011) are in agreement with the concept of crop residue management, increasing soil organic matter and nutrient cycling as reported by Delgado and Follett (2002) (Figure 4.10). To reduce losses of phosphorus and nitrogen we need to implement best conservation practices such as terraces, contour planting, buffers, and improved management of nutrients. Without removal of crop residue, losses of nitrogen will be lower (Meki et al. 2011). These studies from Meki et al. (2011) are in agreement with the concept of precision harvesting (Cruse and Herndl 2009; Delgado and Berry 2008) and are also in agreement with the precision conservation approach of Berry et al. (2003) to reduce losses of nutrients from fields.

Another recent study that evaluated the effects of removal of crop residue on erosion and nutrient loss was conducted by Thomas et al. (2011). They conducted modeling evaluations of the potential to remove crop residue in a biofuel cropping system and found that for the different soils, as the rate of crop residue removal increased, the rate of agrochemical and nutrient loss increased for both conventional and no-till systems. The losses of sediment and agrochemicals were significantly lower with the no-till system than the conventional system. There was a significant variability of erosion losses and agrochemical losses by soil, showing again the need to account for spatial and temporal variability of soils and/or management systems before deciding how much crop residue can be removed from a soil system (Mekki et al. 2011; Thomas et al. 2011).

Roberts (2007) reported that these geotechnologies can be used to improve nutrient management to apply the 4 Rs. These studies from Meki et al. (2011) and Thomas et al. (2011) clearly show that the 4 Rs will not be enough for minimizing the nutrient losses of a system, especially if the system is a conventional system with a higher rate of crop residue removal. Berry et al. (2003, 2005) reported that we can use spatial technologies to increase conservation effectiveness and apply

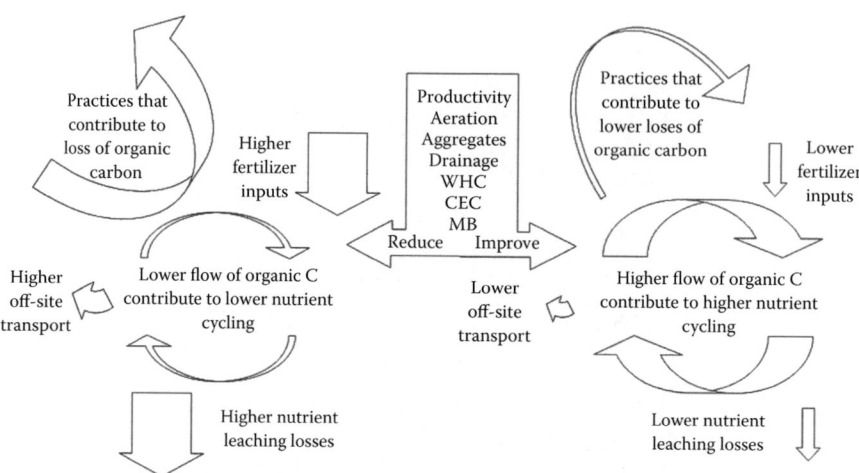

FIGURE 4.10 Effects of organic carbon on nutrient cycling and productivity. CEC, cation exchange capacity; MB, microbial biomass; WHC, water holding capacity. (From Delgado, J. A., and R. F. Follett, *J. Soil Water Conserv.* 57:455–464, 2002.)

precise conservation practices at the right place. Cox (2005) simplified the Berry et al. (2003) precision conservation concept as applying the right conservation practice, at the right place, at the right time, and at the right scale (the 4 Rs for conservation).

We could join these two concepts—the 4 Rs for nutrient management (Roberts 2007) and the 4 Rs for conservation (Cox 2005)—to establish that we need 7 Rs for nutrient management and conservation. If we are to increase conservation effectiveness and nutrient use efficiency and minimize the losses of sediment and nutrients to the environment, we need to apply the right product (fertilizer), at the right fertilizer rate, with the right method of fertilizer application, with the right conservation practice, and with conservation practice at the right place, and at the right scale of conservation practice, with both the fertilizer and the conservation practice applied at the right time (right product, right rate, right method, right practice, right place, right scale, and right time: the 7 Rs of nutrient management and conservation).

Only by applying precision farming together with precision conservation will we be able to minimize the nutrient losses from fields (Berry et al. 2003; Cox 2005; Meki et al. 2011; Roberts 2007; Thomas et al. 2011). It takes hundreds of years to form an inch of soil (SSSA 2013) and we can lose it in a thunderstorm, with erosion reducing soil fertility and moving nutrients out of the field. We need to apply the 7 Rs of nutrient management and conservation to maximize conservation and minimize off-site transport of nutrients.

In summary, Roberts (2007) defined the right product as the fertilizer source with the right balance of nitrogen, phosphorus, potassium, and other nutrients to meet the crop's needs. The right rate is defined as the correct amount of fertilizer to apply, accounting for realistic yields and after subtracting any nutrient sources already available to the crop; the right time is the synchronization, to the extent possible, of nutrient fertilizer application with crop demand; and the right place is using the appropriate method of application to apply nutrients where they can be used (Roberts 2007).

The Roberts (2007) definition of right place also mentions that conservation tillage, buffer strips, cover crops, and irrigation management are other practices that can help keep the nutrients at the right place. However, the right conservation practices, at the right time, right place, and right scale as defined by Cox (2005) have not been in a prominent position in the 4 Rs of nutrient management. For example, when Murrell et al. (2009) reviewed in detail the right place as defined by Roberts (2007), conservation practices were not discussed in the document, and the right place was defined as applying the fertilizer and incorporating it within the soil close to the root zone.

There is a need to give the 4 Rs of conservation (Cox 2005) a more prominent position in the 4 Rs of nutrient management; conservation practices were mentioned in the Roberts (2007) definition of right place but were not included in the definition for right place defined by Murrell et al. (2009). The 7 Rs of nutrient management and conservation listed above merge these principles of precision conservation and precision farming. We need to apply the 7 Rs of nutrient management and conservation: right product, right rate, right method, right practice, right place, right scale, and right time.

In the 7 Rs, the right product (fertilizer) and right fertilizer rate are the same as defined by Roberts (2007). The right method of fertilizer application under the 7 Rs is similar to the definition of the right place from Murrell et al. (2009); for example, incorporating applied fertilizer, or banding fertilizer close to the root system is using the right method of fertilizer application (Murrell et al. 2009). Another example of the right method could also be split applications of nitrogen with fertigation. In this new definition, the right time is applicable to nutrient management as well as conservation practices. For nutrient management, the definition of right time is similar to the Roberts (2007) definition.

An example of the right time for conservation practice application is planting a cover crop at the right time after harvesting a vegetable crop that leaves no crop residue to prevent wind or water erosion and to scavenge residual soil nitrate that is available to leach, and killing the cover crop at the right time to improve nutrient management and conservation for the following crop. The right conservation practice will be the application of the correct conservation practice as needed to minimize soil erosion considering spatial and temporal variability. An example would be using no till, minimum till, or strip till management where applicable and viable. If the field is cultivated, then another example would be application of conservation practices as needed to reduce erosion such as terraces, contour strips, contour furrows, and use of buffers at the field boundary considering areas where concentrated flows may require wider buffers.

Conservation practices at the right place will account for spatial and temporal variability of flows of water and nutrients. For example, right place would include application of minimum tillage for highly erodible soils, no removal of crop residue for highly erodible soils, or other examples discussed in this chapter where conservation practices can be applied at the right place using site-specific geotechnologies (e.g., nutrient traps). Another example of right place would be implementing grass waterways in areas identified as critical areas that will be subjected to higher rates of water flow and that can potentially develop ephemeral gullies.

Conservation practices at the right scale involves not only a given field application of a conservation practice, but also application of a conservation practice at the field border and other areas of the farm where nutrients can be captured to reduce the off-site transport of nutrients from the farm. Examples include application of buffers, riparian buffers, nutrient traps, sediment ponds, controlled drainage, and other practices discussed in this chapter using site-specific data to increase conservation effectiveness. All of these conservation practices contribute to the capture of nutrients, soil, and water moving out of the field and out of the farm.

We need to use the 7 Rs of nutrient management and conservation to start thinking more holistically about nutrient management and to view nutrient management not only as a field management option, but to also realize that there are conservation and management practices that can be applied in the field, at the border of the field, and/or within the farm with precision conservation to increase conservation effectiveness and reduce off-site nutrient transport from the farm. This chapter is proposing that conservation practices need a prominent place in the Rs definition. This is in agreement with previous precision conservation concepts and efforts (Berry et al. 2003, 2005; Delgado and Berry 2008; Tomer 2010; Tomer et al. 2013).

The National Resource Inventory has been conducted since 1982 and is reporting that the average soil erosion for cultivated croplands has been reduced from about 9.9 Mg ha^{-1} in 1982 to about 6.7 Mg ha^{-1} in 2007. However, there have been reports that climate change is occurring and that it is increasing the potential for soil erosion with extreme events (SWCS 2003, 2007; Walthall 2012; also see statement from the American Meteorological Society at http://www.ametsoc.org

/policy/2012climatechange.html about climate change). It has been estimated that for every increase of 1% in total rain, the average rate of erosion will increase by 1.7% (Nearing 2001; Nearing et al. 2004; Pruski and Nearing 2002a,b). These changes in climate and precipitation patterns are creating several knowledge gaps in our understanding of the effects of a changing climate and soil erosion processes and additional knowledge gaps as far as our capability to assess how the changing climate is impacting the current soil erosion rates (Delgado et al. 2013). For example, using high-resolution data, Angel et al. (2005) assessed changes in annual and seasonal rainfall erosivities from 1972 to 2002. They found that major cropland areas of the United States, from Florida, to Texas, to Minnesota, to the East Coast, had increasing trends in erosivities, showing nationwide trends of higher rainfall erosivities. It is very important that these and other knowledge gaps about the effects of a changing climate on erosion rates are addressed (Delgado et al. 2013). The findings from Angel et al. (2005) and SWCS (2003, 2007) are also in agreement with recent regional reports that we are losing ground with climate change, extreme events, and higher erosion rates from widespread ephemeral gullies (Cox and Hug 2012; Cox et al. 2012). Management with precision conservation and precision farming is needed to conserve soil quality and to minimize off-site transport of nutrients (Berry et al. 2003, 2005).

The practice of precision harvesting, where crop residue is left in certain areas of the field, could be a great precision conservation tool/approach to reducing erosion and losses of sediment and nutrients from the areas of the field with higher erosion potential. This is in agreement with Cruse and Herndl (2009), who reported that crop residue removal rates should not be abusive to soil and water resources so that soil and water quality could be conserved. They reported the need to understand the effect of crop residue removal rates on soil erosion potential, that crop residue should only be removed at rates that are acceptable, and the need to consider the spatial variability across the landscape. Johnson et al. (2007) reported on the importance of crop residue for soil quality and that harvest of crop residue from straw and stover should not occur unless soil quality protection has been achieved. Crop residue is important; it provides many ecosystem services, it is strongly correlated with good soil quality properties and sustainability (Cruse and Herndl 2009; Doran and Jones 1996; Johnson et al. 2010; Karlen et al. 2009; Lal 2004; Newman et al. 2010), and it can help us adapt to a changing climate (Delgado et al. 2011, 2013; Lal et al. 2011).

4.6.6 DRAINAGE MANAGEMENT

Precision conservation concepts can be applied to identify water and nutrient flows across the landscape and best locations to strategically place a series of drainage management practices to reduce the flows of these nutrients (Strock et al. 2010). The conservation effectiveness of different drainage management practices can be improved using geotechnologies with precision conservation (Delgado and Berry 2008; Strock et al. 2010). Some examples of the different drainage management practices that can be improved or that can be strategically applied across the landscape are controlled drainage, water storage, drainage water design, side inlets, reactive barriers, vegetative filter strips, wetlands, and agronomic management (Strock et al. 2010). Drainage management has been shown to be a tool that can be used to reduce the losses of nutrients from the system (Evans and Skaggs 2004; Gilliam et al. 1979; Skaggs et al. 2012; Steenvoorden 1985; Strock et al. 2007; Vadas et al. 2007). Management of the water table has been shown to reduce NO_3-N losses by increasing losses via denitrification (Evans and Skaggs 2004; Gilliam et al. 1979; Skaggs et al. 2012; Steenvoorden 1985).

Skaggs et al. (2012) reviewed a series of published studies about drainage management and recent studies conducted across different sites showing controlled drainage techniques can contribute to reduced losses of nitrate. The average reduction in nitrate leaching losses was 50% for studies conducted across the Midwest (Illinois, Indiana, Iowa, and Ohio), North Carolina, Ontario, and Sweden. Out of these 20 studies, in only one study was the yield reduced; for 10 studies there was no effect on yield, and in seven studies, yields increased. It has been reported that nitrogen losses from the upper Mississippi watershed are contributing to hypoxia in the Gulf of Mexico. These recent studies are showing that controlled drainage management systems can be used to reduce the transport of nitrate, so there is potential to use controlled drainage systems in site-specific areas to control the nutrient outflow from the regions of Illinois (Cooke and Verma 2012), Indiana (Adeuya et al. 2012), and Iowa (Helmers et al. 2012; Jaynes 2012).

4.6.7 PRECISION CONSERVATION AGRICULTURE

Although the definition of precision conservation is technologically based, it acknowledges that there is potential to use precision conservation in low-tech agricultural systems and/or natural resource conservation in the application of spatial conservation practices (Berry et al. 2003, 2005). Jenrich (2011) reported on the potential to use precision conservation agriculture in small farms in sub-Saharan Africa, and described the agricultural systems in the region, which are in sync with the concept of low-tech approaches such as nondigital, non-GIS maps and the use of survey methods to manage spatial and temporal variation for precision conservation. Farmers in these settings described by Jenrich (2011) can increase precision by using field survey methods as simple as using cords in the field to mark planting rows, determine where to dig the holes to plant the seeds (done by hand), and ensure the holes are the same distance apart. This low-tech approach can help to increase precision for these smallholder farmers.

Smallholder farmers in the region of sub-Saharan Africa have, on average, very low yields. For example, data presented by Jenrich (2011) shows average corn yields ranging from about 0.2 to 1.4 Mg ha^{-1}. These low yields are due to uncontrollable and controllable factors. Some of the controllable factors that are contributing to these low yields are poor timing of operations, imprecise farming practices such as uneven plant populations, uneven cultivation, and uneven fertilization.

Precision conservation agriculture with precise land systems using simple, low-tech surveys such as described above increases the plant populations and yields. The use of this low-tech but precise system also helps increase the precision of hand application of fertilizer, and when used together with conservation agriculture, the yields can be doubled or tripled.

Precision conservation agriculture can help smallholder farmers in sub-Saharan Africa in the management of not only spatial variability, but also the management of temporal variability by better synchronizing the planting time with the best time to grow the corn. There are smallholder farmers in this region who are using this type of precision conservation and are experiencing reduced costs and increased productivity (Jenrich 2011). This is applicable to other regions where smallholder farmers are having low yields and there is potential to increase these yields.

4.7 PRECISION CONSERVATION FOR MANAGEMENT OF NONFIELD AREAS

4.7.1 BUFFERS

Buffers used at the edge of the fields are an effective vegetative barrier that can reduce off-site transport of sediments and nutrients (Dosskey et al. 2002; Fares et al. 2010; Hey et al. 2005; Lowrance et al. 2000). The spatial variability in flows from out of the field was observed in studies conducted during the 1980s (Dillaha et al. 1986, 1989). Dillaha et al. (1986, 1989) noticed that the flows were concentrated in parts of the buffer area, reducing the efficiency of this conservation practice because spatial variability contributed to the generation of concentrated flows that the buffer design was not able to handle.

Buffers are efficient tools that can trap soil and nutrients, stopping transport off the field, but field and plot studies have shown that the efficiency of the buffers is reduced due to the concentration of flows. Not managing the spatial variability of flows allows the creation of concentrated flows and reduces the effectiveness of this conservation practice. That is why it is important to consider spatial and temporal variability when applying the 7 Rs of nutrient management and conservation: right product (fertilizer), right rate, right method, right practice, right place, right scale, and right time. Not considering the right scale or right shape will decrease the efficiency of efforts to reduce the off-site transport of nutrients.

Field and plot studies have shown that buffers that are subjected to this spatially concentrated flow effect are not as effective in trapping sediment and nutrients such as nitrogen and phosphorus (Daniels and Gilliam 1996; Dickey and Vanbderholm 1981; Dillaha and Hayes 1991; Dillaha et al. 1988, 1989; Fabis et al. 1993; Tomer et al. 2007). Topography and soil factors such as slope, land management, and soil type affect the runoff flows and contribute to the concentration of flows in some areas of the field. Current technologies where we can collect very site-specific slope and topographic information can help us to be very accurate in measuring these topographic and slope characteristics. With these technologies it is much easier to predict and assess these variable flow patterns, which facilitates identification of the site-specific positions to place the buffers and improvement of the design with regard to variable width buffers (Dosskey et al. 2002, 2005). Fares et al. (2010) reported that variable widths of buffers could be a good approach that allows for site-specific assessment of properties and conditions to design more effective buffers, but that there is still a need for additional research to assess the advantages and disadvantages of variable width buffers. Dosskey et al. (2007) used the Revised Universal Soil Loss Equation (RUSLE), the vegetative filter strip model (VFSMOD), and geographical information to determine the best position of buffers across the landscape.

Buffers can even be used as tools to protect groundwater. For example, the area of southeast Minnesota has over 8300 sinkholes that could be impacted if runoff carrying nutrients from the crop field areas moves to these sinkholes. Smith et al. (2006) tested the effects of different widths of buffers on the potential to reduce environmental impacts, and they found that pollution can be reduced by about 80% when sinkholes are protected with buffers that are 30 m wide. Buffers

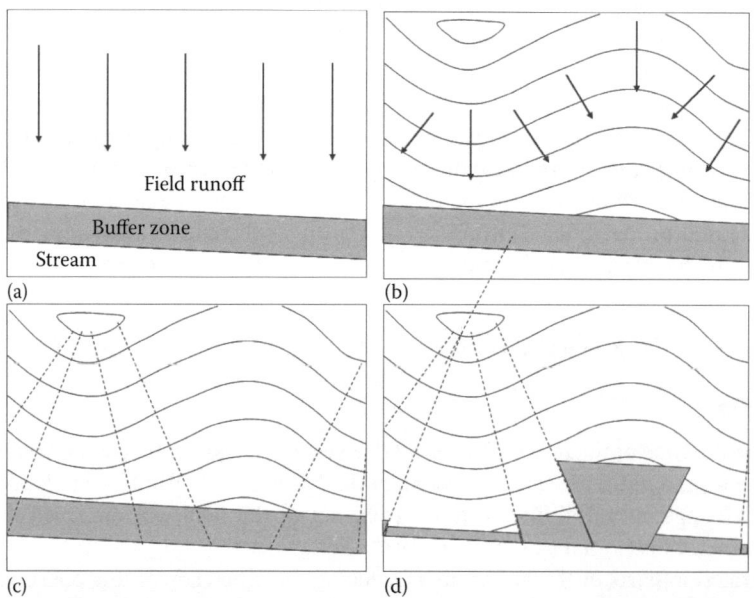

FIGURE 4.11 Diagrams of crop-field runoff patterns, topographic contours, and alternative buffer designs: (a) uniform runoff flow to a uniform-width buffer, (b) nonuniform runoff flow to a uniform-width buffer, (c) nonuniform runoff areas and the corresponding uniform-width buffer locations to which they flow, and (d) nonuniform runoff areas and the corresponding variable-width buffer areas to which they flow. Both (a) and (d) yield an approximately constant level of pollutant filtering along the entire length of the buffer. (From Dosskey, M. G. et al., *J. Soil Water Conserv.* 62:349–354, 2005.)

wider than 30 m have also been reported to positively impact groundwater quality by reducing the potential for NO_3-N leaching (Smith et al. 2006).

Precision conservation technologies can be used to increase the conservation effectiveness of vegetative barriers. The analysis of site-specific topographical areas with real-time kinematic (RTK) integrated with mathematical models in a GIS can produce more accurate changes in elevation across a field and help identify the spatial properties of the soil at the field and its respective hydrological properties to help develop a more precise flow map across the topography. This in turn can help identify the best sites to place the buffers with respect to the spatial and temporal variability and contribute to increased conservation effectiveness by designing better buffers that have greater potential for reducing off-site soil and nutrient transport (Dosskey et al. 2002, 2005) (Figure 4.11).

Dosskey et al. (2002) reported on the correlation between the sediment trapping efficiency and the buffer-to-area ratio, and that as the buffer-to-area ratio increased the sediment-trapping efficiency of the buffer increased. Since there is spatial variability of flows across a field, there is also spatial variability in the loss of nutrients across a field, and since the flows are concentrated in some specific areas, as the flows of water move out of the system, they do so at a higher rate of water flow and off-site nutrient transport because the buffer efficiency is reduced and the buffer cannot handle the points where the flow is concentrated. By using a precision conservation approach and identifying the areas where the flows will be concentrated, we can improve the buffer with a variable width buffer design, keep more nutrients in the field, and reduce the off-site transport of nutrients (Dosskey et al. 2002, 2005) (Figure 4.11).

4.7.2 RIPARIAN BUFFERS

Riparian buffers are also an effective vegetative barrier that can reduce the off-site transport of sediments and nutrients and protect water quality. Riparian buffers are vegetation located close to water bodies such as streams, lakes, or wetlands; they provide great benefits to the environment by reducing the transport of sediments and nutrients to water bodies (Bolton et al. 1991; Fares et al. 2010; Lowrance et al. 1997; Mayer et al. 2005; Mitsch et al. 1999; Tomer et al. 2003, 2007). A natural riparian buffer can have diverse vegetation, and with ecoengineering we could design a riparian buffer with diverse vegetation (eg., warm season and cool season grasses), which is an advantage in using riparian buffers as a tool to manage some of the effects of the spatial and temporal flows (Tomer et al. 2007). Using ecoengineering concepts to design a riparian buffer with diverse vegetation can provide an advantage in capturing nutrients by matching the growing period of a given type of vegetation planted in the riparian buffer to the time period when the nutrient flows are occurring, to catch the nutrients. This approach for some site-specific locations in the landscape can increase the potential for greater recovery of nutrients and increase the efficiency of the riparian buffer, reducing environmental impact (Tomer et al. 2007).

These riparian buffers, which are also called vegetative filter strips, can remove nitrogen from belowground flows (Tomer et al. 2007). The belowground preferential flow can depend on many factors, but similar to the spatial variability observed for surface runoff, topography, soil type for the soil profile, and other factors, can contribute to concentrated belowground flows across some areas of the riparian buffers, which could decrease the efficiency of the vegetative filter strip (Mayer et al. 2005). Riparian buffers are also good tools to reduce the environmental impacts of nutrients and to trap nutrients, and we can use the Riparian Ecosystem Management Model (REMM) to evaluate the effects of buffers of different shapes and soil depths (rooting depths) (Lowrance et al. 2000).

Tomer et al. (2007) studied the spatial flows of nutrients, and there is the potential to use spatially different vegetation to manage the times of larger flows or higher nutrient transport to the riparian buffer. They found that the higher phosphorus concentrations in the belowground flows occurred when the switchgrass was not in its growing stage. They suggested that planting a grass that has a much higher evapotranspiration rate when the flows are accumulating at the fastest rate will better reduce the potential negative impacts of the higher flows, since it will evapotranspire the water at a higher rate when the accumulation is highest. They reported that there is the potential to use a cool-season grass that can grow when the warm-season grass is not growing if the accumulation is highest during the times when the warm-season grass is not growing. The data from Tomer et al. (2013) shows the potential precision conservation approaches that could be used to manage spatial and temporal variability of nutrient flows from the field, and that by considering site-specific, variable information, the conservation effectiveness of riparian buffers can be increased.

4.7.3 NUTRIENT TRAPS AND SEDIMENT PONDS

Nutrient traps are a new advance in nutrient management/precision conservation that can be used to reduce the off-site transport of nutrients. There is potential to strategically place denitrification traps across the landscape to reduce the movement of nitrate-nitrogen (Hunter 2001). Hunter (2001) proposed that we can use plumes of oil as barriers deep in the soil profile to increase the denitrification of nitrate. The wall of oil injected in the ground can form a barrier to clean the water; as the water moves through the barrier, the carbon source in the oil increases the denitrification potential and the nitrate is predominantly reduced to N_2. Hunter (2013) tested pilot-scale vadose zone biobarriers in cattle corrals in Colorado by considering the flow pathways of nitrates at the site and placing the biobarriers strategically at locations where they could intercept and capture nitrates to prevent off-site transport. He found that the nitrate was being removed at rates of 96%, 83%, and 72% efficiency for sawdust, soybean oil, and sawdust plus soybean oil biobarriers, respectively, and that the nutrient traps could last for many years before the substrate would need to be replaced. This concept of using denitrification barriers to protect groundwater can also be used to protect surface waters. A similar approach was used by Greenan et al. (2006) and Jaynes et al. (2008) that showed reduction of nitrate leaching using denitrification traps in the Midwest.

As far as phosphorus surface runoff, there is also the potential to place nutrient traps at strategic positions in the landscape considering the variability of topography and soils and how they affect the variable flow in order to catch phosphorus (Penn et al. 2007, 2014). Surface waters can also be protected with phosphorus-absorbing material that can be used to reduce the movement of phosphorus (Penn et al. 2007, 2014). Spatial geographic information with RTK and modeling can also be used to determine water flow and phosphorus transport and to apply these traps strategically to capture phosphorus in order to reduce the off-site transport of nutrients and conserve water quality.

Tomer et al. (2007) studied the spatial flows of nutrients, and found that there is the potential to use sediment ponds that can be strategically placed in a watershed to reduce the movement of nutrients. This conservation practice can be used to trap nutrients and reduce their transport, which can affect water quality down the watershed (Lowrance et al. 2007). Different areas of the watershed have variable source pollution and due to variable hydrology may produce more pollutants than other areas of the watershed (Qiu et al. 2007). Precision conservation can be used to assess the variable flows and variable hydrology to manage the variable source pollution in a watershed, and the strategic placement of nutrient traps and sediment ponds within a watershed can help capture some of the nutrient runoff (Qiu et al. 2007).

4.7.4 WETLANDS

Wetlands can also be used to clean surface waters and reduce the transport of nutrients. Hey (2002) and Hey et al. (2005) reported that strategically placed wetlands across the landscape can be used as nutrient harvest farms to remove nitrate from waters and as tools to clean water, and that ecological engineering concepts can be used to establish these wetlands. New geotechnological tools can be used to assess the water and nutrient flows in a watershed to make strategic decisions about where to place these precision conservation practices in order to increase conservation effectiveness (Berry et al. 2003; Hey et al. 2005; Mitsch and Day 2006; Tomer 2010; Tomer et al. 2013).

4.8 CONNECTING NUTRIENT FLOWS AT A WATERSHED LEVEL FOR ECOSYSTEM SERVICES

Tomer et al. (2013) used precision geotechnologies to conduct a watershed analysis of the flows of water, sediment, and nutrients from the contributing field areas of the Lime Creek watershed, Illinois, and to develop precision conservation recommendations to increase conservation effectiveness. In developing a precision conservation plan for the whole watershed, Tomer et al. (2013) made recommendations about where to strategically place riparian buffers and what type of riparian buffer should be used considering spatial and temporal information to improve the management of the flow and the capacity to remove nutrients being transported from the watershed. In their analysis they identified the site-specific locations where the placement of riparian buffers would increase water and nutrient removal. The riparian buffers were divided into locations where (1) increased nutrient uptake was needed, (2) diverse vegetation was needed to maintain a longer growing period, (3) a buffer with higher sediment trapping capacity was needed, (4) deeper-rooted vegetation was needed, and (5) stabilization of banks was needed (Figure 4.12). The Tomer et al. (2013) analysis shows that geotechnologies can be used to assess the variable and temporal flows to determine not only where to place the riparian buffers but what type of riparian buffers to use to increase conservation effectiveness across the watershed.

The Tomer at al. (2013) analysis made recommendations about where to strategically place conservation practices to reduce the transport of nutrients from the watershed (Figure 4.13). Geospatial information that considers the spatial variability of topography, soils, water tables, and other factors can be used to apply precision conservation by strategically placing at site-specific locations: (1) controlled drainage practices where they can more effectively remove nitrate, (2) grass waterways at critical areas where the concentrated flow of nutrients and water is thought to be occurring to help trap sediments and nutrients, (3) wetlands and vegetative buffers to trap water, sediments, and nutrients and also serve as bioreactors to further remove nitrate, and (4) riparian buffers with diverse vegetation to allow removal of nutrients over a longer period.

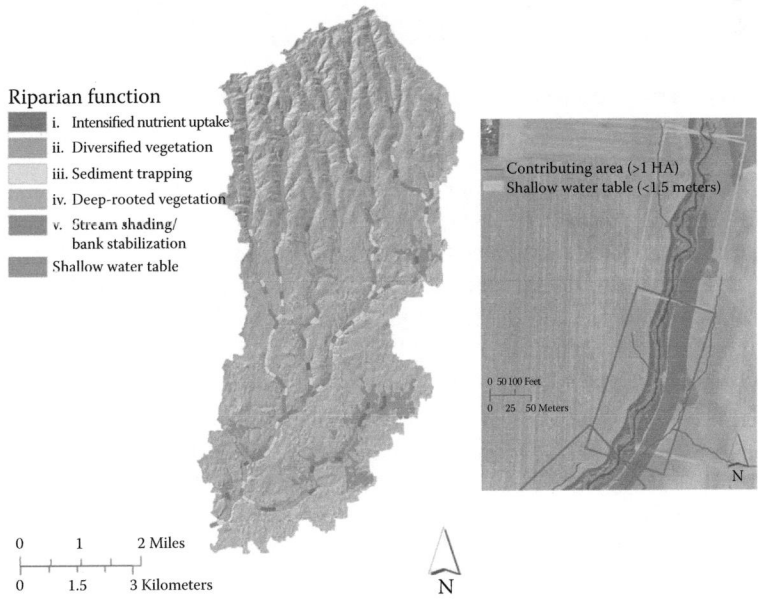

FIGURE 4.12 Distribution of potential riparian buffer functions in Lime Creek and inset showing runoff pathways, shallow water table areas, and riparian segments used for classification. (From Tomer, M. D. et al., *J. Soil Water Conserv.* 68:113A–120A, 2013.)

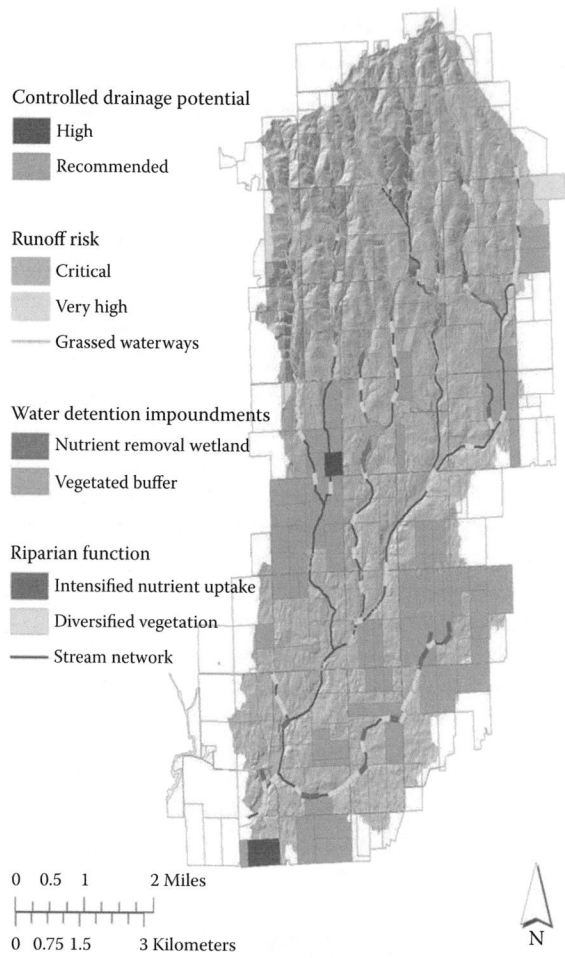

Controlled drainage potential
High
Recommended

Runoff risk
Critical
Very high
Grassed waterways

Water detention impoundments
Nutrient removal wetland
Vegetated buffer

Riparian function
Intensified nutrient uptake
Diversified vegetation
Stream network

0   0.5   1   2 Miles

0 0.75 1.5   3 Kilometers

N

FIGURE 4.13 One possible conservation planning scenario for Lime Creek. (From Tomer, M. D. et al., *J. Soil Water Conserv.* 68:113A–120A, 2013.)

There are new geospatial tools that can be used to connect the nutrient flows from the field to the watershed level and that can evaluate the capacity of conservation practices to provide ecosystem services (e.g., remove nutrients and reduce transport of sediments). There are different tools that can be used to assess the transport of nutrients (e.g., nitrogen, phosphorus) and/or soil particles. We can use precision farming and precision conservation practices at a field level to increase nutrient use efficiency and reduce the transport of nutrients from the field. We can use site-specific ecotechnologies with precision conservation to increase conservation effectiveness at a watershed level. Since nitrogen losses from agricultural systems are impacting natural systems, there is the potential to use precision conservation practices to reduce the environmental impacts of nutrient losses and to trade the savings (reductions in nutrient losses) in ecosystem markets (Delgado 2012; Delgado et al. 2008, 2010; Saleh et al. 2011).

Lal et al. (2012) discussed examples of evolving nutrient trading markets and how these systems could potentially be implemented, and discussed several examples of markets at the time from Connecticut, New York, Pennsylvania, Ohio, and Oregon. There are geospatial tools that can track nutrients from their application in the field to their transport out of the field across the watershed (Saleh et al. 2011). With these new nitrogen trading tools, we are shifting from the concept of what have been called nonpoint sources to increasing our capacity to assess the transport of nutrients at various points across the watershed (Delgado 2012).

There is the potential to use modeling tools to assess the effects of management practices on nitrogen losses (Delgado et al. 2008, 2010; Saleh et al. 2011). There are important nitrogen loss pathways (leaching, denitrification, ammonia volatilization, and emissions of nitrous oxide) that need to be captured by these modeling tools to assess the effects of management practices on nitrogen losses. There are geospatial technologies that can be used for precision conservation to identify hot spots, assess how conservation practices can help reduce the losses and transport of nutrients, and possibly even to trade the savings (reductions in nutrient losses) in future water-and air-quality markets (Delgado et al. 2008, 2010; Hey 2002; Hey et al. 2005; Ribaudo et al. 2005; Saleh et al. 2011).

The geospatial nitrogen trading tool concept works in much the same way as a bank account (Delgado et al. 2008, 2010). If the implementation of the best management practices reduced nitrogen losses by leaching, ammonia volatilization, and/or emissions of N_2O at a site-specific location of the field, then nitrogen has been saved (or, to follow the analogy, there is money in the bank account) that could be traded. If the practices did not lead to any savings in nitrogen (i.e., there was no reduction in nitrogen losses), then there are no savings (or, to follow the analogy, no money in the bank account) to trade (Delgado et al. 2008, 2010).

Spatial nitrogen trading tools integrate management information, soil and plant nitrogen and water balances, soil physical and chemical properties, and other layers of information across the field and even calculate carbon sequestration equivalents from reductions of N_2O emissions. Saleh et al. (2011) developed the Nutrient Tracking Tool (NtrT) that can be used similarly to track spatially across the field not only losses of nitrogen, but also losses of sediments and phosphorus, and the savings can be traded in water- and/or air-quality markets. Another tool that has geospatial capabilities to assess carbon sequestration and emission of trace gases is the Carbon Management Evaluation Tool–Voluntary Reporting (COMET-VR), which can provide information that could be useful for trading savings generated by conservation practices in air-quality markets (Brown et al. 2010). The concept of using site-specific precision conservation considering spatial and temporal variability to assess the effects of conservation practices on transport of nutrients (e.g., losses of reactive nitrogen) can be applied to future trading in air-and/or water-quality markets (e.g., nitrogen trading [NO_3-N, N_2O], carbon sequestration, sediments, and phosphorus).

4.9 PRECISION REGULATION

Cox (2005), reviewing precision conservation concepts, coined the term precision regulation. In his different presentations, he described precision regulation as the capacity to implement precision conservation to increase the efficiency of conservation practices across the watershed (Cox 2005; King et al. 2009). Cox and Hug (2012) reported that implementation of innovative conservation practices with precision regulation could potentially target the hot spots in a watershed by identifying the risky practices that generate a disproportionate amount of pollution, such as those that lead to ephemeral gully erosion, which creates a direct pipeline for mud, fertilizer, and manure to flow into streams and rivers. King et al. (2009) reported that precision regulation can be used to focus voluntary programs using precision conservation to put regulations in place that work. Cox and Hug (2012) also reported that precision regulation does not necessarily need to be mandatory, and it could be voluntary by running model analyses, identifying hot spot areas, and assigning voluntary conservation programs that are focused on these hot spot areas. This is in agreement with Delgado (2012), who reported that these new ecotechnologies with geospatial capabilities are allowing us to move beyond the concept of nonpoint sources to increasing our capacity to assess the transport of nutrients from the field to the watershed.

4.10   RESEARCH AND DEVELOPMENT PRIORITIES

There have been great advances in precision conservation during the last decade but knowledge gaps still exist, and research and development of new tools is needed. Although we now have tools that are being used by private industry in agricultural fields and natural areas (http://www.agrentools .com/blog/) that can quickly assess spatial variability in erosion and/or placement of conservation practices, research is still needed to continue advancement and improvement of these tools for continued improvement in management to increase conservation effectiveness. There is a need for additional research on how to continue expanding the use of GPS, GIS, RS, and modeling tool technologies to improve nutrient management and conservation, especially with new and emerging technologies (e.g., drones and mobile technologies such as smartphone applications). These technologies can be used to do surveys in the field (Figure 4.14), improve nutrient application rates with sensors (Figure 4.15), improve irrigation management (Figure 4.16), and collect aerial information with unmanned aerial vehicles (drones) (Figure 4.17). We must also continue to train personnel in the use of these new tools and technologies. Examples of training personnel can be found in the private industry (e.g., Agreen) (http://precisionconservation.com/agrens-conservation-planning-tools/) and with public organizations such as the Minnesota Department of Agriculture (MDA) (http://www .mda.state.mn.us/protecting/cleanwaterfund/toolstechnology/precisionconsinit.aspx). The MDA is an example of a state organization that is managing spatial variability with precision conservation approaches to protect water quality. The MDA has been supporting the training of personnel with workshops to help conservation practitioners, nutrient managers, and other experts learn how to use digital terrain analysis to help conservation professionals to find the right conservation practices in the right places to improve water quality. This MDA model could be considered as an example for other states and/or countries. These technology transfer efforts will also contribute to the identification of additional research and development priorities as conservation practitioners try to identify the best conservation options to apply using the 7Rs of nutrient management and conservation.

Additional research on how to assess the advantages and disadvantages of using precision conservation to determine the best practice for a site-specific location is needed. For example, Fares

FIGURE 4.14   Soil scientist field-mapping soils utilizing GPS technology in Washington County, VA. (From NRCS website: http://photogallery.nrcs.usda.gov/.)

FIGURE 4.15   Light reflectance sensors mounted on the front of an applicator are used to measure nitrogen deficiency so that nitrogen can be applied as needed (at a variable rate) when the applicator moves through the field. (From ARS website: http://www.ars.usda.gov/is/graphics/photos/.)

FIGURE 4.16   A variable-rate center-pivot irrigation system at Bushland, TX. Technician Luke Britten (left) and agricultural engineer Susan O'Shaughnessy (right) adjust wireless infrared thermometers in the field while technician Brice Ruthardt (center) uses a neutron gauge for soil water measurements. (From ARS website: http://www.ars.usda.gov/is/graphics/photos/.)

FIGURE 4.17 At the Jornada Experimental Range in Las Cruces, NM, a team of scientists prepares to launch an unmanned aerial vehicle from a catapult. The mission is to survey vegetation on the ground in studies of vegetation changes over time. In the foreground are engineering technician Craig Winters (left) and pilot Dave Thatcher (right). Other researchers are in the ground control station. (From ARS website: http://www .ars.usda.gov/is/graphics/photos/.)

et al. (2010) reported the need to assess the advantages and disadvantages of precision conservation practices such as variable width buffers in reducing the off-site transport of nutrients, sediment, soil organic matter, and other agrochemicals. Since precision conservation can be used for climate change adaptation, there are also research gaps in terms of how we can use site-specific conservation practices to adapt to a changing climate while taking spatial and temporal variability into account, especially with an increase in frequency of extreme events such as drought and large precipitation events.

Long-term studies to assess how precision conservation is increasing conservation effectiveness, increasing and/or maintaining yields, and increasing economic returns for farmers when compared with no conservation practices (and/or conservation practices that do not account for spatial and temporal variability) are also needed. We do not have the luxury of letting the hot spot areas of the field continue to lose soils at higher erosion rates, and we need to target these hot spot areas with site-specific conservation practices to minimize the erosion at these sites and the transport of soil and nutrients off-site. Additional research is needed on how crop rotations and/or variable crop residue (precision harvesting) can be managed across the field considering variable spatial properties of the soil, variable erosion rates, and variable yields in order to maintain or increase soil organic matter levels (carbon sequestration) needed for long-term sustainability and soil quality/soil health at any given site.

There is a need to expand research on the economic and environmental benefits of precision conservation application in natural areas and how precision conservation applied in these areas increases the environmental and economic benefits of ecosystem services.

The potential to use precision conservation with riparian forests, sediment ponds, wetlands, and other conservation practices to reduce the transport of sediments, increase wildlife conservation, reduce transport of agrochemicals, reduce (net) emissions of trace gases (carbon dioxide [CO_2], methane [CH_4], N_2O), reduce costs of dredging, reduce the cost of water quality conservation (e.g., the cost to remove nitrate from agricultural sources), and achieve other long-term ecosystem benefits needs to be researched. Models capable of simulating the effects of these practices in natural areas need to be developed and validated. There is a need to continue calibrating, validating, and improving models that connect surface flows of sediments, nutrients, and agrochemicals, emissions and/or sinks of atmospheric gases, and subsurface flows from leaching and tiles.

4.11 SUMMARY AND CONCLUSIONS

Cox (2010) reported that under business as usual, the environmental impacts of nutrient losses from agriculture will not be resolved and that precision conservation and precision regulation are two mechanisms to reduce the environmental impacts of nutrient losses. This is in agreement with the recently released U.S. Government Accountability Office (USGAO) report (2013), which states that more than 40 years after the establishment of the Clean Water Act, as of August 2013, the Environmental Protection Agency (EPA)'s assessment of national water quality found that about 50% of the assessed waters in the United States did not meet the established standards for fishing, swimming, or drinking, with 67% of assessed lake acres and 53% of the assessed miles of rivers impaired, and more of the assessed water bodies not meeting water quality standards than ever before. King et al. (2009) and Cox and Hug (2012) reported that precision regulation does not necessarily need to be forced and that it could be voluntary by running model analyses, identifying hot spot areas, and assigning conservation programs that are focused on these hot spot areas. A recent USGAO report (2013) found that after four decades of incentives for conservation of water quality, there are still persistent water quality problems. The Berry et al. (2003, 2005) concept that precision conservation can help reduce nutrient transport to protect water quality is an important alternative that could be considered.

This chapter has attempted to emphasize the importance of precision conservation for increasing nutrient use efficiency, reducing off-site transport of nutrients from fields, and reducing the transport of nutrients across watersheds. This chapter suggests that the 4 Rs of precision farming for nutrient management (the right product, at the right rate, at the right time, and at the right place) may not be enough. We need 7 Rs for nutrient management and conservation to increase conservation effectiveness and nutrient use efficiency and to minimize the losses of sediment and nutrients from the field. This chapter suggests that by applying the right product (fertilizer), at the right fertilizer rate, with the right method of fertilizer application, with the right conservation practice, and with conservation practice at the right place, and at the right scale of conservation practice, with both the fertilizer and the conservation practice applied at the right time (applying the 7 Rs of nutrient management and conservation: right product, right rate, right method, right practice, right place, right scale, and right time), we will minimize nutrient losses from the field and across the watershed.

REFERENCES

Adeuya, R., N. Utt, J. Frankenberger, L. Bowling, E. Kladivko, S. Brouder, and B. Carter. 2012. Impacts of drainage water management on subsurface drain flow, nitrate concentration, and nitrate loads in Indiana. *J. Soil Water Conserv.* 67:474–484.

Al-Abbas, A. H., R. Barr, J. D. Hall, F. L. Crane, and M. F. Baumgardner. 1974. Spectra of normal and nutrient-deficient maize leaves. *Agron. J.* 66:16–20.

Al-Kaisi, M. M., R. W. Elmore, J. G. Guzman, H. M. Hanna, C. E. Hart, M. J. Helmers, E. W. Hodgson, A. W. Lenssen, A. P. Mallarino, A. E. Robertson, and J. E. Sawyer. 2013. Drought impact on crop production and the soil environment: 2012 experiences from Iowa. *J. Soil Water Conserv.* 68:19A–24A.

Andraski, B. J., and B. Lowery. 1992. Erosion effects on soil water storage, plant water uptake, and corn growth. *Soil Sci. Soc. Am. J.* 56:1911–1919.

Angel, J. R., M. A. Palecki, and S. E. Hollinger. 2005. Storm precipitation in the United States. Part II: Soil erosion characteristics. *J. Appl. Meteorol.* 44:947–959.

Associated Press. 2013. Record nitrate levels from farm runoff are tainting Iowa rivers. http://www.omaha .com/news/record-nitrate-levels-from-farm-runoff-are-tainting-iowa-rivers/article_e8e6e87c-215b-57fc -aa2c-565db01d4c6a.html (accessed May 22, 2014).

Auffhammer, M. 2011. Agriculture: Weather dilemma for African maize. *Nature Clim. Change* 1:27–28. http://www.nature.com/nclimate/journal/v1/n1/full/nclimate1061.html (accessed May 22, 2014).

Bakker, M. M., G. Govers, and M. D. A. Rounsevell. 2004. The crop productivity-erosion relationship: An analysis based on experimental work. *Catena* 57:55–76.

Baligar, V. C., N. K. Fageria, and Z. L. He. 2001. Nutrient use efficiency in plants. *Commun. Soil Sci. Plant Anal.* 32:921–950.

Basso, B., D. Cammarano, C. Fiorentino, and J. T. Ritchie. 2013. Wheat yield response to spatially variable nitrogen fertilizer in Mediterranean environment. *Europ. J. Agronomy* 51:65–70.

Bausch, W. C., and J. A. Delgado. 2003. Ground based sensing of plant nitrogen status in irrigated corn to improve nitrogen management, pp. 151–163. In *Digital Imaging and Spectral Techniques: Applications to Precision Agriculture and Crop Physiology*, T. VanToai, D. Major, M. McDonald, J. Schepers, and L. Tarpley (eds.). ASA Special Publication 66. Madison, WI: ASA, CSSA, SSSA.

Bausch, W. C., and K. Diker. 2001. Innovative remote sensing techniques to increase nitrogen use efficiency of corn. *Commun. Soil Sci. Plant Anal.* 32:1371–1390.

Bausch, W. C., and H. R. Duke. 1996. Remote sensing of plant nitrogen status in corn. *Trans. ASAE* 39:1869–1875.

Bausch, W. C., H. R. Duke, and C. J. Iremonger. 1996. Assessment of plant nitrogen in irrigated corn. In *Proceedings of the 3rd International Conference on Precision Agriculture*, June 23–26, 1996, Minneapolis, Minnesota, P. C. Robert, R. H. Rust, and W. E. Larson (eds.), pp. 23–32. Madison, WI: ASA.

Berry, J. K., J. A. Delgado, R. Khosla, and F. J. Pierce. 2003. Precision conservation for environmental sustainability. *J. Soil Water Conserv.* 58:332–339.

Berry, J. K., J. A. Delgado, F. J. Pierce, and R. Khosla. 2005. Applying spatial analysis for precision conservation across the landscape. *J. Soil Water Conserv.* 60:363–370.

Biermacher, J. T., F. M. Epplin, B. W. Brorsen, J. B. Solie, and W. R. Raun. 2006. Maximum benefit of a precise nitrogen application system for wheat. *Prec. Agric.* 7:193–204.

Bjorneberg, D. L., D. T. Westermann, and J. K. Aase. 2002. Nutrient losses in surface irrigation runoff. *J. Soil Water Conserv.* 57:524–529.

Blackmer, T. M., J. S. Schepers, G. E. Varvel, and E. A. Walter-Shea. 1996. Nitrogen deficiency detection using reflected shortwave radiation from irrigated corn canopies. *Agron. J.* 88:1–5.

Bolton, S. M., T. J. Ward, and R. A. Cole. 1991. Sediment-related transport of nutrients from southwestern watersheds. *J. Irrig. Drain. Eng.* 117:736–747.

Briggs, J. A., T. Whitwell, and M. B. Riley. 1999. Remediation of herbicides in runoff water from container plant nurseries utilizing grassed waterways. *Weed Technol.* 13:157–164.

Brown, J., J. Angerer, S. W. Salley, R. Blaisdell, and J. W. Stuth. 2010. Improving estimates of rangeland carbon sequestration potential in the US Southwest. *Rangeland Ecol. Manag.* 63:147–154.

Bucks, D. A., T. W. Sammis, and G. L. Dickey. 1990. Irrigation for arid areas. In *Management of Farm Irrigation Systems*, G. J. Hoffman, T. A. Howell, and K. H. Solomon (eds.), pp. 499–548. Saint Joseph: ASAE.

Cambardella, C. A., T. B. Moorman, J. M. Novak, T. B. Parkin, D. L. Karlen, R. F. Turco, and A. E. Konopka. 1994. Field–scale variability of soil properties in central Iowa soils. *Soil Sci. Soc. Am. J.* 58:1501–1511.

Chang, J., D. E. Clay, C. G. Carlson, S. A. Clay, D. D. Malo, R. Berg, J. Kleinjan, and W. Wiebold. 2003. Different techniques to identify management zones impact nitrogen and phosphorus sampling variability. *Agron. J.* 95:1550–1559.

Cho, J., G. Vellidis, D. D. Bosch, R. Lowrance, and T. Strickland. 2010. Water quality effects of simulated conservation practice scenarios in the Little River Experimental watershed. *J. Soil Water Conserv.* 65:463–473.

Cooke, R., and S. Verma. 2012. Performance of drainage water management systems in Illinois, United States. *J. Soil Water Conserv.* 67:453–464.

Cox, C. 2005. Precision conservation professional. *J. Soil Water Conserv.* 60:134A.

Cox, C. 2010. Agriculture and nutrient pollution. Illinois Nutrient Summit, September 13–14, 2010. Environmental Working Group. http://www.epa.state.il.us/water/nutrient/presentations/craig_cox.pdf (accessed May 22, 2014).

Cox, C., and A. Hug. 2012. Murky waters: Farm pollution stalls cleanup of Iowa streams. Washington, DC: Environmental Working Group.

Cox, C., A. Hug, and N. Bruzelius. 2012. *Losing Ground.* Washington, DC: Environmental Working Group. http://static.ewg.org/reports/2010/losingground/pdf/losingground_report.pdf (accessed July 17, 2014).

Cruse, R. M., and C. G. Herndl. 2009. Balancing corn stover harvest for biofuels with soil and water conservation. *J. Soil Water Conserv.* 64:286–291.

Dabney, S. M., J. A. Delgado, and D. W. Reeves. 2001. Using winter cover crops to improve soil and water quality. *Comm. Soil Sci. Plant Anal.* 32:1221–1250.

Daniels, R. B., and J. W. Gilliam. 1996. Sediment and chemical load reduction by grass and riparian filters. *Soil Sci. Soc. Am. J.* 60:246–251.

Davatgar, N., M. R. Neishabouri, and A. R. Sepaskhah. 2012. Delineation of site specific nutrient management zones for a paddy cultivated area based on soil fertility using fuzzy clustering. *Geoderma* 173–174:111–118.

Davies, D. B., and R. Sylvester-Bradley. 1995. The contribution of fertilizer nitrogen to leachable nitrogen in the UK: A review. *J. Sci. Food Agric.* 68:399 406.

De Paz, J. M., J. A. Delgado, C. Ramos, M. J. Shaffer, and K. K. Barbarick. 2009. Use of a new Nitrogen Index—GIS assessment for evaluation of nitrate leaching across a Mediterranean region. *J. Hydrol.* 365:183–194.

Delgado, J. A. 1999. NLEAP simulation of soil type effects on residual soil nitrate-nitrogen in the San Luis Valley and potential use for precision agriculture. In *Proceedings of the Fourth International Conference on Precision Agriculture*, July 19–22, 1998, St. Paul, MN, P. C. Robert, R. H. Rust, and W. E. Larson (eds.), pp. 1367–1378. Madison, WI: ASA.

Delgado, J. A. 2001. Use of simulations for evaluation of best management practices on irrigated cropping systems. In *Modeling Carbon and Nitrogen Dynamics for Soil Management*, M. J. Shaffer, L. Ma, and S. Hansen (eds.), pp. 355–381. Boca Raton, FL: Lewis Publishers.

Delgado, J. A. 2002. Quantifying the loss mechanisms of nitrogen. *J. Soil Water Conserv.* 57:389–398.

Delgado, J. A. 2010. Crop residue is a key for sustaining maximum food production and for conservation of our biosphere. *J. Soil Water Conserv.* 65:111A–116A.

Delgado, J. A. 2012. Nitrogen trading tool. In *Encyclopedia of Environmental Management*, S. E. Jorgensen (ed.), pp. 1772–1784. New York: Taylor & Francis.

Delgado, J. A., and R. F. Follett. 2002. Carbon and nutrient cycles. *J. Soil Water Conserv.* 57:455–464.

Delgado, J. A., and W. Bausch. 2005. Potential use of precision conservation techniques to reduce nitrate leaching in irrigated crops. *J. Soil Water Conserv.* 60:379–387.

Delgado, J. A., and J. K. Berry. 2008. Advances in precision conservation. *Adv. Agron.* 98:1–44.

Delgado, J. A., M. Shaffer, and M. K. Brodahl. 1998. New NLEAP for shallow and deep rooted rotations: Irrigated agriculture in the San Luis Valley of south central Colorado. *J. Soil Water Conserv.* 53:332–337.

Delgado, J. A., R. T. Sparks, R. F. Follett, J. L. Sharkoff, and R. R. Riggenbach. 1999. Use of winter cover crops to conserve water and water quality in the San Luis Valley of south central Colorado. In *Soil Quality and Soil Erosion*. R. Lal (ed.), pp. 125–142. Boca Raton, FL: CRC Press.

Delgado, J. A., R. R. Riggenbach, R. T. Sparks, M. A. Dillon, L. M. Kawanabe, and R. J. Ristau. 2001a. Evaluation of nitrate-nitrogen transport in a Potato-barley rotation. *Soil Sci. Soc. Am. J.* 65:878–883.

Delgado, J. A., R. J. Ristau, M. A. Dillon, H. R. Duke, A. Stuebe, R. E. Follett, M. J. Shaffer, R. R. Riggenbach, R. T. Sparks, A. Thompson, L. M. Kawanabe, A. Kunugi, and K. Thompson. 2001b. Use of innovative tools to increase nitrogen use efficiency and protect environmental quality in crop rotations. *Commun. Soil Sci. Plant Anal.* 32:1321–1354.

Delgado, J. A., R. Khosla, W. Bausch, D. G. Westfall, and D. J. Inman. 2005. Nitrogen fertilizer management based on site specific management zones reduces potential for NO_3-N leaching. *J. Soil Water Conserv.* 60:402–410.

Delgado, J. A., M. J. Shaffer, H. Lal, S. McKinney, C. M. Gross, and H. Cover. 2008. Assessment of nitrogen losses to the environment with a Nitrogen Trading Tool (NTT). *Comput. Electron. Agric.* 63:193–206.

Delgado, J. A., R. R. Riggenbach, R. T. Sparks, M. A. Dillon, L. M. Kawanabe, and R. J. Ristau. 2001a. Evaluation of nitrate-nitrogen transport in a potato-barley rotation. *Soil Sci. Soc. Am. J.* 65:878–883.

Delgado, J. A., S. J. Del Grosso, and S. M. Ogle. 2010a. ^{15}N Isotopic crop residue cycling studies suggest that IPCC methodologies to assess N_2O-N emissions should be reevaluated. *Nutr. Cycling Agroecosyst.* 86:383–390.

Delgado, J. A., C. M. Gross, H. Lal, H. Cover, P. Gagliardi, S. P. McKinney, E. Hesketh, and M. J. Shaffer. 2010b. A new GIS nitrogen trading tool concept for conservation and reduction of reactive nitrogen losses to the environment. *Adv. Agron.* 105:117–171.

Delgado, J. A., P. M. Groffman, M. A. Nearing, T. Goddard, D. Reicosky, R. Lal, N. R. Kitchen, C. W. Rice, D. Towery, and P. Salon. 2011. Conservation practices to mitigate and adapt to climate change. *J. Soil Water Conserv.* 66:118A–129A.

Delgado, J. A., M. A. Nearing, and C. W. Rice. 2013. Conservation practices for climate change adaptation. *Adv. Agron.* 121:47–115.

Dickey, E. C., and D. H. Vanderholm. 1981. Vegetative filter treatment of livestock feedlot runoff. *J. Environ. Qual.* 10:279–284.

Dillaha, T. A., and J. C. Hayes. 1991. *A Procedure for the Design of Vegetative Filter Strips.* Washington, DC: USDA Natural Resources Conservation Service.

Dillaha, T. A., J. H. Sherrard, and D. Lee. 1986. *Long-Term Effectiveness and Maintenance of Vegetative Filter Strips.* Bulletin 153. Blacksburg, VA: Virginia Water Resources Research Center, Virginia Polytechnic Institute and State University.

Dillaha, T. A., J. H. Sherrard, D. Lee, S. Mostaghimi, and V. O. Shanholtz. 1988. Evaluation of vegetative filter strips as a best management practice for feed lots. *J. Water Pollution Control Fed.* 60:1231–1238.

Dillaha, T. A., R. B. Reneau, S. Mostaghimi, and D. Lee. 1989. Vegetative filter strips for agricultural nonpoint source pollution control. *Trans. ASAE* 32:513–519.

Doran, J. W., and A. J. Jones. 1996. Methods for assessing soil quality. SSSA Special Publication 49. Madison, WI: SSSA.

Dosskey, M. G., M. J. Helmers, D. E. Eisenhauer, T. G. Franti, and K. D. Hoagland. 2002. Assessment of concentrated flow through riparian buffers. *J. Soil Water Conserv.* 57:336–343.

Dosskey, M. G., D. E. Eisenhauer, and M. J. Helmers. 2005. Establishing conservation buffers using precision information. *J. Soil Water Conserv.* 62:349–354.

Dosskey, M. G., M. J. Helmers, and D. E. Eisenhauer. 2007. An approach for using soil surveys to guide the placement of water quality buffers. *J. Soil Water Conserv.* 61:344–354.

Dubrovsky, N. M., K. R. Burow, G. M. Clark et al. 2010. *Nutrients in the Nation's Streams and Groundwater, 1992–2004.* Reston, VA: U.S. Geological Survey.

Evans, R. O., and R. W. Skaggs. 2004. Development of controlled drainage as a BMP in North Carolina. In *Proceedings of the Eighth National Drainage Symposium*, Sacramento, California, March 21–24, 2004. St. Joseph, MI: American Society of Agricultural Engineers.

Evans, R. G., J. LaRue, K. C. Stone, and B. A. King. 2013. Adoption of site-specific variable rate sprinkler irrigation systems. *Irrig. Sci.* 31:871–887.

Fabis, J., M. Bach, and H. G. Frede. 1993. Vegetative filter strips in hilly areas of Germany. In *Integrated Resource Management and Landscape Modification for Environmental Protection: Proceedings of the International Symposium*, December 13–14, 1993, Chicago. J. K. Mitchell (ed.), pp. 81–88. St. Joseph, MI: ASA.

Fares, A., M. Safeeq, A. Kimoto, and A. Dogan. 2010. Use of buffers to reduce sediment and nitrogen transport to surface water bodies. In *Advances in Nitrogen Management for Water Quality*. J. A. Delgado, and R. F. Follett (eds.), pp. 282–312. Ankeny, IA: SWCS.

Ferguson, R. B., C. A. Gotway, G. W. Hergert, and T. A. Peterson. 1996. Soil sampling for site-specific nitrogen management. In *Proceedings of the 3rd International Conference on Precision Agriculture*, June 23–26, 1996, Minneapolis, Minnesota. P. C. Robert, R. H. Rust, and W. E. Larson (eds.), pp. 13–22. Madison, WI: ASA.

Fiener, P., and K. Auerswald. 2003. Effectiveness of grassed waterways in reducing runoff and sediment delivery from agricultural watersheds. *J. Environ. Qual.* 32:927–936.

Fixen, P. E., and F. B. West. 2002. Nitrogen fertilizers: Meeting contemporary challenges. *AMBIO: A Journal of the Human Environment* 31:169–176.

Fleming, K. L., D. G. Westfall, D. W. Wiens, L. E. Rothe, J. E. Cipra, and D. F. Heermann. 1999. Evaluating farmer developed management zone maps for precision farming. In *Proceedings of the 4th International Conference on Precision Agriculture*, P. C. Robert, R. H. Rust, and W. E. Larson (eds.), pp. 335–343. Madison, WI: ASA.

Fleming, K. L., D. F. Heermann, and D. G. Westfall. 2004. Evaluation soil color with farmer input and apparent soil electrical conductivity for management zone delineation. *Agron. J.* 96:1581–1587.

Fletcher, D. A. 1991. A national perspective. In *Managing Nitrogen for Groundwater Quality and Farm Profitability*, R. F. Follett, D. R. Keeney, and R. M. Cruse (eds.), pp. 9–18. Madison, WI: SSSA.

Follett, R. F., and B. A. Stewart (eds.). 1985. *Soil Erosion and Crop Productivity*. Madison, WI: ASA.

Fox, R. H., W. P. Piekielek, and K. E. Macneal. 1996. Estimating ammonia volatilization losses from urea fertilizers using a simplified micrometeorological sampler. *Soil Sci. Soc. Am. J.* 60:596–601.

Franzen, D. W., L. Reitmeier, J. F. Giles, and A. C. Cattanach. 1999. Aerial photography and satellite imagery to detect deep soil nitrogen levels in potato and sugarbeet. In *Proceedings of the 4th International Conference on Precision Agriculture*, P. C. Robert, R. H. Rust, and W. E. Larson (eds.), pp. 281–290. Madison, WI: ASA.

Freney, J. R., J. R. Simpson, and O. T. Denmead. 1981. Ammonia volatilization. *Ecol. Bull.* 33:291–302.

Fu, Q., Z. Wang, and Q. Jiang. 2010. Delineating soil nutrient management zones based on fuzzy clustering optimized by PSO. *Math. Comput. Model.* 51:1299–1305.

Galzki, J. C., A. S. Birr, and D. J. Mulla. 2011. Identifying critical agricultural areas with 3-meter LiDAR elevation data for precision conservation. *J. Soil Water Conserv.* 66:423–430.

Gilliam, J. W., R. W. Skaggs, and S. B. Weed. 1979. Drainage control to diminish nitrate loss from agricultural fields. *J. Environ. Qual.* 8:137–142.

Goolsby, D. A., W. A. Battaglin, B. T. Aulenbach, and R. P. Hooper. 2001. Nitrogen input to the Gulf of Mexico. *J. Environ. Qual.* 30:329–336.

Gotway, C. A., R. B. Ferguson, and G. W. Hergert. 1996. The effects of mapping and scale on variable rate fertilizer recommendations for corn. In *Proceedings of the 3rd International Conference on Precision Agriculture*, P. C. Robert, R. H. Rust, and W. E. Larson (eds.), pp. 321–330. ASA, Madison, WI.

Greenan, C. M., T. B. Moorman, T. C. Kaspar, T. B. Parkin, and D. B. Jaynes. 2006. Comparing carbon substrates for denitrification of subsurface drainage water. *J. Environ. Qual.* 35:824–829.

Hall, M. D., M. J. Shaffer, R. M. Waskom, and J. A. Delgado. 2001. Regional nitrate leaching variability: What makes a difference in Northeastern Colorado. *J. Am. Water Resour. Assoc.* 37:139–150.

Hallberg, G. R. 1989. Nitrate in ground water in the United States. In *Nitrogen Management and Ground Water Protection*, R. F. Follett (ed.), pp. 35–74. Amsterdam: Elsevier Science B.V.

Helmers, M., R. Christianson, G. Brenneman, D. Lockett, and C. Pederson. 2012. Water table, drainage, and yield response to drainage water management in southeast Iowa. *J. Soil Water Conserv.* 67:495–501.

Hergert, G. W., R. B. Ferguson, and C. A. Gotway. 1996. The impact of variable rate N applications on N use efficiency of furrow irrigated corn. In *Proceedings of the 3rd International Conference on Precision Agriculture*, P. C. Robert, R. H. Rust, and W. E. Larson (eds.), pp. 85–94. Madison, WI: ASA.

Hey, D. L. 2002. Nitrogen farming: Harvesting a different crop. *Restor. Ecol.* 10:1–10.

Hey, D. L., L. S. Urban, and J. A. Kostel. 2005. Nutrient farming: The business of environmental management. *Ecol. Eng.* 24:279–287.

Hornung, A., R. Khosla, R. Reich, and D. G.Westfall. 2003. Evaluation of site-specific management zones: Grain yield and nitrogen use efficiency. In *Precision Agriculture,* J. Stafford, and A. Werner (eds.), pp. 297–302. Wageningen, The Netherlands: Wageningen Academic Publishers.

Hu, C., J. A. Delgado, X. Zhang, and L. Ma. 2005. Assessment of groundwater use by wheat (Triticum aestivum L.) in the Luancheng Xian region and potential implications for water conservation in the northwestern North China plain. *J. Soil Water Conserv.* 60:80–88.

Hunter, W. J. 2001. Remediation of drinking water for rural populations. In *Nitrogen in the Environment: Sources, Problems, and Management*, R. F. Follett and J. T. Hatfield (eds.), pp. 433–453. Amsterdam: Elsevier Science B.V.

Hunter, W. J. 2013. Pilot-scale vadose zone biobarriers removed nitrate leaching from cattle corral. *J. Soil Water Conserv.* 68:52–59.

Inman, D., R. Khosla, and D. G. Westfall. 2005. Nitrogen uptake across site-specific management zones in irrigated corn production systems. *Agron. J.* 97:169–176.

IPCC (Intergovernmental Panel on Climate Change). 2007. Summary for policymakers. In *Climate Change 2007: The Physical Science Basis*. Contribution of Working Group I to the Fourth Assessment Report of the Intergovernmental Panel on Climate Change, S. Solomon, D. Qin, M. Manning, Z. Chen, M. Marquis, K. B. Averyt, M. Tignor, and H. L. Miller (eds.). Cambridge, UK: Cambridge University Press.

Jaynes, D. B. 2012. Changes in yield and nitrate losses from using drainage water management in central Iowa, United States. *J. Soil Water Conserv.* 67:485–494.

Jaynes, D. B., T. C. Kaspar, T. B. Moorman, and T. B. Parkin. 2008. In situ bioreactors and deep drain-pipe installation to reduce nitrate losses in artificially drained fields. *J. Environ. Qual.* 37:429–436.

Jaynes, D. B., T. C. Kaspar, and T. S. Colvin. 2011. Economically optimal nitrogen rates of corn: Management zones delineated from soil and terrain attributes. *Agron. J.* 103:1026–1035.

Jenrich, M. 2011. Potential of precision conservation agriculture as a means of increasing productivity and incomes for smallholder farmers. *J. Soil Water Conserv.* 66:171A–174A.

Johnson J. M. F., M. D. Coleman, R. Gesch, A. Jaradat, R. Mitchell, D. Reicosky, and W. W. Wilhelm. 2007. Biomass-bioenergy crops in the United States: A changing paradigm. Americas *J. Plant Sci. Biotech.* 1:1–28.

Johnson, J. M. F., D. L. Karlen, and S. S. Andrews. 2010. Conservation considerations for sustainable bioenergy feedstock production: If, what, where, and how much? *J. Soil Water Conserv.* 65:88A–91A.

Jokela, W. E., and G. W. Randall. 1989. Corn yield and residual soil nitrate as affected by time and rate of nitrogen application. *Agron. J.* 81:720–726.

Juergens-Gschwind, S. 1989. Ground water nitrates in other developed countries (Europe)–relationships to land use patterns. In *Nitrogen Management and Ground Water Protection*, R. F. Follett (ed.), pp. 75–138. Amsterdam: Elsevier Science B.V.

Karlen, D. L., R. Lal, R. F. Follett, J. Kimble, J. Hatfield, J. Miranowski, C. A. Cambardella, A. Manale, R. P. Anex, and C. W. Rice. 2009. Crop residues: The rest of the story. *Environ. Sci. Technol.* 43:8011–8015.

Khosla, R., and M. M. Alley. 1999. Soil specific nitrogen management on Mid-Atlantic Coastal Plain soils. *Better Crops with Plant Food.* No. 3. Norcross, GA: Potash and Phosphorus Institute.

Khosla, R., K. Fleming, J. Delgado, T. Shaver, and D. Westfall. 2002. Use of site-specific management zones to improve nitrogen management for precision agriculture. *J. Soil Water Conserv.* 57:513–518.

King, E. S., C. Cox, and W. L. Baker. 2009. An urgent call to action: Nutrient Innovations Task Group report. http://water.epa.gov/learn/training/wacademy/upload/2009_12_01_slides.pdf (accessed May 22, 2014).

Knowles, N., M. D. Dettinger, and D. R. Cayan. 2006. Trends in snowfall versus rainfall in the western United States. *J. Climate* 19:4545–4559.

Koch, B., R. Khosla, M. Frasier, and D. G. Westfall. 2003. Economic feasibility of variable-rate nitrogen application in site specific management, pp. 107–112. In *Proceedings of the Western Nutrient Management Conference*, Vol. 5. March 6–7, 2003, Salt Lake City, UT.

Laflen, J. M., and T. S. Colvin. 1981. Effect of crop residue on soil loss from continuous row cropping. *Trans. Am. Soc. Agric. Eng.* 24:605–609.

Lal, R. 1995. Global soil erosion by water and carbon dynamics. In *Soils and Global Change*, R. Lal, J. M. Kimle, E. Levine, and B. A. Stewart (eds.), pp. 131–140. Boca Raton, FL: Lewis Publishers.

Lal, R. 1998. Soil erosion impact on agronomic productivity and environmental quality. *Crit. Rev. Plant Sci.* 17:319–464.

Lal, R. 2004. Soil carbon sequestration impacts on global climate change and food security. *Science* 304:1623–1627.

Lal, R., J. A. Delgado, P. M. Groffman, N. Millar, C. Dell, and A. Rotz. 2011. Management to mitigate and adapt to climate change. *J. Soil Water Conserv.* 66:276–285.

Lal, R., J. A. Delgado, J. Gulliford, D. Nielsen, C. W. Rice, and R. S. Van Pelt. 2012. Adapting agriculture to drought and extreme events. *J. Soil Water Conserv.* 67:162A–166A.

Legg, J. O., and J. J. Meisinger. 1982. Soil nitrogen budgets. In *Nitrogen in Agricultural Soils, Agronomy*, Vol. 22, pp. 503–566. F. J. Stevenson, J. M. Bremner, R. D. Hauck, and D. R. Kenney (eds.). Madison, WI: ASA, CSAA, SSSA.

Lobell, D. B., M. Bänziger, C. Magorokosho, and B. Vivek. 2011. Nonlinear heat effects on African maize as evidenced by historical yield trials. *Nature Clim. Change* 1:42–45. http://www.nature.com/nclimate/journal/v1/n1/full/nclimate1043.html (accessed May 22, 2014).

Lowrance, R. R., L. S. Altier, J. D. Newbold et al. 1997. Water quality functions of riparian forest buffers in Chesapeake Bay watersheds. *Environ. Manage.* 21:687–712.

Lowrance, R., L. S. Altier, R. G. Williams, S. P. Inamdar, J. M. Sheridan, D. D. Bosch, R. K. Hubbard, and D. L. Thomas. 2000. REMM: The riparian ecosystem management model. *J. Soil Water Conserv.* 55:27–34.

Lowrance, R., J. M. Sheridan, R. G. Williams, D. D. Bosch, D. G. Sullivan, D. R. Blanchett, L. M. Hargett, and C. M. Clegg. 2007. Water quality and hydrology in farm-scale coastal plain watersheds: Effects of agriculture, impoundments, and riparian zones. *J. Soil Water Conserv.* 62:65–76.

Luck, J. D., T. G. Mueller, S. A. Shearer, and A. C. Pike. 2010. Grassed waterway planning model evaluated for agricultural fields in the western coal field physiographic region of Kentucky. *J. Soil Water Conserv.* 65:280–288.

Madison, R. J., and J. O. Brunett. 1985. Overview of the occurrence of nitrate in ground water of the United States, pp. 93–105. In *National Water Summary 1984*. USGS Water-Supply Paper 2275. Washington, DC: U.S. Government Printing Office.

Masek, T. J., J. S. Schepers, S. C. Mason, and D. D. Francis. 2001. Use of precision farming to improve application of feedlot waste to increase use efficiency and protect water quality. *Commun. Soil Sci. Plant Anal.* 32:1355–1369.

Mayer, P. M., S. K. Reynolds, and T. J. Canfield. 2005. *Riparian Buffer Width, Vegetative Cover, and Nitrogen Removal Effectiveness: A Review of Current Science and Regulations.* Ada, OK: National Risk Management Research Laboratory, US Environmental Protection Agency.

McMurtrey, J. E., III, E. W. Chappelle, M. S. Kim, J. J. Meisinger, and L. A. Corp. 1994. Distinguishing nitrogen fertilization levels in field corn (Zea mays L.) with actively induced fluorescence and passive reflectance measurements. *Remote Sens. Environ.* 47:36–44.

Meisinger, J. J., and G. W. Randall. 1991. Estimating N budgets for soil-crop systems. In *Managing Nitrogen for Groundwater Quality and Farm Productivity*, R. F. Follett, D. R. Keeney, and R. M. Cruse (eds.), pp. 85–124. Madison, WI: ASA.

Meki, M. N., J. P. Marcos, J. D. Atwood, L. M. Norfleet, E. M. Steglich, J. R. Williams, and T. J. Gerik. 2011. Effects of site-specific factors on corn stover removal thresholds and subsequent environmental impacts in the Upper Mississippi River Basin. *J. Soil Water Conserv.* 66:386–399.

Milburn, P., J. E. Richards, C. Gartley, T. Pollock, H. O'Neill, and H. Bailey. 1990. Nitrate leaching from systematically tiled potato fields in New Brunswick, Canada. *J. Environ. Qual.* 19:448–454.

Mitsch, W. J., and J. W. Day. 2006. Restoration of wetlands in the Mississippi-Ohio-Missouri (MOM) River Basin: Experience and needed research. *Ecol. Eng.* 26:55–69.

Mitsch, W. J., J. W. Day, Jr., J. W. Gilliam, P. M. Groffman, D. L. Hey, G. W. Randall, and N. Wang. 1999. *Reducing Nutrient Loads, Especially Nitrate-Nitrogen, to Surface Water, Ground Water, and the Gulf of Mexico. Topic 5 Report for the Integrated Assessment on Hypoxia in the Gulf of Mexico.* NOAA Coastal Ocean Program Decision Analysis Series No. 19. Silver Spring, MD: NOAA Coastal Ocean Program.

Montgomery, D. R. 2007. Soil erosion and agricultural sustainability. *Proc. Natl. Acad. Sci. U.S.A.* 104:13268–13272.

Moore, I. D., G. J. Burch, and D. H. Mackenzie. 1988. Topographic effects on the distribution of surface soil water and the location of ephemeral gullies. *Trans. Am. Soc. Agric. Eng.* 31:1098–1119.

Mosier, A., D. Schimel, D. Valentine, K. Bronson, and W. Parton. 1991. Methane and nitrous oxide fluxes in native, fertilized and cultivated grasslands. *Nature (London)* 350:330–332.

Mosier, A. R., W. J. Parton, D. W. Valentine, D. S. Ojima, D. S. Schimel, and J. A. Delgado. 1996. CH_4 and N_2O fluxes in the Colorado shortgrass steppe. 1. Impacts of landscape and nitrogen addition. *Global Biogeochem. Cycles* 10:387–399.

Mueller, T. G., H. Cetin, R. A. Fleming, C. R. Dillon, A. D. Karathanasis, and S. A. Shearer. 2005. Erosion probability maps: Calibrating precision agriculture data with soil surveys using logistic regression. *J. Soil Water Conserv.* 62:462–468.

Murrell, T. S., G. P. Lafond, and T. J. Vyn. 2009. Know your fertilizer rights: Right place. *Crops & Soils* November–December 2009.

Nearing, M. A. 2001. Potential changes in rainfall erosivity in the U.S. with climate change during the 21st century. *J. Soil Water Conserv.* 56:229–232.

Nearing, M. A., F. F. Pruski, and M. R. O'Neal. 2004. Expected climate change impacts on soil erosion rates: A review. *J. Soil Water Conserv.* 59:43–50.

Newman, J. K., A. L. Kaleita, and J. M. Laflen. 2010. Soil erosion hazard maps for corn stover management using National Resources Inventory data and the Water Erosion Prediction Project. *J. Soil Water Conserv.* 65:211–222.

Ortega, R. A., and O. A. Santibanez. 2007. Determination of management zones in corn (*Zea mays* L.) based on soil fertility. *Comput. Electron. Agric.* 58:49–59.

Ortega, R. A., D. G. Westfall, and G. A. Peterson. 1997. Variability of phosphorus over landscapes and dryland winter wheat yields. *Better Crops* 81:24–27.

Papiernik, S. K., M. J. Lindstrom, J. A. Schumacher, A. Farenhorst, K. D. Stephens, T. E. Schumacher, and D. A. Lobb. 2005. Variation in soil properties and crop yield across an eroded prairie landscape. *J. Soil Water Conserv.* 60:388–395.

Parton, W. J., D. S. Schimel, C. V. Cole, and D. S. Ojima. 1987. Analysis of factors controlling soil organic matter levels in Great Plains grasslands. *Soil Sci. Soc. Am. J.* 51:1173–1179.

Peng, S., J. Huang, J. E. Sheehy, R. C. Laza, R. M. Visperas, X. Zhong, G. S. Centeno, G. S. Khush, and K. G. Cassman. 2004. Rice yields decline with higher night temperature from global warming. *Proc. Natl. Acad. Sci.* 101:9971–9975.

Penn, C. J., R. B. Bryant, P. J. A. Kleinman, and A. L. Allen. 2007. Removing dissolved phosphorous from drainage ditch water with phosphorous sorbing materials. *J. Soil Water Conserv.* 6:269–276.

Penn, C., J. McGrath, J. Bowen, and S. Wilson. 2014. Phosphorus removal structures: A management option for legacy phosphorus. *J. Soil Water Conserv.* 69:51A–56A.

Pennock, D. J. 2005. Precision conservation for co-management of carbon and nitrogen on the Canadian prairies. *J. Soil Water Conserv.* 62:396–401.

Peoples, M. B., J. R. Freney, and A. R. Mosier. 1995. Minimizing gaseous losses of nitrogen. In *Nitrogen Fertilization in the Environment*, P. E. Bacon (ed.), pp. 565–602. New York: Marcel Dekker Inc.

Peralta, N. R., and J. L. Costa. 2013. Delineation of management zones with soil apparent electrical conductivity to improve nutrient management. *Comput. Electron. Agric.* 99:218–226.

Pike, A. C., T. G. Mueller, A. Schorgendorfer, S. A. Shearer, and A. D. Karathanasis. 2009. Erosion index derived from terrain attributes using logistic regression and neural networks. *Agron. J.* 101:1068–1079.

Pruski, F. F., and M. A. Nearing. 2002a. Runoff and soil loss responses to changes in precipitation: A computer simulation study. *J. Soil Water Conserv.* 57:7–16.

Pruski, F. F., and M. A. Nearing. 2002b. Climate-induced changes in erosion during the 21st century for eight U.S. locations. *Water Resour. Res.* 38(12):1298. doi:10.1029/2001WR000493

Qiu, Z., M. T. Walter, and C. Hall. 2007. Managing variable source pollution in agricultural watersheds. *J. Soil Water Conserv.* 62:115–122.

Quine, T. A., and Y. Zhang. 2002. An investigation of spatial variation in soil erosion, soil properties, and crop production within an agricultural field in Devon, United Kingdom. *J. Soil Water Conserv.* 57:55–64.

Randall, G. W., J. A. Delgado, and J. S. Schepers. 2008. Nitrogen management to protect water resources. In *Nitrogen in Agricultural Systems*, Agronomy Monograph 49. J. S. Schepers, and W. R. Raun (eds.), pp. 907–940. Madison, WI: SSSA.

Rangely, W. R. 1987. Irrigation and drainage in the world. In *Water and Water Policy in World Food Supplies: Proceedings of the Conference*, May 25–30, 1985, W. R. Jordan (ed.), pp. 29–35. College Station, TX: Agriculture and Mining University Press.

Raun, W. R., and G. V. Johnson. 1999. Improving nitrogen use efficiency for cereal production. *Agron. J.* 91:357–363.

Raun, W. R., J. B. Solie, G. V. Johnson, M. L. Stone, R. W. Mullen, K. W. Freeman, W. E. Thompson, and E. V. Lukina. 2002. Improving nitrogen use efficiency in cereal grain production with optical sensing and variable rate application. *Agron. J.* 94:815–820.

Redulla, C. A., J. L. Havlin, G. J. Kluitenberg, N. Zhang, and M. D. Schrock. 1996. Variable nitrogen management for improving groundwater quality. In *Proceedings of the 3rd International Conference on Precision Agriculture*, P. C. Robert, R. H. Rust, and W. E. Larson (eds.), pp. 85–94. Madison, WI: ASA.

Ribaudo, M. O., R. Heimlich, and M. Peters. 2005. Nitrogen sources and Gulf hypoxia: Potential for environmental credit trading. *Ecol. Econ.* 52:159–168.

Ribaudo, M., J. Delgado, L. Hansen, M. Livingston, R. Mosheim, and J. Williamson. 2011. *Nitrogen in Agricultural Systems: Implications for Conservation Policy*. Economic Research Report No. 127. Washington, DC: USDA-ERS.

Roberts, T. L. 2007. Right product, right rate, right time and right place... the foundation of best management practices for fertilizer. In *Fertilizer Best Management Practices: General Principles, Strategies for Their Adoption and Voluntary Initiatives vs Regulations*. Paper presented at the IFA International Workshop on Fertilizer Best Management Practices, Brussels, Belgium, March 7–9, 2007, 29–32. Paris: International Fertilizer Industry Association.

Roberts, D. F., R. B. Ferguson, N. R. Kitchen, V. I. Adamchuk, and J. F. Shanahan. 2012. Relationships between soil-based management zones and canopy sensing for corn nitrogen management. *Agron. J.* 104:119–129.

Robertson, D. M., G. E. Schwarz, D. A. Saad, and R. B. Alexander. 2009. Incorporating uncertainty into the ranking of SPARROW model nutrient yields from Mississippi/Atchafalaya River Basin watersheds. *J. Am. Wat. Res. As.* 45:535–549.

Sadler, E. J., R. G. Evans, K. C. Stone, and C. R. Camp. 2005. Opportunities for conservation with precision irrigation. *J. Soil Water Conserv.* 62:371–379.

Sadler, E. J., W. C. Bausch, N. R. Fausey, and R. B. Ferguson. 2010. Water management: A key to reducing nitrogen losses. In *Advances in Nitrogen Management for Water Quality*, J. A. Delgado, and R. F. Follett (eds.), pp. 38–60. Ankeny, IA: SWCS.

Saleh, A., O. Gallego, E. Osei, H. Lal, C. Gross, S. McKinney, and H. Cover. 2011. Nutrient Tracking Tool—A user-friendly tool for calculating nutrient reductions for water quality trading. *J. Soil Water Conserv.* 66:400–410.

Sawchik, J., and A. P. Mallarino. 2007. Evaluation of zone soil sampling approaches for phosphorus and potassium based on corn and soybean response to fertilization. *Agron. J.* 99:1564–1578.

Scharf, P. C., J. P. Schmidt, N. R. Kitchen, K. A. Sudduth, S. Y. Hong, J. A. Lory, and J. G. Davis. 2002. Remote sensing for N management. *J. Soil Water Conserv.* 57:518–524.

Scharf, P. C., D. K. Shannon, H. L. Palm, K. A. Sudduth, S. T. Drummond, N. R. Kitchen, L. J. Mueller, V. C. Hubbard, and L. F. Oliveira. 2011. Sensor-based nitrogen applications out-performed producer-chosen rates for corn in on-farm demonstrations. *Agron. J.* 103:1683–1691.

Schepers, J. S., D. D. Francis, M. Vigil, and F. E. Below. 1992a. Comparison of corn leaf nitrogen and chlorophyll meter readings. *Commun. Soil Sci. Plant Anal.* 23:2173–2187.

Schepers, J. S., T. M. Blackmer, and D. D. Francis. 1992b. Predicting N fertilizer needs for corn in humid regions: Using chlorophyll meters. In *Predicting Nitrogen Fertilizer Needs for Corn in Humid Regions,* B. R. Bock and K. R. Kelley (eds.), pp. 105–114. Proceedings of the Soil Science Society of America Symposium, Denver, CO, October 28, 1991. Muscle Shoals, AL: Tennessee Valley Authority Bulletin.

Schepers, J. S., T. M. Blackmer, and D. D. Francis. 1998. Chlorophyll meter method for estimating nitrogen content in plant tissue. In *Handbook on Reference Methods for Plant Analysis,* Y. P. Kalra (ed.), pp. 129 135. Boca Raton, FL: CRC Press.

Schepers, A., J. F. Shanahan, M. A. Liebig, J. S. Schepers, S. Johnson, and A. Luchiari. 2004. Delineation of management zones that characterize spatial variability of soil properties and corn yields across years. *Agron. J.* 96:195–203.

Schimel, D. S., M. A. Stillwell, and R. G. Woodmansee. 1985. Biogeochemistry of C, N, and P in a soil catena of the shortgrass steppe. *Ecology* 66:276–282.

Schimel, D. S., W. J. Parton, F. J. Adamsen, R. G. Woodmansee, R. L. Senft, and M. A. Stillwell. 1986. The role of cattle in the volatile loss of nitrogen from a shortgrass steppe. *Biogeochemistry* 2:39–52.

Schumacher, J. A., T. C. Kaspar, J. C. Ritchie, T. E. Schumacher, D. L. Karlen, E. R. Ventris, G. M. McCarty, T. S. Colvin, D. B. Jaynes, M. J. Lindstrom, and T. E. Fenton. 2005. Identifying spatial patterns of erosion for use in precision conservation. *J. Soil Water Conserv.* 62:355–362.

Secoges, J. M., W. M. Aust, J. R. Seiler, C. A. Dolloff, and W. A. Lakel. 2013. Streamside management zones affect movement of silvicultural nitrogen and phosphorus fertilizers to piedmont streams. *J. Appl. For.* 37:26–35.

Segura, C., G. Sun, S. McNulty, and Y. Zhang. 2014. Potential impacts of climate change on soil erosion vulnerability across the conterminous United States. *J. Soil Water Conserv.* 69:171–181.

Shaffer, M. J., and Delgado, J. A. 2002. Essentials of a national nitrate leaching index assessment tool. *J. Soil Water Conserv.* 57:327–335.

Shanahan, J. F., J. S. Schepers, D. D. Francis, G. E. Varvel, W. W. Wilhelm, J. S. Tringe, M. R. Schlemmer, and D. J. Major. 2001. Use of remote sensing imagery to estimate corn grain yield. *Agron. J.* 93:583–589.

Shanahan, J. F., N. R. Kitchen, W. R. Raun, and J. S. Schepers. 2008. Responsive in-season nitrogen management for cereals. *Comput. Electron. Agric.* 61:51–62.

Sharpe, R. R., and L. A. Harper. 1995. Soil, plant, and atmospheric conditions as they relate to ammonia volatilization. *Fert. Res.* 42:149–153.

Skaggs, R. W., N. R. Fausey, and R. O. Evans. 2012. Drainage water management. *J. Soil Water Conserv.* 67:167A–172A.

Smith, T. A., D. L. Osmond, and J. W. Gilliam. 2006. Riparian buffer and nitrate removal in a lagoon-effluent irrigated agricultural area. *J. Soil Water Conserv.* 61:273–281.

Snyder, C. S. 2012. Are Midwest corn farmers over-applying fertilizer N? *Better Crops* 96:3–4.

Snyder, C. S., and T. W. Bruulsema. 2007. *Nutrient Use Efficiency and Effectiveness in North America: Indices of Agronomic and Environmental Benefit.* Norcross, GA: IPNI.

Srivastava, K. P., and I. D. Moore. 1989. Application of terrain analysis to land resource investigations of small catchments in the Caribbean. In *Proceedings of the 20th International Conference of the Erosion Control Association,* pp. 229–249. Vancouver, BC, Canada. Steamboat Springs, CO: International Erosion Control Association.

SSSA (Soil Science Society of America). 2013. How do soils form? Soils matter, get the scoop! August 29, 2013. http://soilsmatter.wordpress.com/2013/08/29/soil-formation/ (accessed May 22, 2014).

Stanhill, G., V. Kalkofi, M. Fuchs, and Y. Kagan. 1972. The effects of fertilizer applications on solar reflectance from a wheat crop. *J. Agric. Res.* (Israel) 22:109–118.

Steenvoorden, J. H. A. M. 1985. Nutrient leaching losses following application of farm slurry and water quality considerations in the Netherlands. In *Efficient Land Use of Sludge and Manure,* A. Dam Kofoed, J. H. Williams, and P. L'Hermite (eds.), pp.168–176. Proceedings of Round-Table Seminar on Efficient Land Use of Sludge and Manure, Commission European Communities Environmental Research Program, held in Brorup-Askov, Denmark, June 25–27, 1985. New York: Elsevier Applied Science Publication.

Stone, M. L., J. B. Solie, W. R. Raun, R. W. Whitney, S. L. Taylor, and J. D. Ringer. 1996. Use of spec-tral radiance for correcting in-season fertilizer nitrogen deficiencies in winter wheat. *Trans. ASAE* 39:1623–1631.

Strock, J. S., C. J. Dell, and J. P. Schmidt. 2007. Managing natural processes in drainage ditches for nonpoint source nitrogen control. *J. Soil Water Conserv.* 62:188–196.

Strock, J. S., P. J. A. Kleinman, K. W. King, and J. A. Delgado. 2010. Drainage water management for water quality protection. *J. Soil Water Conserv.* 65:131A–136A.

SWCS (Soil and Water Conservation Society). 2003. Conservation implications of climate change: Soil erosion and runoff from cropland. A report from the Soil and Water Conservation Society. Ankeny, IA: SWCS.

SWCS (Soil and Water Conservation Society). 2007. Planning for extremes: A report from a Soil and Water Conservation Society workshop held in Milwaukee, Wisconsin, November 1–3, 2006. Ankeny, IA: SWCS.

Thomas, M. A., B. A. Engel, and I. Chaubey. 2011. Multiple corn stover removal rates for cellulosic biofuels and long-term water quality impacts. *J. Soil Water Conserv.* 66:431–444.

Thorne, C. R., L. W. Zezenbergen, E. H. Grissinger, and J. B. Murphey. 1986. Ephemeral gullies as sources of sediment. In *Proceedings of the 4th Federal Interagency Sedimentation Conference,* Vol. 1, pp. 301–309. Washington, DC: Government Printing Office.

Tomer, M. D. 2010. How do we identify opportunities to apply new knowledge and improve conservation effectiveness? *J. Soil Water Conserv.* 65:261–265.

Tomer, M. D., D. E. James, and T. M. Isenhart. 2003. Optimizing the placement of riparian practices in a water-shed using terrain analysis. *J. Soil Water Conserv.* 58:198–206.

Tomer, M. D., T. B. Moorman, J. L. Kovar, D. E. James, and M. R. Burkart. 2007. Spatial patterns of sediment and phosphorous in a riparian buffer in western Iowa. *J. Soil Water Conserv.* 62:329–338.

Tomer, M. D., S. A. Porter, D. E. James, K. M. B. Boomer, J. A. Kostel, and E. McLellan. 2013. Combining precision conservation technologies into a flexible framework to facilitate agricultural watershed plan-ning. *J. Soil Water Conserv.* 68:113A–120A.

Tribe, D. 1994. *Feeding and Greening the World, the Role of Agricultural Research.* Wallingford, UK: Commonwealth Agriculture Bureau.

Tucker, C. J. 1979. Red and photographic infrared linear combinations for monitoring vegetation. *Remote Sens. Environ.* 8:127–150.

Udawatta, R. P., P. P. Motavalli, and H. E. Garrett. 2004. Phosphorus loss and runoff characteristics in three adjacent agricultural watersheds with claypan soils. *J. Environ. Quality* 33:1709–1719.

USDA, National Agricultural Statistic Service (NASS), Census of Agriculture. 2009. *2008 Farm and Ranch Irrigation Survey*, Table 4.

USGAO (United States Government Accountability Office). 2013. Clean Water Act: Changes needed if key EPA program is to help fulfill the nation's water quality goals. Washington, DC: GAO.

Vadas, P. A., M. S. Srinivasan, P. J. A. Kleinman, J. P. Schmidth, and A. L. Allen. 2007. Hydrology and ground-water nutrient concentrations in a ditch-drained agroecosystem. *J. Soil Water Conserv.* 62:178–188.

Walthall, C. L., J. Hatfield, P. Backlund et al. 2012. Climate change and agriculture in the United States: Effects and adaptation. USDA Technical Bulletin 1935. Washington, DC: USDA-ARS.

Williams, J. D., D. S. Long, and S. B. Wuest. 2011. Capture of plateau runoff by global positioning system–guided seed drill operation. *J. Soil Water Conserv.* 66:355–361.

Wood, G. A., G. Thomas, and J. C. Taylor. 1999. Developing calibration techniques to map crop variation and yield potential using remote sensing. In *Proceedings of the 4th International Conference on Precision Agriculture*, P. C. Robert, R. H. Rust, and W. E. Larson (eds.), pp. 1367–1378. Madison, WI: ASA.

Xin-Zhong, W., L. Guo-Shun, H. Hong-Chao, W. Zhen-Hair, L. Qing-Hua, L. Xu-Feng, H. Wei-Hong, and L. Yan-Tao. 2009. Determination of management zones for a tobacco field based on soil fertility. *Comput. Electron. Agric.* 65:168–175.

Zhang, Y., M. Hernandez, E. Anson, M. A. Nearing, H. Wei, J. J. Stone, and P. Heilman. 2012. Modeling cli-mate change effects on runoff and soil erosion in southeastern Arizona rangelands and implications for mitigation with conservation practices. *J. Soil and Water Conserv.* 67:390–405.

5 Smallholder Management of Diverse Soil Nutrient Resources in West Africa
Economics and Policy Implications

Ezra D. Berkhout, Angelinus C. Franke, and Tahirou Abdoulaye

CONTENTS

5.1 INTRODUCTION

"The high efficiency of low external inputs on crop production in semi-arid West Africa is for an important part the result of precision farming" (Brouwer and Bouma 1997). On one hand, chemical fertilizers are expensive, discouraging their use, while on the other hand, many soils are poorly responsive to higher doses of fertilizer use. Brouwer and Bouma (1997) thus posit that what is purchased is applied in a highly efficient manner. In this chapter we further expand on this premise and illustrate that their argument does not stand in isolation. West African smallholder farmers practice the art of closely matching scarce resources, mainly soils, labor, or fertilizer, so that the net gains are highest. The empirical literature reviewed in this chapter suggests that many smallholders are particularly good at this form of precision agriculture. For many, low use of external inputs within their specific contexts and low production levels are an efficient strategy, an insight that echoes the "poor but efficient hypothesis" originally introduced by Schultz (1964).

There exists widespread agreement on the need to raise agricultural productivity in West Africa. Yields are well below their theoretical potential and attainable levels (Pingali and Heisey 1999; van Ittersum et al. 2013) and a combination of adequate technologies and policies are needed to enhance production (Ruben et al. 2001, 2007). There equally exists a clear understanding that past blanket interventions have been largely unsuccessful due to the lack of incorporating heterogeneity (IAASTD 2009). Heterogeneity exists at multiple levels: at the country level, implying comparative advantages between countries and regions, but also at the smallest field level. Such within-farm variability in soil fertility, or the variability between farms within a small geographic area, is sometimes greater than the mean variation across districts (Poulton et al. 2006; Brouwer cited in Poulton et al. 2006). As a result, responses to new technologies and fertilizer differ across fields, with the least fertile fields often being unresponsive. Some recent studies suggests that the poorest households more frequently own such fields (Marenya and Barrett 2007). For them, use of fertilizer or other inputs, such as labor for timely crop management, remain economically unattractive under current conditions (Tittonell and Giller 2013).

On one hand, technological development is slowly paying attention to such on-farm diversity. On the other hand, much of the current fertilizer policies primarily aim to increase its use, ignoring much of the on-farm diversity and largely bypassing the poorest groups. For many smallholders, low output from agriculture cannot be attributed to inefficiently low levels of external input use; rather, as we discuss in this chapter, it is a result of the inefficiency of current input use. What is thus needed are policies and technologies that facilitate a diverse group of farmers to better match their soil characteristics with more appropriate inputs and technologies, with a long-term goal of enhancing the production capacity of their soils. Thus, stimulating input use efficiency also provides for a sound base to profitably expand input use.

In this chapter we further build the foundation for, and elaborate on, these insights. In Section 5.2 we introduce a theoretical economic framework to understand the driver for production decisions, particularly the rate by which fertilizer is used, and how these are influenced by wider socio-economic conditions. We also discuss the impact of variation in production, mostly due to weather conditions, and how this affects the efficient input level and the speed by which farmers are able to discern it.

In Section 5.2.1 we turn to reviewing the empirical literature, investigating the rates by which farmers apply fertilizer endogenous to field-level plot characteristics and how this impacts crop yields. Very few studies have investigated this proposition in detail in larger samples. We draw in some detailed studies from East Africa to illustrate how substantial variation in fertilizer use among smallholders signals underlying variation in its returns.

As argued, these findings have major implications for the development of technologies and policies, to which we turn our attention in Section 5.2.2. We review and evaluate a diverse set of policies including fertilizer subsidies and evaluate how these affect the efficiency by which farmers use fertilizers. In the second half of this section we turn our attention to novel policies, particularly ones that could spur the design and production of different fertilizer blends as well as those that could enhance the uptake of integrated soil fertility management techniques. Further design and ground-testing of such novel policies requires close cooperation between governments, research organizations, and the private sector, something we call for in Section 5.2.3.

5.2 A FRAMEWORK TO MODEL FARMER DECISION-MAKING WHEN SOILS ARE DIVERSE

We introduce a framework that allows us to investigate production decisions of a typical rural African farming household. Our discussion largely follows standard microeconomic theory (Varian 1993; Bardhan and Udry 1999) with a focus on diverse soil fertility levels across fields and its implication for optimal levels of fertilizer use. First, in Section 5.2.1, we introduce the analysis at the field level, aiming to understand primary production relations. Second, in Section 5.2.2, we extend these insights by incorporating the wider economic and institutional environment in which farmers operate. In Section 5.2.3 we provide a brief overview of the theory related to processes of learning-by-doing, with a focus on learning optimal levels of fertilizer use, and in Section 5.2.4 we discuss the empirical estimation of heterogeneous responses to fertilizer use.

5.2.1 Field-Level Production Economics

Crop yields respond to fertilizer as captured by a standard concave production function with diminishing returns to increasing fertilizer application rates. Figure 5.1 depicts a maize grain response to nitrogen fertilization in northern Nigeria, showing an initial steep response to fertilizer application. Crop yields become less responsive, eventually flattening out to a yield of 1.15 Mg ha^{-1}. This is a Von Liebig production relationship, whereby crop yield increase when approaching the horizontal asymptote is no longer limited by nitrogen.

Mathematically, crop yield y_a can be described as a function of fertilizer input x, giving $y_a(x)$ (Figure 5.2). This function assumes all other relevant inputs are fixed. The maximum yield in the function equals y_a^*. At this point an overall increase in crop yield can only be achieved by relieving constraints other than those tackled by fertilizer application (e.g., by increasing the supply of other nutrients or water, or protecting the crop against biotic pressures). This underlies the difference with $y_b(x)$, the second production function shown in Figure 5.2. An equally valid interpretation of the difference between $y_a(x)$ and $y_b(x)$, and more useful within the scope of our investigation, is the depiction of crop yield response in two fields, field a and field b, differing in soil fertility, through differences in soil organic matter, soil structure, and the availability of macro- and micronutrients. Marginal responses to equal amounts of fertilizer may well differ across both fields, suggesting slopes may differ next to different maximal production ceilings y_a^* and y_b^*.

We use Figure 5.2 to illustrate the most efficient level of fertilizer at a field and how and why this typically differs between fields. An important assumption is that farmers' decisions are guided by a single objective, namely maximizing field-level profits, something we discuss in Section 5.2.2. Economic theory provides that the most optimal level of fertilizer use occurs when the marginal

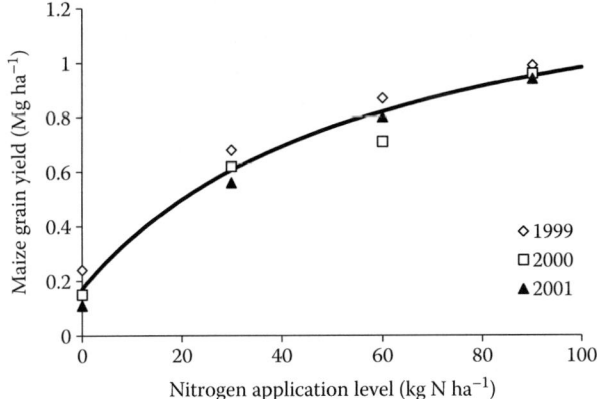

FIGURE 5.1 Response to nitrogen fertilizer application (with a blanket application of phosphorus and potassium) on degraded soils (shallow, stony outer fields) in northern Nigeria (average of eight sites in southern Kaduna State) over 3 years. The highest economic returns to nitrogen fertilizer were realized at around 30 kg N ha^{-1}, well below recommended fertilizer rates. (From Franke et al., *Experimental Agriculture* 40: 463–79, 2004.)

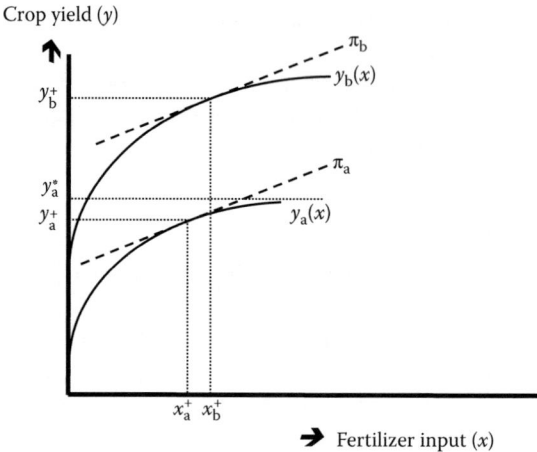

FIGURE 5.2   Identifying the most efficient level of fertilizer use.

returns equal the marginal costs. The most efficient level of fertilizer use is thus guided by the shape of the production function, the unit price of the crop p_y, and the unit price of fertilizer p_x. In Figure 5.2, the most efficient levels of fertilizer use are x_a^+ or x_b^+ for production functions $y_a(x)$ and $y_b(x)$, respectively. Below we provide the intuitive explanation to these facts. We refer the reader to a microeconomics textbook (e.g., Varian 1993) for its mathematical exposition.

The two dashed lines, π_a and π_b, tangent to the production functions are isoprofit lines. They are based on combinations of fertilizer input x and crop output y that provide the same level of profit. The first-order derivatives of these isoprofit lines equal the ratio between the input and the output price, $\frac{p_x}{p_y}$. Moreover, one isoprofit line placed above another in the figure indicates larger profits, logically indicating that profits for any given level of input are higher on the more fertile field b. Hence, the maximization problem a farmer faces has the graphical equivalent of finding the highest isoprofit curve within the production possibility of each field. This point is reached when the production function and the isoprofit curve are tangent to one another, at levels x_a^+ and x_b^+, respectively. Now consider a farmer who applies an input level $x < x_a^+$. The associated profit would be the line, tangent to π_a and π_b, which cuts through the point $(x, y_a(x))$. Since this line will lie below π_a, the profit level it reflects is also lower than π_a. Hence, a farmer can safely increase fertilizer application to x_a^+ and simultaneously increase field-level profits to π_a. The reverse argument holds for the case when $x > x_a^+$. At these levels fertilizer use is excessive, and the associated profit levels are again lower than the maximal π_a.

The efficient levels of input use change when prices do. An increase in input prices implies a reduction of its use when, *ceteris paribus*, the marginal benefits do not change. Conversely, an increase in the output price warrants, *ceteris paribus*, an increased used of inputs. *Ceteris paribus* implies that all other production factors, including inputs such as organic fertilizers and labor, remain unchanged. Most often this does not hold, and a price change leads farmers to reassess all inputs being used, thereby accounting for substitution and complementary effects.

The efficient levels of fertilizer use, x_a^+ and x_b^+, on field a and b only coincide when functions differ through a one-to-one upward shift. It is widely understood that higher levels of soil organic matter combined with inorganic fertilizers offer synergetic effects. Such a synergy would be captured by an outward shift of $y_b(x)$ relative to $y_a(x)$ similar to the one shown in Figure 5.1, and not by a one-to-one upward shift.

The implication is that optimal inorganic fertilizer use x_b^+ is typically larger at the more fertile fields. This theoretical insight provides an empirical test for efficiency among smallholders. More

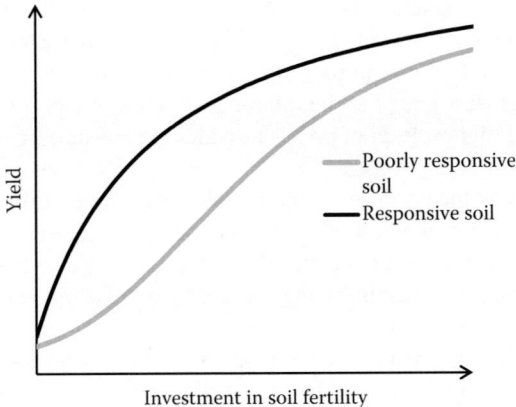

FIGURE 5.3 Theoretical yield response to investments in soil fertility of a responsive and a poorly responsive soil.

efficient farmers apply more inorganic inputs on their more fertile fields (i.e., those fields that display stronger marginal responses to inputs). A more practical implication is that a blanket recommendation of fertilizer use only rarely coincides with optimal field-level fertilizer use x_a^+ or x_b^+. The latter is, of course, not problematic if the blanket recommendation is sufficiently close to optimal field-level use, implying minor variation in soil fertility and fertilizer responses across fields. This, however, appears not to hold in many instances across West Africa, something we return to in Section 5.3.

Soils that have no or a very poor response to fertilizer applications despite the application of recommended agronomic management are often referred to as nonresponsive soils. The sites underlying Figure 5.1 were poorly responsive to nitrogen applications given that attainable maize yields in this area equal 5–7 Mg grain ha^{-1}, while observed yields leveled off approaching 1.15 Mg ha^{-1} despite considerable nitrogen, phosphorus, and potassium applications. Soils may be nonresponsive because of soil chemical, physical, or biological limitations such as missing nutrients not supplied by the fertilizer, shallow soils, water logging, low pH, soil toxicities, soilborne diseases, and so forth. Although such soils can often be made *responsive* with appropriate management, this requires sizeable and prolonged investments in soil fertility, for instance repeated applications of organic inputs, liming, or deep plowing. While a responsive soil is expected to give a standard concave production function in response to investments in soil fertility, the production function of a nonresponsive soil may be better captured by an asymmetric S shape (Figure 5.3). An S-shaped response curve is unlikely to occur when considering a single nutrient input as in Figure 5.1, but if multiple interacting inputs are required for a response, an S-shaped response curve is well possible. The slope of the S curve is rather flat at low levels of investments, indicating a poor return on these investments. Farmers on poorly responsive soils who are unable to sufficiently invest in soil fertility to reach the steeper part of the S curve continue to receive poor returns to their investments and are thus caught in a soil-fertility-based poverty trap (Tittonell et al. 2013). Very fertile soils may also be nonresponsive to investments in soil fertility, but this is not often the case among smallholders.

5.2.2 Household-Level Production Economics

The key assumption underlying the insights presented in Section 5.2.1 is that farmers maximize profits. African smallholders, however, only rarely do so. To understand why profit maximization is not the dominant production objective of most African smallholders (Singh et al. 1986; Bardhan and Udry 1999), one needs to realize that African farm households are both producers and consumers. Consumption and production decisions in smallholder households can only be considered to be taken independent from each other if farmers face near-perfect input and output markets.

Let us first consider the production side. A "pure" producer, as we assumed in Section 5.2.1, is interested in achieving the highest level of profits. Under perfect market conditions, a producer can always vary the input and output mix in order to select the profit-maximizing strategy. We thereby not only consider inorganic fertilizer, but the full range of relevant inputs including labor and capital as well. Subsequently, the highest level of profits provides the consuming household with the largest budget to purchase consumption goods.

Consumer theory rests on the premise that households maximize a utility function conditional on a budget constraint. The utility function reflects consumer preferences and includes food items, but often also goods such as leisure. As mentioned, the budget constraint is based on the profit achieved in farm production, and since it is maximal, the achieved level of utility will also be the highest level possible given a farmer's resources.

Even though this framework treats production and consumption decisions independently, farmers may still consume their own farm produce, just as they consume part of their own labor endowment as leisure. The key point, however, is that when markets are available and function perfectly, smallholders make these decisions based on common market prices, and consumption decisions are not influenced by production activities, and vice versa.

However, this assumption frequently breaks down. Instead of equating the marginal value of product to marginal costs, most farm households equate marginal value of product to marginal utility (e.g., Singh et al. 1986; De Janvry et al. 1991; Bardhan and Udry 1999), which reflects the existence of transaction costs, driving a wedge between consumer and producer prices. These costs not only include direct transport and marketing costs, but also costs associated with weak institutions such as a lack of law enforcement, ill-functioning labor or credit markets, or price asymmetries. High transaction costs lead to a market failure with the trade in a specific product ceasing by some groups.

Virtually all rural smallholders in the developing world face market imperfections of some sort (see Bardhan and Udry 1999 for an overview of the literature). The existence of transaction costs, sometimes substantial, in Africa is well documented (Omamo 1998; Renkow et al. 2004) and tend to be more severe in more isolated areas (e.g., Moser et al. 2009). There, farmers are unable to reap the benefits of lower world market prices for major food crops and forego opportunities to produce more remunerative products. These, in turn, are valued locally at prices lower than true market clearing prices. Demand and supply for agricultural products is mostly inelastic in isolated areas, with farmers unable to fully respond to opportunities created by price changes (Barrett 2008b). Transaction costs are typically lower in well-connected regions, but could still be severe for particular groups. Here, vulnerable groups could still face major difficulties in accessing labor or credit markets. The exact response to market imperfections differs across regions and households (De Janvry et al. 1991; Barrett 2008b) and how, for instance, the use of inorganic fertilizer is affected remains ambiguous. Farmers equate internal shadow costs and prices, which could exceed or fall below local market prices, determining whether they are net buyers, net sellers, or remain autarkical.

Consider the presence of transaction costs of inorganic fertilizer use. If a household is a net seller of an output, a reduction in the transaction costs makes inorganic fertilizer less costly, but the effect on output remains ambiguous when ignoring complementary markets, for example, for labor. If they are substitutes, farmers may expand production and substitute scarce expensive labor for cheaper fertilizer. When the substitution effect is sizeable, farmers could well switch to become net buyers of a product and reallocate labor to more remunerative activities. Such an effect would, however, not be observed if fertilizer and labor are complements, for example, to suppress increased weed growth or to apply fertilizer precisely. Market imperfections equally affect farmer use of organic fertilizers. The most vulnerable groups, if excluded from labor markets, possess very high opportunity costs of labor, limiting the use of labor-intensive manure and compost.

Finally, farmers' decisions are influenced by risk aversion or environmental objectives (e.g., Berkhout et al. 2010) and heterogeneity in such objectives could result from market imperfections. Economic theory predicts that increases in risk aversion displayed by farmers lower the efficient rate of application (e.g., Hardaker et al. 1997), something that has often been cited as an explanation for low fertilizer use in the (semi-) arid regions of Africa (e.g., Morris et al. 2007). Furthermore, market imperfections are considered to aggravate price risks (Barrett 2008b); however, how risk and risk aversion are exactly incorporated in day-to-day decision-making by smallholders remains to be fully understood (Just et al. 2010).

5.2.3 LEARNING ABOUT THE MOST EFFICIENT LEVEL OF FERTILIZER USE

So far, we assumed farmers to possess full information on the shape of the underlying production curves on each field. In other words, farmers can assess with close precision the most efficient levels of input on each field. Such an assumption may occasionally be accurate for industrialized European or U.S. agriculture, with precise computer monitoring of returns to input use even within a field, but this does not hold for African smallholders. Instead farmers probably have an imprecise idea of the shape of the production function, whereby each cropping season generates additional information that allows them to update their beliefs about the return to fertilizer and underlying production relationships (e.g., Conley and Udry 2010 for a review on the empirical literature on this topic). This process is commonly called learning-by-doing or Bayesian learning. We briefly present some key insights into this concept, with particular focus on learning about optimal fertilizer use. We refer the interested reader to Jackson (2008) for a more formal exposition of the mathematics behind Bayesian learning.

If, and how quickly, a farmer is able to discern the exact impact of an input, say fertilizer, on output roughly depends on two factors. First, the shape of the production function plays a role; specifically, the magnitude of output changes resulting from input changes. Next, one needs to consider exogenous factors such as the weather or biotic stresses such as pests and diseases. If the variation in rainfall is high and the effect of fertilizer use on output is only small, it becomes peculiarly difficult for a farmer to discern the impact of fertilizer use on crop yields. Rainfall levels swing from high to low over the seasons, and so will crop yields, irrespective of fertilizer use. A farmer primarily attributes increases in output to more favorable weather conditions and requires a long string of observations to discriminate between the small effects caused by fertilizer applications and the larger weather effects. Before a farmer reaches this point, however, he or she may already have abandoned the idea of using fertilizer even though its application could have been marginally profitable. The opposite situation occurs when the variation in rainfall is only small and the effect of fertilizer use on output is more pronounced. Then, a farmer quickly observes the yield-increasing effects of fertilizer, since yield variation is much less influenced by variations in rainfall.

Farmers may thus refrain from using fertilizer on some or all of their fields because they never identify the profitable level of fertilizer use. Such an explanation could be particularly realistic on the most marginal fields, where returns to fertilizer are intrinsically small and the effect of rainfall variation more pronounced (e.g., due to less water-holding capacity due to lower soil organic matter content).

A learning farmer could benefit from favorable social architectures and institutions that facilitate social learning, which is an extension of the concept of learning-by-doing. If the learning farmer has access to observations from others producing under comparable conditions and/or the learning farmer has full information on input differences, these observations equally serve the learning farmer to update his or her beliefs. We refer to other studies (Bandiera and Rasul 2006; Conley and Udry 2010; Barham et al. 2014) for empirical examples on social learning.

Finally, fertilizer response curves are likely to vary from season to season on a single field due to changing weather conditions, variable biotic pressures, build-up or degradation of soil fertility, and so forth. Figure 5.4 illustrates the wide seasonal variability of maize crop response to nitrogen (N) applications in response to rainfall variation and a buildup of soil phosphorus reserves on an initially highly phosphorus deficient soil in northern Nigeria. It serves to illustrate the difficulty a farmer faces in distinguishing between yield increases caused by fertilizer application and those

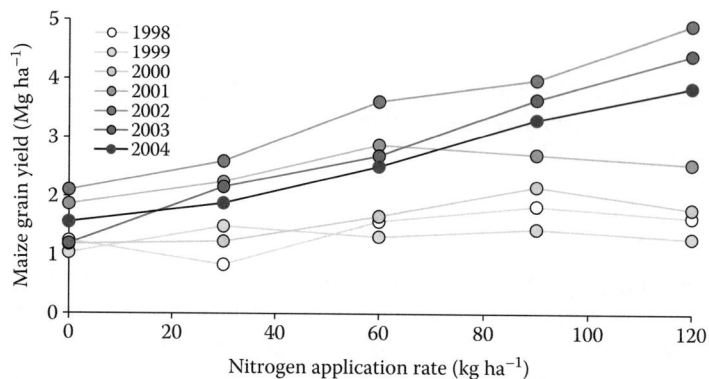

FIGURE 5.4 Seasonal variation in grain yield response to nitrogen fertilizer applications in a single-site trial with maize under monoculture in northern Nigeria (Zaria) over 7 years. (After Franke et al. 2008a). Potassium was annually applied at a rate of 33 kg ha^{-1}, phosphorus at 20 kg P ha^{-1} in 1998–2001, and increased to 40 kg P ha^{-1} and 28 kg S ha^{-1} in 2002–2004.

caused by increased rainfall. Moreover, the marginal response to fertilizer differs under different rainfall regimes, also implying that the efficient level of fertilizer application differs from year to year. Observe the differences in yields between Figures 5.1 and 5.4, both displaying research findings from the same region, further illustrating the influence of small-scale differences in plot characteristics on crop yields.

5.2.4 EMPIRICAL CONSIDERATIONS WHEN ACCOUNTING FOR DIVERSITY IN SOIL FERTILITY

The rate by which smallholders apply fertilizer is an endogenous choice conditioned by, among other factors, field-level soil fertility, household characteristics, and variables reflecting market imperfections. Moreover, the marginal responses of fertilizer use are heterogeneous across different types of fields. The true impact of these factors, particularly (1) by how much inorganic fertilizer and field-level soil fertility contribute to crop yield and (2) by how much field-level soil fertility explains actual use of inorganic fertilizer, often goes unnoticed in empirical economic studies investigating smallholder decisions.

Such studies using econometric techniques on larger samples often do not control for variation in soil fertility and explain crop yields solely from physical and observable inputs. Many estimation results on the marginal effects of fertilizer application are therefore biased, and most likely overestimate its true impact. Farmers may also decide to apply fertilizers only on some fertile fields, which need to be accounted for to avoid selection bias. Furthermore, many studies implicitly assume that the marginal fertilizer responses are homogenous across a sample, which again seems unrealistic given the information in previous sections. The severity of such biases depends in part on the scope of a study. For instance, the homogeneity assumption will, with everything else being equal, yield an unbiased estimate for the average marginal response to fertilizer across the sample but be biased for individual observations. It depends on the specific focus of the researchers whether this is problematic, but it will often be when the focus rests on developing an effective policy intervention to stimulate fertilizer use among smallholders.

Empirical statistical studies that aim to understand smallholder endogenous fertilizer application and its impact on crop production should therefore not rely on common ordinary least squares (OLS) techniques. As we discuss below, Suri (2011) finds substantial discrepancy between common OLS estimates and econometric techniques that control for heterogeneous responses. Similarly, using techniques from spatial econometrics, Florax et al. (2002) suggests OLS or instrumental variable

(IV) regressions to be largely inadequate for understanding endogenous farmer adoption decisions in the midst of ubiquitous heterogeneity in soil responses to technologies. Only a handful of studies have addressed the link between soil fertility, endogenous fertilizer application rates, heterogeneous marginal responses, and crop yield explicitly. These form the base of our review in the next section.

5.3 REVIEWING THE LITERATURE

Empirical studies have only recently begun to investigate heterogeneous use of fertilizer across smallholders in Africa in larger samples. Their overall number remains limited; only few focus on West Africa, and nearly all originate from Kenya. Given the dearth of studies from West Africa and the important insights that the Kenya studies deliver on farmer responses to diverse soil fertility responses, we will discuss and compare these in our review.

A group of companion papers (Florax et al. 2002; Gandah et al. 2003a,b; Voortman et al. 2004) provide detailed insights in on-farm diversity in soil fertility resources, farmer nutrient management practices, and millet yields in Niger. Millet yields are well explained (81%) by spatial diversity in soil properties and application of macronutrients, particularly nitrogen, something which is only feasible on the younger cover sands. Crop yields on these soils are already higher and have more favorable soil chemistry (Florax et al. 2002). On the marginal older cover sands, inorganic fertilizer application makes little sense (Florax et al. 2002; Voortman et al. 2004).

Millet grain yields, soil carbon, and soil phosphorus decline significantly with the distance to the homestead. High transport costs and risk of animal theft prevent farmers from using the little manure they have available on more distant fields (Gandah et al. 2003b). Such soil fertility gradients are also widely observed across Africa (e.g., Tittonell et al. 2005; Zingore et al. 2007; Tittonell and Giller 2013). In some instances Nigerian farmers incorporate crop residues into the top soil. Women thresh millet in the shade of trees and choose different trees in different seasons (Gandah et al. 2003b). Acacia species (*Faidherbia albida*), as pictured in Photograph 5.1, provide shade to livestock in the dry season but discard leaves in the raining season. Light competition is minimal, while soils benefit from nutrients supplied by urine and manure (Brouwer and Bouma 1997). These papers not only illustrate that soil conditions change markedly across small distances, but that farmers adopt nutrient management strategies that explicitly incorporate this diversity. Such findings also emerge from other studies, most notably a few from Kenya.

PHOTOGRAPH 5.1 Millet crop yield gradient. Millet yields are higher directly under the Acacia tree (*Faidherbia albida*) due to nutrient deposits from leaves, manure, and urine (Sahelian Agro-Ecological Zone, Niger).

Two studies disentangle the relation between diverse soil fertility resources, particularly organic matter content, fertilizer use, and maize yields (Marenya and Barrett 2009a,b). The authors explicitly acknowledge that fertilizer use is an endogenous choice based on field soil fertility, and a standard ordinary least squares regression of fertilizer use on maize yields will produce inconsistent and biased insights in the on-farm response to fertilizers. The authors circumvent the econometric bias and find that on roughly one-third of the fields, strongly degraded and carrying low organic carbon content, marginal productivity of fertilizers are very low, rendering its use unprofitable. Indeed farmers mostly refrain from fertilizer application on such fields. Photograph 5.2 illustrates this point, depicting strong heterogeneity in response to P-fertilizer, even within a single field.

In a related study, the authors point out that the most marginal and degraded fields are most often owned by the poorest farmers (Marenya and Barrett 2007). They postulate that findings in previous studies, attributing differences in fertilizer use to differences in wealth, most probably reflect a spurious relationship. The authors argue that the real explanation of low fertilizer use rests with differences in field characteristics across farmers in different wealth categories. Suri (2011) reaches a similar conclusion, albeit not accounting for soil fertility differences explicitly, after estimating an econometric model that explicitly accounts for heterogeneous responses to inputs. The estimation results show that hybrid seed and fertilizer adoption decisions by farmers are fully rational, with only those smallholders adopting for whom net returns are positive.

Similar variation in returns is documented in West Africa. Beaman et al. (2013) conduct a policy experiment among a large sample of rice farmers in Mali. Rather than observing farmer use of fertilizer, free fertilizer handouts are provided to a randomly selected group of farmers. Since this provision is not associated with preexisting differences in soil fertility, selection effects cancel out. As expected, the provision of free fertilizers leads to a substantial increase in fertilizer use, while complementary inputs such as labor and herbicides are crowded in. While the value of outputs increases significantly, profits do not when the costs of fertilizers are factored in. The variation in profits across the sample is large, with a 90% confidence interval on the rates of return ranging from −0.48 to 0.46. Hence, for a substantial portion of farmers, fertilizer use at the rates they applied in the experiment is unprofitable. Although not explicitly investigated, it is likely that in part the variation in returns is explained by the variation in field-level soil fertility.

Duflo et al. (2008) illustrate that official fertilizer recommendations in Kenya are too high for many smallholders. Basing their insights on an experimental design with random subsets receiving fertilizer or serving as a control, they equally observe that for 13.5% of the farmers in their sample any quantity of inorganic fertilizer use is economically unattractive. Duflo et al. (2008) do not take

PHOTOGRAPH 5.2   Soybean crop yield gradient. Within-field variability in soybean yield (Tropical Forest Agro-Ecological Zone, Sierra Leone).

explicit account of differences in field-level fertility (although they do control for household differences), making it impossible to link the variation in returns to such factors. The authors suggest that negative experiences with the officially recommended rate could be one reason for the overall low levels of fertilizer use. In a companion paper (Duflo et al. 2011), the authors investigate a behavioral economics explanation based on impatience and suggest that targeting farmers with the sale of fertilizer vouchers right after the crop harvest could boost fertilizer use in subsequent seasons.

More studies suggest that official fertilizer recommendations are too high, including in countries in West Africa. In northern Nigeria (Figure 5.1), for example, economically optimum fertilizer rates are much lower than recommended fertilizer rates (90–120 kg N ha^{-1}). This insight is equally observed by Sheahan, Black, and Jayne (2013), albeit again in Kenya. Using nationwide data they estimate profitability of fertilizer use across various regions. They equally control for differences in soils, including variables to reflect major soil types, but do not find any significant interaction between soil types and nitrogen application in fertilizers. Next to the possibility that interactions indeed may not exist, the authors hypothesize that the soil classifications included are too broad and do not capture the more relevant local diversity in soils in villages and households.

Across the board, most farmers apply fertilizer at rates close to economic optima. Some room for optimization still exists. Farmers in the high-potential western highlands could increase profits by reducing fertilizer use, suggesting other nutrients have become limiting, while there still appears to be room for expanding fertilizer use in the arid lowlands. Given the near economic optima of application rates, the authors concur with Marenya and Barrett (2009a,b) that policies to expand fertilizer use are probably ineffective if other aspects of soil fertility (like increasing organic carbon content) are not also moved along.

Many African smallholders may use little inputs, but these are still efficient, an insight originally posited by Schultz (1964). This review shows that many farmers follow a form of precision agriculture, closely matching field- and farm-level use of fertilizer and other inputs such as labor with the actual marginal returns. Farmers understand the diversity of soil resources well and how this affects crop yields. Such insights are also documented in studies from other regions in Africa (Kimiti et al. 2007; Mairura et al. 2007). Moreover, they manage limited resources in efficient nutrient management strategies. The studies reviewed suggest that smallholders apply larger quantities of inorganic fertilizer on more fertile soils due to strong complementarities, small supply, and high transport costs to outward fields. In both West (Florax et al. 2002) and East Africa (Sheahan et al. 2013) researchers found that within the current set of constraints on the various soil types, only limited opportunities to increase the efficiency of inorganic fertilizer use exist. Application of nutrients, both from organic and inorganic sources, is minimal or nonexistent on poorer soils (Marenya and Barrett 2009a,b), and some evidence exists that such fields are more frequently owned by the poorest households (Marenya and Barrett 2007).

The existence of poverty traps perpetuated by a weak soil fertility base has been hypothesized both in the economic literature (Antle et al. 2006; Barrett 2008a) and the agronomic literature (Tittonell and Giller 2013). It rests on the assumption that multiple equilibria exist with regard to investments in the soil fertility base of a field. If the economic productivity on a field drops below a certain threshold, investment in long-term maintenance of soils becomes inefficient and maximizing short-term production through nutrient mining becomes economically efficient. Conversely, fields above the threshold warrant sustained investment in its soil fertility base. The first, and still limited, insights reviewed above suggest the existences of poverty traps, but more research is still desired, particularly in connection to policy interventions that could support smallholders out of these poverty traps (see Section 5.2.3).

Most of the studies reviewed did not explicitly include price differences. That is not to imply that prices do not play a role. In the developed world a substantial increase in the practice of precision agriculture has been attributed to rising prices of fertilizer (Harris et al. 2008), encouraging farmers to cut excess and wasteful use of inorganic fertilizers. At the same time, some argue that the low prices of major crops in part explain the lack of productivity growth in agriculture in Africa (Fuglie

2010), while higher transport costs to remote regions partly explain the low use of inorganic fertilizers (Jayne et al. 2003).

Thus, costs and benefits of fertilizer application, either market-based or internal opportunity costs, are what constitute the prime incentives for farmers to use fertilizers to replenish their soils, indirectly increasing the efficiency of current fertilizer use, and to increase agricultural output. This, in principle, opens up mechanisms for policy makers to influence African smallholder production decisions and if well-tailored, such policies can enhance investment in soils and sustainably spur agricultural development. Nevertheless, it is not intuitively clear which strategies are most effective, especially considering that varying marginal responses to fertilizer are ubiquitous. Whether the poorest farmers, possibly caught in low soil fertility poverty traps, can be assisted out of their precarious situation deserves specific attention.

5.4 SPURRING AGRICULTURAL DEVELOPMENT AMIDST UBIQUITOUS DIVERSITY IN SOIL RESOURCES

In the remainder of this chapter we discuss potentially effective interventions, with a focus on enhancing the resource efficiency of African smallholders. In Section 5.4.2 we discuss the room for expanding the use of inorganic fertilizer, while Section 5.4.3 discusses ways to enhance efficiency of inorganic fertilizer use. Section 5.4.4 focuses on mechanisms that can be used to encourage replenishment of soil fertility, particularly soil organic matter. First, however, we review a number of macro-level interventions that ubiquitously improve the incentives for all smallholders.

5.4.1 CREATING AN ECONOMICALLY STABLE PRODUCTION ENVIRONMENT

Boserup (1965) posited a theory of agricultural intensification and outlined how population growth and resulting land scarcity invoke practices of agricultural intensification. Maintaining the soil fertility base of land under cultivation becomes economically efficient when opportunities for land expansion and/or shifting cultivation are exhausted. Infrastructural development thereby aids the intensification process by providing affordable access to necessary inputs.

Such associations are observed in the West African subregion (Abdoulaye and Lowenberg-DeBoer 2000; Harris and Yusuf 2001; de Ridder et al. 2004; Aune and Bationo 2008). Infrastructural development provides a much-needed stimulus to intensify production, partially through a reduction in transaction costs. It is estimated that 15% of the retail price of inorganic fertilizer in northern Nigeria represents domestic transport costs (Gregory and Bumb 2006). These costs may be substantially higher in the landlocked countries of West Africa. Furthermore, infrastructural improvements may lead to a more timely supply of fertilizers in production areas (Yanggen et al. 1998).

Improvements in rural infrastructure have been found to be an effective mechanism to stimulate agricultural development in China and Asia (Fan and Zhang 2004; Zhang and Fan 2004). Better roads reduce transaction costs and close the wedge between farmgate and world market prices with welfare effects accruing to both vulnerable and wealthier households. Net sellers benefit from higher prices of marketed surplus and net consumers are able to purchase subsistence crops at lower prices and shift their own productive capacity to goods for which they display comparative advantages. This could entail a shift toward the production of high-value, and often perishable, crops such as vegetables. In both cases the underlying value of land would increase, making investments in the soil fertility base more remunerative, while better infrastructure tends to reduce price risks.

A number of early studies documented the relevance of market imperfections on smallholder decisions in Nigeria, Sierra Leone, and Senegal (Singh et al. 1986; Goetz 1992). Insights from this region remain scarce, although there is no reason to believe transaction costs and market imperfections are markedly different in West Africa as compared to other regions. The effects on investments in soils, however, remains ambiguous.

For well-connected regions, the final step on the ladder of agricultural intensification could logically be the substitution of labor-intensive organic manure for affordable and labor-extensive inorganic fertilizers (Boserup 1965; McIntire et al. 1992). However, the effects of infrastructural improvements on the application of organic matter remain less clear. Various studies from West Africa document increased rates of organic fertilizer application, including compost, manure, and city waste, close to urban areas such as Kano in Nigeria (e.g., Lewcock 1995; Harris and Yusuf 2001; de Ridder et al. 2004).

Several studies point to the effects of road-building on the agricultural labor supply. Remunerative nonfarm employment opportunities imply that opportunity costs of labor are higher than returns from agriculture, making labor-intensive investments in the soil base, such as the application of organic fertilizers, less likely (Lichtenberg 2006). In Cameroon, road upgrades enabled farmers to better reap benefits from employment in the nonagricultural sector, suggesting that isolation keeps farmers mostly trapped in poverty with agriculture as their dominant means of income (Gachassin et al. 2010). Ruijs et al. (2004), considering regional trade in Burkina Faso, point out that the benefits associated with a physical road upgrade without addressing various institutional weaknesses will only be small. Neither study suggests that improved access significantly increases the underlying value of land coupled with increased investments in the soil fertility base.

Reviewing the literature on transaction costs in East and South Africa, Barrett (2008b) observes that insights into the best mechanisms to stimulate smallholder participation in markets are not well known. He concludes that while investments in infrastructure are an important means of reducing marketing costs, complementary macro-, meso-, and micro-level interventions are still needed. Historically, macropolicies included sectorwide policies such as import tariffs and guaranteed prices for outputs. Many of these policies have favored the urban population more than rural areas and made investments in agriculture and soils relatively less attractive. Many African governments reversed these perverse incentives, but such a trend is much less apparent in West African countries such as Nigeria or Cote d'Ivoire (Anderson and Masters 2009). In the following sections we turn our attention to meso- and micro-level policies, with a particular focus on enhancing the efficiency of fertilizer use.

5.4.2 Augmenting the Use of Inorganic Fertilizers

Policies to stimulate fertilizer use among smallholders have traditionally included the promotion of recommended application rates and often the subsidization of fertilizer prices, with an ultimate goal to raise output. Most studies suggest these policies have been ineffective in reaching this goal.

First, blanket fertilizer recommendations ignore much of the actual on-farm diversity observed and only rarely coincide with the economically optimal fertilizer rate at the field level. They are typically based on trials carried out on research stations, sometimes in the 1970s and 1980s (Gregory and Bumb 2006), and changes across time and space render such recommendations ineffective for many smallholders. Our review suggests that most farmers understand this discrepancy well, with few smallholders following official recommendations. Still the Abuja declaration on fertilizer use led to rather generic advice on augmenting fertilizer use across Africa by 8 to 50 kg of nutrients per hectare (African Union 2006). Such recommendations are inefficient (Gandah et al. 2003b; Suri 2011) and governments should focus instead on building extension services that are better equipped to deal with diversity in production circumstances (Banful et al. 2010).

Next, many African governments have experimented with fertilizer subsidies. Historically, such subsidies, implemented through parastatal agencies, were tailored mostly toward enhanced production of commercial commodities such as cotton and cocoa (Anderson and Masters 2009). Many subsidies for export commodities were abolished during the era of structural adjustment programs. Nevertheless, many African countries reinstated input subsidies, inorganic fertilizer, and improved seeds mostly for major food crops. Popular support for fertilizer subsidies remains high, not least because they entail a large, and often politically motivated, transfer of funds into rural areas. This holds in the well-documented case of Malawi (Dorward and Chirwa 2011; Resnick et al. 2012) but is also true in West Africa as well, such as in Ghana and Nigeria (Banful 2011). This creates substantial fiscal burdens to developing economies.

A subsidy reduces the costs of using the fertilizer, shifting the most efficient level of use upward and thereby augmenting production. The jury is still out on the question of whether fertilizer subsidies are an effective policy instrument. A blanket subsidy on fertilizer prices proportionally benefits those (wealthier) farmers who are using fertilizer in larger quantities. In other words, a fertilizer subsidy particularly benefits wealthier farmers who cultivate relatively fertile fields. Blanket subsidies have therefore largely been replaced with subsidies targeting vulnerable households. Dorward and Chirwa (2011) suggest that the current Malawi subsidy program, using a system of locally redeemable vouchers coupled with a monitoring system, has been effective in improving food security and poverty reduction among the most vulnerable farmers. Several Nigerian states, such as Kano and Taraba, equally subsidize fertilizer through redeemable vouchers. While program participants apply more fertilizer at lower prices, the quality of the fertilizer supplied appears to be lower and often supplied untimely (Liverpool-Tasie et al. 2010). Others (Holden and Lunduka 2012) claim the Malawi voucher scheme did not benefit the poor any better than random distribution of inputs and elite capture in the system remains a major worry. In Nigeria a substantial portion of agricultural subsidies is now distributed through mobile phones, potentially lowering the risk of misappropriation (Okafor and Malizu 2013). Nevertheless, the impact of such programs on farmers who do not own mobile phones remains undocumented.

More worries remain. Instances are rife where overzealous governments engaged in the distribution of fertilizers themselves, stifling private sector engagement (Dorward and Chirwa 2011; Jayne et al. 2013). In Nigeria, fertilizer subsidies differ across states and there exists substantial leakage from subsidized states to unsubsidized states (Banful et al. 2010), while fertilizer subsidies have been claimed to crowd out 19% to 35% of fertilizer purchases from private dealers (Diao et al. 2013).

A generic reduction in the price of inorganic fertilizers changes the most efficient input mix through substitution and complementary effects but does not create a financial incentive to apply fertilizer more efficiently. If failures in concurrent complementary markets such as the labor market exist, farmers remain unresponsive to fertilizer price reductions. If inorganic and organic fertilizers are strong complements, a subsidy could trigger farmers to apply more organic fertilizers. The empirical literature (Dorward and Chirwa 2011) suggests, however, that farmers see both types of fertilizer as substitutes with a fertilizer subsidy rendering the use of organic fertilizers or biological nitrogen fixation by legumes less attractive (also observed in, e.g., Franke et al. 2006).

Sometimes the indivisibility of fertilizer packages for sale further compound the lack of fertilizer use. Many smallholders would not only lack the financial means to buy fertilizer in quantities of, say, 50 kg, but would also be better off economically by using smaller quantities. Various programs have been implemented that stimulate sales of fertilizers in smaller quantities, most notably the Farmer Input promotions (FIPS) program in Kenya (Poulton et al. 2006), allowing farmers to purchase quantities closer to their economic optima. Nevertheless, repackaging fertilizer into smaller quantities adds cost to retail prices. Retailers charge a premium of up to 15% on small packages in Nigeria (Gregory and Bumb 2006).

While policies thus focused on making fertilizers more affordable, some claim that high prices per se are not a main cause of low usage. Duflo et al. (2011) suggest a behavioral explanation based on impatience. After harvest, smallholders allocate crop income to various activities, and may have run out of savings when demands for fertilizers arise in the next cropping season. Their research suggests an efficient intervention is to encourage farmers to buy vouchers directly after harvest, which can be redeemed for fertilizer in the subsequent cropping season. Clearly, a role exists for smart policy design based on insights from behavioral economics as is becoming more common in various other sectors. Yet insights from behavioral economics in African smallholder contexts are still few and far between, and before actual policies can be designed and implemented, replications and additional research in different contexts would be desired. Still, Duflo et al. (2011) equally observe that about half of the farmers in their sample refrain from using fertilizer in the improved policy design.

As we have shown, farmers by and large have an understanding of optimal application rates of fertilizer blends currently for sale. The reduction of prices, sales of microdoses, and bypassing behavioral constraints help to augment fertilizer use for some, but for many smallholders purchase and application of fertilizer remains inefficient. The policies discussed so far focus on augmenting fertilizer use and crop production rather than on improving the efficiency of fertilizer use. Two avenues could achieve the latter. First, fertilizer blends may not be appropriate given farmer contexts. Second, better-targeted policies are needed that stimulate the concurrent uptake of organic fertilizer, particularly among the poorest groups of smallholders.

5.4.3 Enhancing the Efficiency of Inorganic Fertilizers Directly

A change in application methods can improve the efficiency of currently applied fertilizer quantities. Mobile nutrients such as nitrogen especially benefit from a split application (two or three times in the season), better aligning nutrient availability with plant requirements and lessening the risk of nutrient losses during the season. Spot application (placing the fertilizer at the base of each plant stand) can also improve fertilizer use efficiency (relative to broadcasting). Such methods are currently being advocated (Tabo et al. 2011; Twomlow et al. 2011). Both methods are more labor-intensive and whether yield gains warrant additional labor input has, to our knowledge, not yet been investigated.

As well, the composition of the fertilizer blends can be improved to better fit specific soil condition and crop needs. Use of inorganic fertilizers without paying attention to the full spectrum of soil fertility including micronutrients, has been cited to explain the stagnation of crop yields under green revolution technologies in South Asia (Ladha et al. 2003; Aggarwal et al. 2004 cited in Keyzer et al. 2009). Research in Nigeria and Togo points toward stagnating crop yields on continuously cropped fields due to decline in macronutrients (nitrogen, phosphorus, and potassium) but also due to other limiting factors such as magnesium or sulfur (Vanlauwe et al. 2005; Nziguheba et al. 2009). Clearly, the composition of fertilizer blends matters. The long-term use of the same inorganic fertilizer blend leads to mining of those nutrients that are not included, while the proportion of the nutrients that are included may not be optimal. Instead of policies that augment the use of current fertilizers, a policy that promotes the diffusion of new and better-targeted fertilizer blends may well enhance production efficiency and/or production across the subregion.

In West Africa various fertilizer blends are being marketed, with differences occurring mostly between countries. Country-specific differences exist in the share of nutrients contained in cotton fertilizer blends. Harmonization of these blends lower retail prices due to economies of scale (Gregory and Bumb 2006). It is, however, not clear whether harmonized blends, considering varying production conditions, would truly benefit farmers.

Whether the current fertilizer blends for sale are also the most appropriate is largely unknown not only to policy makers but equally to farmers and fertilizer companies. Farmers typically do not know whether critical nutrients are missing in their fields and the vast majority of smallholders have no access to means of testing their soils for missing nutrients. However, this only partially explains the small demand for soil tests. Even if reliable and affordable soil tests are available, the generated information is of little use when markets for missing nutrients are unavailable. In addition, farmers may still need to learn-by-doing and identify the most efficient application rate in successive seasons.

On the supply side of the market, fertilizer companies could choose to invest and market different types of fertilizer blends. A wider range in blends could imply lower volumes for different products and create an upward pressure on unit prices for each type of fertilizer. Higher production costs to accommodate a wider range of nutrients may further aggravate a price increase. On the other hand, better-matched fertilizer blends may boost overall fertilizer demand and the price effect remains ambiguous.

Private companies, however, have little incentive to offer a wider range of fertilizer blends or complementary nutrients as long as information on the specific nutrients remains missing. Companies could choose to invest in on-farm soil testing, but generating such knowledge is costly, next to the costs of developing new fertilizer blends in itself. As well, competitors may free ride on the information one company generates, which deters many companies from investing in such innovation in the first place. A classical market failure is the result, something that can be alleviated through well-designed government coordination.

A specific role would be to stimulate information gathering on the diversity in soil fertility, limiting nutrients for different crops and associated responses to different types of fertilizer. Such an exercise has recently taken place in Korea, whereby fertilizer blends were better aligned to local soil characteristics (OECD 2012). Ideally such information is stored and maintained in databases publicly available to farmers and agricultural input dealers. Such a continuous exercise involves cooperation between farmers, government, and research organizations as well as the private sector, thereby potentially experimenting with new participatory methods of fertilizer delivery. The envisioned outcome of such an exercise would be a substantial increase in the efficiency of fertilizer application that better tailors field- and plant-level nutrient demand with the market supply of various nutrients. Detailed statistical analysis aids to better understand the relationships between inputs, soils, and productivity. Research activities should be targeted to a limited number of representative sites, allowing for the deduction of generalizable insights across a wider region (Florax et al. 2002).

In such a scenario room may still exist for fertilizer subsidies. Morris et al. (2007) suggests a role for market smart subsidies, specifically subsidies that stimulate a nascent fertilizer industry and stimulate experimentation and learning among farmers. Both arguments would hold in the case described above. Effects of crowding out would be minimal since new fertilizer blends are not yet being marketed. Second, as described in Section 5.2.3, application of a new input implies a phase of experimentation and fine-tuning among farmers, something that invokes various opportunity costs. A subsidy would temporarily lower the costs of experimentation. Recent experiences from the subsidy program in Malawi further help to design such a program in the most effective way possible.

Nevertheless, as the review in Section 5.3 showed, the choice not to apply inorganic fertilizers can often be directly attributed to the weak soil fertility base; for instance, its soil organic matter content and/or the availability of complementary micronutrients. A major effort is thus needed to incentivize the restoration of such soils by smallholders. How this can be accomplished is, however, no trivial task.

5.4.4 Enhancing the Efficiency of Inorganic Fertilizers Indirectly by Stimulating the Use of Organic Matter

A strong macroeconomic case exists (von Braun et al. 2012) for investing and safeguarding the productive base of soils, for example, through the application of increased amounts of organic amendments. Organic fertilizers such as manure and compost increase the soil organic matter content of soils, favorably change the soil structure, and increase the water and nutrient retention capacity of the soils. These factors increase the marginal productivity of inorganic fertilizers, something that is widely documented in the agronomic literature (e.g., Heathcote 1970; Jones 1971, 1976; Iwuafor et al. 2001; Vanlauwe et al. 2001, 2002; Franke et al. 2008a) and is equally perceived by smallholders in (West) Africa (Florax et al. 2002; Gandah et al. 2003a,b; Nkonya et al. 2004; Marenya and Barrett 2007, 2009a).

Yet supply of organic fertilizers in West Africa tends to be small and inelastic. Low external input agriculture relying solely on manure and compost to replenish soils requires substantial amounts of biomass per unit of cropland, either from on-farm production or through reliance on large pasture areas (McIntire and Powell 1993). The nutrient content of manure, particularly with respect to nitrogen and phosphorus, is relatively low, partially because of inefficient livestock feeding strategies. Sustaining plant growth without mining soil nutrients would necessitate sizeable amounts of manure application of up to 6 to 10 Mg ha^{-1} (Abdoulaye and Sanders 2006). Such quantities, or the underlying biomass requirements, are unavailable to farmers in the (semi-)arid areas of West Africa (Reddy 1988; Hayashi et al. 2012). As a corollary, studies suggest that at average fertilizer prices, the use of inorganic fertilizers to maintain soil nutrients is most efficient (McIntire and Powell 1993).

The small and inelastic supply of manure adds explanation to the observations that organic and inorganic fertilizer do not appear to be complementary inputs in its standard economic definition. This would imply that a lower price of either product leads to an increased demand for the other, which could be a mechanism to stimulate uptake of organic fertilizer use. We are not aware of studies that explicitly investigated cross-price elasticities between organic and inorganic fertilizer, but evidence from countries that experimented with fertilizer subsidies does not suggest a concurrent increase in use of organic fertilizers or uptake of integrated soil fertility management (ISFM) techniques. Rather, the various syntheses suggest cheaper inorganic fertilizers substitute organics and therefore call for additional focus on promoting the uptake of ISFM and organic fertilizer (Poulton et al. 2006; Dorward and Chirwa 2011).

Furthermore, manure and compost application is costly. Transport and field application involve substantial labor and/or capital investments (Photograph 5.3), leading farmers to apply most of these inputs on fields close to the homestead (Gandah et al. 2003b; Tittonell et al. 2005). Application of inputs on outer lying fields can also be less attractive due to risk of crop theft and associated higher monitoring costs. Such investments are prohibitive to many smallholders (Photograph 5.4) (Harris and Yusuf 2001; Franke et al. 2010), particularly during the planting season when labor demand for agricultural activities peaks. In many parts of West-Africa farmers apply manure when the opportunity costs of labor are low, which occurs in the dry season weeks before the onset of rains (Harris and Yusuf 2001). As a result, much of the available nitrogen in manure will have volatized when the cropping season starts. Equally worrying is that losses of manure are greatest among the poorest smallholders, for whom the on-farm supply of manure is already small (Rufino 2008). Cheaper strategies to apply organic matter include the corralling of animals or threshing the grain harvest within fields (Florax et al. 2002; Gandah et al. 2003a) on a rotational basis.

These limitations are most acute for the poorest smallholders who may face the highest opportunity costs of labor and are being constantly pressed to secure incomes from the limited resources they possess. Problematically, the beneficial effects of organic matter application are delayed, only becoming manifest after a number of seasons (e.g., Ridder and Keulen 1990; Franke et al. 2008b), further disadvantaging the poorest with high time discount rates. Low yields imply a small supply

PHOTOGRAPH 5.3   Transport of manure to farmer field (Northern Guinea Savannah, Nigeria).

PHOTOGRAPH 5.4 Manure application. Peaks in labor demand lead farmers to apply manure several weeks before the start of the rains and field preparation activities (Northern Guinea Savannah, Nigeria).

of crop residues, which some poor farmers may sell if they do not possess their own animals. These factors could further underlie possible low soil fertility poverty traps, as discussed in Section 5.2.1 Strategies designed to enhance the application of organic fertilizers need to address both the high costs and limited supply and rest with a combination of both technologies and policies. We start with discussing the role of technologies.

In recent years the documentation of heterogeneity in production contexts in the various regions in Africa has gained ground (e.g., Tittonell et al. 2005; Giller et al. 2006; Berkhout 2009; Franke et al. 2014). Such work has in part been inspired by the realization that the development of blanket technological recommendations has not proved effective to promote agricultural growth (IAASTD 2009), an insight that mimics our discussion on ineffective blanket fertilizer recommendations. Development of technological interventions now explicitly incorporates on-farm (within village or region) diversity (Tittonell et al. 2005) and focus on combining the use of organic and inorganic fertilizer in production systems that rotate cereals with nitrogen-fixing legumes (Place et al. 2003; Sanginga et al. 2003). Research focuses on increasing crop biomass production available for animal consumption, and eventually production of organic fertilizers, without adversely affecting food grain production (Kristjanson et al. 2005). Another focus rests with increasing the efficiency by which such biomass is converted into animal manure (Franke et al. 2010). The suggestion that soil fertility mining can be arrested through reliance on cereal-legume rotations in combination with crop-livestock integration has been long known. In fact, very similar production systems were promoted by colonial administrators in Africa as long ago as the 1920s (Sumberg 1998).

However, the unfavorably high labor costs associated with manure application remain unresolved with the exception of the promotion of mechanization. Organic fertilizers remain bulky, although the development of Biochar, inspired by the black soils of the Amazons (Barrow 2012), could prove to be more labor-efficient. Others have called for research that could increase the quality of the current manure supply (Harris and Yusuf 2001) while changing on-field corralling strategies, which is less haphazard but better planned, and could be a low-cost alternative to increase the efficiency of manure allocation across fields (Florax et al. 2002).

Despite technological advances, uncertainty with respect to the availability of sufficient quantities remains. Policies that temporarily stimulate an increased supply could be desirable. We broadly distinguish three groups of potential policies. First, macropolicies could stimulate the use of organic inputs by altering input and output prices. Second, policies could increase the local supply of biomass. Third, policies could serve to enhance the use of organic inputs from external sources.

The precarious situation of some smallholders and their inability to pursue much-needed investment in their soils has led some to observe that such groups should perhaps be assisted with an exit from agriculture (Fan et al. 2013). They could then be assisted to pick up remunerative employment in nonagricultural sectors while the remaining farmers can increase the scale of their farming operations. The application of organic fertilizer, or adoption of other ISFM techniques, could be inhibited by economies of scale. For example, the use of draft animals or tractors to transport and apply organic material only makes economic sense at certain scales. A smallholder operating on a larger scale, less pressed by day-to-day subsistence requirements, could well display a lower time discount rate, thereby accepting a longer period for investments in soils to materialize (Nkonya et al. 2004).

Ethical issues aside, a smooth exit from agriculture would only be feasible if realistic alternative employment opportunities exist outside agriculture. A concomitant requirement would be the existence of an adequate system of land tenure, with effective institutions to monitor and enforce title rights and the ability to transfer these formally. Uncertainty about land tenure is thus often linked to the paucity of on-farm investments in soils (see Nkonya et al. 2004 for an overview of findings), but empirical evidence on the relationship between land tenure and land investments is ambiguous (von Braun et al. 2012). Differences between various types of schemes could explain the observed ambiguity, since participatory and low-cost schemes to certify land tenure have been found to raise soil conservation investments effectively (Holden et al. 2009). Furthermore, in many instances trusted and effective local institutions governing land rights exist (Knowler 2004).

Others have called for macro-level policies that temporarily support domestic agricultural income, such as levying taxes on food imports (Koning et al. 2001; Heerink 2005). Such policies artificially raise the output price and provide a stimulus to increase the use of organic and inorganic inputs. Increased production also augments the availability of biomass for the production of organic fertilizers. On the other hand, inelastic supply responses due to transaction costs and missing or imperfect markets, particularly in more isolated areas, lessen the effectiveness of such programs (De Janvry et al. 1991). Moreover, many vulnerable smallholders are net buyers of imported food crops, and the overall welfare effects of such policies could be negative.

Policies could also enhance the incentives for farmers to restore degraded lands, for example, through the adoption of agroforestry practices. Such practices increase the local supply of biomass (e.g., using nitrogen-fixating tree species, which has been facilitated by two types of policies: payments for environmental services (PES) and public work schemes.

A variety of successful PES schemes has led to enhanced sustainable land management ranging from arresting the Great Dust Bowl in the United States (Lal et al. 2012) to restoring degraded rangelands in Central America (Pagiola et al. 2004). PES has been advocated as a way in which smallholders could benefit from the global market for carbon credits (Lal et al. 2012; Marenya et al. 2012). Whether they would be eligible is uncertain, since it is not clear whether such practices store atmospheric carbon for sufficiently long periods. Use of PES would be impractical if one moves beyond directly observable field characteristics to, for example, actual organic matter application rates.

As well, public work schemes (food for work, or cash for work) run in various countries in Africa. Such programs have proved effective (Bezu and Holden 2008; Andersson et al. 2011) in providing basic income to the marginalized groups in society. They can be used to deliver public goods such as road maintenance or for investments in sustainable land use (Holden et al. 2006), for example, through measures arresting soil erosion or implementing agroforestry practices. Nevertheless, caution should be exercised, as top-down implemented public work schemes could also push up rural wages (Khera 2011) and lead to an overall increase in production costs to farmers.

Finally, a role could exist for the private sector to engage in the supply of organic fertilizers. Although sources of biomass and organic fertilizer are often in short supply locally, substantial supplies exist elsewhere, notably from urban centers in West Africa as well as manure from intensive livestock rearing. Smart financial incentives and appropriate regulation could play a role in making these sources available to African smallholders.

Markets for organic fertilizers exist in some areas, notably in areas with high levels of agricultural intensification close to major urban areas such as Kano in Nigeria (Harris and Yusuf 2001). Similarly, various types of urban waste are collected and used in agriculture (Lewcock 1995). More recently, the collection of urban waste and its conversion into compost for agricultural use has been taken up by a private company in Lagos, Nigeria (World Bank 2006). The linkage to carbon-emission trading scheme enhances the business case for this company, and more generally the provision of start-up subsidies to such companies could increase the supply of organic inputs for use in agriculture. Similar incentives could play a role in importing organic fertilizers based on animal manure from, for instance, European countries. Trade in animal manure that is treated subject to relatively strict public health norms is common from The Netherlands to other countries within the European Union (EU) (CBS 2013), but is nonexistent in countries outside the EU. Economies of scale in processing and transport of organic fertilizers could thereby lower the price that farmers pay. Nevertheless, subsidizing the production or import of organic fertilizers runs at the risk of perverse externalities. These could include increased deforestation locally or the conversion of food products.

An important question remains whether such policies can provide sufficient incentives for the poorest farmers. Even though the policies discussed lower the costs of applying organic inputs, this reduction may not be sufficient for the poorest. Much like subsidies on inorganic fertilizers, a policy instrument can be devised that subsidizes the use of organic fertilizers, entailing the specific

TABLE 5.1
Potential Policies to Enhance the Use of Organic Inputs

Type of Intervention	Main Advantages	Disadvantages and Potential Externalities
Land tenure reforms	• More secure land rights could lead to increased investments in the soil fertility base • Trade in land and scale increases could make ISFM technologies more attractive	• Much uncertainty remains if, how, and when the listed advantages can be achieved
Macro-level price policies	• More favorable output prices enhance the use of organic and inorganic inputs	• Lower costs of organic fertilizers for all, but lacks a specific targeting for the most marginal farmers
Payment for environmental services	• Remunerates farmers for applying soil conservation measures that are easily observable, such as agroforestry practices	• Expensive to monitor if focus on actual use of organic fertilizer • Does not address high transport costs • Potentially subject to elite capture
Public work schemes	• Reduces the transport costs of organic fertilizers to farmers • Provides incomes to the most marginalized farmers • May increase biomass available for organic fertilizers • Ability to learn from experiences in various public work schemes	• Potentially subject to elite capture
Incentivize companies that supply organic fertilizers, for example, through temporary tax breaks	• Could be used to unlock the supply of biomass sources from elsewhere (urban waste or animal manure from other continents) • Could be coupled with subsidies to lower transport costs	• Subsidized organic fertilizers could lead to perverse externalities, such as increased deforestation or conversion of food products • Lower costs of organic fertilizers for all, but lacks a specific targeting for the most marginal farmers

targeting of marginal groups, and the distribution of vouchers that can be redeemed at local agro-dealers. Nevertheless, worries about elite capture remain and a subsidy may crowd out the farmers' own supplies of manure and compost.

The most effective examples of PES and public work schemes rely on existing social structures and institutions. Public work schemes are best designed in a participative manner (Holden et al. 2006). Still, schemes that propagate payments for environmental services are prone to elite capture (OECD 2013 citing Topa et al. 2009), and one can easily imagine that services delivered in public work schemes proportionally benefit wealthier groups.

This list of policy measures, as summarized in Table 5.1, is not exhaustive, nor are they exclusive. A combination of some measures may deliver results that equally benefit the poorest. How this can be achieved needs to be the scope of further investigation through a joint effort by farmers, researchers, governments, and the private sector.

5.5   CONCLUSION AND THE WAY FORWARD

West African soil fertility resources are characterized by substantial diversity, waging diverse responses of crop yields to the application of external inputs such as inorganic fertilizer. Missing complementary nutrients, low organic matter content, adverse soil chemical properties, biotic factors, soil structural factors, or a combination of these, lead to unresponsiveness to external input application on some fields. Many farmers appear keenly aware of this diversity and vary fertilizer application across fields close to economic optima. Our review on the ways by which West African smallholders manage diverse soil fertility resources adds further credence to the premise that farmers are poor but efficient.

This comprehension also signals that smallholders respond to incentives conditioned to their local contexts, creating possibilities by policy makers to influence smallholder production, and if well-designed, steering them toward greater investments in soil fertility resources and higher productivities. This was the focus of the second half of this chapter. The provision of a stable macroeconomic environment coupled with investments in rural infrastructure and institutions, including the provision of extension services catering to heterogeneous information demands from farmers, should remain a cornerstone for policy makers. However, these policies in and of themselves are unlikely to be sufficient for scaling up production. This particularly holds true for the most marginal fields.

Policies to augment the use of inorganic fertilizers, notably subsidies, have mostly been marred by adverse effects. Instead the major emphasis should rest with policies that enhance fertilizer use efficiency. These include mechanisms directly stimulating fertilizer efficiency, such as the design and distribution of better-suited fertilizer blends, next to efforts to indirectly stimulate fertilizer use through the restoration of marginal and degraded fields. Temporary fertilizer subsidies may still be required to facilitate the development and uptake of novel fertilizer blends. This could take the form of direct support to fertilizer producers or targeted subsidies to agricultural producers. In similar ways private sector actors could be stimulated to provide organic fertilizers. Direct transfers to smallholders, for example through PES or cash-for-work programs, may further support the restoration of degraded lands. Technologies should improve the efficiency by which biomass resources are recycled on-farm, next to development and uptake of innovations such as biochar.

Just like technological recommendations, single-policy instruments do not fully apprehend the diversity across much of Africa, for example in biophysical production conditions. Instead, specific sets of policies should be devised for different regions (Ehui and Pender 2005; Ruben et al. 2007). These should further account for diversity at the lowest scale, exhibiting responsive and unresponsive fields possibly more frequently owned by more vulnerable groups. Nevertheless, our review also reveals a paucity of insights into the most effective technologies and policies or combinations thereof. Currently no straight answers can be provided to the question as to which combination of technologies and policies works best in a specific region.

Concerted efforts are therefore required by researchers, governments, and the private sector. First, a major effort should rest with increasing the availability of information on diversity in soils. A fair amount of agronomic research has been conducted and is currently ongoing in West Africa. Findings on differential crop responses to nutrient applications, as well as missing nutrients, are not currently stored in portals where such information is easily accessible either to farmers or to fertilizer producers. Governments should take an active lead in unlocking such information in national or regional databases, both for identifying areas that lack attention from researchers as well as for identifying patterns, for example on missing micronutrients, emerging from such data.

Second, a major effort should be made to experiment with various technologies and policies or combinations thereof in different settings. Novel fertilizer blends require ground proofing with various groups of farmers and subsequent scaling up of such blends could require the temporary provision of subsidies. PES schemes could prove to be an effective mechanism for land use restoration, but whether this actually works, and which level of payments actually suffice, are not known. Hence, policy experiments need to simultaneously bring insights into the costs and benefits on which the various actors, both the different types of smallholders as well as the private sector, base their decisions. Such efforts help to uncover insights into best-bet approaches and also help in identifying the right monetary incentives by which actors move to more desirable outcomes. As well, generated information should be unlocked to the general public for maximal impact, for example, by sharing insights across countries.

REFERENCES

Abdoulaye, T., and J. Lowenberg-DeBoer. 2000. Intensification of Sahelian Farming Systems: Evidence from Niger. *Agricultural Systems* 64: 67–81.

Abdoulaye, T., and J. H. Sanders. 2006. New Technologies, Marketing Strategies and Public Policy for Traditional Food Crops: Millet in Niger. *Agricultural Systems* 90: 272–92.

African Union. 2006. Abuja Declaration on Fertilizer, Africa Fertilizer Summit. Abuja.

Aggarwal, P. K., P. K. Joshi, J. S. I. Ingram, and R. K. Gupta. 2004. Adapting Food Systems of the Indo-Gangetic Plains to Global Environmental Change: Key Information Needs to Improve Policy Formulation. *Environmental Science & Policy* 7: 487–98.

Anderson, K., and W. A. Masters. 2009. *Distortions to Agricultural Incentives in Africa*. Washington, DC: World Bank Publications.

Andersson, C., A. Mekonnen, and J. Stage. 2011. Impacts of the Productive Safety Net Program in Ethiopia on Livestock and Tree Holdings of Rural Households. *Journal of Development Economics* 94: 119–26.

Antle, J. M., J. J. Stoorvogel, and R. O. Valdivia. 2006. Multiple Equilibria, Soil Conservation Investments, and the Resilience of Agricultural Systems. *Environment and Development Economics* 11: 477–92.

Aune, J. B., and A. Bationo. 2008. Agricultural Intensification in the Sahel–the Ladder Approach. *Agricultural Systems* 98: 119–25.

Bandiera, O., and I. Rasul. 2006. Social Networks and Technology Adoption in Northern Mozambique. *The Economic Journal* 116: 869–902.

Banful, A. B., E. Nkonya, and V. Oboh. 2010. Constraints to Fertilizer Use in Nigeria: Insights from Agricultural Extension Service. *IFPRI Discussion Paper 01010*. Washington, DC: International Food Policy Research Institute (IFPRI).

Banful, A. B. 2011. Old Problems in the New Solutions? Politically Motivated Allocation of Program Benefits and the "New" Fertilizer Subsidies. *World Development* 39: 1166–76.

Bardhan, P., and C. R. Udry. 1999. *Development Microeconomics*. New York: Oxford University Press.

Barham, B. L., J.-P. Chavas, D. Fitz, V. Ríos-Salas, and L. Schechter. 2014. Risk, Learning, and Technology Adoption. *Agricultural Economics* 46: 11–24.

Barrett, C. B. 2008a. Poverty Traps and Resource Dynamics in Smallholder Agrarian Systems, pp. 17–40. In *Economics of Poverty, Environment and Natural-Resource Use*. Dordrecht, The Netherlands: Springer.

Barrett, C. B. 2008b. Smallholder Market Participation: Concepts and Evidence from Eastern and Southern Africa. *Food Policy* 33: 299–317.

Barrow, C. J. 2012. Biochar: Potential for Countering Land Degradation and for Improving Agriculture. *Applied Geography* 34: 21–28.

Beaman, L., D. Karlan, B. Thuysbaert, and C. Udry. 2013. Profitability of Fertilizer: Experimental Evidence from Female Rice Farmers in Mali. *American Economic Review* 103: 381–86.

Berkhout, E. D. 2009. Decision-Making for Heterogeneity: Diversity in Resources, Farmers' Objectives and Livelihood Strategies in Northern Nigeria. PhD thesis, Wageningen University, The Netherlands.

Berkhout, E. D., R. A. Schipper, A. Kuyvenhoven, and O. Coulibaly. 2010. Does Heterogeneity in Goals and Preferences Affect Efficiency? A Case Study of Farm Households in Northern Nigeria. *Agricultural Economics* 41: 265–73.

Bezu, S., and S. Holden. 2008. Can Food-for-Work Encourage Agricultural Production? *Food Policy* 33: 541–49.

Boserup, E. 1965. *The Conditions of Agricultural Growth: The Economics of Agrarian Change under Population Pressure*. London: Allen and Unwin.

Brouwer, J., and J. Bouma. 1997. Soil and Crop Growth Variability in the Sahel: Highlights of Research (1990–95) at Icrisat Sahelian Center. *Information Bulletin* No. 49.

CBS. Getransporteerde Mest Naar Herkomst En Bestemming. 2013. Centraal Bureau voor de Statistiek–Statistics Netherlands, http://statline.cbs.nl/StatWeb/publication/?VW=T&DM=SLNL&PA=71461NED&D1=0-7&D2=0&D3=1-2&D4=a&HD=140410–1545&HDR=T&STB=G1,G2,G3.

Conley, T. G., and C. R. Udry. 2010. Learning about a New Technology: Pineapple in Ghana. *American Economic Review* 100: 35–69.

De Janvry, A., M. Fafchamps, and E. Sadoulet. 1991. Peasant Household Behavior with Missing Markets: Some Paradoxes Explained. *Economic Journal* 101: 1400–17.

de Ridder, N., H. Breman, H. van Keulen, and T. J. Stomph. 2004. Revisiting a "Cure against Land Hunger": Soil Fertility Management and Farming Systems Dynamics in the West African Sahel. *Agricultural Systems* 80: 109–31.

Diao, X., A. Kennedy, F. Cossar et al. 2013. Evidence on Key Policies for African Agricultural Growth. *IFPRI Discussion Paper 01242*. Washington, DC: International Food Policy Report (IFPRI).

Dorward, A., and E. Chirwa. 2011. The Malawi Agricultural Input Subsidy Programme: 2005/06 to 2008/09. *International Journal of Agricultural Sustainability* 9: 232–47.

Duflo, E., M. Kremer, and J. Robinson. 2008. How High Are Rates of Return to Fertilizer? Evidence from Field Experiments in Kenya. *American Economic Review* 98: 482–88.

Duflo, E., M. Kremer, and J. Robinson. 2011. Nudging Farmers to Use Fertilizer: Theory and Experimental Evidence from Kenya. *American Economic Review* 101: 2350–90.

Ehui, S., and J. Pender. 2005. Resource Degradation, Low Agricultural Productivity, and Poverty in Sub-Saharan Africa: Pathways out of the Spiral. *Agricultural Economics* 32: 225–42.

Fan, S., and X. Zhang. 2004. Infrastructure and Regional Economic Development in Rural China. *China Economic Review* 15: 203–14.

Fan, S., J. Brzeska, M. Keyzer, and A. Halsema. 2013. From Subsistence to Profit. *IFPRI Food Policy Report 26*. Washington, D.C: International Food Policy Research Institute (IFPRI).

Florax, R. J. G. M., R. L. Voortman, and J. Brouwer. 2002. Spatial Dimensions of Precision Agriculture: A Spatial Econometric Analysis of Millet Yield on Sahelian Coversands. *Agricultural Economics* 27: 425–43.

Franke, A. C., S. Schulz, B. Oyewole, and S. Bako. 2004. Incorporating Short-Season Legumes and Green Manure Crops into Maize-Based Systems in the Moist Guinea Savanna of West Africa. *Experimental Agriculture* 40. 463–79.

Franke, A. C., J. Ellis-Jones, G. Tarawali et al. 2006. Evalauting and Scaling-Up Integrated Striga Hermonthica Control Technoligies among Farmers in Northern Nigeria. *Crop Protection* 25: 868–78.

Franke, A. C., G. Laberge, B. Oyewole, and S. Schulz. 2008a. A Comparison between legume Technologies and Fallow, and Their Effects on Maize and Soil Traits, in Two Distinct Environments of the West African Savannah. *Nutrient Cycling in Agroecosystems* 82: 117–35.

Franke, A. C., S. Schulz, B. Oyewole, J. Diels, and O. Tobe. 2008b. The Role of Cattle Manure in Enhancing On-Farm Productivity, Macro- and Micro-Nutrient Uptake, and Profitability of Maize in the Guinea Savanna. *Exprimental Agriculture* 44: 313–28.

Franke, A. C., E. D. Berkhout, E. N. O. Iwuafor et al. 2010. Does Crop-Livestock Integration Lead to Improved Crop Production in the Savanna of West Africa? *Experimental Agriculture* 46: 439–55.

Franke, A. C., G. J. van den Brand, and K. E. Giller. 2014. Which Farmers Benefit Most from Sustainable Intensification? An Ex-Ante Impact Assessment of Expanding Grain Legume Production in Malawi. *European Journal of Agronomy* 58: 28–38.

Fuglie, K. 2010. Accelerated Productivity Growth Offsets Decline in Resource Expansion in Global Agriculture. *Amber Waves* 8: 46–51.

Gachassin, M., B. Najman, and G. Raballand. 2010. Roads Impact on Poverty Reduction—A Cameroon Case Study. *World Bank Policy Research Working Paper* 5209.

Gandah, M., J. Bouma, J. Brouwer, P. Hiernaux, and N. Van Duivenbooden. 2003a. Strategies to Optimize Allocation of Limited Nutrients to Sandy Soils of the Sahel: A Case Study from Niger, West Africa. *Agriculture, Ecosystems & Environment* 94: 311–19.

Gandah, M., J. Brouwer, P. Hiernaux, and N. Van Duivenbooden. 2003b. Fertility Management and Landscape Position: Farmers' Use of Nutrient Sources in Western Niger and Possible Improvements. *Nutrient Cycling in Agroecosystems* 67: 55–66.

Giller, K. E., E. C. Rowe, N. de Ridder, and H. van Keulen. 2006. Resource Use Dynamics and Interactions in the Tropics: Scaling up in Space and Time. *Agricultural Systems* 88: 8–27.

Goetz, S. J. 1992. A Selectivity Model of Household Food Marketing Behavior in Sub-Saharan Africa. *American Journal of Agricultural Economics* 74: 444–52.

Gregory, D. I., and B. Bumb. 2006. *Factors Affecting Supply of Fertilizer in Sub-Saharan Africa.* Agriculture and Rural Development Department, World Bank.

Hardaker, J. B., R. B., Hiurne, R. B. M. Anderson, and G. Lien. 1997. *Coping with risk in agriculture.* CABI Publishing, Willingford, UK.

Harris, F., and M. A. Yusuf. 2001. Manure Management by Smallholder Farmers in the Kano Close-Settled Zone, Nigeria. *Experimental Agriculture* 37: 319–32.

Harris, J. M., K. Erickson, J. Dillard et al. 2008. Agricultural Income and Finance Outlook. ed. Economic Research Service. US Department of Agriculture, Washington, DC.

Hayashi, K., N. Matsumoto, E. T. Hayashi et al. 2012. Estimation of Nitrogen Flow within a Village-Farm Model in Fakara Region in Niger, Sahelian Zone of West Africa. *Nutrient Cycling in Agroecosystems* 92: 289–304.

Heathcote, R. 1970. Soil Fertility under Continuous Cultivation in Northern Nigeria. *Experimental Agriculture* 6: 229–37.

Heerink, N. 2005. Soil Fertility Decline and Economic Policy Reform in Sub-Saharan Africa. *Land Use Policy* 22: 67–74.

Holden, S., C. B. Barrett, and F. Hagos. 2006. Food-for-Work for Poverty Reduction and the Promotion of Sustainable Land Use: Can It Work? *Environment and Development Economics* 11: 15–38.

Holden, S. T., K. Deininger, and H. Ghebru. 2009. Impacts of Low-Cost Land Certification on Investment and Productivity. *American Journal of Agricultural Economics* 91: 359–73.

Holden, S. T., and R. W. Lunduka. 2012. Who Benefit from Malawi's Targeted Farm Input Subsidy Program? *Forum for Development Studies* 40: 1–25.

IAASTD 2009. International Assessment of Agricultural Knowledge, Science and Technology for Development (IAASTD): Global Report. B. D. McIntyre, H. P. Herren, J. Wakhungu, and R. T. Watson (eds.). IAASTD, Washington, DC.

Iwuafor, E., K. Aihou, J. Jaryum et al. 2001. On-Farm Evaluation of the Contribution of Sole and Mixed Applications of Organic Matter and Urea to Maize Grain Production in the Savannah, pp. 185–97. In *Integrated Plant Nutrient Management in Sub-Saharan Africa: From Concept to Practice*, B. Vanlauwe, J. Diels, N. Sanginga, and R. Merckx (eds.). Wallingford, CT: CABI.

Jackson, M. O. 2008. *Social and Economic Networks*. Princeton, NJ: Princeton University Press.

Jayne, T. S., J. Govereh, M. Wanzala, and M. Demeke. 2003. Fertilizer Market Development: A Comparative Analysis of Ethiopia, Kenya, and Zambia. *Food Policy* 28: 293–316.

Jayne, T. S., D. Mather, N. Mason, and J. Ricker-Gilbert. 2013. How Do Fertilizer Subsidy Programs Affect Total Fertilizer Use in Sub-Saharan Africa? Crowding Out, Diversion, and Benefit/Cost Assessments. *Agricultural Economics* 44: 687–703.

Jones, M. J. 1971. The Maintenance of Soil Organic Matter under Continuous Cultivation at Samaru, Nigeria. *The Journal of Agricultural Science* 77: 473–82.

Jones, M. J. 1976. The Significance of Crop Residues to the Maintenance of Fertility under Continuous Cultivation at Samaru, Nigeria. *The Journal of Agricultural Science* 86: 117–25.

Just, D. R., S. V. Khantachavana, and R. E. Just. 2010. Empirical Challenges for Risk Preferences and Production. *Annual Review of Resource Economics* 2: 13–31.

Keyzer, M., W. van Veen, R. Voortman et al. 2009. Nutrient Shortages and Agricultural Recycling Options Worldwide, with Special Reference to China. 17th Annual Conference of the European Association of Environmental and Resource Economists. Amsterdam.

Khera, E. (ed.). 2011. *The Battle for Employment Guarantee*. New Delhi, India: Oxford University Press.

Kimiti, J. M., A. O. Esilaba, B. Vanlauwe, and A. Bationo. 2007. Participatory Diagnosis in the Eastern Drylands of Kenya: Are Farmers Aware of Their Soil Fertility Status, pp. 961–68. In *Advances in Integrated Soil Fertility Management in Sub-Saharan Africa: Challenges and Opportunities*, A. Bationo, B. Waswa, J. M. Kihara, and J. Kimetu (eds.). Dordrecht, The Netherlands: Springer.

Knowler, D. J. 2004. The Economics of Soil Productivity: Local, National and Global Perspectives. *Land Degradation & Development* 15: 543–61.

Koning, N., N. Heerink, and S. Kauffman. 2001. Food Insecurity, Soil Degradation and Agricultural Markets in West Africa: Why Current Policy Approaches Fail. *Oxford Development Studies* 29: 189–207.

Kristjanson, P., I. Okike, S. Tarawali, B. Singh, and V. Manyong. 2005. Farmers' Perceptions of Benefits and Factors Affecting the Adoption of Improved Dual-Purpose Cowpea in the Dry Savannas of Nigeria. *Agricultural Economics* 32: 195–210.

Ladha, J. K., D. Dawe, H. Pathak et al. 2003. How Extensive Are Yield Declines in Long-Term Rice-Wheat Experiments in Asia? *Field Crops Research* 81: 159–80.

Lal, R., U. Safriel, and B. Boer. 2012. Zero Net Land Degradation: A New Sustainable Development Goal for Rio+ 20. A Report Prepared for the Secretariat of the United Nations Convention to Combat Desertification. Bonn, Germany: United Nations Convention to Combat Desertification (UNCCD).

Lewcock, C. 1995. Farmer Use of Urban Waste in Kano. *Habitat International* 19: 225–34.

Lichtenberg, E. 2006. A Note on Soil Depth, Failing Markets and Agricultural Pricing: Comment. *Journal of Development Economics* 81: 236–43.

Liverpool-Tasie, L., A. B. Banful, and B. Olaniyan. 2010. An Assessment of the 2009 Fertilizer Voucher Program in Kano and Taraba, Nigeria. IFPRI Nigeria Strategy Support Program Working Paper 17.

Mairura, F. S., D. N. Mugendi, J. I. Mwanje, J. J. Ramisch, and P. K. Mbugua. 2007. Assessment of Farmers' Perceptions of Soil Quality Indicators within Smallholder Farms in the Central Highlands of Kenya, pp. 1035–46. In *Advances in Integrated Soil Fertility Management in Sub-Saharan Africa: Challenges and Opportunities*, A. Bationo, B. Waswa, J. M. Kihara, and J. Kimetu (eds.). Dordrecht, The Netherlands: Springer.

Marenya, P. P., and C. B. Barrett. 2007. Household-Level Determinants of Adoption of Improved Natural Resources Management Practices among Smallholder Farmers in Western Kenya. *Food Policy* 32: 515–36.

Marenya, P., E. Nkonya, W. Xiong, J. Deustua, and E. Kato. 2012. Which Policy Would Work Better for Improved Soil Fertility Management in Sub-Saharan Africa, Fertilizer Subsidies or Carbon Credits? *Agricultural Systems* 110: 162–72.

Marenya, P. P., and C. B. Barrett. 2009a. Soil Quality and Fertilizer Use Rates among Smallholder Farmers in Western Kenya. *Agricultural Economics* 40: 561–72.

Marenya, P. P., and C. B. Barrett. 2009b. State-Conditional Fertilizer Yield Response on Western Kenyan Farms. *American Journal of Agricultural Economics* 91: 991–1006.

McIntire, J., D. Bourzat, and P. Pingali. 1992. *Crop-Livestock Integration in Sub-Saharan Africa*. Washington, DC: World Bank.

McIntire, J., and J. Powell. 1993. African Semi-Arid Tropical Agriculture Cannot Grow without External Inputs. Paper presented at the International Conference on Livestock and Sustainable Nutrient Cycling in Mixed Farming Systems of Sub-Saharan Africa, Addis Ababa, 1993.

Morris, M., V. A. Kelly, R. J. Kopicki, and D. Byerlee. 2007. Fertilizer Use in African Agriculture: Lessons Learned and Good Practice Guidelines. *Directions in Development: Agriculture and Rural Development*. Washington, DC: World Bank.

Moser, C., C. Barrett, and B. Minten. 2009. Spatial Integration at Multiple Scales: Rice Markets in Madagascar. *Agricultural Economics* 40: 281–94.

Nkonya, E., J. Pender, P. Jagger et al. 2004. Strategies for Sustainable Land Management and Poverty Reduction in Uganda. *IFPRI Research Report 133*. Washington, DC: International Food Policy Research Report (IFPRI).

Nziguheba, G., B. K. Tossah, J. Diels et al. 2009. Assessment of Nutrient Deficiencies in Maize in Nutrient Omission Trials and Long-Term Field Experiments in the West African Savanna. *Plant and Soil* 314: 143–57.

OECD. 2012. Food and Agriculture. *OECD Green Growth Studies*. Paris: OECD.

OECD. 2013. Putting Green Growth at the Heart of Development. *OECD Green Growth Studies*. Paris: OECD.

Okafor, G. O., and C. F. Malizu. 2013. New Media and Sustainable Agricultural Development in Nigeria. *New Media and Mass Communication* 20: 66–73.

Omamo, S. W. 1998. Transport Costs and Smallholder Cropping Choices: An Application to Siaya District, Kenya. *American Journal of Agricultural Economics* 80: 116–23.

Pagiola, S., P. Agostini, J. Gobbi et al. 2004. Paying for Biodiversity Conservation Services in Agricultural Landscapes. *Environment Department Papers 96*. Washington, DC: World Bank.

Pingali, P. L., and P. W. Heisey. 1999. Cereal Crop Productivity in Developing Countries: Past Trends and Future Prospects. Economics Working Paper 99–03 CIMMYT Mexico.

Place, F., C. B. Barrett, H. A. Freeman, J. J. Ramisch, and B. Vanlauwe. 2003. Prospects for Integrated Soil Fertility Management Using Organic and Inorganic Inputs: Evidence from Smallholder African Agricultural Systems. *Food Policy* 28: 365–78.

Poulton, C., J. Kydd, and A. Dorward. 2006. Increasing Fertilizer Use in Africa: What Have We Learned. Agriculture and Rural Development Discussion Paper 25. Washington, DC: World Bank, Agriculture and Rural Development Department.

Reddy, K. C. 1988. Rapport De L'agronomie Générale, Campagne, 1987. INRAN/DRA, Kollo, Niger.

Renkow, M., D. G. Hallstrom, and D. D. Karanja. 2004. Rural Infrastructure, Transactions Costs and Market Participation in Kenya. *Journal of Development Economics* 73: 349–67.

Resnick, D., F. Tarp, and J. Thurlow. 2012. The Political Economy of Green Growth. UNU-WIDER Working Paper 2012/11. Helsinki, Finland: United Nations University–World Institute for Development Economics Research (UNU-WIDER).

Ridder, N., and H. Keulen. 1990. Some Aspects of the Role of Organic Matter in Sustainable Intensified Arable Farming Systems in the West-African Semi-Arid-Tropics (Sat). *Fertilizer Research* 26: 299–310.

Ruben, R., A. Kuyvenhoven, G. Kruseman, D. Lee, and C. Barrett. 2001. Bioeconomic Models and Ecoregional Development: Policy Instruments for Sustainable Intensification, pp. 115–33. In *Tradeoffs or Synergies? Agricultural Intensification, Economic Development and the Environment*, D. R. Lee and C. B. Barrett (eds.). Wallingford, CT: CABI.

Ruben, R., J. Pender, and A. Kuyvenhoven. 2007. Sustainable Poverty Reduction in Less-Favoured Areas: Problems, Options and Strategies, pp. 1–62. In *Sustainable Poverty Reduction in Less-Favoured Areas*, R. Ruben, J. Pender, and A. Kuyvenhoven (eds.). Wallingford, CT: CABI.

Rufino, M. C. 2008. Quantifying the Contribution of Crop-Livestock Integration to African Farming. PhD thesis, Wageningen University, The Netherlands.

Ruijs, A., C. Schweigman, and C. Lutz. 2004. The Impact of Transport and Transaction Cost Reductions on Food Markets in Developing Countries: Evidence for Tempered Expectations for Burkina Faso. *Agricultural Economics* 31: 219–28.

Sanginga, N., K. E. Dashiell, J. Diels et al. 2003. Sustainable Resource Management Coupled to Resilient Germplasm to Provide New Intensive Cereal-Grain-Legume-Livestock Systems in the Dry Savanna. *Agriculture, Ecosystems & Environment* 100: 305–14.

Schultz, T. W. 1964. *Transforming Traditional Agriculture*. Vol. 3, Studies in Comparative Economics. New Haven: Yale University Press.

Sheahan, M., R. Black, and T. S. Jayne. 2013. Are Kenyan Farmers Under-Utilizing Fertilizer? Implications for Input Intensification Strategies and Research. *Food Policy* 41: 39–52.

Singh, I., L. Squire, and J. Strauss. 1986. *Agricultural Household Models: Extensions, Applications, and Policy*. Baltimore: Johns Hopkins University Press.

Sumberg, J. 1998. Mixed Farming in Africa: The Search for Order, the Search for Sustainability. *Land Use Policy* 15: 293–317.

Suri, T. 2011. Selection and Comparative Advantage in Technology Adoption. *Econometrica* 79: 159–209.

Tabo, R., A. Bationo, B. Amadou et al. 2011. Fertilizer Microdosing and "Warrantage" or Inventory Credit System to Improve Food Security and Farmers' Income in West Africa, pp. 113–21. In *Innovations as Key to the Green Revolution in Africa*, A. Bationo, B. Waswa, J. M. Okeyo, F. Maina, and J. M. Kihara (eds.). Dordrecht, The Netherlands: Springer.

Tittonell, P., and K. E. Giller. 2013. When Yield Gaps Are Poverty Traps: The Paradigm of Ecological Intensification in African Smallholder Agriculture. *Field Crops Research* 143: 76–90.

Tittonell, P., A. Muriuki, C. J. Klapwijk et al. 2013. Soil Heterogeneity and Soil Fertility Gradients in Smallholder Farms of the East African Highlands. *Soil Science Society of America Journal* 77: 525–38.

Tittonell, P., B. Vanlauwe, P. A. Leffelaar, E. C. Rowe, and K. E. Giller. 2005. Exploring Diversity in Soil Fertility Management of Smallholder Farms in Western Kenya: I. Heterogeneity at Region and Farm Scale. *Agriculture, Ecosystems & Environment* 110: 149–65.

Topa, G., A. Karsenty, C. Megevand, and L. Debroux. 2009. The Rainforests of Cameroon: Experience and Evidence from a Decade of Reform. *Directions in Development; Environment and Sustainable Development*. Washington, DC: World Bank.

Twomlow, S., D. Rohrbach, J. Dimes et al. 2011. Micro-Dosing as a Pathway to Africa's Green Revolution: Evidence from Broad-Scale on-Farm Trials, pp. 1101–13. In *Innovations as Key to the Green Revolution in Africa*, A. Bationo, B. Waswa, J. M. Okeyo, F. Maina, and J. M. Kihara (eds.). Dordrecht, The Netherlands: Springer.

van Ittersum, M. K., K. G. Cassman, P. Grassini et al. 2013. Yield Gap Analysis with Local to Global Relevance—A Review. *Field Crops Research* 143: 4–17.

Vanlauwe, B., K. Aihou, S. Aman et al. 2001. Maize Yield as Affected by Organic Inputs and Urea in the West African Moist Savanna. *Agronomy Journal* 93: 1191–99.

Vanlauwe, B., J. Diels, N. Sanginga, and R. Merckx (eds.). 2002. *Integrated Plant Nutrient Management in Sub-Saharan Africa: From Concept to Practice*: Wallingford, CT: CABI.

Vanlauwe, B., J. Diels, N. Sanginga, and R. Merckx. 2005. Long-Term Integrated Soil Fertility Management in South-Western Nigeria: Crop Performance and Impact on the Soil Fertility Status. *Plant and Soil* 273: 337–54.

Varian, H. R. 1993. *Intermediate Microeconomics: A Modern Approach*. New York: Norton.

von Braun, J., N. Gerber, N. Mirzabaev, and E. Nkonya. 2012. The Economics of Land Degradation. *Global Soil Week 2012*. Berlin, Germany.

Voortman, R. L., J. Brouwer, and P. J. Albersen. 2004. Characterization of Spatial Soil Variability and Its Effect on Millet Yield on Sudano-Sahelian Coversands in Sw Niger. *Geoderma* 121: 65–82.

World Bank. 2006. Nigeria: Earthcare Solid Waste Composting Project. Carbon Finance Unit. World Bank, Washington, DC.

Yanggen, D., V. A. Kelly, T. Reardon, and A. Naseem. 1998. Incentives for Fertilizer Use in Sub-Saharan Africa: A Review of Empirical Evidence on Fertilizer Response and Profitability. MSU International development working paper 70. Michigan State University, East Lansing, MI.

Zhang, X., and S. Fan. 2004. How Productive Is Infrastructure? A New Approach and Evidence from Rural India. *American Journal of Agricultural Economics* 86: 492–501.

Zingore, S., H. K. Murwira, R. J. Delve, and K. E. Giller. 2007. Soil Type, Management History and Current Resource Allocation: Three Dimensions Regulating Variability in Crop Productivity on African Smallholder Farms. *Field Crops Research* 101: 296–305.

6 Soil Fertility Assessed by Infrared Spectroscopy

Du Changwen, Ma Fei, Lu Yuzhen, and Zhou Jianmin

CONTENTS

6.1 INTRODUCTION

Soil is a very important resource in the earth's surface, which provides a medium for the growth of plant, animals, and microbes. Furthermore, soil fertility plays a vital role in the growth of these living things and is closely linked with the production of food, fuel, feed, and fiber. Therefore, assessment of soil fertility is routine work in soil management and crop production. However, laboratory analysis-based determination of soil properties is time-consuming and costly, and is often not suitable in modern agriculture. Infrared spectroscopy can be used as an alternative technique for this purpose in soil analysis, and transmission and reflectance methods are already widely used. However, the transmission method is usually used in soil qualitative analysis, while reflectance can be used in soil quantitative analysis, and most soil-related research is focused on reflectance spectroscopy. The objective of this chapter is to describe the applicability of these techniques for site-specific management of soil fertility with specific reference to application of precision agriculture to smallholder farming in China.

The techniques of infrared reflectance spectroscopy, including diffuse reflectance spectroscopy and total attenuated reflectance spectroscopy, are involved in soil quantitative analysis. Excellent performance of predicting soil carbon (C) and nitrogen (N) content using infrared reflectance spectra is observed, and in most cases, the predictions of soil phosphorous (P), potassium (K), calcium (Ca), magnesium (Mg), sulfur (S), and some other microelements are satisfactory. Besides the prediction of soil nutrients, soil water, soil clays, soil microbes, and so forth are also characterized and evaluated using infrared reflectance spectroscopy. In recent years, infrared photoacoustic spectra have been used in soil analysis; this method is more convenient in sample pretreatment and spectra recording, and the recorded soil spectra contain more useful information than that generated by conventional reflectance spectroscopy. Although currently the application of infrared photoacoustic spectroscopy in soil analysis is limited, it has great potential for the future, and better performance will be demonstrated in the characterization of soil fertility. The application of infrared spectroscopy in soil fertility largely depends on spectra analysis. Partial least square (PLS) and artificial neural network (ANN) are two widely used mathematical tools in the prediction of soil properties, and more mathematical tools combined models will further benefit prediction capabilities and use in precision agriculture. In order to make full use of soil infrared spectra, a soil spectra library construction is urgently needed so that soil fertility can be quickly evaluated by combining suitable mathematical models, which will promote sustainable agriculture through adoption of precision agriculture based on these soil analytical techniques.

6.1.1 Definition of Soil Fertility

Soil, occurring in the upper few centimeters (50 to 150 cm) of the Earth's surface at the interface between the atmosphere, biosphere, hydrosphere, and geosphere, plays a vital role in issues of food production and ecological environment, and soil, the foundation for most terrestrial life, has unrivaled complexity. Soil contains minerals, organic matter, and uncountable numbers of organisms, as well as varying amounts of air and water, which provide support for life (Wilding and Lin 2006). A single gram of soil usually contains tens to thousands of millions of fungi and bacteria, plus thousands of diverse plant and animal species; soil is both an ecosystem in itself and a critical part of the larger terrestrial ecosystem (Uphoff et al. 2006). From the earliest perceptions of soils as the organic enriched surface layer to today's pedologic horizonation of profiles, there is a rich history of beliefs and understanding of this vital life-sustaining resource (Richard 2006). This biologically active, structured porous medium, called the *pedosphere*, mediates most of the biogeophysical and chemical interactions among the land, its surface and groundwaters, and the atmosphere.

As we know, yields are difficult to maintain with a sandy soil, whereas the yields are much better in clay soils, in which the difference of this soil property is expressed with soil fertility. Soil fertility is a commonly used concept in soil science, and it is the function of soil properties, including soil nutrients, soil moisture, soil mineral, and soil organic matter (Desbiez et al. 2004), and in different conditions the limited factor in soil fertility varied, (Cardoso and Kuyper 2006). Therefore, soil fertility is very comprehensive, which can't be measured directly but can be evaluated by some other soil properties (Bautista-Cruz et al. 2007). Considering sustainable agriculture (both economic and environmental aspects), soil fertility can be defined as the ability of a soil to serve as a suitable substrate on which plants can grow and develop in a sustainable way (Izac 2003; Adjei-Nsiah et al. 2007). Fertile soils facilitate root development, supply water, air, and nutrients to plants, have little soil erosion, and do not have pest and disease burdens that result in catastrophic impacts on the plants. Ultimately, fertile soil needs a well-balanced combination of soil mineral and soil organic matter that results in good production. Yet soil fertility is also a function of gases, liquids, organic matter, and myriad organisms in soils. As the wooden bucket principal states, water capacity is decided by the shortest board rather than the longest one, and soil fertility can be looked on as the water capacity in the wooden bucket, which is limited by the poorest soil property. Hence, soil fertility requires a biologically framed understanding of the soil system for making its management more productive and sustainable.

6.1.2 Soil Fertility and Food Production

With an increasing population on Earth, feeding people and protecting the ecological environment are critical problems faced by most countries. Maintaining soil fertility is an essential way to solve these problems, but doing so requires preservation of its organic matter, physical properties, water, and nutrient levels. In most cases, the leading factor involved in soil fertility is nutrient status and deficiency of soil organic matter (Alfaia et al. 2004). Recent research has provided evidence that soil fertility is declining in many farmed areas, and one reason for this is the high amount of input of chemical fertilizers. Another significant driver is tillage-related soil erosion, which dramatically affects the soil organic matter stocks along with associated nutrients (Gobeille et al. 2006). Nutrient budgets are especially threatening for the fertility of soils whose nutrient stocks are already small, such as sandy soils or acid soils with low organic matter contents. The nutrient balance of a site can improve soil structure by reducing unproductive nutrient losses from erosion and leaching and by increasing nutrient inputs through nitrogen fixation, water-holding capacity, and crop rooting volume, as well as increasing biological activity in the soil by providing biomass and a suitable microclimate. However, a better understanding of the interactions among crops, soils, and microbes has also helped to keep expectations at a realistic level and to recognize what agroecosystems can and cannot achieve (Izac 2003).

Obviously, soil fertility is very important to agricultural production, thus it is necessary to know how to evaluate the fertility of soil in order to know what measures can be taken to maintain it or to delay its declining rate.

6.1.3 Soil Fertility and Soil Nutrients

Mineral elements are usually the principal factor in the assessment of soil fertility since they are the essential materials needed in plant growth, while soil organic matter plays an indirect role in soil fertility, which shows as capacity that holds nutrient and water. The mineral portion of soil, which differs from system to system in its chemical composition and its physical characteristics, has long been the focus of most soil science research. These mineral elements exist in different-sized soil particles, classified (from large to small) as sand, silt, or clay. The mineral composition of soil establishes its physical properties, and it influences and is influenced by the life forms that are present. The availability of soil nutrients and the supply capacity of soil nutrients are largely dependent on soil water and soil organic matter (Uphoff et al. 2006), thus soil water and soil organic matter are also closely related to soil fertility. Soil water is usually about a quarter of soil volume, although the actual amount can vary greatly over time and between soil systems. With too little water, soil systems become desiccated, and with too much, they are saturated. Air in well-aggregated soil can be another quarter of the volume, containing oxygen, hydrogen, nitrogen, and carbon in gaseous forms. The more pore space within the soil, the greater will be its capacity for holding both water and air that benefit plants as well as other flora and fauna in the soil. For any given soil porosity, the amounts of water and air are usually inversely related. As for soil organic matter, it usually occupies between 1% and 6%, although it can be higher. This organic materials category encompasses non-living organic matter that is derived from the growth, reproduction, death, and decomposition of plants, animals, and microbes and exists in the soil as humus or as other inanimate material, and an immense variety of living flora and fauna, referred to collectively as the soil biota. The organic portion of soils includes both soil organisms and the various biological substances and processes that animate soil systems. The connection between the mineral and organic components of soil systems is intimate, converging at the smallest scale of soil structure and function in what are called clay-humus complexes. At the next higher level of structure, in microaggregates, inert and living materials are practically fused. Although this is well known, the biological dimensions of soil systems are too often regarded more as secondary or intervening variables rather than as central and determining factors.

Therefore, soil nutrients level and organic matter content are the main soil properties used in the evaluation of soil fertility, and the conventional option to determine these soil properties is mainly based on chemical methods in which sample pretreating, such as soil extraction and soil digestion, and sample processing, such as colorization, are needed (Faithfull 2002). Obviously, chemical methods provide a useful tool for understanding the soil, but most of these methods are time-consuming and costly, which makes them unsuitable for fast or *in situ* evaluation of soil quality as well as the analysis of mass soil samples that are needed in modern agriculture (McCarty and Reeves 2006; Ortega and Santibanez 2007). Therefore, instrumental methods based on rapid sensing of soil are needed for highly efficient food production.

6.1.4 INFRARED SPECTROSCOPY IN SOIL ANALYSIS

Soil sensing emerges as a new discipline in soil science that can provide abundant information for soil management. Remote sensing is widely used in soil sensing; however, it is well known that it is difficult to obtain soil fertility information using remote sensing due to large prediction errors resulting from the plant cover above soil. Alternatively, proximal soil sensing is widely used in soil analysis, in which infrared spectroscopy is highly recommended. Infrared spectroscopy has advantages over some of the conventional techniques of soil analysis: they are rapid, timely, and less expensive, and hence are more efficient when a large number of analyses and samples are required (McCarty and Reeves 2006; Nanni and Dematte 2006). Moreover, spectroscopic techniques do not require expensive and time-consuming sample preprocessing or the use of (environmentally harmful) chemical extractants. Infrared spectroscopy may, in some instances, be more straightforward than conventional soil analysis and also be more accurate (McCauley et al. 1993; Viscarra Rossel et al. 2001, 2006). One other advantage is the potential adaptability of the techniques for *in situ* field use (Viscarra Rossel and McBratney 1998). These are particularly important advantages now that

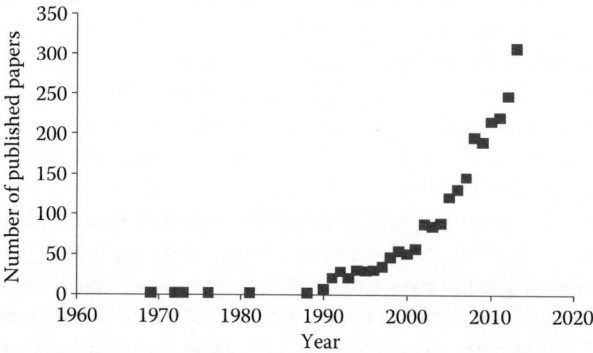

FIGURE 6.1   Number of published papers related with soil infrared spectroscopy. The papers were searched using keywords of soil and infrared spectroscopy in the Web of Science.

there is an increasing global need for larger amounts of good quality, inexpensive spatial soil data to be used in environmental monitoring and precision agriculture (Mouazen et al. 2007).

Papers on soil analysis using infrared spectroscopy have been published since the 1970s (Figure 6.1) and have increased since the 1990s; even more rapid growth has been observed in the last 10 years, and this tendency will most likely continue. These applications cover a large discipline including soil chemistry, soil physics, soil biology et al. Therefore, infrared spectroscopy has been widely used and applied in soil analysis with great potential, which will play an important role in modern agriculture.

6.2 TECHNIQUES OF INFRARED SPECTROSCOPY

6.2.1 Absorption of Infrared Spectroscopy

Infrared spectroscopic techniques are highly sensitive to both organic and inorganic phases of the soil (i.e., soil organic matter and soil clay), making them useful in the agricultural and environmental sciences. Intense fundamental molecular frequencies related to soil components occur in the mid-infrared (MIR) between wavelengths of 2500 and 25,000 nm. The visible and near infrared portions of the electromagnetic spectrum are highly used (Viscarra Rossel et al. 2006). Infrared spectroscopy is a technique based on the vibrations of atoms in a molecule. An infrared spectrum is commonly obtained by passing infrared radiation through a soil sample and determining what fraction of the incident radiation is absorbed at a particular energy. The energy at which any peak in an absorption spectrum appears corresponds to the frequency of a vibration of a part of a sample molecule, which makes infrared spectroscopy an alternative method in soil evaluation (Dematte et al. 2004).

6.2.2 Methods of Infrared Spectroscopy

6.2.2.1 Infrared Transmission Spectroscopy

Transmission spectroscopy is the oldest and most straightforward infrared method. This technique is based on the absorption of infrared radiation at specific wavelengths as it passes through a sample. It is possible to analyze samples in liquid, solid, or gaseous forms when using this approach (Stuart 2004).

Liquid samples can be run neat or by dissolving in a solvent. The sample concentration and path length should be selected to obtain the transmittance in the range of 15%–70% in order to get a good infrared (IR) spectrum. This will correspond to about 0.02 mm cell thick in the case of most neat liquids and concentration of 10% and cell length of 0.1 mm in the case of most solutions. The solvent selected must be transparent in the region of interest. Neat liquids can be analyzed between salt plates made of sodium chloride (NaCl) or potassium bromide (KBr). Non- or low volatility liquids can be analyzed by placing a drop of the sample onto specially prepared thin polyethylene (or other) polymer substrates. These supports are cheap and disposable. They absorb IR only in well-known, narrow bands, which depend on the substrate materials. These absorptions can be accounted for using the clean substrate as a background.

Solid samples, such as a soil sample, can be prepared for IR analysis by the pellet technique. This technique is based on the fact that a sample in KBr powder can be compressed under pressure with or without vacuum to form transparent disks. In this technique, a solid sample of approximately 2–3 mg is allowed to mix with about 0.2–1 g of KBr (which is transparent to IR). The mixture is thoroughly ground in a mortar, then pressed in a pellet die under a pressure of about 6000–10,000 psi to obtain a transparent disk. Good dispersion of the sample in KBr is critical. It should be pointed out that bands near 3448 and 1639 cm^{-1} from moisture often appear in the spectra obtained by this technique. One should avoid moisture by, for example, freeze-drying the sample as needed. The KBr disk can be prepared with a Mini-Press accessory. Pastes and other semisolids are routinely analyzed with the help of an attenuated total reflectance (ATR) attachment (Linker et al. 2004).

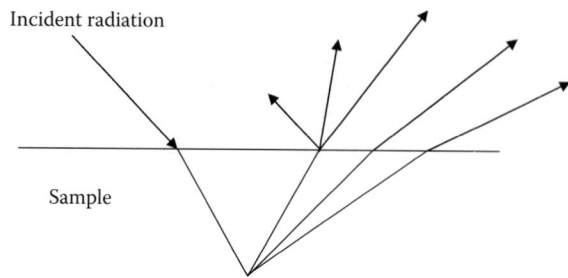

FIGURE 6.2   Illustration of diffuse reflectance.

6.2.2.2   Infrared Diffuse Reflectance Spectroscopy

In external reflectance, the energy that penetrates one or more particles is reflected in all directions and this component is called diffuse reflectance. In the diffuse reflectance (infrared) technique, commonly called DRIFT, the DRIFT cell reflects radiation to the powder and connects the energy reflected back over a large angle. Diffusely scattered light can be collected directly from material in a sampling cup or, alternatively, from material collected by using an abrasive sampling pad. Figure 6.2 illustrates diffuse reflectance from the surface of a sample.

Kubelka and Munk developed a theory describing the diffuse reflectance process for powdered samples that relate the sample concentration to scattered radiation intensity (Kubelka and Munk 1931).

6.2.2.3   Infrared Attenuated Total Reflectance Spectroscopy

Fourier transform infrared attenuated total reflectance spectroscopy (FTIR-ATR) utilizes the phenomenon of total internal reflection (Figure 6.3). A beam of radiation entering a crystal will undergo total internal reflection when the angle of incidence at the interface between the sample and crystal is greater than the critical angle, where the latter is a function of the refractive indices of the two surfaces. The beam penetrates a fraction of wavelength beyond the reflecting surface and when a material that selectively absorbs radiation, such as soil, is in close contact with the reflecting surface, the beam loses energy at the wavelength where the material absorbs. The resultant attenuated radiation is measured and plotted as a function of wavelength by the spectrometer and gives rise to the absorption spectral characteristics of the sample.

The crystal used in ATR cells are made from materials that have low solubility in water and are of a very high refractive index. Such materials include zinc selenide, germanium, and thallium iodide. Different designs of ATR cells allow both liquid and solid samples to be examined. It is also possible to set up a flow-through ATR cell by including an inlet and outlet in the apparatus. This allows for continuous flow of soil solutions through the cell and permits spectral changes to be monitored over time.

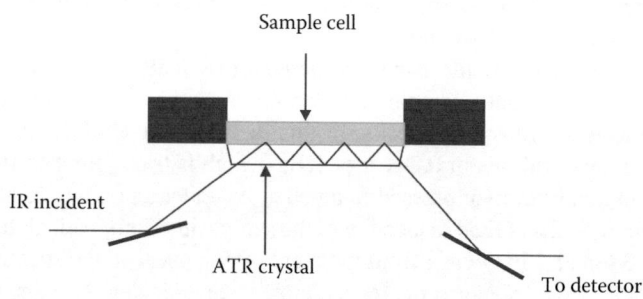

FIGURE 6.3   Schematic of a typical attenuated total reflectance cell.

In most cases, a zinc-selenide-based ATR cell is used in analysis, and this ATR cell is not suitable for direct soil analysis due to difficulty for a well touching between the soil sample and Zinc selenide crystal. Thus, soil paste was prepared in soil analysis (Linker et al. 2004, 2005), but water absorption is strong, the obtained information is limited, and the useful wavenumber range is in 1000–1500 cm⁻¹, which can be applied in direct nitrate determination (Shao et al. 2014). Direct soil testing using ATR requires a diamond crystal, which needs a higher input in ATR cell.

6.2.2.4 Infrared Photoacoustic Spectroscopy

Fourier transform infrared photoacoustic spectroscopy (FTIR-PAS) is a relatively new infrared technique that is based on photoacoustic theory. The photoacoustic phenomenon was invented in 1880 by Alexander Graham Bell while experimenting with a photophone. After this accidental discovery, several experiments with solid, liquid, and gas samples were conducted; however, the phenomenon remained in the background until the advent of the microphone (Ryczkowski 2010). Until the 1970s, a quantitative derivation was presented for the acoustic signal in a photoacoustic cell in terms of the optical, thermal, and geometric parameters of the system, giving a theoretical foundation for the technique of photoacoustic spectroscopy (Rosencwaig and Gersho 1976), and this major improvement made it a very competitive and widely used method.

FTIR-PAS is based on the absorption of electromagnetic radiation by molecules (McClelland et al. 2002). Nonradioactive relaxation processes (such as collisions with other molecules) lead to local warming of the soil sample matrix. Pressure fluctuations are then generated by thermal expansion, which can be detected by a very sensitive microphone (Figure 6.4) (Du et al. 2007). The resulting spectrum differs from both equivalent transmission and reflectance spectra (both diffusion reflectance spectrum and total attenuated reflectance spectrum) because the technique detects nonradioactive transitions in the sample. PAS spectra can be obtained for materials that are optically opaque, such as soil samples; the absorption is measured directly, and the acoustic wave amplitude is proportional to the absorption and the concentration of the absorbing gas (e.g., air, argon, helium [He]; He is mostly preferred for its high thermal sensibility). A spectrum is obtained by rationing the detector signal to that generated with a totally absorbing material such as carbon black (McClelland et al. 2002). The detected signal in photoacoustic spectroscopy is proportional to the target concentration and can be used with highly absorbing soil samples without any pretreatment. Photoacoustic (PA) measurements are unique in that they depend directly on the energy absorbed by the sample rather than on what is transmitted or reflected; using mathematical tools, such as partial least squares (PLS), FTIR-PAS has found wide application (Armenta et al. 2006; Zoltan et al. 2011).

Two types of FTIR-PAS measurements are commonly undertaken. The first involves signal detection undertaken when the FTIR spectrometer is operating in its conventional rapid scan mode. For spectrometers equipped with a Michelson interferometer operating in this mode, the modulation

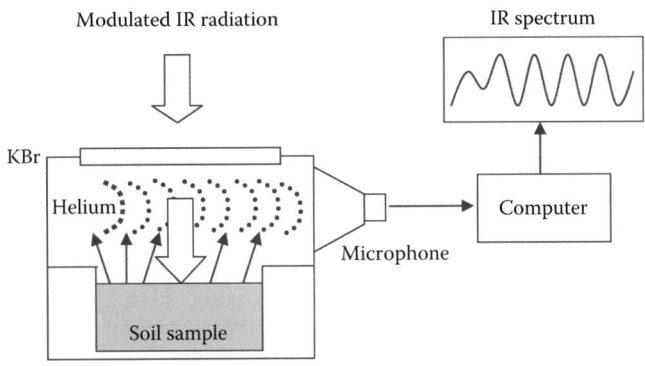

FIGURE 6.4 Testing of a typical photoacoustic spectroscopy cell.

frequency is proportional to the wavenumber; the moving mirror in the interferometer provides the audio-frequency modulation necessary for the generation of a PA signal (Michaelian 2010). To achieve a constant modulation frequency, which is independent of the wavenumber, the interferometer should be operated in a step-scan mode. Spectra obtained in this mode do not exhibit the wavenumber-dependent frequency effects that affect spectra in rapid scan mode. In a conventional PA cell, the acoustic resonator and microphone play important roles in signal production (Miklos et al. 2001). In recent decades, quartz-enhanced PA cells and cantilever-type PA cells were developed to obtain PAS spectra, and the signal amplitude, Q-factor, and signal-to-noise ratio were improved (Wu et al. 2015), thus the sensibility was remarkably enhanced (Dong et al. 2010; Koskinen et al. 2010). However, these PA cells were popularly used in gas monitoring (as a gas sensor) (Grave et al. 2015), and the applications are limited for solid and liquid samples (Hirschmann et al. 2010). A PA cell can be improved through modification of the resonator and microphone, such as a quartz-enhanced PA cell and cantilever-type PA cell.

FTIR-PAS has several unique advantages over conventional infrared spectroscopy techniques. For example, it needs minimum sample pretreatment, and soil particle size and high infrared radiation absorption by soil samples have little influence on spectral recording (Du and Zhou 2007). Subsequently, better repeatability associated with FTIR-PAS can result in improved accuracy and/ or precision of recorded soil attributes. Using multivariate statistic analysis, such as PLS, FTIR-PAS made it a wide application with quantitative purpose in solid substance (Wahls et al. 2000; Bjarnestad and Dahlman 2002; Irudayaraj et al. 2002; Armenta et al. 2006).

6.3 ASSESSMENT OF SOIL FERTILITY BY INFRARED SPECTROSCOPY

6.3.1 Infrared Spectra-Based Soil Qualitative Analysis

6.3.1.1 Characterization of Soil Components

The soil mid-infrared spectrum (400–4000 cm^{-1}) can be approximately divided into four regions and the nature of a group frequency may generally be determined by the region in which it is located (Figure 6.5). The fundamental vibrations in the 2500–4000 cm^{-1} region are generally due to O–H, C–H, and N–H stretching. O–H stretching produces a broad band that occurs in the range of 3700–3600 cm^{-1}. By comparison, N–H stretching is usually observed between 3400 and 3300 cm^{-1}. The absorption is generally much sharper than O–H stretching and may, therefore, be differentiated. C–H stretching bands occur in the range of 3000–2850 cm^{-1}. Triple-bond stretching absorptions fall in the 2000–2500 cm^{-1} region because of the high force constants of the bonds. C–C bonds absorb between 2300 and 2050 cm^{-1}, while the nitrile group (C–N) occurs between 2200 and 2300 cm^{-1}. The principal bands in the 1500–2500 cm^{-1} regions are due to C=C and C=O stretching. Carbonyl stretching is one of the easiest absorptions to recognize in an infrared spectrum, and it occurs in the 1650–18,030 cm^{-1} region. C=C stretching is much weaker and occurs at around 1650 cm^{-1}. It has been assumed so far that each band in an infrared spectrum can be assigned to a particular deformation of the molecule, the movement of a group of atoms, or the bending or stretching of a particular bond. Many vibrations may vary by hundreds of wavenumbers, and these combine most bending and skeletal vibrations, which absorb in the 650–1500 cm^{-1} region, referred to as the fingerprint region.

The absorptions observed in the near-infrared (NIR) regions (4000–13,000 cm^{-1}) are overtones or combination of fundamental stretching bands that occur in the 1700–3000 cm^{-1} region. The bands in the NIR are often overlapped, making them less useful versus mid-infrared region for qualitative analysis. However, there are important differences between the NIR positions of different functional groups and these differences can often be exploited for quantitative analysis.

Soil clay minerals had abundant absorptions in mid-infrared region (Figure 6.6), and each clay mineral showed specific absorption, which can play a role as a fingerprint in specific identification (Du et al. 2007, 2008a). Kaolin (1:1 clay mineral) and montmorillonite (2:1 clay mineral) are

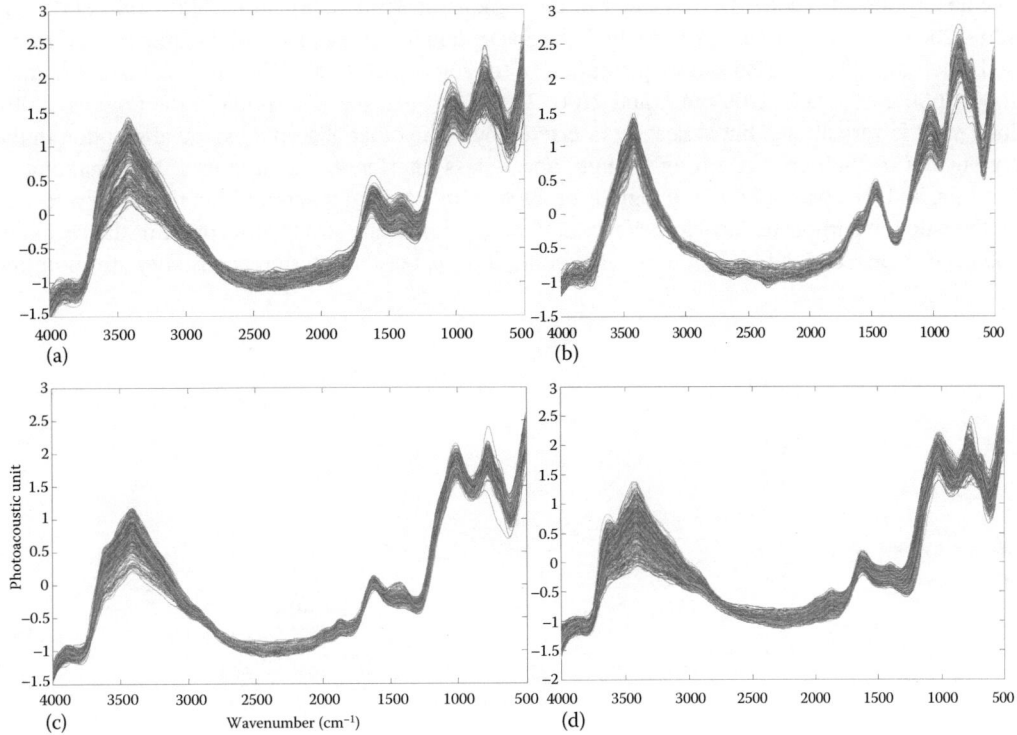

FIGURE 6.5 Mid-infrared photoacoustic spectra of typical farmland soils in China. (a) Black soil from north China ($n = 647$), (b) fluvo-aquic soil from middle China ($n = 221$), (c) paddy soil from east China ($n = 1635$), and (d) red soil from south China ($n = 1648$).

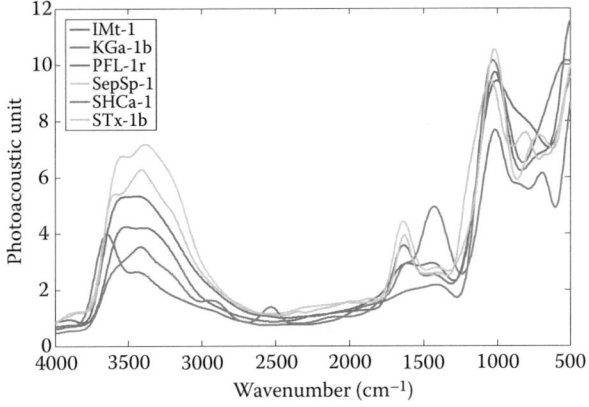

FIGURE 6.6 FTIR-PAS spectra of typical clay minerals in soil. Source clays were purchased from clay minerals society. IMt-1 = illite, KGa-1b = kaolin, PFL-1 = attapulgite, SepSp-1 = sepiolite, SHCa-1 = hectorite, STx-1b = montmorillonite.

two popular clay types commonly encountered in the investigated soils. Clear absorptions are visible in several spectral regions, and in particular around 2800–3700 cm^{-1}, 2200–2600 cm^{-1}, 1800–2100 cm^{-1}, and 900–1600 cm^{-1}. The absorptions of bentonite are demonstrated in the regions of 2800–3700 cm^{-1} (O–H stretching), 1500–1800 cm^{-1} (C=O stretching), and 800–1200 cm^{-1} (fingerprint region), in which the absorption in the regions of 2800–3700 cm^{-1} is a wide band.

The absorptions of kaolin are indicated in the regions of 3500–3700 cm^{-1}, 1500–2000 cm^{-1}, and 800–1200 cm^{-1} (Si–O stretching), in which the absorption in the regions of 800–1200 cm^{-1} is strong. Soil calcium carbonate also shows absorption in the regions of 2900–3100 cm^{-1}, 2300–2600 cm^{-1}, 1000–1600 cm^{-1}, 1600–1700 cm^{-1}, and 2100–2200 cm^{-1}, and the absorption in the regions 1000–1600 cm^{-1} is very strong, but is heavily interfered by some other absorptions; the absorption in the regions 2300–2600 cm^{-1} is strong enough, and is less interfered, which is useful in quantitative analysis, and the absorption in the regions of 2900–3100 cm^{-1} might come from the water attached on the calcium carbonate particle surface. Soil water indicates a strong absorption in the region of 1600–1700 cm^{-1} and 2900–3600 cm^{-1} (Du et al. 2008b), but the interference is very strong in the

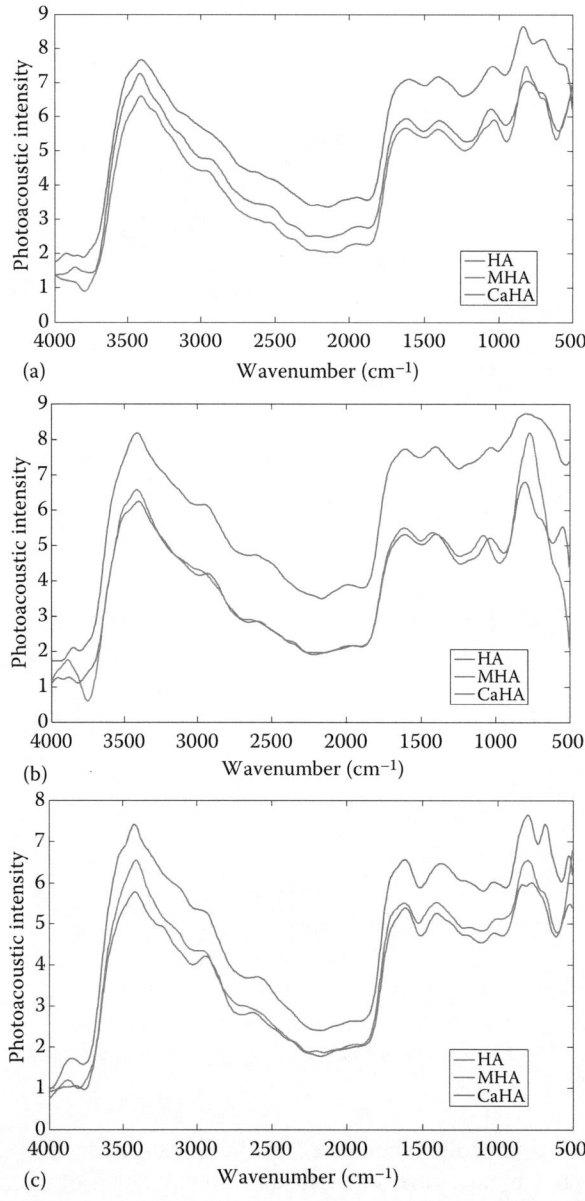

FIGURE 6.7 Mid-infrared photoacoustic spectra of different humic substances from soil. (a) Humic substances from Elliott, (b) humic substances from Pahokee, and (c) humic substances from Loenardite.

region of 2900–3600 cm^{-1}. In addition, besides the characteristics of soil components absorptions soil properties are usually interrelated (Cohen et al. 2005; Du 2012), which makes each property directly or indirectly determined by infrared spectroscopy.

Soil humic substances also contribute to soil spectra (Figure 6.7). Different humic substances have different spectra appearances, especially between humic acid (HA) and other humic substances mobile humic acid (MHA) and recalcitrant calcium humate (CaHA), and the humic substance from different soil varies in component and structure, which provides fundamentals for soil qualitative and quantitative analysis.

6.3.1.2 Soil Identification

Soil types are determined by soil components, and each soil component has a specific characterization of the infrared spectrum (Figures 6.6 and 6.7). Spectral signatures of soils are defined by their reflectance, or absorbance, as a function of wavelength. Under controlled conditions, the signatures are due to electronic transitions of atoms and vibrational stretching and bending of structural groups of atoms in molecules and crystals. The fundamental vibrations of most soil materials can be found in the mid-infrared region, with overtones and combinations found in the NIR region. Soil minerals such as different clay types have very distinct spectral signatures in the infrared region because of strong absorption of the overtones of SO_4^{2-}, SO_3^{2-}, and OH^- and combinations of fundamental features, for example, water (H_2O) and carbon dioxide (CO_2) (Figure 6.6) (Brown et al. 2006; Du et al. 2007). Therefore, it is possible to have a soil identification based on soil infrared spectrum.

Infrared spectroscopy is a well-established technique for the identification of chemical compounds and/or specific functional groups in compounds, and thus is a useful tool for soil applications (Johnston and Aochi 1996; Haberhauer and Gerzabek 2001). In particular, reflectance spectroscopy can be used for nondestructive assessment of soil and crop physical and biochemical properties (Chang et al. 2001; Dunn et al. 2002; Shepherd and Walsh 2002; Cozzolino and Moron 2003, 2006; Shepherd et al. 2003). Although the NIR range (800–2500 nm) is still the most widely used, mid-infrared spectroscopy is becoming increasingly common due to the specificity of the absorbance bands in that spectral range (Stuart 1997). In particular, mid-infrared attenuated total reflectance (ATR) spectroscopy can be used for fast and simple determination of nitrate concentration in water and soil pastes (Shaviv et al. 2003; Linker et al. 2004). Linker et al. (2005, 2006) also showed that mid-infrared ATR spectroscopy could be used to identify major types of agricultural soils based primarily on absorbance bands associated with characteristic soil constituents (e.g., calcium carbonate, clay minerals, and possibly organic constituents), and such identification of soil types led to the significant improvement of ATR-based determination of nitrate in soil pastes (Linker et al. 2006). FTIR-PAS is another spectral technique that can be used for the identification of constituents in complex systems (McClelland et al. 2002). A major advantage of photoacoustic spectroscopy is that it is suitable for highly absorbing solid samples without any special pretreatment. With respect to soil analysis, this is a major advantage compared to transmittance measurements that require time-consuming preparation of KBr pellets or the ATR configuration that requires a saturated soil paste and suffers from interferences associated with the presence of water (Linker et al. 2004), and very good classification performances were achieved, with correct classification rates of the validation samples typically above 95% (Du et al. 2008a; Linker 2008).

6.3.2 Infrared Spectra-Based Soil Quantitative Analysis

Infrared spectroscopy has been widely used in the soil quantitative analysis, and most of soil properties have been well predicted (Table 6.1).

6.3.2.1 Soil Nutrients

Soil nutrients, such as C, N, P, K, S, Ca, and microelements, play the most important role in soil fertility. Usually the nutrient contents are determined through laboratory analysis; however, many

TABLE 6.1

Adsorbed Water Vibrations in the Middle Infrared Spectra

Frequency (cm⁻¹)	Vibration Mode	Water Type
3750–2900	Stretching OH	OH groups on surface or at specific crystallographic sites and liquid water
3280	Stretching OH Symmetric, $\nu 1$	Liquid
3490	Stretching OH, Symmetric, $\nu 1$	Liquid
3300–3000	Stretching OH; hexameric and more complex clusters	Adsorbed water
3430	Stretching OH	Bulk water
3150	Stretching OH	Water firmly bound to a specific site,
3050	Stretching OH	perhaps with a cluster structure
3250	Stretching OH	First spectrum: mix of bulk and adsorbed water components
3135	Stretching OH	Adsorbed water before vacuum pumping
3100	Stretching OH	Adsorbed water under vacuum pumping

of the existing methods of soil analysis are resource-intensive, and do not lend themselves to the use of a large number of samples (Ludwig et al. 2002). It is possible to evaluate the nutrient content using infrared spectroscopy (Chang et al. 2001; Pirie et al. 2005; Verma and Deb 2007a,b), and the technique of infrared reflectance spectroscopy could be a faster, cheaper, and more objective way to evaluate soil nutrients (Daniel et al. 2003; Brown et al. 2006). Soil properties are usually inter-related, which makes the predictions of most soil properties possible (Cohen et al. 2005; Rinnan and Rinnan 2007). Viscarra Rossel et al. (2006) provided a review of some literature comparing quantitative predictions of soil nutrients using various multivariate techniques and reflectance spectra response in the infrared region of the spectrum. The content of soil C and N are mainly studied because they are more sensitive to infrared radiation. The calibration coefficients (R^2) are in the region of 0.80–0.98, and the root mean square errors (RMSE) are very satisfactory in fast evaluation of soil fertility (Reeves et al. 2001; McCarty et al. 2002; Cozzolino and Moron 2006; Stevens et al. 2008).

Soil C can be spectrally measured with a reasonable level of accuracy depending on the type of instrument and environmental conditions, with RMSE ranging from 1 to 15 g C kg⁻¹ (Ludwig et al. 2002; Brown et al. 2006; Stevens et al. 2006). The calibration coefficient (R^2) is more than 0.8 (Zimmermann et al. 2007; Leach et al. 2008; Wetterlind et al. 2008a), and the size of the soil particle has a strong influence on the calibration and validation (Barthes et al. 2008a). C mineralization is also studied using diffuse reflectance spectra and it is useful in evaluating C storage potential in soils (Mutuo et al. 2006). Infrared spectra of soil C source materials (i.e., humic acids, fulvic acids, and their interaction products) has provided much information about their characterization and determination (Byler et al. 1987; Francioso et al. 2007), which will benefit the evaluation of soil fertility.

Soil nitrate concentration can be directly measured using FTIR-ATR through the correlation between nitrate concentration and the vibration band around 1350 cm⁻¹ (Verma and Deb 2007b). Shaviv et al. (2003) showed that MIR spectroscopy using either standard ATR crystal can be used for direct determination of nitrate concentration in water, soil extracts, or soil pastes. By applying a straightforward chemometric approach, Linker et al. (2004) improved the determination accuracy and overcame some of the interferences associated with direct measurements in soil pastes. However, this correlation between soil nitrate concentration and the infrared absorption band is soil-dependent, due mostly to varying contents of carbonate (Linker et al. 2004, 2005; Jahn et al. 2006). Linker et al. (2005) suggested the use of a two-stage method that can be summarized as follows: (1) determination of the soil type by comparing the so-called fingerprint region of the spectrum 800–1200 cm^{-1} to a reference spectral library, and (2) determination of the nitrate concentration using the model corresponding to this soil type. This soil identification approach led to determination errors significantly lower than those reported earlier (Shaviv et al. 2003; Linker et al. 2004), and determination errors range from 6.2 to 13.0 mg /kg, depending on the soil type, with the lowest errors for light sandy soils. These determination errors are appreciably smaller than those obtained using a single model calibrated using all the data (Linker et al. 2006).

For the prediction of the other soil nutrients including P, K, and microelements, the calibration results are not stable, which are pending to the variability and capacity of calibration set (Janik et al. 1998; Ludwig et al. 2002; Cozzolino and Moron 2003; Brown et al. 2006), and it also showed that NIR was not a good tool for P and K prediction with R, 0.47 and 0.68, and SEP, 33.70 and 26.54, respectively (He et al. 2007), and future research should be addressed to build calibrations for open populations (Terhoeven-Urselmansa et al. 2008).

Infrared reflectance spectroscopy is usually used in soil quantitative analysis, but it has certain limitations, especially in the sample pretreatment. Recently, infrared photoacoustic spectroscopy was used in soil quantitative analysis, and a better calibration result for soil C, N, P, and K was observed (Du and Zhou 2007; Du et al. 2008c). This technique does not need sample pretreatment, and a fast and *in situ* monitoring of soil nutrients can be attained, which makes it a promising method in the evaluation of soil fertility.

Using infrared spectroscopy combined with Geographic Information System (GIS) and statistical methods, the N, P, K, and soil organic matter (SOM) spatial variability within the field can be obtained (Odlare et al. 2005; Christy 2008; Wetterlind et al. 2008a), and their distribution maps can be drawn (He et al. 2005) in which soil nutrient status can be directly indicated. The reference maps for the predicted and measured values of N and OM were almost the same, unlike with P and K, due to the unsuccessful prediction of these constituents. A phosphorus sensing system could be developed using diffuse reflectance of soil for soil P testing (Bogrekci and Lee 2005a; Maleki et al. 2006; Mouazen et al. 2007), and based on this technique, spectral OM and N maps could be drawn, in which the variability could be well represented (Figure 6.8) and would be useful in precision agriculture (Bogrekci and Lee 2005b). The maps derived from the infrared spectra data are promising, and the potential for developing a cost-effective strategy to map soil from infrared spectra data at the farm scale is considerable (Wetterlind et al. 2008b).

6.3.2.2 Soil Clays

Absorptions by water bonds associated with clay content and other bonding associated with clay type provide the opportunity to use infrared spectra for quantifying clay information in soil. Using infrared spectroscopy, research on air-dried ground soil samples has shown predictions of soil clay content with R^2 values ranging from 0.56 to 0.91 and RMSE ranging from 23 to 11 g kg^{-1} (Ben-Dor and Banin 1995; Janik et al. 1998; Shepherd and Walsh 2002; Islam et al. 2003; Sorensen and Dalsgaard 2005; Brown et al. 2006). An infrared spectra data-based calibration model for determination of clay in soil was developed and tested in practice. The model showed ruggedness, linearity, and stable prediction error over the calibrated content range (2%–26% clay). The uncertainty of the method was <40% higher than the reproducibility standard deviation of the reference method for clay content below 26%. As seen in many other cases, the prediction error was dependent on the content range calibrated. When the range was extended from 2% to 74% clay, the estimated prediction error increased to 3.4%. However, a standard deviation (SD)/RMSE ratio of 4.7 demonstrated a high correlation between NIR spectral data and reference data (Sorensen and Dalsgaard 2005). Visible NIR diffuse reflectance spectra was capable of predicting soil clay content *in situ* at varying water contents, in which the RMSE was 61 g kg^{-1} (Waiser et al. 2007).

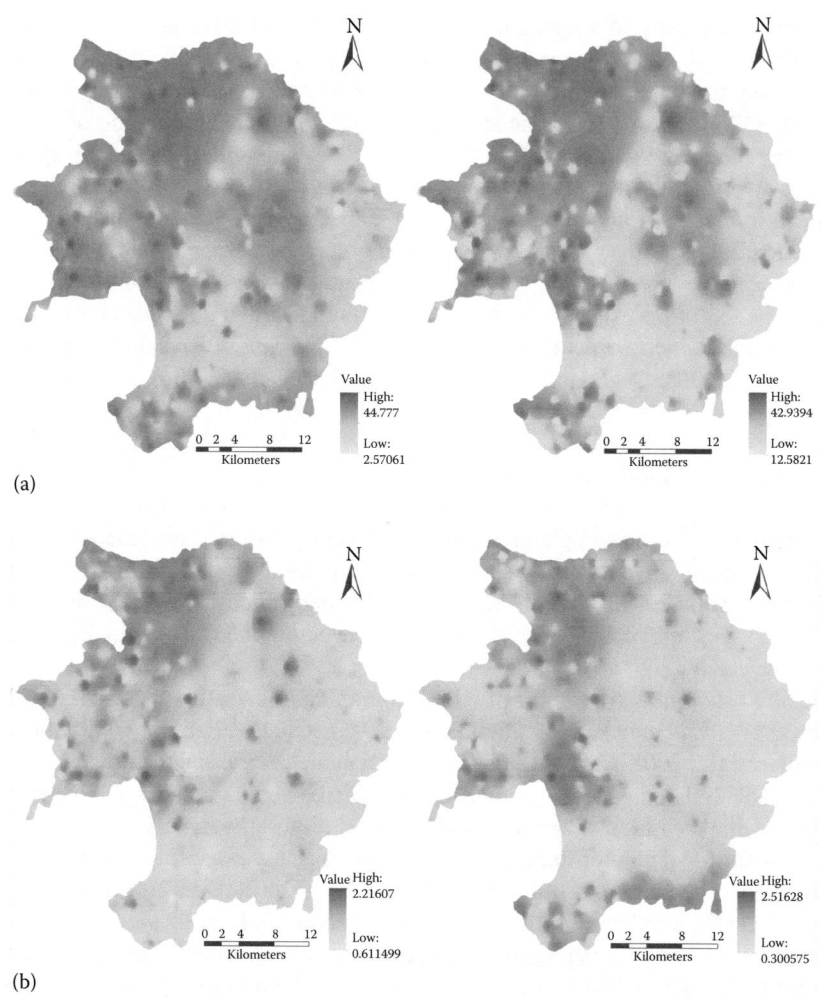

FIGURE 6.8 Actual (left) and partial least-squares predicted (right) soil organic matter and nitrogen content in Lishui County, Jiangsu Province, China. (a) Organic matter content in soil and (b) total N in soil.

Figures 6.6 and 6.7 show the photoacoustic spectra of major soil components (inorganic clays and organic humic substances) (Du et al. 2007). Montmorillonite (standing for smectites) and kaolin, which represent the two clay types most commonly encountered in the investigated soils, have spectra that differ notably, share only the strongest band around 1040 cm^{-1}, and have some overlap around the 3600–3700 cm^{-1} range. Calcium carbonate is easily identifiable with a series of strong and well-defined bands (900, 1450–1550, 2500–2550, and 2800–3000 cm^{-1}), while the overall intensity of the quartz spectrum is much lower and includes only weak bands in the 1000–2000 cm^{-1} region. Comparison shows that the spectra of mineral soils agree well with expectations based on soil composition. For instance, the fluvo-aquic soil, which has a very high calcium carbonate ($CaCO_3$) content and relatively high clay content but cannot contain kaolin, has very strong bands around 1040 cm^{-1} (smectite), 1430 cm^{-1}, and 2520 cm^{-1} ($CaCO_3$), and two smaller peaks around 1600 and 1800 cm^{-1} (smectite and $CaCO_3$). The spectra of soils (Figure 6.6) with very low calcium carbonate content (red soil and paddy soil) are devoid of bands in the 1430 and 2520 cm^{-1} region. With regard to clay soils, the differences are more subtle, but the bands in the range of 1000–1200 cm^{-1}, around 1600 cm^{-1}, and the shoulders that can be seen around 780, 1800, and 3700 cm^{-1} in the spectra of the Chinese Red and paddy soils point to the presence of kaolin or a mixture of kaolin and smectite in these soils. Furthermore, the black soil sample, which is expected to contain only smectite, shows none of the bands that are characteristic to kaolin (780 and 1800 cm^{-1}). Regardless of type, clay content directly determines the amount of hydroscopic water present in air-dried soil. Figure 6.6 shows that the red and fluvo-aquic soils, which have the lowest clay content, have the lowest intensities in the 1600–1650 and 3000–3600 cm^{-1} intervals that include the major water bands. This hints to the potential of utilizing PAS spectroscopy for assessing the amount of hygroscopic water or for indirectly determining characteristics such as clay content or specific surface area of a given soil.

6.3.2.3 Soil Water

Water is essential for plant growth, and as an inevitable consequence of opening their stomata to enable gaseous exchange during photosynthesis, plant transpiration. The soil water content also has a pronounced influence on nutrient uptake from the soil as it affects root growth and the transport of nutrients to the root. Furthermore, soil water influences the availability of oxygen, microbial, and faunal activity, leaching of nutrients and agrochemicals into the subsoil, and swelling and shrinking of certain clay soils. Therefore, soil water is one of the most critical soil components for successful plant growth and land management, particularly in arid lands. Measurement of soil water content can be very beneficial for site-specific irrigation, seeding, and land management. The conventional method to determine water content by oven drying of samples collected from fields is a difficult, costly, and time-consuming procedure. NIR spectroscopy is a proven technique for the measurement of soil water content (Viscarra Rossel and McBratney 1998) because it is fast, nondestructive, and cost-effective, although a NIR instrument is expensive. Soil water content is considered one of the most critical factors affecting the accuracy of NIR models developed for the determination of other soil properties, and it can be successfully measured with NIR spectroscopy (Mouazen et al. 2005, 2006).

Most of soil water is absorbed in organic matter or in the mineral surface, and the respective vibration of adsorbed water in the MIR is listed in Table 6.2 (Richard et al. 2006). Soil water retention is an important property of soil and has a heavy influence on soil fertility. Soil water retention varies widely with soil composition and texture, but measurements are often time-consuming and expensive using traditional laboratory methods since soil water retention is affected by soil density, particle size, mineral and organic composition, and prespace density and distribution. Soil minerals and soil organics are sensitive to infrared spectra, and there is a strong absorption band of water around 1600 cm^{-1} in the infrared reflectance spectrum, which makes it possible to analyze water retention using infrared spectroscopy (Janik et al. 2007), and the determination errors range from 0.01 to 0.02 g water/g dry soil (Linker et al. 2006). Soil particle size shows significant influence on the determination of soil water, and the soil particles that are less than 45 μm provide a reasonable model for estimating the water content of hydrated asteroids (Milliken and Mustard 2007).

6.3.2.4 Soil Microbes

Soil governs plant productivity in terrestrial ecosystems and maintains the equilibrium of biogeochemical cycles through biotransformations (or functions) mediated by living organisms. It has been recognized for many years that microbes are responsible for 80%–90% of these functions (Nannipieri et al. 2003). It is thus of great interest for the sustainability of our environment to assess if the procedures of restoration of sites degraded by changes of land use may allow the soil to partially or totally recover its microbial functions. Microbial-based indicators such as changes in total biomass or in the structure of the total microbial community or of a given group of microorganisms have been often used to describe soil fertility (Schloter et al. 2003).

A selection of microbial functions and species diversity has been made out of the many soil functioning monitoring possibilities. General functional aspects are soil respiration, nitrification, nitrogen fixation, and bacterial DNA synthesis. General diversity aspects include ratio fungi/bacteria,

TABLE 6.2
Review of the Literature Comparing Quantitative Predictions of Various Soil Attributes Using a Multivariate Statistical Technique and Spectral Response in Infrared Regions of the Electromagnetic Spectrum

Soil Attribute	Reference
CEC	Ben-Dor and Banin 1995; Janik et al. 1998; Chang et al. 2001; Shepherd and Walsh 2002; Islam et al. 2003; Du et al. 2011; Waruru et al. 2014
Ca	Ben-Dor and Banin 1995; Janik and Skjemstad 1995; Janik et al. 1998; Chang et al. 2001; Shepherd and Walsh 2002; Cozzolino and Moron 2003; Islam et al. 2003; Du et al. 2013; Soriano-Disla et al. 2013; Liu and Liu 2014; Ramirez-Lopez et al. 2014
Clay	Ben-Dor and Banin 1995; Janik and Skjemstad 1995; Janik et al. 1998; Chang et al. 2001; Shepherd and Walsh 2002; Cozzolino and Moron 2003; Islam et al. 2003; Sorensen and Dalsgaard et al. 2005; Viscarra Rossel et al. 2006; Waiser et al. 2007; Summers et al. 2011; Ge et al. 2014; Soriano-Disla et al. 2014
Cu	Chang et al. 2001; Cozzolino and Moron 2003; Siebielec et al. 2004; Abdi et al. 2012; Soriano-Disla et al. 2013; Wang et al. 2014
EC	Janik et al. 1998; Islam et al. 2003; Viscarra Rossel et al. 2006; Du et al. 2011; Cozzolino et al. 2013; Gholizadeh et al. 2013b
Fe	Janik et al. 1998; Chang et al. 2001; Cozzolino and Moron 2003; Islam et al. 2003; Siebiele et al. 2004; Du et al. 2011; Soriano-Disla et al. 2013
K	Janik et al. 1998; Chang et al. 2001; Cozzolino and Moron 2003; Daniel et al. 2003; Islam et al. 2003; He et al. 2007; Du et al. 2011; Shao and He 2011; Abdi et al. 2012; Liu and Liu 2013; Soriano-Disla et al. 2013; Liu and Liu 2014
Mg	Janik et al. 1998; Chang et al. 2001; Shepherd and Walsh 2002; Cozzolino and Moron 2003; Islam et al. 2003; Mouazen et al. 2010; Soriano-Disla et al. 2013; Liu and Liu 2014
Mn	Janik et al. 1998; Chang et al. 2001; Bertrand et al. 2002; Moron and Cozzolino 2003; Chodak et al. 2004; Soriano-Disla et al. 2013
N	Dalal and Henry 1986; Janik and Skjemstad 1995; Janik et al. 1998; Reeves et al. 1999; Chang et al. 2001; Reeves et al. 2001; Chang and Laird 2001; Martin et al. 2002; Linker 2004; Linker et al. 2005; Linker et al. 2006; He et al. 2007; Shao and He et al. 2011; Abdi et al. 2012; St Luce et al. 2012; Yang et al. 2012; Calderon et al. 2013; Dick et al. 2013; Shi et al. 2013; Vohland et al. 2014
Na	Janik et al.1998; Chang et al. 2001; Islam et al. 2003; Chodak et al. 2004; Cozzolino and Moron et al. 2010; Mouazen et al. 2010; Soriano-Disla et al. 2013
SOM	Dalal and Henry 1986; Ben-Dor and Banin 1995; Janik and Skjemstad 1995; Janik et al. 1998; Masserschmidt et al. 1999; Reeves et al. 1999, 2001; Chang and Laird 2001; Fidêncio et al. 2002; McCarty et al. 2002; Shepherd and Walsh 2002; Daniel et al. 2003; Islam et al. 2003; Viscarra Rossel et al. 2006; Ellerbrock and Gerke 2004; Moron and Cozzolino 2004; Bornemann et al. 2010; Ertlen et al. 2010; Calderon et al. 2011; Yang et al. 2012; Calderon et al. 2013; Gholizadeh et al. 2013a; Ge et al. 2014; Kim et al. 2014; Liu and Liu 2014; Wang et al. 2014
P	Janik et al. 1998; Daniel et al. 2003; Maleki et al. 2006; Viscarra Rossel et al. 2006; Cohen et al. 2007; Du and Zhou 2007; He et al. 2007; Chen et al. 2008; Zornoza et al. 2008; Janik et al. 2009; Du et al. 2011; Shao and He 2011; Abdi et al. 2012
pH	Janik and Skjemstad 1995; Janik et al. 1998; Reeves et al. 1999; Chang et al. 2001; Reeves et al. 2001; Moron and Cozzolino 2002; Shepherd and Walsh 2002; Islam et al. 2003; Zornoza et al. 2008; Dacqui et al. 2010; Du et al. 2011; Gholizadeh et al. 2013b; Vohland et al. 2014
Sand	Janik et al. 1998; Chang et al. 2001; Shepherd and Walsh 2002; Cozzolino and Moron 2003; Islam et al. 2003; Ferraresi et al. 2012; Han et al. 2012; Ge et al. 2014
Silt	Janik et al. 1998; Chang et al. 2001; Shepherd and Walsh 2002; Cozzolino and Moron 2003; Islam et al. 2003; Sorensen and Dalsgaard 2005; Ferraresi et al. 2012
Zn	Chang et al. 2001; Siebielec et al. 2004; Chodak et al. 2007; Wang et al. 2014

mycorrhiza, suppressiveness to pathogens, and catabolic genes. Functions based on narrow diversity, such as nitrification and nitrogen fixation, are most valuable in relation to monitoring adverse influences. The most common approach is to select appropriate microbial functions as relevant indicators of soil functioning. Nevertheless, one of the major limitations is that measurements of microbial functions are often time-consuming and require a large number of soil sample analyses to be representative of a given situation. FTIR is one of the methods that has been successfully used for detecting and identifying microorganisms (Rinnan and Rinnan 2007), especially in food products (Mariey et al. 2001; Irudayaraj et al. 2002; Al-Qadiri et al. 2006). The possibility of using infrared spectroscopy provides many opportunities for understanding both the temporal dynamics and the spatial variability of the recovery of key microbial functions during soil restoration. Some of these studies showed that discrimination was possible not only at the genus level, but also at the species and strain levels (Linker and Tsror 2008). Calibrations were performed between infrared reflectance spectral data and microbial-based indicators using the PLS model, and the microbial functions were precisely predicted (Schimann et al. 2007). Furthermore, identification and speciation of bacterial spores can be made using FTIR-PAS (Thompson et al. 2003), and discrimination was performed at the genus level and at the strain level for five soilborne fungi. For discrimination between the five fungi at the genus level, the success rate for the validation samples ranged from 75% to 89%. For discrimination between the two Colletotrichum strains, the success rate was 78% (Linker and Tsror 2008).

6.3.3 Mathematical Tools in the Treatment of Spectral Data

6.3.3.1 Data Preprocessing

Preprocessing is a very important part of the analysis of spectroscopic data and is defined as any mathematical manipulation of the spectral data prior to primary analysis. There are a number of techniques available, such as normalization, baseline corrections, spectrum smoothing, difference spectrum, and spectral derivatives in the pretreatment of spectra data, which is helpful to both the qualitative and quantitative interpretation of spectra (Beebe 1998; Stuart 2004).

Normalization of a spectral data is accomplished by dividing each absorbance by a constant, which is used to remove systematic variation. Note that normalization may remove important spectral information, and it is suggested to check the normalization effect combing the primary analysis in calibration.

It is common to use a baseline joining the points of lowest absorbance on a peak, preferably in reproducibly flat parts of the absorption line. The absorbance difference between the baseline and the top of the band is then used. Spectra are usually preprocessed with a smoothing filter (first-order Savitzky–Golay filter with a 25-point window). The Savitzky–Golay filter method essentially performs a local polynomial regression to determine the smoothed value for each data point. This method is superior to adjacent averaging because it tends to preserve features of the data such as peak height and width, which are usually washed out by adjacent averaging, and the detail of the smoothing filter is given by Savitzky and Golay (1964).

The most straightforward method of analysis for complex spectra is difference spectroscopy. This technique may be carried out by simply subtracting the infrared spectrum of one component of the system from the combined spectrum to leave the spectrum of the other component. If the interaction between the components results in a change in the spectral properties of either one or both of the components, the changes will be observed in the difference spectra. Spectral subtraction may be applied for the data collected for solutions, such as the infrared ATR spectrum of soil paste. It is necessary to record spectra of both soil paste and water, and water spectra may then be subtracted from the soil paste spectrum, after which the soil spectrum is obtained (Linker et al. 2006). However, water absorptions are very strong under certain circumstances, and in this situation make it difficult to investigate the soil spectrum.

Spectra may also be differentiated, and the benefits of derivatives techniques are twofold. Resolution is enhanced in the first derivative since changes in the gradient are examined, and the second derivative gives a negative peak for each band and shoulder in the absorption spectrum (Mark and Workman 2007). The advantage of derivatization is more readily appreciated for more complex soil spectrum, in which sharp bands are enhanced at the expense of broad ones, which may allow for the selection of a suitable peak.

6.3.3.2 Prediction Model

The Beer-Lambert law is used to do quantitative analysis; however, it cannot be directly used in soil quantitative analysis. The infrared spectra of soil have many overlapping peaks, and isolation of absorption should be made (Linker et al. 2006; Du et al. 2007). Therefore, a multivariate calibration method, such as principal components regression (PCR), PLS, and ANN, is necessary in soil analysis. The techniques of PLS and ANN are explained next, respectively.

A general form of the PLS model is expressed as

$$\begin{aligned} X &= TP^T + E \\ Y &= UQ^T + F \end{aligned}$$ (6.1)

where X is the variable predictor matrix (absorbance), Y is the variable response matrix (soil properties), T and U are the X-scores and Y-scores matrices, P and Q are the X-loading and Y-loading matrices, and E and F are the X-residual and Y-residual matrices. The coordinates of the sample in a coordinate system defined by the principal components (PCs) are called scores. The loading vectors are the bridge between the variable space and the PC space. The loadings provide information about how much each variable contributes to each PC. In the case here, T contains information about the samples and P contains information about the wavenumber. The detailed PLS algorithm in the PLS analysis is well described by Blanco et al. (2000) and Wold et al. (2001).

With numerous and correlated X-variables there is a substantial risk for overfitting (i.e., getting a well-fitting model with little or no predictive power). Hence, a strict test of the predictive significance of each PLS component is necessary, stopping when components start to be nonsignificant (Wold et al. 2001). Cross-validation (CV) is a practical and reliable way to test this predictive significance and has become the standard in PLS analysis. Basically, CV is performed by dividing the data in a number of groups and then developing a number of parallel models from reduced data with one of the groups deleted. After developing a model, differences between actual and predicted Y-values are calculated for the deleted data. The sum of squares of these differences is computed and collected from all the parallel models to form the predictive residual sum of squares, which estimates the predictive ability of the model. Ratio of standard error to prediction error (RPD) is an important statistical parameter used to evaluate the calibration models. In agricultural application, RPD > 3 was considered acceptable and RPD > 5 excellent (Malley et al. 1999). However, there is no critical level of RPD for the infrared analysis in soil science, and acceptable values depend on the intended application of the predicted values. Three categories based on RPD in the ranges >2, 1.4–2.0, and <1.4 were used to indicate decreasing reliability of predicting (Chang et al. 2001). Dunn et al. (2002) and Pirie et al. (2005) reported the similar results of suitable limits for RPD: <1.6, poor; 1.6–2.0, acceptable; and >2.0, excellent. In this study the RPD values in optimized PLS models were acceptable, and comprehensively, the PLS models were excellent for soil organic matter, soil-available N and P, and they were relatively poor for soil-available K. Comparing with research results of Pirie et al. (2005), the predicting ability of soil properties using photoacoustic spectra-based PLS modeling, both in near-infrared and mid-infrared region, was satisfactory enough comparing with reflectance spectra.

Since soil is a complex mixture and soil nutrient content is related through many soil components, multivariate calibration techniques were used to extract related information in the FTIR-PAS spectra. PLS regression was used to develop a correlation between the PAS spectra and the soil nutrient content in the soil samples. Du et al. (2008a) demonstrated results of the leave-one-out cross-validation calibration using different PLS factor numbers varied from 2 to 9. For each soil property, the calibration error kept decreasing to near zero and the calibration coefficient kept increasing to near 1. However, the validation error initially became smaller, then turned larger and larger, which meant that modeling involving too many PLS factors would lead to overfitting. The PLS factor number could be selected with the lowest validation error where the calibration error and calibration coefficient were still good enough. Four PLS factors that explained 96.98% variance of input vectors and output vectors were selected for available N and organic matter, and the RPD values were 5.27 and 3.48, respectively; five PLS factors that explained 98.58% variance of input vectors and output vectors were selected for available P and available K, and the RPD values were 5.51 and 3.85, respectively (Du et al. 2008a).

ANN is typically organized in layers where these layers are made up of a number of interconnected nodes that contain an activation function. Input vectors are presented to the network via the input layer that communicates to one or more hidden layers where the actual processing is done via a system of weighted connections. ANN allows one to estimate relationships between one or several input variables called independent variables or descriptors and one or several output variables called dependent variables or responses. Information in an ANN is distributed among multiple cells (nodes) and connections between the cells (weights) (Despagne and Massart 1998). Most ANNs contain some form of learning rule that modifies the weights of the connections according to input patterns that it is presented with. There are many different kinds of learning rules used by neural networks; in back-propagation neural networks (BP-ANN) learning is a supervised process that occurs with each cycle of epoch (i.e., each time the network is presented with a new input pattern) through a forward activation flow of inputs and the backward error propagation of weight adjustment (Ramadan et al. 2005). There are many variations of the back-propagation algorithm. The simplest implementation of back-propagation learning updates the network weights and biases in the direction in which performance function decreases most rapidly, the negative of the gradient. One iteration of this algorithm can be written as

$$x_{k+1} = x_k - a_k g_k \tag{6.2}$$

where x_k is a vector of current weights and biases, g_k is the current gradient, and a_k is the learning rate. In this work, gradient descent with momentum is applied and the performance function is the mean square error, the average squared error between the network outputs, and the actual output. For the basic gradient descent algorithm, the weights and biases are moved in the direction of the negative gradient of the performance function. Gradient descent with momentum often provides faster convergence because momentum allows a network to respond not only to the local gradient but also to recent trends in the error surface. Momentum can also help the network to overcome a shallow local minimum in the error surface and settle down at or near the global minimum. Momentum can be added to back-propagation learning by making weight changes equal to the sum of a fraction of the last weight change and the new change suggested by the back-propagation rule. The magnitude of the effect that the last weight change is allowed to have is mediated by a momentum constant that can be any number between 0 and 1. When the momentum constant is 0, the weight change is based solely on the gradient. When the momentum constant is 1, the new weight change is set to equal the last weight change and the gradient is simply ignored. The performance of the network was also tested by reducing the dimension of the input vectors before the training process (Sun et al. 2003).

ANN was implemented to estimate soil organic matter, phosphorus, and potassium content, and satisfactory results were attained (Daniel et al. 2003). An effective procedure for performing this operation is the principal component analysis (PCA), which can reduce input data (Despagne et al. 1998). This technique has three effects: it orthogonalizes the components of the input vectors (so

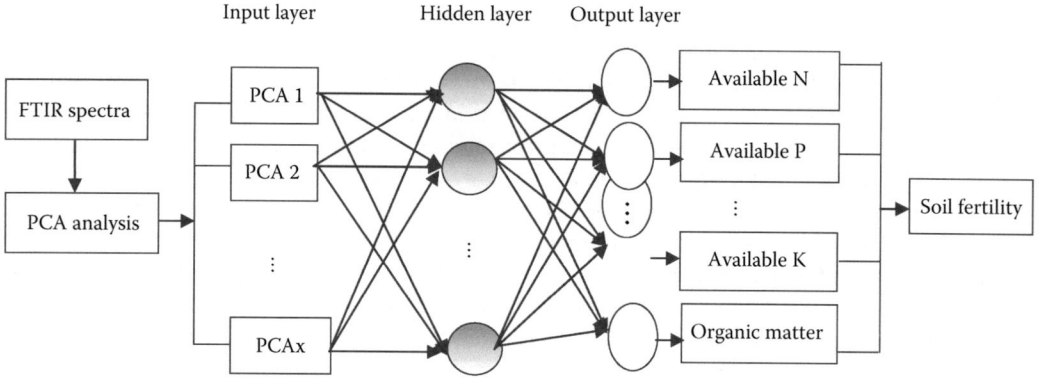

FIGURE 6.9   Schematic diagram of three-layer BP ANN model in the evaluation of soil fertility.

that they are uncorrelated with each other), it orders the resulting orthogonal components (principal components) so that those with the largest variation come first, and it eliminates those components that contribute the least to the variation in the data set. Application of the PC scores instead of the original variables as the net inputs leads to efficient reduction of the net architecture and usually gives better prediction of the *Y*-variables. PC scores of soil mid-infrared photoacoustic spectra were used as the input of an ANN model, and soil properties as well as soil fertility could be successfully predicted (Figure 6.9) (Du et al. 2007, 2008a).

6.3.3.3   Model Verification

Leave-one-out cross-validation was widely used to determine the number of factors to retain in the calibration models. In this instance, 30 bilinear factors were tested. To select the optimal cross-validated calibration model, we computed the RMSE of predictions. Generally, the model with the lowest RMSE is selected.

$$RMSE = \sqrt{\frac{\sum_{i=1}^{n} (\hat{y}_i - y_i)^2}{n}} \qquad (6.3)$$

where $\hat{y}_i$ = values of the predicted variable and y_i = values of the actual values.

The calibration models were independently validated against the soil data. The procedure employed for the quantification of prediction biases and errors was also that of leave-one-out cross-validation. The validation predictions involved computing each calibration model using $n - 1$ soil samples and predicting the soil property of the sample removed. Explicitly, the procedure entailed removing a soil sample from the prediction set, then computing the calibration model using the $n - 1$ samples and predicting the soil property of the sample removed. This procedure was repeated for all samples and accordingly, all soil properties. The mean error (ME) was used to quantify bias, the RMSE (their precision may then be easily inferred), and determination coefficients (R^2) (Viscarra Rossel et al. 2006).

A key requirement for empirical modeling is that validation samples be similar to calibration samples; or, put another way, to build a global empirical soil characterization model we would need a calibration library that spanned the range of possibilities for soil composition (Dardenne et al. 2000). It was computed that 5.2×10^9 carefully selected calibration samples would be required to span the global soil compositional space, and a far more reasonable calibration size for our tropical soil model, 5.9×10^9 samples. Tropical soils are, on the whole, compositionally much less diverse than less weathered temperate soils and should therefore be more amenable to empirical modeling approaches (Brown et al. 2006). In regions with uniform parent material (e.g., loess deposits), it was expected to construct reliable calibrations with a limited number of samples. Further research is needed to test whether local calibration procedures (Berzaghi et al. 2000) could help to reduce the size of calibration sets for regional or watershed applications. However, parent materials like glacial till with a range of primary and secondary minerals might well require large calibration datasets even in geographically restricted areas.

6.4   SOIL INFRARED SPECTRUM INFORMATION SYSTEM

Stable crop production as well as protection and enhancement of the global environment require the development of innovative new methodologies to assess the spatial and temporal variability of soil fertility and soil properties. In particular, spectroscopic techniques such as infrared photoacoustic spectroscopy offer the potential to quickly and inexpensively characterize soils relative to standard laboratory techniques. In the FTIR-PAS-based soil analysis, the calibration set is the most important aspect, and the analyzed results will strongly depend on the selected soil number and soil type. Given the compositional diversity of soils, enough independent samples would be required to construct a complete global, empirical soil library for calibration purposes.

At present, with a relatively limited soil spectral library, expanding the soil spectral library is necessary by scanning previously characterized state and national soil archives (Brown et al. 2006), and it is important to consider the number of samples that are needed to adequately describe the soil variability in the region in which the library is to be used (Viscarra Rossel et al. 2008). Du has initially composed a soil information system based on the soil FTIR-PAS spectra library (Figure 6.10) (Du 2010), and cloud technology was applied in the system, which provided an alternative platform for soil analysis. The soil spectra library, which was stored in cloud, is one of the most important components in this software; however, the soil spectra data is far from enough in general applications. It is proposed that global libraries should be used in conjunction with local calibration samples and with easy-to-measure auxiliary predictors such as nutrient content, clay content, organic matter, and pH.

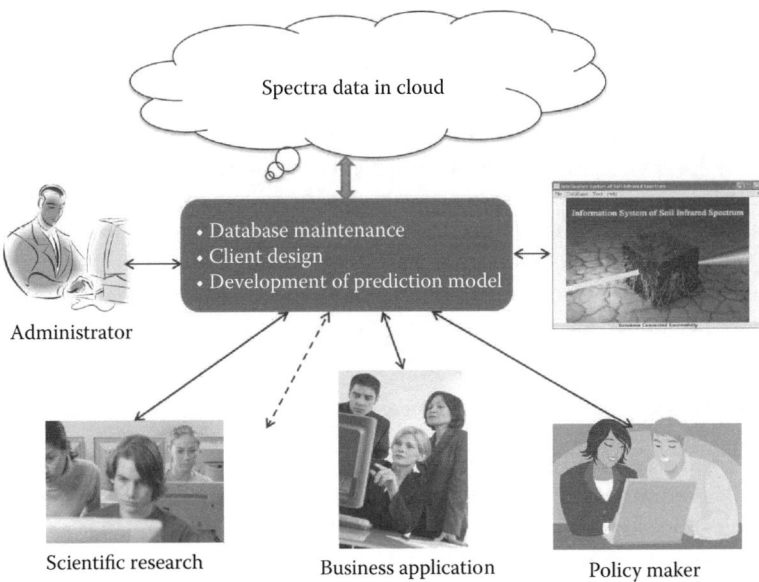

FIGURE 6.10   Application of the information system of the soil infrared spectrum combining cloud technology.

Developments in theoretical soil spectroscopy and spectral processing techniques should also improve predictions while reducing calibration requirements. We anticipate that the future development of soil infrared spectroscopy and the expansion of soil spectral libraries will support the assessment of soil fertility at a scale and resolution not previously possible.

6.5 RESEARCH HIGHLIGHTS IN THE FUTURE

6.5.1 Construction of a Soil Infrared Spectra Library

Stable crop production as well as protection and enhancement of the global environment require the development of innovative new methodologies to assess the spatial and temporal variability of soil fertility and soil properties. In particular, spectroscopic techniques such as infrared spectroscopy offer the potential to quickly and inexpensively characterize soils relative to standard laboratory techniques. Given the compositional diversity of soils, enough independent samples would be required to construct a complete global, empirical soil library for calibration purposes. However, since we are able to construct useful predictive models, fundamental and partially independent soil-spectral relationships may reduce the required number of calibration samples to a manageable number. At present, with a relatively limited soil spectral library, expanding the soil-spectral library is necessary by scanning previously characterized state and national soil archives (Brown et al. 2006), and it is important to consider the number of samples that are needed to adequately describe the soil variability in the region in which the library is to be used (Viscarra Rossel et al. 2008). This work, necessarily involving multiple labs, will require that we address calibration transfer between spectrometers of the same and different types.

It is proposed that global libraries should be used in conjunction with local calibration samples and with easy-to-measure auxiliary predictors such as clay content, organic matter, and pH. Developments in theoretical soil spectroscopy and spectral processing techniques should also improve predictions while reducing calibration requirements. We anticipate that the future development of soil infrared spectroscopy and the expansion of soil-spectral libraries will support the assessment of soil fertility at a scale and resolution not previously possible.

6.5.2 Mathematic Tools in Spectral Data Mining

MIR photoacoustic spectroscopy is a very promising technique for the analysis of soil with higher absorbance, and the characteristics of main soil components can be shown in the FTIR-PAS spectra (Du et al. 2007, 2008a,b). The merits of this technique make it possible to construct a soil spectral library besides reflectance spectroscopy. Through calibration with soil fertility related properties and crop production, a soil fertility parameter extracted from soil spectra using mathematical tools can be reached, which will be helpful in the practice of fertilization. However, the application of infrared photoacoustic spectroscopy in soil analysis is very limited, and more soil types and soil samples involved calibrations and verifications should be made to connect soil fertility and soil spectra, which needs global collaboration.

6.6 CONCLUSIONS

Soil fertility is usually depicted by soil properties such as soil nutrients level, soil organic matter, and soil water. Therefore, the related soil properties should be determined first, and then a model based on the soil properties can be developed to evaluate soil fertility under site-specific conditions. The conventional technique of soil fertility evaluation is based on laboratory analysis and is time-consuming and costly, which is not suitable for precision agriculture for many determinations. Thus, infrared spectroscopy provides an alternative technique for the evaluation of soil fertility and its applications to precision agriculture.

Infrared transmission and reflectance spectroscopy are both useful in soil analysis. Infrared transmission spectroscopy is commonly used in soil qualitative analysis, such as identification of soil organic matter, and reflectance spectroscopy can be used in soil quantitative analysis. Reflectance spectroscopy, including diffuse spectroscopy and attenuated total spectroscopy, is widely used in soil analysis, but sometimes the prediction error is high, and sample pretreatment is also required. Infrared photoacoustic spectroscopy is a new technique used in soil analysis that does not need sample pretreatment, and more useful information can be obtained in the photoacoustic spectra. The merits of FTIR-PAS technique makes it promising in the evaluation of soil fertility in the future for application in precision agriculture.

Spectral analysis is very important in the evaluation of soil fertility using infrared spectroscopy. Since the interference of multi-components in the soil the multi-calibration should be involved to extracting the needed information of soil fertility in the soil spectra. PLS and ANN are two important statistic methods to reach this purpose. Enough soil sample number and soil variance are also needed in a good multivariate calibration of soil fertility. Library of soil infrared spectra including at least thousands of soil samples should be constructed, which will be a useful information system in the evaluation of soil fertility. This spectral information system will provide fast and in situ evaluation of soil fertility, which will benefit the sustainable agriculture.

ACKNOWLEDGMENTS

This work was supported by the National Natural Scientific Foundation of China (40871113) and National Basic Research Program of China (2015CB150403).

REFERENCES

Abdi, D., G.F. Tremblay, N. Ziadi, G. Belanger, L.E. Parent. 2012. Predicting soil phosphorus-related properties using near-infrared reflectance spectroscopy. *Soil Science Society of America Journal* 76: 2318–2326.

Adjei-Nsiah, S., T.W. Kuyper, C. Leeuwis, M.K. Abekoe, K.E. Giller. 2007. Evaluating sustainable and profitable cropping sequences with cassava and four legume crops: Effects on soil fertility and maize yields in the forest/savannah transitional agro-ecological zone of Ghana. *Field Crops Research* 103: 87–97.

Alfaia, S.S., G.A. Ribeiro, A.D. Nobre, L. Flávio, J. Luizão. 2004. Evaluation of soil fertility in smallholder agroforestry systems and pastures in western Amazonia. *Agriculture Ecosystems and Environment* 102: 409–414.

Al-Qadiri, H.M., M. Lin, A.G. Cavinato, B.A. Rasco. 2006. Fourier transform infrared spectroscopy, detection and identification of *Escherichia coli* O157:H7 and *Alicyclobacillus* strains in apple juice. *International Journal of Food Microbiology* 111: 73–80.

Armenta, S., J. Moros, S. Garrigues, M. LaGuardia. 2006. Direct determination of Mancozeb by photoacoustic spectroscopy. *Analytica Chimica Acta* 567: 255–261.

Barthes, B.G., D. Brunet, E. Hien, F. Enjalric, S. Conche, G.T. Frescheta, R.J. Toucet-Louri. 2008. Determining the distributions of soil carbon and nitrogen in particle size fractions using near-infrared reflectance spectrum of bulk soil samples. *Soil Biology & Biochemistry* 40: 1533–1537.

Bautista-Cruz, A., R. Carrillo-Gonzalez, M.R. Arnaud-Vinas, C. Robles, F. de Leon-Gonzàlez. 2007. Soil fertility properties on Agave angustifolia Haw. Plantations. *Soil & Tillage Research* 96: 342–349.

Beebe, K.R. 1998. *Chemometrics: A pratical guide.* John Wiley & Sons Inc., New York, 26–52.

Ben-Dor, E., A. Banin. 1995. Near-infrared analysis as a rapid method to simultaneously evaluate several soil properties. *Soil Science Society of America Journal* 59: 364–372.

Bertrand, I., L.J. Janik, R.E. Holloway, R.D. Armstrong, M.J. McLaughlin. 2002. The rapid assessment of concentrations and solid phase associations of macro- and micronutrients in alkaline soils by mid-infrared diffuse reflectance spectroscopy. *Australian Journal of Soil Research* 40: 1339–1356.

Berzaghi, P., J.S. Shenk, M.O. Westerhaus. 2000. LOCAL prediction with near infrared multi-product databases. *Journal of Near Infrared Spectroscopy* 8: 1–9.

Bjarnestad, S., O. Dahlman. 2002. Chemical compositions of hardwood and softwood pulps employing photoacoustic Fourier transform infrared spectroscopy in combination with partial least-squares analysis. *Analytical Chemistry* 74: 5851–5858.

Blanco, M., J. Coello, H. Iturriaga, S. Maspoch, J. Pages. 2000. NIR calibration in non-linear systems: Different PLS approaches and artificial neural networks. *Chemometrics and Intelligent Laboratory Systems* 50: 75–82.

Bogrekci, I., W.S. Lee. 2005a. Spectral phosphorus mapping using diffuse reflectance of soils and grass. *Biosystems Engineering* 91: 305–312.

Bogrekci, I., W.S. Lee. 2005b. Spectral soil signatures and sensing phosphorus. *Biosystems Engineering* 92: 527–533.

Bornemann, L., G. Welp, W. Amelung. 2010. Particulate organic matter at the field scale: Rapid acquisition using mid-infrared spectroscopy. *Soil Science Society of America Journal* 74: 1147–1156.

Brown, D.J., K.D. Shepherd, M.G. Walsh, M.D. Mays, T.G. Reinsch. 2006. Global soil characterization with VNIR diffuse reflectance spectroscopy. *Geoderma* 132: 273–290.

Byler, D.M., W.V. Gerasimowicz, H. Susi, M. Schnitzer. 1987. FT-IR spectra of soil constituents: Fulvic acid and fulvic acid complex with ferric ions. *Applied Spectroscopy* 41: 1428–1430.

Calderon, F. J., J.B. Reeves, H.P. Collins, E.A. Paul. 2011. Chemical differences in soil organic matter fractions determined by diffuse-reflectance mid-infrared Spectroscopy. *Soil Science Society of America Journal* 75: 568–579.

Calderon, F., M. Haddix, R. Conant, K. Magrini-Bair, E. Paul. 2013. Diffuse reflectance Fourier transform mid-infrared spectroscopy as a method of characterizing changes in soil organic matter. *Soil Science Society of America Journal* 77: 1591–1600.

Cardoso, I.M., T.W. Kuyper. 2006. Mycorrhizas and tropical soil fertility. *Agriculture, Ecosystems and Environment* 116: 72–84.

Chang, C.W., D.A. Laird, M.J. Mausbach, C.R. Hurburgh Jr. 2001. Near-infrared reflectance spectroscopy-principal components regression analysis of soil properties. *Soil Science Society of America Journal* 65: 480–490.

Chen, P.F., L.Y. Liu, J.H. Wang, T. Shen, A.X. Lu, C.J. Zhao. 2008. Real-time analysis of soil N and P with near infrared diffuse reflectance spectroscopy. *Spectroscopy and Spectral Analysis* 28: 295–298.

Chodak, M., P. Khanna, B. Horvath, F. Beese. 2004. Near infrared spectroscopy for determination of total and exchangeable cations in geologically heterogeneous forest soils. *Journal of Near Infrared Spectroscopy* 12: 315–324.

Chodak, M., M. Niklinska, F. Beese. 2007. Near-infrared spectroscopy for analysis of chemical and microbiological properties of forest soil organic horizons in a heavy-metal-polluted area. *Biology and Fertility of Soils* 44: 171–480.

Christy, C.D. 2008. Real-time measurement of soil attributes using on-the-go near infrared reflectance spectroscopy computers and electronics in agriculture. *Journal of Computers and Electronics in Agriculture* 61: 10–19.

Cohen, M.J., J.P. Prenger, W.F. DeBusk. 2005. Visible-near infrared reflectance spectroscopy for rapid, nondestructive assessment of wetland soil quality. *Journal of Environmental Quality* 34: 1422–1434.

Cohen, M.J., J. Paris, M.W. Clark. 2007. P-sorption capacity estimation in southeastern USA wetland soils using visible/near infrared VNIR reflectance spectroscopy. *Wetlands* 27: 1098–1111.

Cozzolino, D., A. Moron. 2003. The potential of near-infrared reflectance spectroscopy to analyse soil chemical and physical characteristics. *Journal of Agricultural Sciences* 140: 65–71.

Cozzolino, D., A. Moron. 2006. Potential of near-infrared reflectance spectroscopy and chemometrics to predict soil organic carbon fractions. *Soil & Tillage Research* 85: 78–85.

Cozzolino, D., A. Moron. 2010. Influence of soil particle size on the measurement of sodium by near-infrared reflectance spectroscopy. *Communications in Soil Science and Plant Analysis* 41: 2330–2339.

Cozzolino, D., W.C. Cynkar, R.G. Dambergs, N. Shah, P. Smith. 2013. In situ measurement of soil chemical composition by near-infrared spectroscopy: A tool toward sustainable vineyard management. *Communication in Soil Science and Plant Analysis* 44: 1610–1619.

Dacqui, L.P., A. Pucci, L.J. Janik. 2010. Soil properties prediction of western Mediterranean islands with similar climatic environments by means of mid-infrared diffuse reflectance spectroscopy. *European Journal of Soil Science* 61: 865–876.

Dalal, R. C., R. J. Henry. 1986. Simultaneous determination of moisture, organic-carbon, and total nitrogen by near-infrared reflectance spectrophotometry. *Soil Science Society of America Journal* 50(1): 120–123.

Daniel, K.W., N.K. Tripathi, K. Honda. 2003. Artificial neural network analysis of laboratory and in situ spectra for the estimation of macronutrients in soils of Lop Buri Thaland. *Australian Journal of Soil Research* 41: 47–59.

Dardenne, P., G. Sinnaeve, V. Baeten. 2000. Multivariate calibration and chemometrics for near infrared spectroscopy: Which method? *Journal of Near Infrared Spectroscopy* 8: 229–237.

Dematte, J.A.M., R.C. Campos, M.C. Alves, P.R. Fiorio, M.R. Nanni. 2004. Visible–NIR reflectance: A new approach on soil evaluation. *Geoderma* 121: 95–112.

Desbiez, A., R. Matthewsa, B. Tripathi, J. Ellis-Jones. 2004. Perceptions and assessment of soil fertility by farmers in the mid-hills of Nepal. *Agriculture, Ecosystems and Environment* 103: 191–206.

Despagne, F., B. Walczak, D. Massart. 1998. Transfer of calibrations of near-infrared spectra using neural networks. *Applied Spectroscopy* 52: 732–745.

Despagne, F., D. Massart. 1998. Neural networks in multivariate calibration. *Analyst* 123: 157–178.

Dick, W.A., B. Thavamani, S. Conley, R. Blaisdell, A. Sengupta. 2013. Prediction of beta-glucosidase and beta-glucosaminidase activities, soil organic C, and amino sugar N in a diverse population of soils using near infrared reflectance spectroscopy. *Soil Biology & Biochemistry* 56: 99–104.

Dong, L., A.A. Kosterev, D. Thomazy, F.K. Tittel. 2010. QEPAS spectrophones: Design, optomization, and performance. *Applied Physics* B 100: 627–635.

Du, C.W. 2010. Soil infrared information system. v1.0, software certificate No. 2010R11L027920.

Du, C.W. 2012. *Soil infrared photoacoustic spectroscopy and its application.* China Scientific Press, Beijing.

Du, C.W., J.M. Zhou. 2007. Prediction of soil available phosphorus using Fourier transform infrared photoacoustic spectroscopy. *Chinese Journal of Analytical Chemistry* 35: 119–122.

Du, C.W., R. Linker, A. Shaviv. 2007. Characterization of soils using photoacoustic mid-infrared spectroscopy. *Applied Spectroscopy* 61: 1063–1067.

Du, C.W., R. Linker, A. Shaviv. 2008a. Soil identification with Fourier transform infrared photoacoustic spectroscopy. *Geoderma* 143: 85–90.

Du, C.W., J.M. Zhou, H.Y. Wang, X.Q. Chen, A.N. Zhu, J.B. Zhang. 2008b. Study on the soil mid-infrared photoacoustic spectroscopy. *Spectroscopy and Spectral Analysis* 28: 1242–1245.

Du, C.W., J.M. Zhou, H.Y. Wang, X.Q. Chen, A.N. Zhu, J.B. Zhang. 2008c. Determination of soil properties using infrared photoacoustic spectroscopy using techniques of partial least square PLS. *Vibrational Spectroscopy* 49: 32–37.

Du, C.W., J. Deng, J.M. Zhou, H.Y. Wang, X.Q. Chen. 2011. Characterization of greenhouse soil properties using mid-infrared photoacoustic spectroscopy. *Spectroscopy Letters* 44: 359–368.

Du, C.W., Z.Y. Ma, J.M. Zhou, K.W. Goyne. 2013. Application of mid-infrared photoacoustic spectroscopy in monitoring carbonate content in soils. *Sensors and Actuators B-Chemical* 188: 1167–1175.

Dunn, B. W., H.G. Beecher, G.D. Batten, S. Ciavarella. 2002. The potential of near-infrared reflectance spectroscopy for soil analysis—A case study from the Riverine Plain of south eastern Australia. *Australian Journal of Experimental Agriculture* 42: 607–614.

Ellerbrock, R.H., H.H. Gerke. 2004. Characterizing organic matter of soil aggregate coatings and biopores by Fourier transform infrared spectroscopy. *European Journal of Soil Science* 55: 219–228.

Ertlen, D., D. Schwartz, M. Trautmann, R. Webster, D. Brunet. 2010. Discriminating between organic matter in soil from grass and forest by near-infrared spectroscopy. *European Journal of Soil Science* 61: 207–216.

Faithfull, N.T. 2002. *Methods in agricultural chemical analysis.* CABI Publishing, Wallingford, Oxon, UK, 57–104.

Ferraresi, T.M., W.T.L. da Silva, L. Martin-Neto, P.M. da Silveira, B.E. Madari. 2012. Infrared spectroscopy in determination of soil texture. *Revista Brasileira De Ciencia Do Solo* 36: 1769–1777.

Fidêncio, P.H., R.J. Poppi, J.C. Andrade. 2002. Determination of organic matter in soil using near-infrared spectroscopy and partial least squares regression. *Communication in Siol Science and Plant Analysis* 33: 1607–1615.

Francioso, O., E. Ferrari, M. Saladini, D. Montecchio, P. Gioacchini, C. Ciavatta. 2007. TG–DTA, DRIFT and NMR characterization of humic-like fractions from olive wastes and amended soil. *Journal of Hazardous Materials* 149: 408–417.

Ge, Y. F., J. A. Thomasson, C. L. S. Morgan. 2014. Mid-infrared attenuated total reflectance spectroscopy for soil carbon and particle size determination. *Geoderma* 213: 57–63.

Gholizadeh, A., L. Boruvka, M. Saberioon, R. Vasat. 2013. Visible, near-infrared, and mid-infrared spectroscopy applications for soil assessment with emphasis on soil organic matter content and quality: State-of-the-art and key issues. *Applied Spectroscopy* 67: 1349–1362.

Gholizadeh, A., M.A.M. Soom, M.M. Saberioon, L. Boruvka. 2013. Visible and near infrared reflectance spectroscopy to determine chemical properties of paddy soils. *Journal of Food Agriculture & Environment* 11: 859–866.

Gobeille, A., J. Yavitt, P. Stalcup, A. Valenzuela. 2006. Effects of soil management practices on soil fertility measurements on Agave tequilana plantations in Western Central Mexico. *Soil & Tillage Research* 87: 80–88.

Grave, R.A., R. Nicoloso, P. Cassol, C. Aita, J. Correa, M. Dalla Costa, D. Fritz. 2015. Short-term carbon dioxide emission under contrasting soil disturbance levels and organic amendments. *Soil & Tillage Research* 146: 184–192.

Haberhauer, G., M. Gerzabek. 2001. FTIR-spectroscopy of soils-characterization of soil dynamic processes. *Trends in Applied Spectroscopy* 3: 103–109.

Han, L., D. Li, W. Fang, Y. Wang, R. Gaussoin. 2012. Analysis of soil chemical properties of sand-based turfgrass rootzone using Fourier transform-infrared spectroscopy. *Communications in Soil Science and Plant Analysis* 43: 2709–2721.

He, Y., H.Y. Song, A.G. Pereira, A.H. Gómez. 2005. A new approach to predict N, P, K and OM content in a loamy mixed soil by using near infrared reflectance spectroscopy. In *Advances in Intelligent Computing*, eds., D.S. Huang, X.-P. Zhang, G.-B. Huang, Springer, Berlin, 859–867.

He, Y., M. Huang, A. Garcia, A. Hernandez, H. Song. 2007. Prediction of soil macronutrients content using near-infrared spectroscopy. *Computers and Electronics in Agriculture* 58: 144–153.

Hirschmann, C.B., J. Uotila, S. Ojala, J. Tenhunen, R.L. Keiski. 2010. Fourier transform infrared photoacoustic multicomponent gas spectroscopy with optical cantilever detection. *Applied Spectroscopy* 64: 293–297.

Irudayaraj, J., H. Yang, S. Sakhamuri. 2002. Differentiation and detection of microorganisms using fourier transform infrared photoacoustic spectroscopy. *Journal of Molecular Structure* 606: 181–188.

Islam, K., B. Singh, A. McBratney. 2003. Simultaneous estimation of several soil properties by ultra-violet, visible, and near-infrared reflectance spectroscopy. *Australian Journal of Soil Research* 4: 1101–1114.

Izac, A.M. 2003. Economic aspects of soil fertility. Management and agroforestry practices. In *Trees, crops and soil fertility*, eds., G. Schroth, F.L. Sinclair, CABI Publishing, New York, 2003, 13–20.

Jahn, B.R., R. Linker, S. Upadhyaya, A. Shaviv, D. Slaughter, I. Shmulevich. 2006. Mid-infrared spectroscopic determination of soil nitrate content. *Biosystems Engineering* 94: 505–515.

Janik, L. J., J. O. Skjemstad. 1995. Characterization and analysis of soils using midinfrared partial least-squares .2. correlations with some laboratory data. *Australian Journal of Soil Research* 33(40): 637–650.

Janik, L.J., R. Merry, J. Skjemstad. 1998. Can mid infrared diffuse reflectance analysis replace soil extractions? *Australian Journal of Experimental Agriculture* 38: 681–696.

Janik, L.J., S. Forrester, A. Rawson. 2009. The prediction of soil chemical and physical properties from mid-infrared spectroscopy and combined partial least-squares regression and neural networks PLS-NN analysis. *Chemometrics and Intelligent Laboratory Systems* 97: 179–188.

Janik, L.J., R.H. Merry, S.T. Forrester, D.M. Lanyon, A. Rawson. 2007. Rapid prediction of soil water retention using mid-infrared spectroscopy. *Soil Science Society of American Journal* 71: 507–514.

Johnston, C.T., Y. Aochi. 1996. Fourier transform infrared and raman spectroscopy. In *Methods of soil analysis*, Part 3, eds., J.M. Bartels, J.M. Bigham, Soil Science Society of America, Inc., American Society of Agronomy, Inc., Madison, WI, 269–321.

Koskinen, V., J. Fonsen, K. Roth, J. Kauppinen. 2007. Cantilever enhanced photoacoustic detection of carbon dioxide using a tunable diode laser source. *Applied Physics B* 86: 451–454.

Kim, I., R. R. Pullanagari, M. Deurer, R. Singh, K. Y. Huh, B. E. Clothier. 2014. The use of visible and near-infrared spectroscopy for the analysis of soil water repellency. *European Journal of Soil Science* 65(30): 360–368.

Kubelka, P., F. Munk. 1931. Ein beitrag zur optik der farbanstriche. *Zeitschrift für technische Physik* 12: 593–601.

Leach, C.J., T. Wagner, M. Jones, S. Juggins, A. Stevenson. 2008. Rapid determination of total organic carbon concentrations in marine sediments using Fourier transform near-infrared spectroscopy FT-NIRS. *Organic Geochemistry* 39: 910–914.

Linker, R. 2004. Waveband selection for determination of nitrate in soil using mid-infrared attenuated total reflectance spectroscopy. *Applied Spectroscopy* 58: 1277–1281.

Linker, R. 2008. Soil classification via mid-infrared spectroscopy. In *Computer and computing technologies in agriculture*, Vol. 2, ed., D. Li, IFIP International Federation for Information Processing, a Springer Series in Computer Science, Boston, 1137–1146.

Linker, R., L. Tsror. 2008. Discrimination of soil-borne fungi using Fourier transform infrared attenuated total reflection spectroscopy. *Applied Spectroscopy* 62: 302–305.

Linker, R., A. Kenny, A. Shaviv, L. Singher, L. Shmulevich. 2004. FTIR/ATR nitrate determination of soil pastes using PCR, PLS and cross-validation. *Applied Spectroscopy* 58: 516–520.

Linker, R., I. Shmulevich, A. Kenny, A. Shaviv. 2005. Soil identification and chemometrics for direct determination of nitrate in soils using FTIR-ATR mid-infrared spectroscopy. *Chemosphere* 61: 652–658.

Linker, R., M. Weiner, I. Shmulevich, A. Shaviv. 2006. Nitrate determination in soil pastes using attenuated total reflectance mid-infrared spectroscopy: Improved accuracy via soil identification. *Biosystems Engineering* 94: 111–118.

Liu, X. M., J. S. Liu. 2013. Measurement of soil properties using visible and short wave-near infrared spectroscopy and multivariate calibration. *Measurement* 46(10): 3808–3814.

Liu, X.M., J.S. Liu. 2014. Using short wave visible-near infrared reflectance spectroscopy to predict soil properties and content. *Spectroscopy Letters* 47: 729–739.

Ludwig, B., P.K. Khanna, J. Bauhus, P. Hopmans. 2002. Near infrared spectroscopy of forest soils to determine chemical and biological properties related to soil sustainability. *Forest Ecology and Management* 171: 121–132.

Maleki, M.R., L.V. Holm, H. Ramon, R. Merckx, J. Baerdemaeker, A. Mouazen. 2006. Phosphorus sensing for fresh soils using visible and near infrared spectroscopy. *Biosystems Engineering* 95: 425–436.

Malley, D.F., L. Yesmin, D. Wray, S. Edwards. 1999. Application of near-infrared spectroscopy in analysis of soil mineral nutrients. *Communication in Soil Science and Plant Analysis* 30: 999–1012.

Mariey, L., J. Signolle, C. Amiel, J. Travert. 2001. Discrimination, classification, identification of microorganisms using FTIR spectroscopy and chemometrics. *Vibrational Spectroscopy* 26: 151–159.

Mark, H., J. Workman. 2007. *Chemometrics in spectroscopy.* Academic Press, London, 340–378.

Martin, P.D., D.F. Malley, G. Manning, L. Fuller. 2002. Determination of soil organic carbon and nitrogen at the field level using near-infrared spectroscopy. *Canadian Journal of Soil Science* 82: 413–422.

Masserschmidt, I., C. J. Cuelbas, R. J. Poppi, J. C. De Andrade, C.A. De Abreu, C. U. Davanzo. 1999. Determination of organic matter in soils by FTIR/diffuse reflectance and multivariate calibration. *Journal of Chemometrics* 13(13–4):265–273.

McCarty, G.W., J. Reeves, V. Reeves, R. Follett, J. Kim. 2002. Mid-infrared and near-infrared diffuse reflectance spectroscopy for soil carbon measurement. *Soil Science Society of America Journal* 66: 640–646.

McCarty, G.W., J.B. Reeves. 2006. Comparison of near infrared and mid infrared diffuse reflectance spectroscopy for field-scale measurement of soil fertility parameters. *Soil Science* 171: 94–102.

McCauley, J.D., B. Engel, C. Scudder, M. Morgan, P. Elliot. 1993. Assessing the spatial variability of organic matter. ASAE Paper No. 93–1555, American Society of Agricultural Engineers, St. Joseph, MI.

McClelland, J.F., R. Jones, S. Bajic. 2002. Photoacoustic spectroscopy. In *Handbook of vibrational spectroscopy*, Volume II, eds., J.M. Chalmers, P.R. Griffiths, Wiley & Sons, Chichester, UK.

Michaelian, K.H. 2010. Chapter 1, Introduction, In: *Protoacouctic IR Spectroscopy.* Wiley-VCH: Weinheim, Germany, pp. 1–10.

Miklos, A., P. Hess, Z. Bozoki. 2002. Application of acoustic resonators in protoacoustic trace gas analysis and metrology. *Review of Scientific Instruments* 72: 1937–1955.

Milliken, R.E., J. Mustard. 2007. Estimating the water content of hydrated minerals using reflectance spectroscopy II. Effects of particle size. *Icarus* 189: 574–588.

Moron, A., D. Cozzolino. 2002. Application of near infrared reflectance spectroscopy for the analysis of organic C, total N and pH in soils of Uruguay. *Journal of Near Infrared Spectroscopy* 10: 215–221.

Moron, A., D. Cozzolino. 2003. Exploring the use of near infrared reflectance spectroscopy to study physical properties and microelements in soils. *Journal of Near Infrared Spectroscopy* 11: 145–154.

Moron, A., D. Cozzolino. 2004. Determination of potentially mineralizable nitrogen and nitrogen in particulate organic matter fractions in soil by visible and near-infrared reflectance spectroscopy. *Journal of Agricultural Science* 142: 335–343.

Mouazen, A.M., J. Baerdemaeker, H. Ramon. 2005. Towards development of on-line soil moisture sensor using a fiber-type NIR spectrophotometer. *Soil Tillage Research* 80: 171–183.

Mouazen, A.M., R. Karoui, J. Baerdemaeker, H. Ramon. 2006. Characterization of soil water content using measured visible and near infrared spectra. *Soil Science Society of America Journal* 70: 1295–1302.

Mouazen, A.M., M. Maleki, J.D. Baerdemaeker, H. Ramon. 2007. On-line measurement of some selected soil properties using a VIS–NIR sensor. *Soil & Tillage Research* 93: 13–27.

Mouazen, A.M., B. Kuang, J. De Baerdemaeker, H. Ramon. 2010. Comparison among principal component, partial least squares and back propagation neural network analyses for accuracy of measurement of selected soil properties with visible and near infrared spectroscopy. *Geoderma* 158: 23–31.

Mutuo, P.K., K. Shepherd, A. Albrecht, G. Cadisch. 2006. Prediction of carbon mineralization rates from different soil physical fractions using diffuse reflectance spectroscopy. *Soil Biology & Biochemistry* 38: 1658–1664.

Nanni, M.R., J.A.W. Dematte. 2006. Spectral reflectance methodology in comparison to traditional soil analysis. *Soil Science Society of America Journal* 70: 393–407.

Nannipieri, P., J. Ascher, M.T. Ceccherini, L. Landi, G. Pietramellara, G. Renella. 2003. Microbial diversity and soil functions. *European Journal of Soil Science* 54:655–670.

Odlare, M., K. Svensson, M. Pell. 2005. Near infrared reflectance spectroscopy for assessment of spatial soil variation in an agricultural field. *Geoderma* 126: 193–202.

Ortega, R.A., O.A. Santibanez. 2007. Determination of management zones in corn Zea mays L. based on soil fertility. *Computers and Electronics in Agriculture* 58: 49–59.

Pirie, A., B. Singh, K. Islam. 2005. Ultra-violet, visible, near-infrared and mid-infrared diffuse reflectance spectroscopis techniques to predict several soil properties. *Australian Journal of Soil Research* 43: 713–721.

Ramadan, Z., P.K. Hopke, M.J. Johnson, K.M. Scow. 2005. Application of PLS and back-propagation neural networks for the estimation of soil properties. *Chemometrics and Intelligent Laboratory Systems* 75: 23–30.

Ramirez-Lopez, L., K. Schmidt, T. Behrens, B. van Wesemael, J. Dematte, T. Scholten. 2014. Sampling optimal calibration sets in soil infrared spectroscopy. *Geoderma* 226: 140–150.

Reeves, J. B., G. W. McCarty, J. J. Meisinger. 1999. Near infrared reflectance spectroscopy for the analysis of agricultural soils. *Journal of Near infrared Spectroscopy* 7(3): 179–193.

Reeves, J.B., G. McCarty, V. Reeves. 2001. Mid-infrared diffuse reflectance spectroscopy for the quantitative analysis of agricultural soils. *Journal of Agriculture and Food Chemistry* 49: 766–772.

Richard, W.A. 2006. *Concepts of soils. Soils: Basic concept and future challenges*, Cambridge University Press, Cambridge, UK, 2–9.

Richard, T., L. Mercury, F. Poulet, L. Hendecourt. 2006. Diffuse reflectance infrared Fourier transform spectroscopy as a tool to characterise water in adsorption/confinement situations. *Journal of Colloid and Interface Science* 304: 125–136.

Rinnan, R., S. Rinnan. 2007. Application of near infrared reflectance NIR and fluorescence spectroscopy to analysis of microbiological and chemical properties of arctic soil. *Soil Biology & Biochemistry* 39: 1664–1673.

Rosencwaig, A., A. Gersho. 1976. Theory of the photoacoustic effect with solids. *Journal of Applied Physics* 47: 64–66.

Ryczkowski, J. 2010. Infrared protoacoustic spectroscopy in catalysis and surface science. *Applied Surface Science* 256: 5545–5550.

Savitzky, A., M. Golay. 1964. Smoothing and differentiation of data by simplified least squares procedures. *Analytical Chemistry* 36: 1627–1639.

Schimann, H., R. Joffre, J. Roggy, R. Lensi, A. Domenach. 2007. Evaluation of the recovery of microbial functions during soil restoration using near-infrared spectroscopy. *Applied Soil Ecology* 37: 223–232.

Schloter, M., H. Bach, S. Metz, U. Sehy, J. Munch. 2003. Influence of precision farming on the microbial community structure and functions in nitrogen turnover. *Agriculture, Ecosystem and Environment* 98: 295–304.

Shao, Y.N., Y. He. 2011. Nitrogen, phosphorus, and potassium prediction in soils, using infrared spectroscopy. *Soil Research* 49: 166–172.

Shao, Y.Q., C. Du, Y. Shen, F. Ma, J. Zhou. 2014. Rapid determination of n isotope labeled nitrate using Fourier transform infrared attenuated total reflection spectroscopy. *Chinese Journal of Analytical Chemistry* 42: 747–752.

Shaviv, A., A. Kenny, I. Shmulevich, L. Singher, Y. Reichlin, A. Katzir. 2003. IR fiberoptic systems for in situ and real time monitoring of nitrate in water and environmental systems. *Environmental Science and Technology* 37: 2807–2812.

Shepherd, K.D., M. Walsh. 2002. Development of reflectance spectral libraries for characterization of soil properties. *Soil Science Society of America Journal* 66: 988–998.

Shepherd, K.D., P. Palm, C. Gachengo, B. Vanlauwe. 2003. Rapid characterization of organic resource quality for soil and livestock management in tropical agroecosystems using near-infrared spectroscopy. *Agronomy Journal* 95: 1314–1322.

Shi, T.Z., L. Cui, J. Wang, T. Fei, Y. Chen, G. Wu. 2013. Comparison of multivariate methods for estimating soil total nitrogen with visible/near-infrared spectroscopy. *Plant and Soil* 366: 363–375.

Siebielec, G., G. McCarthy, T. Stuczynski, J. Reeves. 2004. Near- and mid-infrared diffuse reflectance spectroscopy for measuring soil metal content. *Journal of Environmental Quality* 33: 2056–2069.

Sorensen, L.K., S. Dalsgaard. 2005. Determination of clay and other soil properties by near infrared spectroscopy. *Soil Science Society of America Journal* 69: 159–167.

Soriano-Disla, J.M., L. Janik, M. McLaughlin, S. Forrester, J. Kirby, C. Reimann. 2013. Prediction of the concentration of chemical elements extracted by aqua regia in agricultural and grazing European soils using diffuse reflectance mid-infrared spectroscopy. *Applied Geochemistry* 39: 33–42.

Soriano-Disla, J.M., L. Janik, R. Rossel, L. Macdonald, M. McLaughlin. 2014. The performance of visible, near-, and mid-infrared reflectance spectroscopy for prediction of soil physical, chemical, and biological properties. *Applied Spectroscopy Reviews* 49: 139–186.

Stevens, A., B. Wesemael, G. Vanderschrick, S. Touré, B. Tychon. 2006. Detection of carbon stock change in agricultural soils using spectroscopic techniques. *Soil Science Society America Journal* 70: 844–850.

Stevens, A., B. Wesemael, H. Bartholomeus, D. Rosillon, B. Tychon, E. Ben-Dor. 2008. Laboratory, field and airborne spectroscopy for monitoring organic carbon content in agricultural soils. *Geoderma* 144: 395–404.

St Luce, M., N. Ziadi, J. Nyiraneza, G.F. Tremblay, B. Zebarth, J. Whalen, M. Laterriere. 2012. Near infrared reflectance spectroscopy prediction of soil nitrogen supply in humid temperate regions of Canada. *Soil Science Society of America Journal* 76: 1454–1461.

Stuart, B. 1997. Biological applications of infrared spectroscopy. In *Analytical chemistry by open learning*, ed., D J. Ando, Wiley & Sons, Chichester, UK.

Stuart, B. 2004. *Infrared spectroscopy: Fundamentals and applications*, John Wiley & Sons, Ltd, Chichester, UK, 25–54.

Summers, D., M. Lewis, B. Ostendorf, D. Chittleborough. 2011. Visible near-infrared reflectance spectroscopy as a predictive indicator of soil properties. *Ecological Indicators* 11: 123–131.

Sun, Y., Y. Peng, Y. Chen, A. Shukla. 2003. Application of artificial neural networks in the design of controlled release drug delivery systems. *Advanced Drug Delivery Reviews* 55: 1201–1215.

Terhoeven-Urselmansa, T., H. Schmidt, R. Joergensen, B. Ludwig. 2008. Usefulness of near-infrared spectroscopy to determine biological and chemical soil properties: Importance of sample pre-treatment. *Soil Biology & Biochemistry* 40: 1178–1188.

Thompson, S.E., N. Foster, T. Johnson, N. Valentine, J. Amonette. 2003. Identification of bacterial spores using statistical analysis of Fourier transform infrared photoacoustic spectroscopy data. *Applied Spectroscopy* 57: 893–899.

Waiser, T.H., L.S. Cristine, D. Brown, C. Hallmark. 2007. In situ characterization of soil clay content with visible near-infrared diffuse reflectance spectroscopy. *Soil Science Society of America Journal* 71: 389–396.

Uphoff, N., A. Ball, E. Fernandes, H. Herren, O. Husson, C. Palm, J. Pretty, N. Sanginga, J. Thies. 2006. Understanding the functioning and management of soil systems. In *Biological approaches to sustainable soil systems*, eds., N. Uphoff, A. S. Ball, E. Fernandes, H. Herren, O. Husson, M. Laing, C. Palm, J. Pretty, P. Sanchez, N. Sanginga, J. Thies, CRC Press, Taylor & Francis Group, Boca Raton, FL, pp. 3–12.

Verma, S.K., M.K. Deb. 2007a. Direct and rapid determination of sulphate in environmental samples with diffuse reflectance Fourier transform infrared spectroscopy using KBr substrate. *Talanta* 71: 1546–1552.

Verma, S.K., M.K. Deb. 2007b. Nondestructive and rapid determination of nitrate in soil, dry deposits and aerosol samples using KBr-matrix with diffuse reflectance Fourier transform infrared spectroscopy DRIFTS. *Analytica Chimica Acta* 582: 382–389.

Viscarra Rossel, R.A., A. McBratney. 1998. Laboratory evaluation of a proximal sensing technique for simultaneous measurement of clay and water content. *Geoderma* 85: 19–39.

Viscarra Rossel, R.A., A.B. McBratney. 2001. A response-surface calibration model for rapid and versatile site-specific lime requirement predictions in south-eastern Australia. *Australian Journal of Soil Research* 39: 185–201.

Viscarra Rossel, R.A., D. Walvoort, A. McBratney, L. Janik, J.O. Skjemstad. 2006. Visible, near infrared, mid infrared or combined diffuse reflectance spectroscopy for simultaneous assessment of various soil properties. *Geoderma* 131: 59–75.

Viscarra Rossel, R.A., Y. Jeon, I. Odeh, A. McBratney. 2008. Using a legacy soil sample to develop a mid-IR spectral library. *Australian Journal of Soil Research* 46: 1–16.

Vohland, M., M. Ludwig, S. Thiele-Bruhn, B. Ludwig. 2014. Determination of soil properties with visible to near- and mid-infrared spectroscopy: Effects of spectral variable selection. *Geoderma* 223: 88–96.

Wahls, M.W.C., E. Ketta, J.C. Leyte. 2000. Depth profiles in coated paper: Experimental and simulated FT-IR photoacoustic difference magnitude spectra. *Applied Spectroscopy* 54: 214–220.

Waiser, T.H., C. Morgan, D. Brown, C. Hallmark. 2007. In situ characterization of soil clay content with visible near-infrared diffuse reflectance spectroscopy. *Soil Science Society of America Journal* 71: 389–396.

Wang, J.J., L.J. Cui, W. Gao, T. Shi, Y. Chen, Y. Gao. 2014. Prediction of low heavy metal concentrations in agricultural soils using visible and near-infrared reflectance spectroscopy. *Geoderma* 216: 1–9.

Waruru, B.K., K. Shepherd, G. Ndegwa. 2014. Rapid estimation of soil engineering properties using diffuse reflectance near infrared spectroscopy. *Biosystems Engineering* 121: 177–185.

Wetterlind, J., B. Stenberg, A. Jonsson. 2008a. Near infrared reflectance spectroscopy compared with soil clay and organic matter content for estimating within-field variation in N uptake in cereals. *Plant and Soil* 302: 317–327.

Wetterlind, J., B. Stenberg, M. Soderstrom. 2008b. The use of near infrared NIR spectroscopy to improve soil mapping at the farm scale. *Precision Agriculture* 9: 57–69.

Wilding, L.P., H. Lin. 2006. Advancing the frontiers of soil science towards a geoscience. *Geoderma* 131: 257–274.

Wold, S., M. Sjostrom, L. Eriksson. 2001. PLS-regression: A basic tool of chemometrics. *Chemometrics and Intelligent Laboratory Systems* 58: 109–130.

Wu, H.P., L. Dong, W. Ren, W. Yin, W. Ma, L. Zhang, S. Jia, F.K. Tittel. 2015. Position effects of acoustic micro-resonator in quartz enhanced photoacoustic spectroscopy. *Sensors and Actuators B-Chemical* 206: 364–370.

Yang, X.M., H. Xie, C. Drury, W. Reynolds, J. Yang, X. Zhang. 2012. Determination of organic carbon and nitrogen in particulate organic matter and particle size fractions of Brookston clay loam soil using infrared spectroscopy. *European Journal of Soil Science* 63: 177–188.

Zimmermann, M., J. Leifeld, J. Fuhrer. 2007. Quantifying soil organic carbon fractions by infrared spectroscopy. *Soil Biology & Biochemistry* 39: 224–231.

Zoltan, B., P. Andrea, S. Gabor. 2011. Photoacoustic instruments for practical applications: Present, potentials, and future challenges. *Applied Spectroscopy Reviews* 46: 1–37.

Zornoza, R., C. Guerrero, J. Mataix-Solera, K. Scow, V. Arcenegui, J. Mataix-Beneyto. 2008. Near infrared spectroscopy for determination of various physical, chemical and biochemical properties in Mediterranean soils. *Soil Biology & Biochemistry* 40: 1923–1930.

7 Use of Variable Rate Application in Soil Fertility Management by Small Farmers
Status, Issues, and Prospects

Yeong Sheng Tey, Mark Brindal, and Chin Ding Lim

CONTENTS

7.1 INTRODUCTION

Small-scale agriculture is the main source of world food production, accounting for up to 70% of food (Fairtrade Foundation 2013). However, most smallholders are confronted with the intertwined issues of poverty and malnutrition. As these issues demand immediate attention, agricultural innovations must address broad income generation and food security needs (Lee 2005). One solution to both issues lies in improved crop yield, which is itself the result of better soil fertility management in precision agriculture (PA).

The USDA (2007) posits that PA "is a management system that is information and technology based, is site specific and uses one or more of the following sources of data: soils, crops, nutrients, pests, moisture, or yield, for optimum profitability, sustainability, and protection of the environment." This concept emphasizes the use of technology to supply specific information for precise decision-making.

The concept of PA addresses one of the major weaknesses of traditional/conventional production practice: soil variability on the farmland is often overlooked. As depicted in Figure 7.1, fertilizers are therefore typically uniformly applied over an entire farm in an attempt to prevent nutritional deficiency and yield loss (Tey and Brindal 2012). These standard application rates result in under- or oversupply of nutrition for crop growth. Without proper prescription, traditional/conventional production practices have rarely achieved maximal crop yield. Targeting to remedy this situation,

FIGURE 7.1 Major weaknesses of traditional/conventional production practice—less information-armed fertilizer application on most small-scale farms. (From ClimateTechWiki, 2014.)

soil fertility management in PA involves the implementation of variable rate application (VRA)—computer-controlled equipment that continually readjusts fertilizer application.

Technologies used to implement VRA include soil sampling analysis, Global Positioning Systems (GPSs), and variable-rate applicators. Figure 7.2 illustrates a variable rate fertilizer application system for precision rice farming in Japan. Sampling data provides the prescription for the particular fertilizers to be applied according to variability of spatial patterns within the farm. A GPS receiver in the "spreader" equipment assists the recognition of a precise location. Assisted by the GPS, a computer-controlled variable-rate applicator varies the types and amounts of fertilizer according

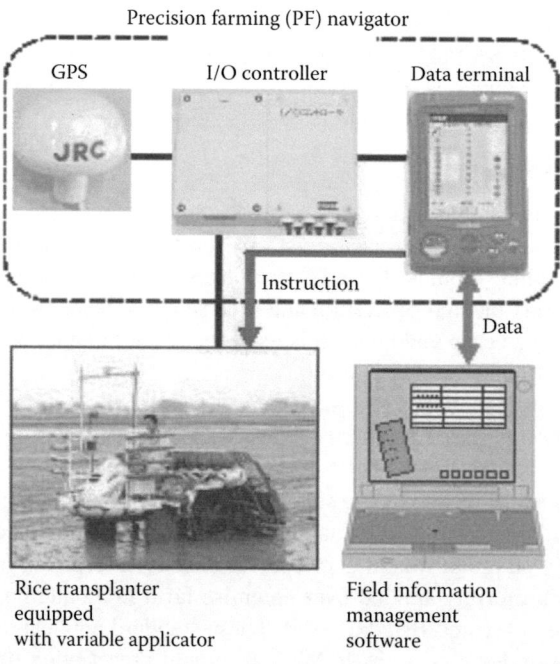

FIGURE 7.2 Variable rate fertilizer application system for precision rice (*Oryza sativa* L.) farming in Japan. PF, precision farming. (From National Agriculture and Food Research Organization, 2014.)

to the information supplied by the stored sampling data analysis. In this manner, VRA helps to improve crop productivity efficiently on what is hoped will be a sustainable basis.

In an effort to encourage the uptake of PA, VRA has been promoted in both developed and developing countries. The package of VRA technologies has been increasingly used on large farms in many developed countries, such as the United States (Fountas et al. 2005), Australia (Robertson et al. 2013), Germany (Reichardt and Jürgens 2009), and Canada (Goddard 2005). Some investments in VRA have also been made by large-scale plantations (e.g., sugarcane) in developing countries (e.g., Brazil, see Palaniswami et al. 2011; Silva et al. 2011). It is obvious that critical questions remain to whether small farmers have made use of VRA and what are their underlying issues. Clearly, in such cases, economies of scale in the potential saving of input costs justify the capital investment costs involved in both the initial soil sampling and the purchase of the technology. The economics of smallholdings represent a different paradigm.

Accordingly, a number of studies have attempted to understand low/variable adoption of PA overall (see the review in Tey and Brindal 2012). In particular, Khanna (2001), Roberts et al. (2002), Roberts et al. (2004), and Robertson et al. (2013) offer insights into the motivations behind VRA. These empirical studies suggest that farmer and household characteristics, farm biophysical characteristics, farm financial/management characteristics, and attributes of the innovations can be used to explain adoptive decision-making. Given these findings, decision-making in relation to VRA use is seen as complex. A specific knowledge base should be built to comprehend both the economic and motivational needs of small farmers.

To assist in this, this chapter aims to review current variable fertilizer practices, identify key characteristics of VRA, and discuss some of the key factors, helping to explain the current state of VRA use, which recent research has suggested as being critical issues among smallholder farmers. We compliment this by documenting some of our personal observations resulting from our interaction with farmers, extension agents, and researchers. It is hoped that this work will prove valuable in encouraging efficient fertilizer application according to spatial need within smallholders' land holding. The information should also have applicability to the promotion of PA in general.

7.2   CURRENT VARIABLE FERTILIZER PRACTICES ON SMALL FARMS

The concept of crop management according to field variability and site-specific conditions is highly applicable to small farmers (Pinstrup-Andersen et al. 1999). This might be considered axiomatic since "precision agriculture" is variously defined as "that kind of agriculture that increases the number of (correct) decisions per unit area of land per unit time with associated net benefits" (McBratney et al. 2005). With or without the involvement of modern equipment, such a concept accentuates the need for improved decision-making in fertilizer application.

7.2.1   TRADITIONAL METHODS

Even in the absence of modern equipment, fertilizers are often applied variably. This is because of the farmers' understanding the fields in their smallholding. As noted by Tittonell and Giller (2012), in sub-Saharan Africa, farmers identify and ascribe different local names to field-specific soil fertility. They also recognize gradients of soil fertility. Based on such knowledge, they tend to plant crops more densely, weed more frequently, and apply less fertilizer (including manure) to the more fertile fields.

In Asian countries, rice (*Oryza sativa* L.) yields are recorded by farmers. All rice farmers are obliged to report their rice production to the local department of agriculture so that governments can be aware of the national food security status. Through that mechanism, an information base exists through which rice farmers can compare their current and previous yields (season or year). Lower productivity implies that more fertilizer input is required to increase crop yields in the next cropping season or year; higher productivity suggests that the same fertilizer practice is likely to help maintain production output in the next season or year. In addition, yield difference between rice paddies are also noted by farmers. Such information is also used as a guide for fertilizer application (i.e., high yielding plots serve as a benchmark, with the implication that low-yielding plots might benefit from higher applications of fertilizer).

In rice farming, leaf color is traditionally used as a natural indicator. It serves as a sign of both the current and potential nutrient issues that the rice crop may be facing. Yellowing leaf tips imply that there is a nutrient imbalance. In that case, soil properties should be checked, i.e., nutrient composition, pH, hydration, and access to sunlight. Sudden leaf shedding or dropping signals a variety of ailments, including environmental change, chemical exposure, pest or disease battle, and insufficient nutrients or hydration. Given various possibilities, rice farmers rely on their experience to interpret the deficiencies giving rise to the leaf color. Correct interpretation of the various signs produces a better site-specific nutrient resolution and/or preventive treatment.

7.2.2 Modern Methods

To facilitate a better interpretative process, leaf color panel charts have been introduced to rice farmers in many countries. These cheap panels (as low as US$1) have provided an acceptable substitute for the more expensive chlorophyll meters. By simply holding a rice leaf against a panel (see Figure 7.3), a farmer identifies the degree of nitrogen supply to the plant (deficiency, adequacy, or excess). They have proved useful as an index to measure the nutritional status and quality of all rice varieties (Yang et al. 2003). Guidelines on their use vary according to local conditions. This results in a site-specific and real-time recommendation on fertilization timing and quantity needed in order to increase productivity. As a simple, effective, and low-cost tool, leaf color charts have become popular and are used extensively on rice farms. Their popularity has motivated further innovations, such as web- and smartphone-based applications. For example, LeafCorder—a US$1 smartphone application—allows Indonesian rice farmers to carry out the same function through digital leaf color images, which are sent to and analyzed by a web-based intelligent system. A similar smartphone application has recently been designed for Filipino (IRRI 2012) and Thai (Intaravanne and Sumriddetchkajorn 2012) rice farmers.

Owing to the popularity of mobile phones, the Pinoy Farmers' Text Center was initiated in 2004. At Php1.00 (US$0.022) per text message, short messaging service (SMS) is one of the cheapest means to get answers on production (including nutrient management) and marketing questions from rice experts in Philippines. From an average of 2000 text messages per month in 2008 (Barroga et al. 2009), the center now responds to more than 3000 text enquiries monthly (The Philippine Star 2012). Armed with such information, rice farmers have been observed to make better decisions: resulting

FIGURE 7.3 Leaf color chart functions as an index to measure the nutrient need of rice plants. (From International Rice Research Institute, 2012.)

in increased production and greater marketability. To date, this SMS service is already used by some 30,000 Filipino rice farmers and this number continues to rise (The Philippine Star 2013).

Launched by the International Rice Research Institute (IRRI) in 2010, Nutrient Manager for Rice (NMRice) is a free information tool that is built upon the principles of site-specific nutrient management for small-scale rice farmers. This innovation is the result of improvement from its predecessors: the Nutrient Management fact sheet in 2008 and the compact disk-based Nutrient Management for Rice in 2009. Farmers and extension workers are now able to obtain site-specific fertilizer guidelines through country-specific web, mobile SMS and smartphone applications (see Figure 7.4).

- Using the web, NMRice collects information by asking 15–17 simple questions about a farmer's rice cultivation conditions. The answers are transmitted to and analyzed by a database which then calculates a fertilizer recommendation detailing the amounts, sources, and timing of fertilizer application for that specific rice field. The web-based NMRice is available to help rice farmers in Bangladesh, Guangdong Province of China, Tamil Nadu State of India, Indonesia, the Philippines, and West Africa. However, it should be noted that it is unfortunate that in those countries many rice farmers and extension workers have limited access to internet.
- In contrast to the issue of lack of internet connectivity, most small-scale rice farmers (including those in the most remote areas) and extension workers possess a mobile phone. Through an interactive voice response system, farmers dial a toll-free number and answer a list of questions pertaining to cultivation conditions by pressing the appropriate keypad numbers. Then, the farmer receives a text message with a guideline matching the site-specific needs and conditions of that farmer. Since its launch in 2011, NMRiceMobile has received more than 10,000 calls from Filipino rice farmers (New Agriculturist 2012). This information tool is also available in Indonesia.

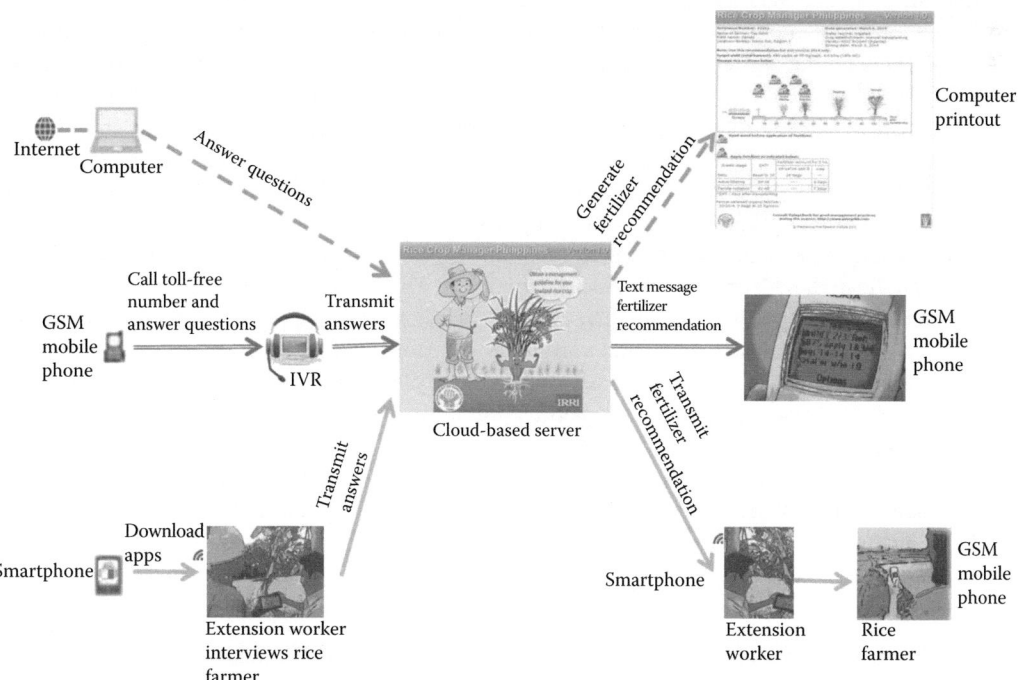

FIGURE 7.4 Flow chart of Nutrient Manager for Rice (NMRice) operations. (Authors' illustration.)

- With the prevalence of smartphones, NMRice now also works using Android operating systems. While complementing its web- and mobile phone-based tools, NMRiceApp is targeted mostly at extension workers. With the help of visuals, extension workers understand questions better and are more able to process the fertilizer recommendation for the farmer comprehensively. NMRiceApp is currently only available in the Philippines.

Both the pilot projects and actual use have demonstrated that NMRice is beneficial. In an Indonesian pilot project, for example, 300 rice farmers across nine provinces have recorded higher yields after modifying their fertilizer practice in line with the recommendation accessed through NMRice. These higher yields have led to a significant increase in net income, exceeding US$100 per hectare per season (New Agriculturist 2012). Such potential is further affirmed through a field survey on the economic impacts of NMRice. In Zaini (2012), the users of NMRice have consistently outperformed the non-users in all studied areas: the net profit of NMRice users is 5.4%–18.1% more than the non-users respectively.

7.2.3 Major Weaknesses of the Traditional and Modern Methods

A major limitation to traditional methods arises because decision-making is based on an individual's heuristics, i.e., their observation, experience, and knowledge (which may be incomplete). While understanding and making use of these factors is never easy, farmers may, in fact, achieve better solutions or a satisfactory result through trial and error. Using such an approach, farmers trial various techniques repeatedly (sometimes with variations) until success is achieved or error is diminished over time. Such a process can be long and challenging, especially when it is attempted in a milieu of environmental variability. For example, the nutrient demand of rice crops varies according to changes in the soil fertility gradient from field to field. In addition, it varies in between dry season, wet season, and the interregnum. Therefore, it becomes inherently difficult for farmers to plan for effective variable application of fertilizers using these traditional methods.

Similarly, the innovations discussed above are not without their own weaknesses. Many farmers do not have any implicit advice on fertilizer application rates. In consequence, farmers develop a total reliance on the system and lack the ability to adapt to changes. Such an issue is partially a result of the structure of the system itself, since it leaves little room for communication on the science inherent in nutrient management. To assist farmers to understand and learn the implications behind the advice generated, there is still a need for educative programs. This can involve interaction between farmers and extension workers. It is in such areas as this that extension services will always have a role in bridging farmers' knowledge gaps.

Nevertheless, it becomes obvious that some small-scale farmers have practiced variable rate fertilizer application with or without modern equipment. This shows that they have adapted the concept of site-specific application to suit their particular context and supports the contention that small farmers know their farm's features and variability (Stafford 2000; Booltink et al. 2001; Maohua 2001). The application of fertilizers is already manually adjusted according to the field-specific understandings or observations. Consequently, small landholding farmers are demonstrably as amenable to potential deviation and would potentially make precise decisions if presented with a package of VRA technologies. Given such predisposition, it is therefore crucial to understand why VRA is not adopted by most smallholders, or whether, because of their understanding of their own landholdings, some farmers feel that they already practice VRA so can gain little from the additional information.

7.3 FACTORS INFLUENCING THE USE OF VRA BY SMALL FARMERS

From the outset, it is important to recognize that small farmers are not a homogenous group. Resources are asymmetrically endowed and distributed across small farms. Such unequal settings are compounded by varied attitudes and capacity among small farmers. Their attitudes differ in many ways, including the belief or evaluation that they have toward the attributes, risks, costs, and benefits of VRA. Their capacity also determines their managerial and financial ability to maintain the function of a system in response to change (Carpenter et al. 2001).

7.3.1   ACQUISITION COSTS AND FINANCIAL CAPACITY

Unlike traditional sustainable agricultural practices, the operation of VRA requires the purchase of equipment and/or subscription of services. Their costs have remained high although the investment costs have reduced in accordance with the increasing availability of VRA equipment and service providers in the market over time (Jochinke et al. 2007). In Australia, for example, about $10,000 is needed for investment in the basic package of VRA (zone analysis, GPS, and variable rate controller); at the other end of the spectrum, nearly $100,000 is essential to acquire the high-end package of VRA (high-level zone analysis, high accuracy GPS, auto-steer vehicles, and multipurpose variable rate controllers) (Robertson et al. 2007). Such acquisition costs become prohibitive to small farmers, especially since few incentives or subsidies have been provided to enhance their affordability.

As VRA is capital intensive, financial capacity is necessarily found to be correlated with the adoption of technology. This is largely because wealthier small farmers (either measured directly by income or asset base) have both greater purchasing power and risk tolerance (Daberkow and McBride 1998). A similar impact is also exhibited by those who have greater ability to raise capital by selling alternative crops or livestock, generating additional incomes from off-farm employment, and who have access to credit and loans (Isgin et al. 2008). However, typically, farmers of smallholdings generally generate lesser incomes than those with large properties. Due to limited assets, they often face difficulty in securing financial support from outside parties (Tey et al. 2014a). Therefore, it is not surprising to observe that most small farmers remain unchanged in their fertilizer application practice or prefer the low-cost variable rate fertilizer application methods outlined above.

7.3.2   PROFITABILITY POTENTIAL AND INVESTMENT BEHAVIOR

Return on investment is a primary concern of farmers. However, it remains uncertain when the economic return of any particular VRA investment is greater than the costs of the purchase of the equipment, the sacrifice of amenity, and additional skilled labor (Tey and Brindal 2012). As noted earlier, the prices of VRA equipment remain high. While such expensive investment does not lead to immediate returns, small landholders have to spend substantial time learning to master its operation, data collection, data analysis, and the resultant interpretation implications for farm variability. Alternatively, these tasks can be undertaken by hired (skilled) labor or consultants, but these also come at additional cost unless provided by an outside funding agency, such as a government. A further complication arises in tandem with the advent of rising input costs. There is invariably fluctuation in market prices. Therefore, inherent uncertainty exists in relation to the potential of VRA (given its extra input costs) to achieve higher levels of profitability.

As demonstrated, the adoption of VRA involves a series of analyses in respect to its economic viability in specific agricultural systems and varying environments. Small landholding farmers are more likely to invest if the inter-temporal, short, and long-term costs are outweighed by the anticipated net benefits of VRA. While such consideration also applies to other agricultural production systems, the considerable monetary investment needed to acquire VRA technology make it particularly cogent in this context. Moreover, small landholding farmers presumably need some time to achieve proficiency since, as they learn by doing, their improvement in the efficient use of VRA operation is time dependent. In juxtaposition, the issues of poverty and food insecurity are pressing and inculcate a short-run perspective in small landholding farmers (Lee 2005).

7.3.3 COMPLEXITY AND MANAGEMENT CAPACITY

Unlike many high-yielding production practices (e.g., hybrid varieties), VRA use is not a straight-forward process (Robertson et al. 2013). It involves a complex series of stages: data collection, interpretation, and judgment. Sampling data, to cite just one example, requires small farmers to make a range of management decisions. They have, from a set of various alternatives, to decide what information they need and where and how it is to be collected. The collected data must be carefully analyzed and interpreted according to individual crop requirements. If a specific nutritional addition (or indeed subtraction in the case of over fertilization with specific elements) is identified to enhance crop production, it is necessary for small farmers to evaluate the relative agronomic advantages and economic returns from the additional fertilizer application to that particular crop. Other complementary VRA technologies (e.g., GPS and variable-rate applicators) present equally challenging adoption scenarios. Consequently, the implementation of VRA remains dependent on and requires the intensive use of management inputs.

Management capacity is widely found in the literature to be associated with the use of VRA. Farmer age and farming experience are often cited as the proxy for shorter experience horizons and better knowledge of field variability (Roberts et al. 2004). However, we assert that the relationship between these interrelated factors and VRA is a blunt and, therefore, inappropriate instrument whose indicators are not necessarily either positive or significant. In contrast, the education level of the household head is expected to be positively associated with the use of VRA since, as demonstrated, it is information intensive. Therefore, management ability is the key to its successful operation (Walton et al. 2008). While improved management starts with general education, most small farmers face greater constraints in acquiring the specific knowledge associated with VRA. They therefore remain constrained to their experience and knowledge in the variable application of fertilizers.

7.3.4 INFORMATION INTENSITY AND SKILLS

The implementation of VRA is facilitated by the use of readily abundant information. It is important to recognize the value of VRA lies in increased information flow as a key to greater overall management efficiency. One should undergo comprehensive training to implement VRA efficiently. As most small farmers are not adequately trained, they face many challenges in respect to their ability to analyze these datasets and interpret their results as the basis for decision-making in soil fertility management. Thus, skills play the key function in VRA.

The role of skills, both as a special characteristic of PA and as a primary constraint, in influencing the use of VRA has been documented. Measures of hired consultant and participation in training program have been shown to affect adoptive decision making. Robert (2002) has noted the range of skill requirements characterizing VRA, and emphasizes that agricultural dealers, consultants, and cooperative personnel are among important potential sources to which might assist farmers adopt more tailored site-specific fertilizer management regimes. The underlying reason is VRA requires new skills that are often missing on farms.

7.3.5 AWARENESS AND ACCESS TO INFORMATION

Information is the key to the education of small landholding farmers and the improvement of management skills in relation to VRA. To begin with, information is necessary in shaping awareness. The equipment facilitating the use of VRA have been available for more than a decade, but many farmers reportedly remain unaware of what PA stands for (Reichardt and Jürgens 2009; Reichardt et al. 2009). VRA technologies are still seen as an innovation. Creating awareness is an essential first step in leading in facilitating subsequent changes in small farmers' attitude and behavior.

Most studies have shown that access to the information and advisory services related to the technological innovation facilitates adoption (Robertson et al. 2013). Typical sources of information include agricultural schools, extension programs, farmer associations, non-governmental organizations, mass media, input suppliers, and other farmers. Access to these sources does not necessarily lead to VRA adoption unless specific information on VRA is provided. However, most of these parties have little or no specific information. For example, suitable teaching materials are lacking at the teaching level: many agricultural professionals have not received training in the theories and/or practice of VRA. Consequently, the relevant information is still unavailable to most smallholding farmers and they remain untrained in the use of VRA equipment.

7.3.6 FARM SIZE AND LAND RIGHTS

Farm size of small landholders, especially as this measure is relevant to crop type, varies and is among the key determinants of VRA adoption (Walton et al. 2008). Farm enlargement, either through leasing or consolidation, is one measure whereby the profitability of the technological application is enhanced (Takacs-Gyorgy 2008). Larger operational scales enable the cost per unit of output generally to decrease as fixed costs and variable costs are spread over more units of output. It is more economical to invest VRA on larger farms: there is an inverse relationship between farm size and the improvement in gross margin required to cover the cost of VRA investment (Robertson et al. 2007). Moreover, larger farms have a greater capacity to absorb risks and spread them over a wide productive base. Consequently, the likelihood of VRA use scales upward in line with farm size growth.

Even when farm holdings are of an identical size, the nature of the tenure must be considered. Part or the whole of the farmland may be alternatively owned or rented by small landholding farmers. Private ownership may not include freehold possession. It may, in fact, refer to leasehold in which smallholders buy from or are given the right by the local government to farm the land for an indeterminate length of time (i.e., a 99-year lease). However, while some might contend that this system of tenure is less certain than freehold, since such leasehold might still be diverted for other purposes under public orders, the same acquisition right is generally applicable even to freehold tenure (Tey et al. 2014b). A greater uncertainty does, however, exist where farmland is rented on a periodic basis (i.e., yearly or monthly).

Unfortunately, a lot of small farmers do not own their farmland. Short-term tenure encourages the leaseholder to farm the land in a manner which returns the maximum in short-term yield even if the additional gains are gleaned at the detriment to the long term sustainability of the land parcel. Such decreased assurance of continuity of access to the same farmland has been shown to affect VRA investment (Roberts et al. 2004). As mentioned earlier, this also means shorter planning horizons for the same farmland, prompting the exploitation of various short-term benefits.

7.4 CONCLUSIONS AND IMPLICATIONS

The application of fertilizer inputs at a notional "averaged" rate is clearly neither economically nor environmentally efficient. Not only because such fertilizers have an input cost to the farmer, but they are often, and especially in the case of developing countries, heavily subsidized by the public purse. Inefficient application therefore represents economic waste. Additionally, since the excess fertilizer is invariably water soluble, it enters either the underground water tables or surface waters. The leaching and runoff often cause environmental degradation, an extreme form of which is toxic environmental blooms, which endanger human and animal life. This represents a dangerous and avoidable environmental inefficiency. Finally, chemical fertilizers are manufactured from finite natural resources, most of which are not renewable in what would be generally acceptable time frames. Profligate current fertilizer use is bought by creating a deficit against the future.

Therefore, the introduction of VRA (e.g., soil sampling analysis, GPS, and variable-rate applicators) is essential to the future of global food production since 70% is produced by smallholding farmers. Inherent in the concept is that farmers will make guided (correct) decisions on fertilizer application using the optimum amounts of the right type of fertilizer according to soil properties and time. When implemented properly, this innovation can harness the agricultural sector to maintain a profitable and sustainable agricultural industry through cost-effective use of inputs (chemicals, labor, and machinery), increased yields, improved food safety, and increasing product value (Robertson et al. 2013).

This paper supports the applicability of the VRA concept to small farms. Fertilizer practice, as we have mentioned, is already somewhat managed according to field variability and site-specific conditions.

The traditional methods used for that purpose include using soil color, soil fertility, and leaf color to infer nutritional status; production records are sometimes used to identify the yield difference between fields and seasons. Any indication of nutrient imbalance or deficiency signals a need for the application of more fertilizer to the field. Such an indicator, however, provides little information to help smallholding farmers in quantifying fertilizer application.

Modern methods, including the use of leaf color charts to measure the nutritional status and quality of rice crops and computer- and mobile phone-based applications are clearly more helpful in achieving efficient resolution of the problems. They have proved user-friendly and their recommendations are easy to understand. They also provide more concrete recommendations on fertilizer need (i.e., both as to quantity and type).

It is clear that both traditional and modern variable rate application methods rationalize the need for fertilizer. This often leads to a cost reduction for inputs which can translate into higher farm profits. However, such application, especially in the absence of *a priori* knowledge, can negatively impact on yields. Consequently, VRA is arguably an essential component to achieve optimum yields.

However, VRA has not been adopted commonly amongst smallholding farms. We have identified that factors contributing to this non-adoption are largely related to the inherent attributes of the innovation and their corresponding endowments.

VRA technologies, as they are implemented in the developed world, are capital intensive and are therefore less affordable to typical smallholding farmers, particularly in developing nations. Not only do they have limited income and assets, they also face difficulty in raising capital. These financial constraints further undermine their capacity to hire consultants, absorb potential failure, and to undertake any associated risks. These circumstances in tandem with the chronic budgetary pressures experienced in many developing countries give rise to important implications for international and national support in the form of financial assistance. At the farm enterprise level, public funding could be offered through a variety of flexible financial instruments (e.g., installment), tax reduction or rebates, subsidies and incentives, and co-investment strategies with other farmers. One logical extension of that strategy lies in funding farmer-centered research, which aims to escalate VRA product development and intensify marketing competition. The resultant (increasing) availability of hopefully cheaper VRA technologies will make innovation more affordable.

Longer term prospective returns of VRA are uncertain, but immediate profits are the main concern of most smallholding farmers. Uncertainty is perceived as risk. Given that smallholding farmers have limited budgets, their degree of risk tolerance is generally perceived as low. One way to reduce any uncertainty is through technical support for the operation of VRA. However, such professional services are largely provided by VRA suppliers and/or consultants at a cost. Since such a service charge is likely to drive smallholding farmers away, it is vital for public extension authorities to train their staff or instigate a specialized committee for that purpose. In particular, the latter structure is dedicated to the provision of complimentary knowledge transfer services: it should be more effective. It could be that the initial results could provide more in the area of cost reduction than in yield improvement. Nevertheless, it would still be a valuable adjunct in assisting farmers to make a profitable beginning and to continue as a guide for farmers to achieve optimum productivity in the long-run.

The operation of VRA is challenging and requires intensive use of management inputs. As a knowledge intensive agricultural system, management ability provides the key to its successful adaptation and adoption. Improved management can only come about through education and training. Such outreach programs should necessarily encourage farmer involvement, enhancing their observational, analytical, experimental, and communications skills. Coupled with that, there should be more effort aimed at simplifying the operation of VRA. One of these might involve farmer participatory research in which smallholding farmers experiment in their own fields, learn, modify (simplify), and adopt VRA and subsequently spread their experience to other farmers. The main advantage of this approach is that smallholding farmers "learn by doing" with operational processes being altered on the basis of their direct experience.

A key feature of VRA operation is that it relies on abundant information inputs. A different set of skills is required to make optimum use of the information. In congruence with the foregoing recommendation, a key implication is the provision of more educational and training programs for the smallholding segment. In addition, technical assistance should also be made convenient and accessible. As such assistance is mostly rendered by professionals (e.g., extension workers, consultants, suppliers, and dealers), separate educator training programs should be developed for them.

Unfortunately, a general awareness of VRA is low among smallholding farmers and information currently available is too general. An obvious need here is to gear up the promotion of PA in general and VRA specifically. It is to be hoped that awareness would lead to more information searches and would arouse positive changes in farmer attitudes and behaviors toward VRA. At the same time, different levels of message (i.e., from shallow to deep) should be tailored specifically for beginners, intermediate, and advanced users and promoters of VRA. Any/all promotional programs should not just target smallholding farmers, but should also permeate through their networking systems (i.e., including farming solution providers).

The issue then is whether it is possible to realize the use of PA and VRA on most smallholding farms.

We have mentioned that conventional economic wisdom implies the aggregation of small holdings to unit sizes at which the investment costs for VRA become economically viable. However, such conclusions ignore many of social and political realities of the developing world: the social dislocation which would result from tens or hundreds or thousands of smallholders and their families being dispossessed of a way of life and livelihood which has been their family heritage for centuries has not been considered. The political implications are especially complicated where such farmers hold their tenure by virtue of a lease and would therefore be disposed, presumably with no right to compensation.

Such wisdom is also based on the supposed efficiency of vast mono-cultural systems, which dominate regions where climatic variability necessitates seasonal rather than continuous production regimes. Since climatic patterns facilitate continuous production, different paradigms are applicable. Indeed, in areas of high rainfall such are experienced throughout the tropics, a complimentary answer to VRA might lie in production diversification to ensure maximum use of fertilizer on the farm sites rather than accepting its inevitable leaching into water courses. Such diversification is more easily facilitated in small intensively managed landholdings than in vast enterprises.

We believe that, if small landholders can be properly informed and educated, one solution might lie in the more utilization of the type of modern variable rate fertilizer methods mentioned previously (of which NMRice is an example). Provided that information can be accurately gleaned, it seems axiomatic that any smallholding can benefit from any user-friendly and easy-to-understand innovations. For instance, a number of rice paddies would require a variable application rate within a particular paddy or, even were this to be the case a general application rate for that paddy, it would be far less wasteful than a general inefficient rate spread over many hectares. Once such a rate is ascertained, very cheap spreaders can be utilized to apply fertilizer and variable rates between the different paddies. This would not only save input costs and be more efficient, but labor requirements would also be no more than are now actually expended. These will result in higher profitability albeit that only a few of these farms would enjoy a productivity increase.

The principal issue is not, therefore, how to turn the developing world into a replication of efficient farming practice in other climatic regions, but to develop PA and VRA in a manner conducive to the climatology and social needs of the agricultural systems in which it must be embedded. The prospect is to utilize the science for efficient production while building on the strengths provided by intense farming utilizing human intelligence. In this manner, hopefully, smallholdings will not become more than traps for poverty and malnutrition.

Traditional farming methodologies have evolved which subconsciously seek to adopt the principle of PA and VRA. Programs such as NMRice seek to build such traditions into a more scientific and effective paradigm. Their widespread adoption is an encouraging beacon, underscoring the importance of tailored and unique methodologies in precision farming promotions.

REFERENCES

Barroga, R. F., Asis, O. R. M., Domingo, O. C., Quilang, E. J. P., and Alvarez, T. G. V. 2009. "Pinoy farmers" text center: Serving rice farmers, caring for the environment through SMS [short messaging system]. Paper presented at the Federation of the Crop Science Society of the Philippines Scientific Conference, Dumaguete City.

Booltink, H., van Alphen, B., Batchelor, W., Paz, J., Stoorvogel, J., and Vargas, R. 2001. Tools for optimizing management of spatially-variable fields. *Agr. Syst.* 70:445–476.

Carpenter, S., Walker, B., Anderies, J. M., and Abel, N. 2001. From metaphor to measurement: Resilience of what to what? *Ecosystems* 4:765–781.

Daberkow, S. G., and McBride, W. D. 1998. Socioeconomic profiles of early adopters of precision agriculture technologies. *Agribusiness* 16:151–168.

Fairtrade Foundation. 2013. Powering up smallholder farmers to make food fair: A five point agenda. http://www.fairtrade.org.uk/includes/documents/cm_docs/2013/F/FT_smallholder%20report_2013_lo-res.pdf (accessed January 2, 2014).

Fountas, S., Blackmore, S., Ess, D., Hawkins, S., Blumhoff, G., Lowenberg-Deboer, J., and Sorensen, C. G. 2005. Farmer experience with precision agriculture in Denmark and the US eastern corn belt. *Precis. Agr.* 6:121–141.

Goddard, T. W. 2005. An overview of precision conservation in Canada. *J. Soil Water Conserv.* 60:456–461.

Intaravanne, Y., and Sumriddetchkajorn, S. 2012. BaiKhao (rice leaf) app: A mobile device-based application in analyzing the color level of the rice leaf for nitrogen estimation, *Proc. SPIE* 8558, Optoelectronic Imaging and Multimedia Technology II, 85580F.

IRRI (International Rice Research Institute). 2012. App to gauge rice crop health via phone image wins award. http://irri-news.blogspot.com/2012/06/app-to-gauge-rice-crop-health-via-phone.html (accessed January 3, 2014).

Isgin, T., Bilgic, A., Forster, D. L., and Batte, M. 2008. Using count data models to determine the factors affecting farmers' quantity decisions of precision farming technology adoption. *Comput. Electron. Agr.* 62:231–242.

Jochinke, D. C., Noonon, B. J., Wachsmann, N. C., and Norton, R. M. 2007. The adoption of precision agriculture in an Australian broadacre cropping system—Challenges and opportunities. *Field Crop. Res.* 104:68–76.

Khanna, M. 2001. Sequential adoption of site-specific technologies and its implications for nitrogen productivity: A double selectivity model. *Am. J. Agr. Econ.* 83:35–51.

Lee, D. R. 2005. Agricultural sustainability and technology adoption: Issues and policies for developing countries. *Am. J. Agr. Econ.* 87:1325–1334.

Maohua, W. 2001. Possible adoption of precision agriculture for developing countries at the threshold of the new millennium. *Comput. Electron. Agr.* 30:45–50.

McBratney, A., Whelan, B., Ancev, T., and Bouma, J. 2005. Future directions of precision agriculture. *Precis. Agr.* 6:7–23.

New Agriculturist. 2012. Phone advice helps rice farmers earn more. http://www.new-ag.info/en/focus/focusItem.php?a=2764 (accessed February 10, 2014).

Palaniswami, C., Gopalasundaram, P., and Bhaskaran, A. 2011. Application of GPS and GIS in sugarcane agriculture. *Sugar Tech* 13:1–6.

Pinstrup-Andersen, P., Pandya-Lorch, R., and Rosegrant, M. W. 1999. World food prospects: Critical issues for the early twenty-first century. 2020 Vision Food Policy Report. Washington, DC: International Food Policy Research Institute.

Reichardt, M., and Jürgens, C. 2009. Adoption and future perspective of precision farming in Germany: Results of several surveys among different agricultural target groups. *Precis. Agr.* 10:73–94.

Reichardt, M., Jürgens, C., Klöble, U., Hüter, J., and Moser, K. 2009. Dissemination of precision farming in Germany: Acceptance, adoption, obstacles, knowledge transfer and training activities. *Precis. Agr.* 10:525–545.

Robert, P. C. 2002. Precision agriculture: A challenge for crop nutrition management. *Plant Soil* 247:143–149.

Roberts, R. K., English, B. C., and Larson, J. A. 2002. Factors affecting the location of precision farming technology adoption in Tennessee. *J. Ext.* 40(1). http://www.joe.org/joe/2002february/rb3.php.

Roberts, R. K., English, B. C., Larson, J. A., Cochran, R. L., Goodman, W. R., Larkin, S. L., Marra, M. C., Martin, S. W., Shurley, W. D., and Reeves, J. M. 2004. Adoption of site-specific information and variable-rate technologies in cotton precision farming. *J. Agr. Appl. Econ.* 36:143–158.

Robertson, M., Isbister, B., Maling, I., Oliber, Y., Wong, M., Adams, M., Bowden, N., and Tozer, P. 2007. Opportunities and constraints for managing within-field spatial variability in Western Australian grain production. *Field Crop. Res.* 104:60–67.

Robertson, M. J., Llewellyn, R. S., Mandel, R., Lawes, R., Bramley, R. G. V., Swift, L., Metz, N., and O'Callagham, C. 2013. Adoption of variable rate fertiliser application in the Australian grains industry: Status, issues and prospects. *Precis. Agr.* 13:181–199.

Silva, C. B., de Moraes, M. A. F. D., and Molin, J. P. 2011. Adoption and use of precision agriculture technologies in the sugarcane industry of São Paulo state, Brazil. *Precis. Agr.* 12:67–81.

Stafford, J. V. 2000. Implementing precision agriculture in the 21st century. *J. Agr. Eng. Res.* 76:267–275.

Takacs-Gyorgy, K. 2008. Economic aspects of chemical reduction on farming: Role of precision farming—Will the production structure change? *Cereal Res. Commun* 36:19–22.

Tey, Y. S., and Brindal, M. 2012. Factors influencing the adoption of precision agricultural technologies: A review for policy implications. *Precis. Agr.* 13:713–730.

Tey, Y. S., Li, E., Bruwer, J., Abdullah, A. A., Brindal, M., Radam, A., Ismail, M. M., and Darham, S. 2014a. Factors influencing the adoption of sustainable agricultural practices in developing countries: A review. *Environ. Eng. Manage. J.* In press.

Tey, Y. S., Li, E., Bruwer, J., Abdullah, A. A., Brindal, M., Radam, A., Ismail, M. M., and Darham, S. 2014b. The relative importance of factors influencing the adoption of sustainable agricultural practices: A factor approach for Malaysian vegetable farmers. *Sustain. Sci.* 9:17–29.

The Philippine Star. 2012. IT program provides info to rice farmers. http://www.philstar.com/agriculture/771824/it-program-provides-info-rice-farmers (accessed February 15, 2014).

The Philippine Star. 2013. PhilRice project receives Arab Gulf dev't award. http://www.philstar.com/science-and-technology/2013/02/28/913855/philrice-project-receives-arab-gulf-devt-award (accessed February 15, 2014).

Tittonell, P., and Giller, K. E. 2012. When yield gaps are poverty traps: The paradigm of ecological intensification in African smallholder agriculture. *Field Crop. Res.* 143:76–90.

USDA (United States Department of Agriculture). 2007. Precision agriculture: NRCS support for emerging technologies. Agronomy Technical Note No. 1. Washington, DC: United States Department of Agriculture.

Walton, J. C., Lambert, D. M., Roberts, R. K., Larson, J. A., English, B. C., Larkin, S. L., Martin, S. W., Marra, M. C., Paxton, K.W., and Reeves, J. M. 2008. Adoption and abandonment of precision soil sampling in cotton production. *J. Agr. Resour. Econ.* 33:428–448.

Yang, W., Peng, S., Huang, J., Sanico, A. L., Buresh, R. J., and Witt, C. 2003. Using leaf color charts to estimate leaf nitrogen status of rice. *Agronomy J.* 95:212–217.

Zaini, Z. 2012. Nutrient Manager for Rice: Indonesia experience with ICT. Paper presented at Syngenta Foundation for Sustainable Agriculture's Beijing Roundtable 2012, Beijing, China.

8 Managing Soil Heterogeneity in Smallholder African Landscapes Requires a New Form of Precision Agriculture

P. Tittonell, R. van Dis, B. Vanlauwe, and K.D. Shepherd

CONTENTS

8.1 INTRODUCTION

Soil spatial heterogeneity is inherent to smallholder African agriculture. It originates from the interaction between geological and geomorphological diversity across landscapes and the effects of land use and management over time. Several studies have documented the causes and impacts of soil heterogeneity on crop productivity in sub-Saharan Africa in the last decade (e.g., Tittonell et al. 2005a,b; Samaké et al. 2005; Mtambanengwe and Mapfumo 2005; Nkonya et al. 2015). A common denominator in these studies is the fact that limited resource availability leads to concentration of organic matter and nutrient inputs in certain areas of the farm or agricultural landscape, resulting in coexisting areas of net accumulation and areas of net depletion. The latter are often more extended than the former. In certain regions, heterogeneity may be created as a strategy to increase resource use efficiency, diversify, or spread risks. Yet in many areas of sub-Saharan Africa, notably in sparsely populated regions, soil heterogeneity may not conform to this specific spatial pattern. This may be the case where particular forms of shifting cultivation still prevail, in sparsely populated regions where no agriculture inputs are used, where perennial crops are the dominant land use, or where physical landscape units are strongly associated with particular forms of land use and management as in the case of irrigated rice cultivation.

Although soil heterogeneity has been well documented, a lesser number of studies proposed ways to deal with heterogeneity in the realm of technical recommendations to improve crop productivity in this region. Evidence from East (e.g., Vanlauwe et al. 2006), West (e.g., Wopereis et al. 2006), and southern Africa (e.g., Zingore et al. 2007a) suggests that soil heterogeneity can greatly influence the response of crops to nutrient additions through mineral and organic fertilizers, the growth and nitrogen fixation potential of legume cover crops (e.g., Chikowo et al. 2006), or the ability of agroforestry trees to prosper and function on the least fertile soils of smallholder landscapes (e.g., Hartemink et al. 1996; Jama et al. 1998). Such soils are often termed non- or poorly responsive, as the observed crop responses to mineral fertilizers tend to be weak (e.g., Tittonell et al. 2008). Any effort toward sustainable intensification of agriculture in the continent needs to deal with the challenge of restoring productivity of such degraded, nonresponsive soils. A form of precision agriculture is urgently needed that targets specific technologies to soil fertility niches within the landscape.

A major challenge in tailoring agricultural technologies and recommendations to address the rehabilitation of nonresponsive soils resides in the recognition and diagnosing of nonresponsiveness. What is the spatial extent of such soils within the agricultural landscape? What are the reasons that render these soils nonresponsive? Are these reasons permanent or transitory? Answering these questions requires extensive assessment of soil condition, often across large areas. Assessments based on soil sampling and chemical analysis are often prohibitive in terms of financial costs and the time required. Infrared and other spectroscopy techniques offer rapid, low-cost alternatives to assess soil condition, either through remote sensing, through sampling and laboratory analysis, or through *in situ* assessments in the field (Bellon-Maurel and McBratney 2011). Yet detailed sampling and measuring operations may be hampered by logistics and/or by the availability of background information such as soil maps and images. The resulting granularity may not be detailed enough to match the scale at which smallholder farmers make decisions. Complementing spectral sensing of soil condition with farmers' perception of soil fertility is then crucial. A few studies from sub-Saharan Africa show potential in matching local indicators of soil quality with formal indicators of soil fertility (e.g., Corbeels et al. 2000; Barrios et al. 2006; Mairura et al. 2007).

The objective of this chapter is to take stock of the knowledge available on soil spatial heterogeneity in sub-Saharan Africa, its causes and consequences, and its assessment through soil analytical techniques and participatory field methods involving farmers. It starts by reviewing studies on soil heterogeneity across diverse African agroecosystems, their causes and their implications for soil management recommendations, and then examines examples of farmer perception of soil fertility and their degree of agreement with other soil indicators, showing how such perceptions influence decision-making and soil responsiveness to management practices. Approaches to improve our predictive ability using new soil-plant diagnostic techniques are finally presented, indicating the potential to integrating these various approaches in comprehensive assessments to delineate soil management recommendation domains.

8.2   SOIL SPATIAL HETEROGENEITY IN SMALLHOLDER LANDSCAPES

Attempts to describe, categorize, and measure soil heterogeneity in smallholder African landscapes use several criteria to represent the effects of soil-landscape variability (or soilscapes as termed by Deckers 2002), the effect of land use and its history, and the effect of current management. At any point in space-time (x, y, t) in a smallholder landscape, soil fertility can be predicted as the interaction of multiple factors that can be summarized in the following conceptual set of models:

Soil fertility$_{(x, y, t)} = f$(Landscape position-soil type; Land use history; Past and current soil management)

Soil management$_{(x, y, t)} = f$(Resource endowment; Perceived soil fertility; Target soil fertility)

The relative importance of one or the other factor in determining soil heterogeneity varies across agroecosystems, largely determined by their context in terms of demography, access to markets, and production potential. For example, in smallholder landscapes situated in the proximity of urban areas, highly populated and well connected to markets, and where intensive cash crop production dominates, soil heterogeneity tends to emerge between farms of different resource endowment or between inherently different soil types on contrasting landscape positions. In less populated and more subsistence-oriented areas, the distance to the homestead can be an important factor in determining soil fertility gradients, as fields closer to the homesteads tend to be managed more intensely but also with greater input of mineral and organic fertilizers. Within a certain community, the production orientation of a farm household (e.g., subsistence- vs. market-oriented) or the type of cropping system (e.g., arable crop rotations vs. perennial banana-coffee grooves) can determine a target level of soil fertility that farmers aim to achieve through their management practices. The interaction of geomorphological factors, historical land management (e.g., access and allocation of resources), and its codependence with farm wealth has been analyzed in several studies (Murage et al. 2000; Tittonell et al. 2005a,b; Samaké et al. 2005; Wopereis et al. 2006; Zingore et al. 2007b; Ebanyat 2009; Tittonell et al. 2010; 2013).

Summarizing our findings in the specialized literature, the most common soil heterogeneity patterns observed in African landscapes are illustrated in Figure 8.1. We are aware that there are other possible spatial patterns that are not properly represented by any of these four models, but such patterns are either site-specific or a variant of these four (e.g., the patterns described by Baijukya et al. 2005 in NW Tanzania and by Alvarez et al. 2014 in the highlands of Madagascar). In various African agroecosystems a decrease in soil fertility is found at increasing distance from the homestead. Such a decrease can be gradual and continuous (Figure 8.1a), as in the classical case described as ring management by Prudencio (1993) in Burkina Faso and by several other authors working in West Africa in the 1980s (e.g., Pieri 1989; Stoop 1987). Fields close to the villages receive more inputs of carbon and nutrients than those located far away (bush fields). The decrease in soil fertility at increasing distances from

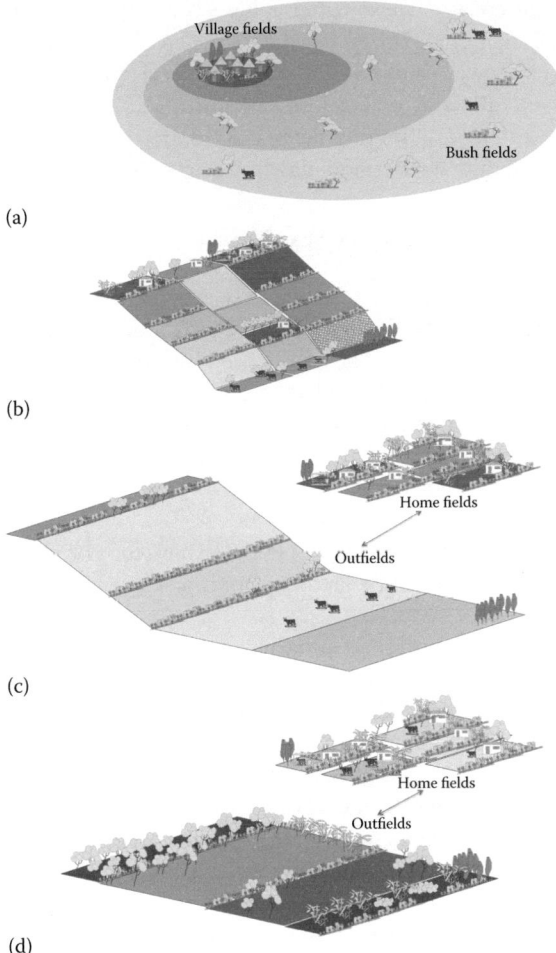

(a)

(b)

(c)

(d)

FIGURE 8.1   Spatial patterns of farmer-induced soil heterogeneity that have been described in sub-Saharan Africa: (a) continuous gradient; (b) discrete gradient; (c) discontinuous gradient; (d) inverse gradient.

the homesteads can exhibit a discrete pattern that results from the differential management of clearly delimited, contiguous fields within a farm perimeter (Figure 8.1b). This pattern, described by Tittonell et al. (2005a,b) for western Kenya, a densely populated region characterized by heavily dissected landscapes, is common in vast areas of the eastern and southern African highlands (Ethiopia, Uganda, Rwanda, Burundi, Tanzania, Zambia, Malawi, and Madagascar). The interaction between farm type in terms of resource endowment and spatial soil variability is a strong one (Tittonell et al. 2010).

In certain areas, communities are organized in villages where each household owns a small area of land (home fields) and a larger field plot or contour in a communally owned cropping area farther away from the homestead (outfields). This pattern is not strictly a gradient, as the variation in soil fertility is discontinuous in space from the fertile home fields to the less fertile outfields (Figure 8.1c). This is one of the most common patterns observed in sub-Saharan Africa and variants of this have been described in East (e.g., Mowo et al. 2006), West (e.g., Samaké et al. 2005), and southern (Zingore et al. 2007) Africa. Farms of higher resource endowment, notably livestock owners, are able to manure and fertilize their soils, both in their home- and their outfields, contributing to soil heterogeneity at landscape level (Zingore et al. 2011). A very conspicuous variant of this pattern is the one often referred to as inverse gradients, which is commonly observed in less densely populated regions and where vast areas of (semi) natural vegetation still remain. The most intensively used fields around the homestead or villages exhibit poorer soil fertility than the cropping plots located farther away (Figure 8.1d). These patterns have been described in coastal and Equatorial West Africa (e.g., Adjei-Nsiah et al. 2004), but they are also common in other sparsely populated regions (e.g., in North Zimbabwe, Baudron et al. 2012). In these agroecosystems, soil fertility maintenance depends largely on fallowing and/or forms of shifting cultivation and in general labor, animal traction and animal manure are limiting factors. An important exception to these four patterns is found in the landscapes dominated by perennial cash crops such as cocoa, oil palm, tea, or coffee.

8.3 CATEGORIZING SOIL HETEROGENEITY

In order to use the patterns described above as the basis for applied research and management recommendations, it is necessary to identify relevant categorization criteria to delimit easily identifiable soil fertility zones in the landscape. The names used by local communities to refer to proximal and distal fields are obviously site-specific and are often associated with some physical reference in the field. Different authors have used land use, cropping intensity, or historical management as categorization criteria for soil variability. However, the two major reference systems used to categorize soil spatial heterogeneity in sub-Saharan Africa were

1. The location of a field or zone with respect to the homesteads or village
2. The location of a field or zone in the landscape or toposequence

Other categorization systems have relied on farmers' perception of soil fertility (e.g., productive and nonproductive fields, Murage et al. 2000). Such categorizations are even more site-specific and not necessarily linked to any physical reference for the recognition of soil fertility situations in the field. Others have used indicators that are easily recognizable by local farmers, such as soil color, texture, structure, and presence/absence of soil life or indicator plant species (Barrios et al. 2001). Carter and Murwira (1995) referred to the various soil condition situations within smallholder farms as soil fertility niches. The authors indicated locations such as old kraal or old boma sites, old termite mounds, areas of concentration of organic resources in home gardens, and charcoal-making sites to be special niches of high fertility concentration, whereas areas of soil disturbance, for example from rill erosion, to be niches of poor soil fertility.

Several studies went beyond describing spatial patterns to the spatial sampling and quantification of soil fertility indicators across heterogeneous landscapes. Table 8.1 provides an overview of published studies that characterized the chemical properties of soils across a large diversity of sub-Saharan

TABLE 8.1

Average Soil Fertility Indicators across Heterogeneous Sub-Saharan African Agroecosystems

Case Study Area	Criteria and Components of the Soil Fertility Gradient	Clay + Silt (%)	pH (H₂O)	SOC (g kg⁻¹)	Total N (g kg⁻¹)	Extractable P (mg kg⁻¹)	Exchangeable Bases (cmol(+) kg⁻¹)			Sources
							K⁺	Mg²⁺	Ca²⁺	
	Historical Field Management (Abandoned Bomas)									
Central Kenya	1–5 years old	19.8	n.a.	18.7	1.7	n.a.	42.2	61.0	50.1	Augustine (2003)
	12–24 years ago	19.8	n.a.	9.8	0.8	n.a.	4.8	24.7	31.2	
	30–39 years ago	19.8	n.a.	3.5	0.2	n.a.	2.5	6.1	14.7	
	Land Use									
Northwest Tanzania	Kibanja Perennial crops	26	5.7 (4.8–6.8)	26 (16–48)	2.2 (2.2–4.2)	123 (10–515)	0.4 (0.1–0.6)	n.a.	4.9 (1.0–10.3)	Baijukya et al. (2005)
	Kikamba Annual crops	34	5.5 (4.5–6.6)	22 (8–46)	1.7 (0.8–3.7)	21 (5–480)	0.2 (0.08–0.4)	n.a.	1.3 (0.8–4.4)	
	Rweya Grasslands	31	5.2 (4.2–5.8)	26 (5–56)	1.3 (0.4–2.1)	13 (5–250)	0.1 (0.04–0.2)	n.a.	0.9 (0.2–1.8)	
	Land Use									
Madagascar	Tanety (hills) Annual crops	58.6	5.15	32.7	2.28	72.5	n.a.	n.a.	n.a.	Alvarez (2012)
	Terraced foothill Rice	56.1	5.91	18.0	1.24	74.3	n.a.	n.a.	n.a.	
	Flood lowland Rice	57.7	5.7	18.0	1.34	64.0	n.a.	n.a.	n.a.	

(Continued)

TABLE 8.1 (CONTINUED)
Average Soil Fertility Indicators across Heterogeneous Sub-Saharan African Agroecosystems

Case Study Area	Criteria and Components of the Soil Fertility Gradient		Clay + Silt (%)	pH (H₂O)	SOC (g kg⁻¹)	Total N (g kg⁻¹)	Extractable P (mg kg⁻¹)	Exchangeable Bases (cmol₍₊₎ kg⁻¹)			Sources
								K⁺	Mg²⁺	Ca²⁺	
Southern Ethiopia	**Field Location**										Elias et al. (1998); Elias and Scoones (1999)
	Enset	Home garden	n.a.	6.9	4.5	2.7	51.2	n.a.	n.a.	n.a.	
	Darkoa	Homestead field	n.a.	6.5	2.65	2.5	21.13	n.a.	n.a.	n.a.	
	Shoka	Outfield	n.a.	5.7	2.65	2.1	5.11	n.a.	n.a.	n.a.	
Sudan–Savannah Zone (e.g., Burkina Faso)	**Field Location**										Smaling and Braun (1996)
	Champs de case	Ring 1	n.a.	7.5 (6.7–8.3)	17 (11–22)	1.4 (0.9–1.8)	110 (20–200)	1.6 (0.4–2.4)	n.a.	n.a.	
	Champs de village	Ring 2	n.a.	6.4 (5.7–7.0)	7.5 (5–10)	0.7 (0.5–0.9)	14.5 (13–16)	0.75 (0.4–1.1)	n.a.	n.a.	
	Champs de brousse	Ring 3	n.a.	6.0 (5.7–6.2)	3.5 (2–5)	0.4 (0.2–0.5)	10.5 (5–16)	0.08 (0.06–0.1)	n.a.	n.a.	
Rwanda	**Distance from Homestead (m)**										Bucagu (2013)
	Home fields	10–30	40	5.5	25.2	2.5	12.5	0.7	n.a.	n.a.	
	Close fields	50–100	40.8	5.2	19.3	1.8	7.9	0.3	n.a.	n.a.	
	Outfields	100–800	38.7	5.2	16	1.5	5.1	0.4	n.a.	n.a.	
Mali	**Distance from Homestead (m)**										Ramisch (1999, 2005)
	Fulawere	10 ± 20	30	5.7	13.2	0.70	10	0.12	n.a.	n.a.	
	Village	80 ± 40	30	5.7	12.4	0.52	8.6	0.25	n.a.	n.a.	
	Hamlet	1340 ± 820	30	6.1	11.5	0.46	11.6	0.22	n.a.	n.a.	

(Continued)

TABLE 8.1 (CONTINUED)
Average Soil Fertility Indicators across Heterogeneous Sub-Saharan African Agroecosystems

Case Study Area	Criteria and Components of the Soil Fertility Gradient		Clay + Silt (%)	pH (H₂O)	SOC (g kg⁻¹)	Total N (g kg⁻¹)	Extractable P (mg kg⁻¹)	Exchangeable Bases (cmol₍₊₎ kg⁻¹)			Sources
								K⁺	Mg²⁺	Ca²⁺	
Northern Togo	**Field Location**										Wopereis et al. (2006)
	Infield	Ring 1	n.a.	7.7	13.4	0.97	48	1.7	1.12	3.9	
	Outfield	Ring 3	n.a.	6.4	6.3	0.51	1.15	0.25	0.56	2.2	
Eastern Uganda	**Historical Field Management (e.g., Kraal)**										Ebanyat (2009)
	Good fields	n.a.	25	6.6	9.3	0.97	19	0.47	0.66	2.1	
	Medium fields	n.a.	23	6.3	6.6	0.69	14	0.37	0.58	1.44	
	Poor fields	n.a.	20	6.1	5.5	0.59	12	0.30	0.53	1.25	
Malawi	**Distance from Homestead (m)**										Kamanga (2011, Chapter 2)
	Home fields	0–50	47	5.4	12	0.8	7.0	n.a.	n.a.	n.a.	
	Middle fields	51–100	39	5.5	9	0.5	4.9	n.a.	n.a.	n.a.	
	Remote fields	>100	35	5.7	7	0.4	3.1	n.a.	n.a.	n.a.	
Mali	**Field Management**										Benjaminsen et al. (2010)
	Gradient 1—most intensively cultivated		n.a.	6.4	6.9	0.27	9.1	3.57	4.75	13.96	
	Gradient 2		n.a.	6.1	6.6	0.23	8.4	2.61	4.36	10.08	
	Gradient 3		n.a.	6.5	8.8	0.39	5.9	3.85	6.04	16.47	
	Gradient 4		n.a.	6.1	8.5	0.35	5.7	2.05	4.77	10.28	
	Gradient 5—least intensively cultivated		n.a.	6.6	12.6	0.67	19.7	2.98	6.15	21.66	
Northern Ethiopia	**Land Use**										Tilahun (2007)
	n.a.	Forest land	40	6.8	8.2	1.05	3.53	2.00	5.02	10.75	
	n.a.	Grazing land	34	6.5	10.7	1.35	3.82	0.93	0.88	3.26	
	n.a.	Cultivated land	35	5.8	5.7	0.55	4.51	0.85	0.81	3.96	

(Continued)

TABLE 8.1 (CONTINUED)
Average Soil Fertility Indicators across Heterogeneous Sub-Saharan African Agroecosystems

Case Study Area	Criteria and Components of the Soil Fertility Gradient		Clay + Silt (%)	pH (H₂O)	SOC (g kg⁻¹)	Total N (g kg⁻¹)	Extractable P (mg kg⁻¹)	Exchangeable Bases (cmol$_{(+)}$ kg⁻¹)			Sources
								K⁺	Mg²⁺	Ca²⁺	
Western Kenya (Aludeka midland region)	**Distance from Homestead (m)**										Tittonell et al. (2005b)
	Home gardens	10 ± 3.6	36.1	5.4	6.9	0.3	2.5	0.28	0.70	2.4	
	Close fields	26 ± 8.6	42.9	5.8	7.5	0.6	5.6	0.44	0.80	3.9	
	Mid-distance fields	54 ± 17	44.3	5.4	8.8	0.6	2.9	0.25	0.90	2.9	
	Remote fields	82 ± 21	39.4	5.2	7.9	0.5	2.3	0.15	0.70	2.3	
Zimbabwe	**Distance from Homestead (m)**										Masvaya et al. (2011)
	Home fields	29 ± 12.7	4.7	5.4	6.9	0.29	16.8	1.98	1.47	5.53	
	Outfields	159 ± 36.4	3.1	5.1	5.4	0.20	9	1.06	1.06	3.15	
Highlands Ethiopia	**Distance from Homestead (m)**										Amede and Taboge (2007)
	Homestead	29 ± 12.7	n.a.	6.7	n.a.	0.23	6.3 (2.5–10)	10.2 (8.6–11.8)	n.a.	6.75	
	Outfields	159 ± 36.4	n.a.	5.9	n.a.	0.17	0.8	3.35 (2.7–4)	n.a.	4.25	

(Continued)

TABLE 8.1 (CONTINUED)
Average Soil Fertility Indicators across Heterogeneous Sub-Saharan African Agroecosystems

Case Study Area	Criteria and Components of the Soil Fertility Gradient		Clay + Silt (%)	pH (H₂O)	SOC (g kg⁻¹)	Total N (g kg⁻¹)	Extractable P (mg kg⁻¹)	K⁺	Mg²⁺	Ca²⁺	Sources
								Exchangeable Bases (cmol₍₊₎ kg⁻¹)			
Sahel of Mali	**Distance from Homestead (m)**										
	n.a.	10	n.a.	8.5	5.4	0.25	8.4	0.07	n.a.	n.a.	Samaké et al. (2005)
	n.a.	100	n.a.	7.2	3.6	0.19	4.5	0.06	n.a.	n.a.	
	n.a.	500	n.a.	6.0	1.2	0.11	2.5	0.02	n.a.	n.a.	
	n.a.	2000	n.a.	5.2	1.0	0.12	2.5	0.01	n.a.	n.a.	
Zimbabwe	**Distance from Homestead (m)**										
	Sandy home field	50	15	5.1	5	0.4	7.2	n.a.	n.a.	n.a.	Zingore et al. (2007a)
	Sandy outfield	100–500	12	4.9	3	0.3	2.4	n.a.	n.a.	n.a.	
	Clayey home field	<50	54	5.6	14	0.8	12.1	n.a.	n.a.	n.a.	
	Clayey outfield	100–500	58	5.4	7	0.5	3.9	n.a.	n.a.	n.a.	

Note: When the results from several farm wealth classes were available, only the middle wealth class was considered; n.a.: not applicable/not available. Average values (and their variation between brackets, when available) are presented as category of soil variability as proposed in the original publications.

African agroecosystems. In most of these case studies soil fertility patterns were ascribed to spatial differences in soil management strategies, although the interaction between management and landscape position cannot be entirely disentangled in most of these examples. The majority of the studies include the distance to homestead (m) as the main driver of variability in soil fertility, and the patterns described were associated with variables such as farmers' risk attitudes (i.e., it is perceived as safer to invest resources in the fields closest to the homesteads), lack of means to transport bulky organic matter (e.g., compost), and limited access to and availability of labor, fertilizer, and animal manure. The actual differences in soil quality attributes between spatial patches on the farm or in the landscape, as well as their spatial distribution, varies widely across case studies (Table 8.1).

For example, the spatial distribution of field types in western Kenya ranges from home gardens to bordering close, mid-distance, and remote fields located from 7 to 100 m away from the homesteads, respectively, and the magnitude of the gradients vary enormously between farm types (see Figure 8.1b). As a result, the average soil fertility indicators do not differ broadly between home and remote fields (note that the term "remote" is used locally by farmers to refer to fields that are poorly accessible due mostly to topography). In such cases, differences between farm types are often wider than between field types. The patterns described in Zimbabwe, distinguishing between home fields and outfields often located at a distance from the homesteads (see Figure 8.1c), tend to exhibit more conspicuous differences in soil indicators between these field types. Large distances from home fields to outfields as well as strong gradients were reported for Rwanda as well, yet these farms include (bordering) close fields in between the other two. In the Sudan-Savannah zone of West Africa, as in the case of southern Burkina Faso, the soil fertility pattern presented in Table 8.1 reproduced approximately the ring model depicted in Figure 8.1a. Soil fertility declined from the homesteads toward the outer ring; ring 1 includes the home field, ring 2 represents the village fields, and ring 3 the bush fields.

However, the categorization of spatial patches as rings remains rather arbitrary and not necessarily consistent between and across the farmer and researcher communities. For example, the rings distinguished by Smaling and Braun (1996) do not correspond exactly to those distinguished by Prudencio (1993), although both studies took place in the same region. The increase in population densities leads to the disappearance of such ring distinction, since new households are built in the outer rings of former ones, creating a new, more complex pattern of spatial heterogeneity (Andrieu et al. 2015; Diarisso et al. 2015). According to Wopereis et al. (2006), soil heterogeneity in these cases occurs mostly due to, for example, abandoned kraals or sandy patches. Two case studies presented in Table 8.1, in central Kenya and eastern Uganda, measured high soil fertility levels on recently abandoned kraals. Another criteria used in the examples of Table 8.1 to categorize spatial soil heterogeneity in the landscape is land use. The example from northwest Tanzania shows a categorization of fields between those used to grow perennial crops like bananas, fields with annual crops, and grasslands (Baijukya et al. 2005). These fields differed widely in terms of extractable phosphorous (P) levels, and secondarily in nitrogen (N) and potassium (K) levels, but not in terms of soil carbon.

Finally, the remaining major criterion to categorize spatial soil heterogeneity that has been used to describe smallholder landscapes in Africa is the position in the landscape, soil-landscape unit, or soilscape. Examples of several published studies are presented in Table 8.2. The distinction of units within a soilscape is determined by the geomorphology of an area and in the majority of case studies the soil fertility gradients relate to the position of a field along a toposequence. In general, a weak decrease in soil fertility is found at the lower slopes or valley-bottom sections of a toposequence, although such decreases were not always significant. This does not apply for extractable P, which tended to be higher at lower landscape positions in several cases. Different positions along the toposequence could be associated with inherently different soil types, originally more or less fertile, but also with their inherent water regime (e.g., excessive runoff on steep slopes, excessive drainage, and poor water-holding capacity in soils of coarser texture, or waterlogging in soils located in valley bottoms) that makes them more or less suitable for cultivation. A major factor that determines land use and management intensity and thereby soil heterogeneity is the accessibility of fields, particularly in mountainous regions, and where access with animals or tractors is barely impracticable.

TABLE 8.2

Average Soil Fertility Indicators for Fields Located in Different Landscape Units across Heterogeneous Sub-Saharan African Soilscapes

Case Study Area	Position of Field along a Toposequence	Clay + Silt (%)	pH (H$_2$O)	SOC (g kg^{-1})	Total N (g kg^{-1})	Extractable P (mg kg^{-1})	Exchangeable Bases (cmol$_{(+)}$ kg^{-1})			Sources
							K$^+$	Mg^{2+}	Ca^{2+}	
Kenya (Meru South)	Upslope	46	5.8	20.3	n.a.	15.6	0.64	1.5	5.6	Tittonell (2008)
	Midslope	48	5.6	19.7	n.a.	17.2	0.54	1.4	4.9	
	Footslope	47	5.5	19.1	n.a.	21.2	0.53	1.3	4.8	
	Valley bottom	51	5.3	20.3	n.a.	24.6	0.56	1.4	4.4	
Rwanda	Upper hill: mountainous	n.a.	5.5	17.3	0.18	2.7	0.28	0.82	5.02	Rushemuka et al. (2014)
	Upper hill: interfluves	n.a.	5.6	11.3	0.85	26	0.51	1.41	4.19	
	Upper hill: shoulder	n.a.	6.5	22.8	0.17	7	0.85	3.55	7.39	
	Hill side	n.a.	4.8	2.4	0.14	1.5	0.09	0.06	0.28	
	Valley bottom (Ibumba)	n.a.	4.3	2.6	0.20	2.5	0.10	0.09	0.49	
Southern Rwanda	Upper slope	39	5.7	16	2.4	n.a.	n.a.	n.a.	4.3	Steiner (1998)
	Middle slope	39	5.1	15	1.6	n.a.	n.a.	n.a.	2.8	
	Lower slope	39	4.9	14	1.6	n.a.	n.a.	n.a.	2.3	
Burkina Faso	Mid-slope	28	6.5	7.1	n.a.	0.8	0.1	0.7	2.1	Stoop (1987)
	Lower slope	39	6	6.1	n.a.	0.5	0.14	0.5	1.2	
	Lowland	59	5.8	9.5	n.a.	0.7	0.14	0.6	2.2	
Uganda	Upper (Erony)	12.0	5.8	4.6	0.5	7.9	0.3	0.6	1.2	Ebanyat (2009)
	Middle (Eitela)	37.0	4.5	6.2	0.7	6.0	0.2	0.5	1.0	
	Middle (Apuuton)	14.0	5.7	5.2	0.6	11.0	0.3	0.6	1.2	
	Bottom (Akao)	10.0	5.5	3.7	0.4	23.1	0.4	0.4	1.0	
South-west Nigeria	Upper slope	44.7	6.1	5.7	0.47	3.81	0.31	0.78	3.09	Salako et al. (2006)
	Middle slope	32.1	6.1	3.8	0.24	3.67	0.13	0.81	2.81	
	Lower slope	26.8	6.2	4.5	0.55	3.55	0.13	0.65	2.67	
Western Niger	Plateau	11.6	5.5	1.7	0.08	2.79	n.a.	n.a.	n.a.	Gandah (1999)
	Upper slope	8	5.7	1.4	0.05	2.15	n.a.	n.a.	n.a.	
	Undulating terraces	7.8	5.7	1.1	0.06	1.7	n.a.	n.a.	n.a.	
	Valley	8.4	5.2	1.2	0.06	2.31	n.a.	n.a.	n.a.	

Note: When results from several soil depths were available, only the upper soil depth was considered; n.a.: not applicable/not available.

8.4 FARMERS' PERCEPTION AND MANAGEMENT OF SOIL FERTILITY

The first visual evidence of the existence of soil fertility gradients in the field is the variation in crop productivity across the landscape. Yet crop productivity is not only the result of soil fertility but of management practices, including nutrient inputs, planting dates, plant population densities, and weeding frequency (Tittonell et al. 2007). Farmers perceive soil quality based on a number of visual cues, but also considering the history of each field, which is usually known to them. They adjust their management practices, when possible, according to such perceptions (Misiko et al. 2009). The rates of organic and mineral fertilizer applications are often associated with soil perceived heterogeneity patterns in different regions of sub-Saharan Africa (Figure 8.2). For example, the best plots and (in most cases the same) plots closest to the homestead receive the highest amounts of N fertilizer. On these fields the organic N input tends to exceed the mineral fertilizer N input, while where mineral fertilizers are used, their N input is higher on the more distant fields. The decrease of fertilizer input associated with worse or remote plots is for the greatest part due to the decrease of organic N input.

   In the studies cited in Figure 8.2, greater organic N input near the homesteads was associated with the presence of cash crops in these fields and the presence of livestock stalled or tethered around the homesteads, with the perception that distant fields are more prone to risks such as damage by marauding livestock or wildlife, or theft, and with the labor required to transport bulky organic materials such as cattle manure to the more distant fields of the farm. These studies and others (e.g., Baijukya et al. 2005; Nkonya et al. 2005; Tittonell and Giller 2013; Diarisso et al. 2015) also document that as a consequence of such management patterns, nutrient balances tend to be positive in the close fields and strongly negative in the distant ones, leading to soil fertility depletion. Farmers' management decisions based on their perceptions tend to reinforce soil heterogeneity when they apply most nutrient resources to the fields perceived to be more fertile (Tittonell et al. 2013). Table 8.3 compiles examples

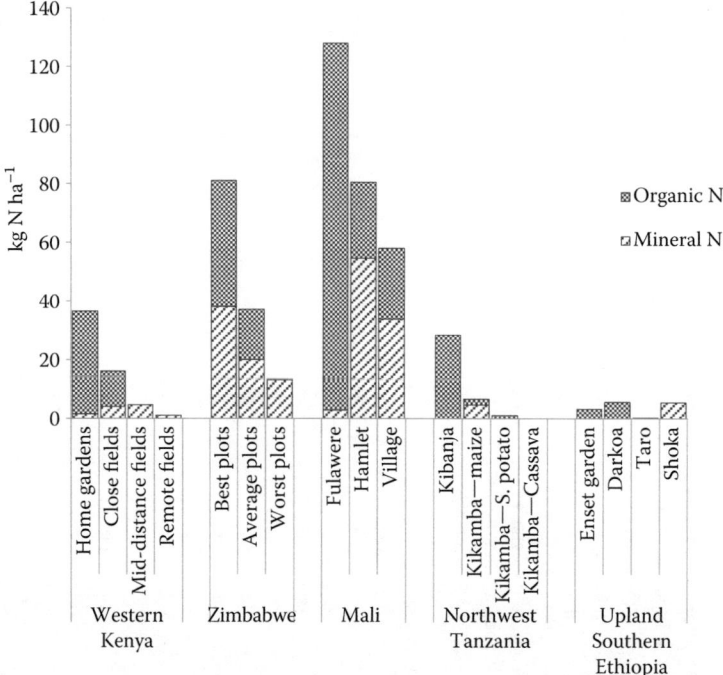

FIGURE 8.2   Mineral and organic N fertilizer input (kg ha⁻¹) along within-farm soil fertility gradients for case studies in Kenya (Tittonell et al. 2005b), Zimbabwe (Zingore et al. 2007a), Mali (Ramisch 2005), Tanzania (Baijukya 2004), and Ethiopia (Elias et al. 1998). When data from different wealth classes were available, only the data from middle-class households were considered.

TABLE 8.3

Indicators Used by Farmers to Distinguish between Soil Quality Classes in Various Smallholder Landscapes of Sub-Saharan Africa

Case Study	Soil Fertility Level					Indicators					Source
Ghana	Fertile fields	Dark soil color	High water holding capacity	Few stones and pebbles present	Located in valley bottom or lower middle slope	Consistently high yields	Fast/high growth rate	Soil is easy to work	Numerous wet worm casts present	Indicator weed: *Chromolaena odorata* with large green leaves	Dawoe et al. (2012)
	Infertile fields	White/pale/light soil	Low water holding capacity	Numerous stones and pebbles present	Located upper slopes/summits	Low yields	Stunted and slow plant growth	Soil is difficult to work	Few worm casts present	Indicator weed: *Chromolaena odorata* with small yellow leaves	
Ethiopia	Reguid (fertile)	Red and brown soil	Heavy texture	Slight stoniness	Location: level (valley bottom)	Maximum and most reliable yield	Deep soil depth	Soil is difficult to work	Intensively cultivated arable land		Corbeels et al. (2000)
	Mehakelay (moderately fertile)	Brown soil	Medium texture	Moderate stoniness	Location: gentle slope (between valley bottom and hills)	Medium yield with slight risk of crop failure	Medium soil depth	Soil is average to work	Some cultivation, also used for pasture		
	Rekik (least fertile)	White and black soil	Light texture	High stoniness	Location: very steep (hilly)	Low yields with high risk of crop failure	Shallow soil depth	Soil is easy to work	Not cultivated		

(Continued)

TABLE 8.3 (CONTINUED)

Indicators Used by Farmers to Distinguish between Soil Quality Classes in Various Smallholder Landscapes of Sub-Saharan Africa

Case Study	Soil Fertility Level	Indicators									Source
Ethiopia	High soil quality (Reguid)	Dark soil color	Texture is clay loam, loamy, loam clay	Deep topsoil depth	Hold moisture well and give and take water easily	High yield	Even growth, matures on time	Easy to work or soil flows and falls apart	Soil stays loose, does not pack	Soil has numerous worm holes and castings, bird behind tillage	Tesfahunegn et al. (2011)
	Medium soil quality (Maekelay)	Brown, gray, or reddish soil color	Too heavy or too light, but no or little problem	Shallow topsoil depth	Soil is drought prone in dry weather	Medium yield	Uneven growth and late to mature	Difficult to work or needs extra passes	Soil has thin hardpan or plow layer	Few worm holes and castings present	
	Low soil quality (Rekik)	Light colored soil	Texture is extremely sandy, clayey, rocky, is a problem	Subsoil exposed or near surface	Soil dries out too fast	Low yield	Stunted growth, never seems to mature	Plow hard or soil never works down	Soil is tight and compacted, cannot get into it, thick hardpan	No casts or holes of worm activity	
Tanzania	Good soil	Black soil color	Cracks during dry season due to high clay content	High water holding capcity	Presence/ vigorous growth of certain plants	Abundance of earthworms	Good crop performance				Mowo et al. (2006)
	Poor soil	Yellow and red colors in soil	Compacted soil	Shallow soil depth	Stunted growth	Presence of rocks and stones	Presence of bracken ferns	Salt visible on soil surface			

(Continued)

TABLE 8.3 (CONTINUED)
Indicators Used by Farmers to Distinguish between Soil Quality Classes in Various Smallholder Landscapes of Sub-Saharan Africa

Case Study	Soil Fertility Level			Indicators				Source
Zimbabwe	Rich field	Red- or gray-colored soil	Relative high clay content	Soils do not dry easily and do not readily wilt crops	Consistently contributing the highest amount of yield	High crop growth and yield responses to external inputs	Exhibit clods on tilling	Presence of islands of termite mounds	Mtambanengwe and Mapfumo (2005)
	Poor field	Light-colored soil	Very sandy soil	Often poor seed emergence due to surface crusting	Crop yields are poor year after year	Low, poor seed emergence, low input response			

TABLE 8.4

Visual Indicators of Soil Quality and Degradation and Their Frequency of Occurrence in Fields Classified by East African Farmers

| Indicator | Category | Fields per Category | | Occurrence within SF Classes (%) | | |
		n	(%)	Poor	Medium	Good
Soil erosion	Sheet	340	17	19	18	13
	Rill	431	22	29	20	13
	Mass	16	1	1	1	1
Hard settings	Temporary	227	92	13	11	9
	Permanent	19	8	1	1	2
Stoniness	0%–5%	1855	93	93	94	94
	5%–25%	72	4	3	4	4
	25%–50%	39	2	2	2	2
	50%–75%	12	1	1	1	0
	>75%	10	1	1	0	0
Slope class	0%–5%	919	46	37	49	55
	5%–10%	442	22	22	22	22
	10%–20%	317	16	16	16	15
	20%–40%	247	12	18	11	7
	> 40%	63	3	6	2	1
Landscape	Upslope	371	19	12	17	33
	Midslope	1423	72	77	74	58
	Footslope	158	8	10	8	5
	Bottomland	36	2	1	1	3
Flooding (occasional/regular)		60	3	3	2	5

Note: 250 randomly selected farmers in Kenya and Uganda classified their fields according to their perceived fertility as poor, medium, or good (*n* = 2607 fields).

of indicators used by farmers to classify soil quality. The indicators chosen by farmers vary from region to region but some are quite consistent across regions. Soil color, texture/structure, depth, workability, landscape position, and average crop yields are most commonly used.

In a study of 250 farms in six districts of Kenya and Uganda farmers were asked to classify all the fields of their farms based on their soil fertility level as good, medium, and poor (Tittonell et al. 2010). Each farmer did this individually, without contrasting them with their neighbor's fields, and the criteria to classify fields also varied from site to site. In general, soil fertility classes perceived by farmers were weakly related to visual indicators of soil degradation and physical impediments and moderately related to the slope of the fields and/or their position in the landscape (Table 8.4). However, the soil quality classes were strongly correlated with soil management decisions and soil chemical indicators (Tittonell et al. 2013). Yet, reducing the cause of poor productivity in distant fields to a problem of poor soil nutrient availability is an oversimplification. Several studies show that crops tend to respond poorly to added nutrients in these fields, as will be discussed in Section 8.5, due to other concomitant causes of soil degradation. Most important, soil degradation reduces the efficiency of nutrient capture and use by crops when fertilizers are applied.

8.5 IMPACT OF SOIL HETEROGENEITY ON SOIL RESPONSIVENESS TO INTERVENTIONS

Heterogeneity in soil fertility status between fields within a single farm or village impacts on the performance of interventions aiming at improving crop productivity. Applying a standard rate and type of fertilizer to crops within a large number of randomly selected farmer fields leads to a wide range of responses, even under optimal crop management practices, including plant spacing and weeding (Figure 8.3). Three broad classes of fields can be distinguished: (1) fertile, less responsive fields, (2) responsive fields in which a strong response to fertilizers is found, and (3) poor, less responsive fields. The latter soils are those where response to fertilizer is limited due to other constraints besides the nutrients contained in the fertilizer, while the former includes homestead fields. For instance, on sandy granitic soils, N use efficiency by maize varied from >50 kg grain kg^{-1} N on the more fertile fields close to homesteads to less than 5 kg grain kg^{-1} N in degraded outfields (Zingore et al. 2007a). In high rainfall zones of Togo, Wopereis et al. (2006) reported average N use efficiencies by maize ranging from 39 kg kg^{-1} in the infields to 22 kg kg^{-1} in the outfields when phosphorus was not applied, and from 40 kg kg^{-1} to 28 kg kg^{-1} when 30 kg ha^{-1} P was applied. In the densely populated region of western Kenya, Vanlauwe et al. (2006) observed that maize response to applied N decreased with distance to the homestead, for instance, in Aludeka from 0.95 kg kg^{-1} to 0.55 kg kg^{-1} relative yield. The limited responsiveness of soils seems to appear predominantly in resource-scarce areas that are densely populated.

These authors ascribed the spatial variability observed in crop responses to fertilizers to several factors. Zingore et al. (2007b) pointed to multiple nutrient deficiencies in the soils of distant fields (especially on sandy soils) such as shortages in K and zinc (Zn). Wopereis et al. (2006) explained such difference by an increased water infiltration capacity and reduced evaporation in infields compared to outfields due to organic household waste and crop residues scattered over the soil surface in the infields. Within a specific soil type, a good proxy for soil fertility status is soil organic matter

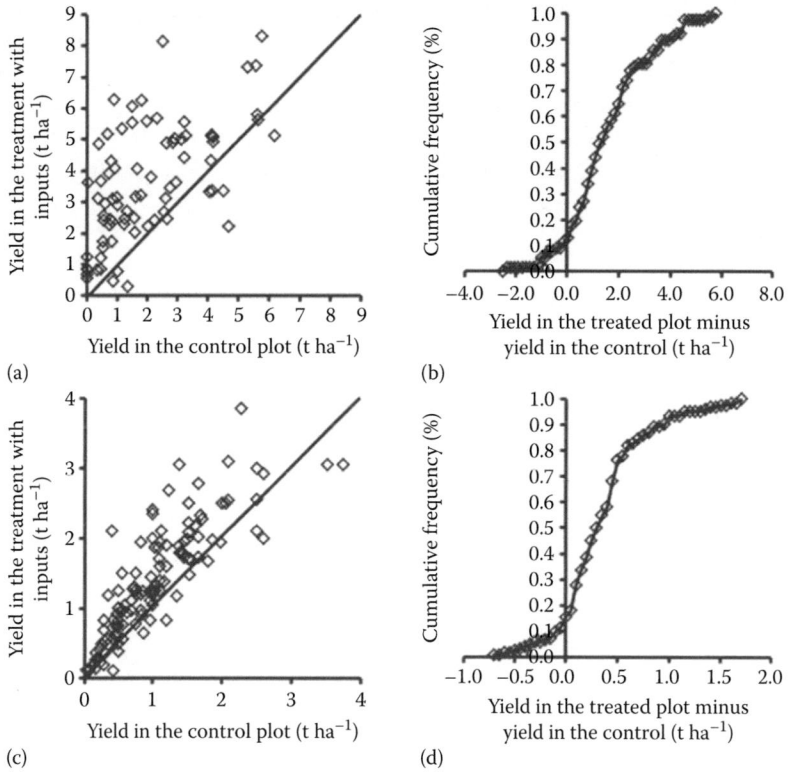

FIGURE 8.3 Yields in the treatments with inputs vs. those without inputs for maize in Kenya (a) and beans in Rwanda (c) and cumulative frequency curves showing the chance to observe specific treatment effect or less for the same data sets (b) and (d). In panels (a) and (c), data points close to the 1:1 line are defined as nonresponsive. (From Vanlauwe et al. [n.d.].)

(SOM) content. SOM contributes positively to specific soil physicochemical properties or processes fostering crop growth, such as cation exchange capacity, soil moisture and aeration, or nutrient stocks. On land where these constraints limit crop growth, a higher SOM content may enhance the demand by the crop for N and consequently increase the fertilizer N use efficiency (Figure 8.4a). On the other hand, SOM also releases available N that may be better synchronized with the demand for N by the plant than fertilizer N. Consequently, a larger SOM pool may result in lower N fertilizer use efficiencies (Figure 8.4b).

Conservation agriculture (CA) is commonly defined around a set of three principles: minimum tillage, soil surface cover, and diversified crop rotations. Since minimal tillage without surface mulch results in depressed yields, and since the most common source of mulching material is crop residues, heterogeneity in soil fertility status within farms and communities is also expected to affect the performance of CA systems (Tittonell et al. 2012). For instance, in central Kenya, Guto et al. (2011) indicated that about 3 t ha^{-1} of maize stover are required to keep a soil cover of minimally 30% at planting (Figure 8.5a), the latter being a requirement to substantially reduce interill soil erosion (Scopel et al. 2004). In a study in western Kenya, in an on-farm experiment across soil fertility gradients, in absence of fertilizer over half of the plots produced maize stover amounts below this threshold (Figure 8.5b). Especially in Aludeka and Shinyalu, strong gradients in maize stover production were observed between plots within farms. These results led Vanlauwe et al. (2014) to argue that strategies for using CA in sub-Saharan Africa must integrate a fourth principle—the appropriate use of fertilizer—to increase the likelihood of benefits of CA for smallholder farmers. However, evidence from other regions where maize is the dominant crop, as well as from cropping systems dominated by other crop species, indicate that the amount of crop residue biomass necessary to achieve 30% soil cover is around 1 t ha^{-1} on average (e.g., Findeling et al. 2003; Naudin et al. 2011; Baudron et al. 2012; Sommer et al. 2014) and that the benefits of CA can also be achieved when including legumes cover crops in the rotation, especially when farmers cannot afford or access mineral fertilizers (Ruzinamhodzi et al. 2012).

Often the analysis of soil heterogeneity is restructured to the topsoil. Soils with subsoil properties unfavorable for root growth and nutrient uptake (e.g., a sandy subsoil with low nutrient stocks

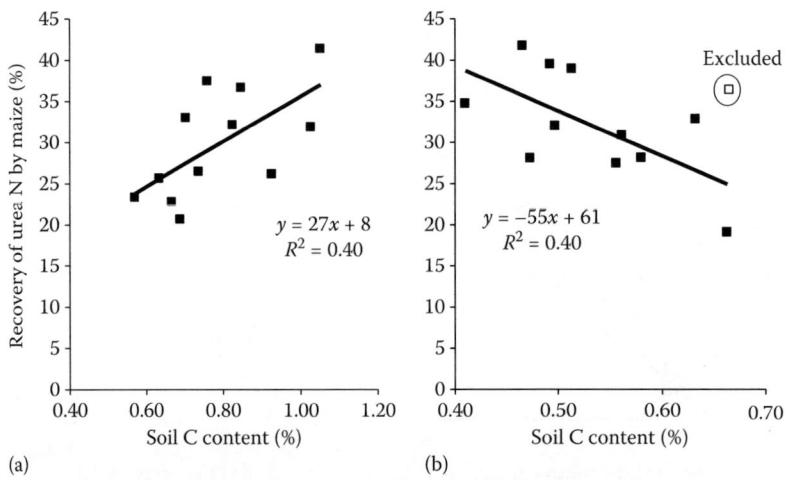

FIGURE 8.4 Observed relationships between recovery of ^{15}N labeled urea N in the maize shoot biomass and the soil organic C content for 12 farmers' fields in (a) Zouzouvou (southern Benin) and (b) Danayamaka (northern Nigeria). Urea was split-applied (one-third at planting, two-thirds at knee height) at 90 kg N ha^{-1} in Zouzouvou and 120 kg N ha^{-1} in Danayamaka. One observation was excluded from the regression analysis for the Danayamaka data. (From Vanlauwe et al. [n.d.].)

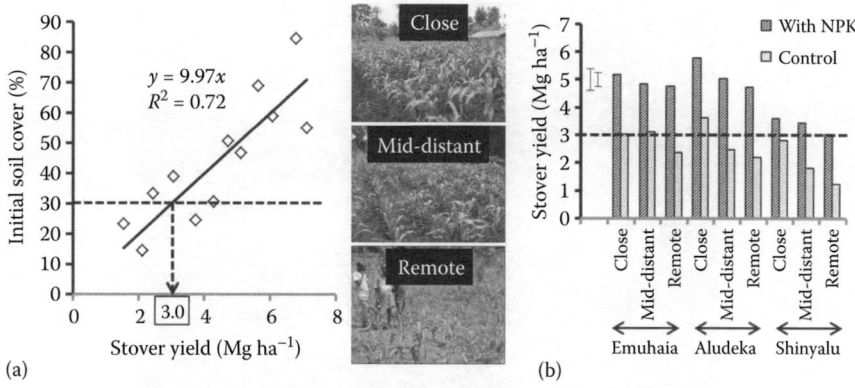

(a)   (b)

FIGURE 8.5   Relationship between maize stover yield and soil cover at the start of the cropping seasons in central Kenya (adapted from Guto et al. 2011) (a), and maize stover yields for different field types in three districts of western Kenya with and without fertilizer (adapted from Vanlauwe et al. 2006) (b). The error bars in (b) are standard errors of the difference to compare stover yields within site, the left bar for the plots with fertilizer, and the right bar for the control plots. The horizontal dashed line represents the 30% soil cover limit required for CA. Photographs in the center give an impression of the maize plots close to harvest on the different field types in central Kenya. (From Vanlauwe et al. [n.d.].)

typical for Arenosols) or a subsoil with high acidity and available Al levels typical for Allisols are often found in close proximity to soils with favorable subsoil properties (e.g., a subsoil with a clay accumulation horizon that contain a large proportion of exchangeable bases on the exchange complex typical for Lixisols) within specific landscapes. Such situations favor or impede the growth of deep-rooting species, including agroforestry species, within farming landscapes. For instance, in a study on alley cropping in southern Benin Republic, *Senna siamea* hedgerows produced significantly less biomass on a Ferralic Cambisol compared to a Rhodic Ferralsol, and separated by a distance of less than 2 km within the same landscape (Aihou et al. 1999). The same species was found to substantially increase the topsoil calcium (Ca) content, effective cation exchange capacity, and soil pH when grown on the latter compared with the former soil, indicating that the hypothesized safety net of trees in a farming system depends partly on the presence of a subsoil of suitable quality (i.e., clay enriched and with high Ca saturation) (Vanlauwe et al. 2005). The same activities demonstrated that *Senna siamea* growing on the Ferralsol was able to rehabilitate a chemically degraded topsoil where maize no longer responded to fertilizer application and restore maize productivity from virtually nil to over 2 t ha^{-1} after a period of 4–5 years (Aihou et al. 1999), thus providing an option for rehabilitating nonresponsive soils in areas with a relatively fertile subsoil.

8.6   SENSING SOIL CONDITION THROUGH SPECTRAL-BASED SURVEILLANCE

The large variation in soil physicochemical fertility over short distances in smallholder systems poses challenges to making soil fertility management recommendations, especially as smallholder farmers are least able to afford soil and plant testing services. Conventional soil maps have limited utility for targeting soil fertility recommendations beyond major variations in parent material or major landforms because they are based on taxonomic mapping units derived from diagnostic horizons rather than from soil properties that directly relate to soil fertility. It is assumed that soil properties are homogeneous within the units whereas the variation of soil fertility levels is often higher within than among mapping units. Furthermore, there are typically a limited number of direct measurements of soil properties taken within map units and the variation in this data is rarely reported. Although there have been attempts to translate soil taxonomic classes into fertility capability classes (Sanchez et al. 2003), it is not possible to derive information such as the prevalence of different soil constraints from such maps.

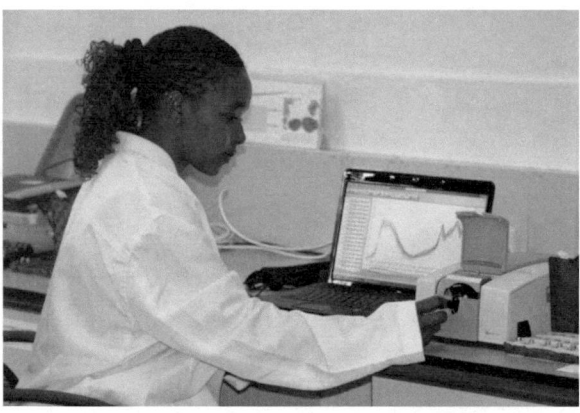

FIGURE 8.6 Portable mid-infrared spectrometer being used for scanning soil samples.

Conventional soil fertility tests rely on extracting and analyzing individual nutrients from soil in an attempt to predict crop response to individual applied nutrients. The conventional tests do not provide an absolute measure of nutrient availability but have to be calibrated using extensive agronomic field trials on different soil types to be able to predict crop response to applied nutrients. They are rarely systematically validated or their success rate documented. Furthermore the soil tests have to be adjusted for different soil types according to factors such as mineralogy, soil texture, and soil organic matter. This is because nutrient supply to plants depends on the interaction of a number of organic-mineral mediated factors that are not captured by the individual nutrient extraction tests (Figure 8.6).

8.6.1 Surveillance Approach

New technological advances in sensing soil condition and digital soil mapping are now making it feasible to map individual soil properties at different scales as opposed to mapping taxonomic units (Sanchez et al. 2009; Herrick et al. 2013). These developments are made possible with the recent availability of (1) accurate georeferencing of soil sampling in the field, (2) rapid, low-cost, and highly reproducible methods for directly characterizing soil properties in the laboratory using infrared spectroscopy (visible, near-infrared, and mid-infrared diffuse reflectance), and (3) multispectral remote sensing data (including unaided aerial vehicles) at medium (30 m) to high spatial resolution (submeter). These advances in sensing soil condition enable a surveillance approach to characterizing and managing soil fertility in much the same way that surveillance systems are used to guide policy and management decisions in public health (Shepherd and Walsh 2007; UNEP 2012; Shepherd et al. 2015).

Infrared spectroscopy (IR) uses a different set of principles than conventional soil fertility tests and provides a single multiple-utility measure of soil production potential and response to management (Shepherd and Walsh 2007; Nocita et al. 2014). With IR, soils can be characterized in a single 30-second measure that requires no chemicals, only light. The shapes of infrared spectra respond to the basic molecular structure of mineral and organic composition of soils and their interactions. It is the organic-mineral composition that determines soil functional properties, including a soil's ability to retain and supply different nutrients and water, nitrogen mineralization capacity, soil charge characteristics, soil structural stability and ability to resist soil erosion, and amount of soil organic carbon in different pools and its protection. Hence IR has been shown to predict a wide range of properties and across a wide range of soils types (Janik et al. 1998; Shepherd and Walsh 2002; Brown et al. 2006; Janik et al. 2007; Shepherd and Walsh 2007; Terhoeven-Urselmans et al. 2010; Bellon-Maurel and McBratney 2011; Soriano-Disla et al. 2014).

8.6.2 Predicting Soil Responsiveness

The IR approach is to calibrate soil and crop responses to management directly to infrared spectra and completely bypass the need for conventional soil tests (Shepherd and Walsh 2007). The same approach applies to IR characterization of plant materials and organic resources (e.g., composts). As an intermediate step, however, until more systematic data on crop and soil management responses is collected, the conventional soil tests are calibrated to infrared spectra. IR does well in predicting soil tests that are strongly related to fundamental soil fertility properties, such as organic carbon, nitrogen, and base status. It does less well in predicting soil tests that measure weakly extractable nutrients (P, K, some micronutrients) (Janik et al. 1998; Shepherd and Walsh 2002; Soriano-Disla et al. 2014), which themselves have variable success in predicting crop response to applied nutrients. A spectral prediction service for Africa, which allows users to upload spectra and obtain predictions of soil properties with prediction uncertainties given, is under beta testing (AfSISa 2015).

To speed up the process of developing calibrations of crop response to spectral tests (and conventional soil tests) for soil fertility recommendations, current research is developing a rapid screening approach using plant growth chambers in the laboratory (Shepherd 2015). Plants are grown in test tubes under controlled conditions in small amounts of soil to which different nutrients are applied. Plant growth potential and response to nutrients can quickly be measured this way and related to infrared spectra of the soils as well as to conventional soil tests. These controlled environment calibrations are then validated against results supplemented with agronomic field trials, which are more expensive to conduct and suffer from many confounding factors such as variable weather. As an intermediate step, current research is also testing the sampling of the existing variation in cultivated landscapes of soil fertility, plant growth (e.g., maize), and plant nutrient status using paired soil and plant samples taken from microplots (e.g., 1 m^2). This research (Muhati 2013) is attempting to establish reference levels for soil and plant tests similar to those done in medicine (Jones and Payne 1997).

IR-based soil fertility analysis has a role for (1) enabling national and subnational soil health mapping of soil fertility constraints and (2) provision of cost-effective soil analysis and advisory services to farmers at the farm-field level. Large area surveillance and mapping applications based on IR have been demonstrated in West Africa (UNEP 2012) and are being applied in the Africa Soil Information Service (AfSISb 2015) and by the Ethiopian Soil Information System (EthioSIS 2015). The World Bank Living Standard Measurement Study and the Ethiopia Central Statistics Agency is testing IR for field-scale characterization of soil fertility linked to long-term household panel surveys (ICRAF 2015). Several studies have used IR to characterize soil fertility in smallholder farming landscapes (Vågen et al. 2006; Awiti et al. 2008; UNEP 2012) and individual fields (Muhati et al. 2011; Tittonell et al. 2005b, 2010, 2013). Commercial soil-testing laboratories are beginning to use IR to provide services to smallholder famers (e.g., Soil Cares 2014). Handheld x-ray fluorescence spectroscopy for plant mineral analysis is also emerging as a promising technology to supplement IR soil and plant N analysis. We expect to see rapid innovation and uptake of spectral-based surveillance and location-specific advisory services over the next 10 years, especially in Africa where existing services are currently limited.

8.7 CONCLUDING REMARKS: PRECISION AGRICULTURE AND SMALLHOLDER FARMING

Targeting soil restoration and/or specific recommendations to increase crop productivity in nonresponsive soils requires major investments to identify the extent and distribution of such soils in Africa and their major limitations. A report published by ISRIC (World Soil Information) in the early 1990s pointed to 494 M ha of African soils being degraded due to human-induced causative factors. Causes such as overgrazing and misconduct of agriculture are often put forward to explain soil degradation (e.g., Oldeman 1994; Kiage 2013). More recently, Vlek et al. (2008) reviewed and reported five major causes of soil degradation in Africa: deforestation, overexploitation, overgrazing, agricultural activities, and (bio) industrial activities. Sandy soils are particularly under stress due to their chemical and physical properties; they represent 13% of the agricultural soils in the continent and constitute a challenge for soil fertility replenishment (Hartemink and Huting 2005). Beyond these coarse, continent-wide assessments, the challenge of targeting specific recommendations to restore productivity in African soils resides in being able to capture the magnitude and impact of spatial soil heterogeneity within farms and landscapes. As soil surveillance technologies progress, the feasibility of sensing such micro- to mesovariability in the field increases. A form of precision agriculture is required that recognizes soil spatial heterogeneity, its causes, and the best soil management technologies to address them. In the context of smallholder farming in Africa, however, precision agriculture approaches need to combine high-tech solutions for soil surveillance with farmers' perception of soil fertility, as this is what ultimately dictates their management decisions, and the development of soil improving interventions that take stock of endogenous innovations, local contexts and trade-offs, and farmers' access to productive resources.

REFERENCES

Adjei-Nsiah S., C. Leeuwis, K. E. Giller, O. Sakyi-Dawson, J. Cobbina, T. W. Kuyper, M. Abekoe and W. Van der Werf (2004). Land tenure and differential soil fertility management practices among native and migrant farmers in Wenchi, Ghana: Implications for interdisciplinary action research. *NJAS Wageningen J. Life Sci.* 52:331–348.

AfSIS. (2015a). Africa Soil Information Service. Spectral Prediction App: Prediction Using Bayesian Additive Regression Trees (BART). http://afsistest.ciesin.columbia.edu/bart_prediction/.

AfSIS. (2015b). Africa Soil Information Service. http://www.africasoils.net.

Aihou, K., N. Sanginga, B. Vanlauwe, O. Lyasse, J. Diels and R. Merckx (1999). Alley cropping in the moist savannah of West Africa: I. Restoration and maintenance of soil fertility in 'terries de barre' soils in Bénin Republic. *Agroforestry Systems* 42, 213–227.

Alvarez, S. (2012). Pratiques de gestion de la biomasse au sein des exploitations familiales d'agriculture-élevage des hauts plateaux de Madagascar: Conséquences sur la durabilité des systèmes, Montpellier SupAgro, CIRAD.

Alvarez S., M. C. Rufino, J. Vayssières, P. Salgado, P. Tittonell, E. Tillard and F. Bocquier (2014). Whole-farm nitrogen cycling and intensification of crop-livestock systems in the highlands of Madagascar: an application of network analysis. *Agricultural Systems* 126: 15–37.

Amede, T. and E. Taboge (2007). Optimizing soil fertility gradients in the Enset (*Ensete ventricosum*) systems of the Ethiopian highlands: Trade-offs and local innovations, pp. 289–297. In *Advances in Integrated Soil Fertility Management in Sub-Saharan Africa: Challenges and Opportunities*, A. Bationo, B. Waswa, J. Kihara and J. Kimetu (eds.), Dordrecht, The Netherlands: Springer.

Andrieu N, J. Vayssieres, M. Corbeels, M. Blanchard, E. Vall and P. Tittonell (2015). From farm scale synergies to village scale trade-offs: Cereal crop residues use in an agro-pastoral system of the Sudanian zone of Burkina Faso. *Agric Syst.* doi:10.1016/j.agsy.2014.08.012.

Augustine, D. J. (2003). Long term, livestock mediated redistribution of nitrogen and phosphorus in an East African savanna. *J. Appl. Ecol.* 40(1):137–149.

Awiti A. O., M. G. Walsh, K. D. Shepherd and J. Kinyamario (2008). Soil condition classification using infrared spectroscopy: A proposition for assessment of soil condition along a tropical forest-cropland chronosequence. *Geoderma* 143:73–84.

Baijukya, F. P. (2004). Adapting to change in banana-based farming systems of northwest Tanzania: The potential role of herbaceous legumes, a PhD thesis, Wageningen, The Netherlands: Wageningen University.

Baijukya, F., N. De Ridder, K. Masuki and K. Giller (2005). Dynamics of banana-based farming systems in Bukoba district, Tanzania: Changes in land use, cropping and cattle keeping. *Agric. Ecosyst. Environ.* 106(4):395–406.

Barrios, E., M. Bekunda, R.J. Delve, A. Esilaba and J. Mowo, (2001). Identifying and classifying local indicators of soil quality. Eastern Africa Version. Participatory Methods for Decision Making in Natural Resource Management. CIAT-SWNM-TSBF-AHI. Nairobi, Kenya.

Barrios E., R. J. Delve, M. Bekunda, J. Mowo, J. Agunda, J. Ramisch, M. T. Trejo and R. J. Thomas (2006). Indicators of soil quality: A south-south development of a methodological guide for linking local and technical knowledge. *Geoderma* 135:248–259.

Baudron, F., P. Tittonell, M. Corbeels, P. Letourmy and K. E. Giller (2012). Comparative performance of conservation agriculture and current smallholder farming practices in semi-arid Zimbabwe. *Field Crops Res.* 132:117–128.

Bellon-Maurel, V. and A. McBratney (2011). Near-infrared (NIR) and mid-infrared (MIR) spectroscopic techniques for assessing the amount of carbon stock in soils—Critical review and research perspectives. *Soil Biol. Biochem.* 43:1398–1410.

Benjaminsen, T. A., J. B. Aune and D. Sidibé (2010). A critical political ecology of cotton and soil fertility in Mali. *Geoforum* 41(4):647–656.

Brown, D., K. D. Shepherd and M. G. Walsh (2006). Global soil characterization using a VNIR diffuse reflectance library and boosted regression trees. *Geoderma* 132:273–290.

Bucagu, C. (2013). Tailoring agroforestry technologies to the diversity of Rwandan smallholder agriculture. PhD thesis, Wageningen, The Netherlands: Wageningen University.

Carter, S., and H. Murwira, (1995). Spatial variability in soil fertility management and crop response in Mutoko Communal Area, Zimbabwe. *Ambio* 24:77–84.

Chikowo, R., P. Mapfumo, P. A. Leffelaar and K. E. Giller (2006). Integrating legumes to improve N cycling on smallholder farms in sub-humid Zimbabwe: Resource quality, biophysical and environmental limitations. *Nutr. Cycl. Agroecosys.* 76:219–231.

Corbeels, M., A. Shiferaw and M. Haile (2000). Managing Africa's Soils, Farmers' knowledge of soil fertility and local management strategies in Tigray, Ethiopia, Vol. 10. London: IIED.

Corbeels, M., A. Shiferaw and M. Haile (2000). *Farmers' knowledge of soil fertility and local management strategies in Tigray, Ethiopia*, IIED-Drylands Programme.

Dawoe, E. K., J. Quashie-Sam, M. E. Isaac and S. K. Oppong (2012). Exploring farmers' local knowledge and perceptions of soil fertility and management in the Ashanti Region of Ghana. *Geoderma* 179–180: 96–103.

Deckers, J. (2002). A system approach to target balanced nutrient management in soilscapes of sub-Saharan Africa, pp 47-61. In: *Integrated Plant Nutrient Management in Sub-Saharan Africa. From concepts to practice*, Vanlauwe, B., Diels, J., Sanginga, N., Merckx, R. (eds.), CAB International, Wallingford, Oxon, UK.

Diarisso, T., M. Corbeels, N. Andrieu, P. Djamen and P. Tittonell (2015). Biomass transfers and nutrient budgets of the agro-pastoral systems in a village territory in south-western Burkina Faso. *Nutr Cycl Agroecosyst*, doi 10.1007/s10705-015-9679-4.

Ebanyat, P. (2009). A road to food? Efficacy of nutrient management options targeted to heterogeneous soilscapes in the Teso farming system, a PhD thesis, Wageningen, The Netherlands: Uganda. Proefschrift Wageningen.

Elias, E. and I. Scoones (1999). Perspectives on soil fertility change: A case study from southern Ethiopia. *Land Degrad. Dev.* 10(3):195–206.

Elias, E., S. Morse and D. G. R. Belshaw (1998). Nitrogen and phosphorus balances of Kindo Koisha farms in southern Ethiopia. *Agric. Ecosyst. Environ.* 71(1–3):93–113.

EthioSIS. (2015). Ethiopia Soil Information System. Agricultural Transformation Agency. http://www.ata.gov .et/highlighted-deliverables/ethiopian-soil-information-system-ethiosis/

Findeling, A., S. Ruy and E. Scopel (2003). Modeling the effects of a partial residue mulch on runoff using a physically based approach. *J. Hydrol.* 275:49–66.

Gandah, M. (1999). Spatial variability and farmer resource allocation in millet production in Niger, a PhD thesis, Wageningen, The Netherlands: Landbouwuniversiteit Wageningen.

Guto, S. N., P. Pypers, B. Vanlauwe, N. de Ridder and K. E. Giller (2011). Socio-ecological niches for minimum tillage and crop-residue retention in continuous maize cropping systems in smallholder farms of central Kenya. *Agron. J.* 103:644–654.

Hartemink, A. E. and J. Huting (2005). Sandy soils in Southern and Eastern Africa: Extent, properties and management. Proceedings *Management of tropical sandy soils for sustainable agriculture, Thailand*, pp. 54–59.

Hartemink, A. E., R. J. Buresh, B. Jama and B. H. Janssen (1996). Soil nitrate and water dynamics in sesbania fallow, weed fallows and maize. *Soil Sci. Soc. Am. J.* 60:568–574.

Herrick, J. E., K. C. Urama, J. W. Karl et al. (2013). The Global Land-Potential Knowledge System (LandPKS): Supporting evidence-based, site-specific land use and management through cloud computing, mobile applications, and crowdsourcing. *J. Soil Water Conserv.* 68(1):5A–12A. doi:10.2489/jswc.68.1.5A.

ICRAF (2015). Improving measurements of agricultural productivity through methodological validation and research (LSMS-ISA). http://www.worldagroforestry.org/research/land-health/projects/wb%20lsms.

Jama, B., R. J. Buresh and F. M. Place (1998). Sesbania tree fallows on phosphorus-deficient sites: Maize yield and financial benefit. *Agron. J.* 90:717–726.

Janik, L. J., R. H. Merry and J. O. Skjemstad (1998). Can mid infrared diffuse reflectance analysis replace soil extractions? *Aust. J. Exp. Agric.* 38:681–696.

Janik, L. J., J. O. Skjemstad, K. D. Shepherd and L. R. Spouncer (2007). The prediction of soil carbon fractions using mid-infrared-partial least square analysis. *J. Aust. Soil Res.* 45(2):73–81.

Jones, R. and B. Payne (1997). *Clinical Investigation and Statistics in Laboratory Medicine*. London: ACB Venture Publications.

Kamanga, B. C. G. (2011). Poor people and poor fields? Integrating legumes for smallholder soil fertility management in Chisepo, central Malawi, PhD thesis, Wageningen, The Netherlands: Wageningen University.

Kiage, L. M. (2013). Perspectives on the assumed causes of land degradation in the rangelands of Sub-Saharan Africa. *Prog. Phys. Geogr.* 37(5):664–684.

Mairura F. S., D. N. Mugendi, J. I. Mwanje, J. J. Ramisch, P. K. Mbugua and J. N. Chianu (2007). Integrating scientific and farmers' evaluation of soil quality indicators in Central Kenya. *Geoderma* 139:134–143.

Masvaya, E., J. Nyamangara, R. Nyawasha, S. Zingore, R. Delve and K. Giller (2011). Effect of farmer resource endowment and management strategies on spatial variability of soil fertility in contrasting agro-ecological zones in Zimbabwe, pp. 1221–1229. In *Innovations as Key to the Green Revolution in Africa*, Vol. 1, A. Bationo, B. Waswa, J. M. Okeyo, F. Maina and J. Kihara (eds.). Dordrecht, The Netherlands: Springer.

Misiko, M., P. Tittonell, K.E. Giller and P. Richards (2009). Strengthening understanding of mineral fertilizer among smallholder farmers in western Kenya. *Agriculture and Human Values* 28:27–38.

Mowo, J. G., B. H. Janssen, O. Oenema, L. A. German, J. P. Mrema and R. S. Shemdoe (2006). Soil fertility evaluation and management by smallholder farmer communities in northern Tanzania. *Agric. Ecosyst. Environ.* 116(1–2):47–59.

Mtambanengwe, F. and P. Mapfumo (2005). Organic matter management as an underlying cause for soil fertility gradients on smallholder farms in Zimbabwe. *Nutr. Cycl. Agroecosyst.* 73:227 – 243.

Muhati, S. I., K. D. Shepherd, C. K. Gachene, M. W. Mburu, R. Jones, G. O. Kironchi and A. Sila (2011). Diagnosis of soil nutrient constraints in small-scale groundnut (*Arachis hyopaea* L.) production systems of Western Kenya using infrared spectroscopy. *J. Agric. Sci. Technol.* A1:111–127.

Muhati, S. I. (2013). Refining fertilizer use recommendations for smallholder maize fields in African landscapes. PhD proposal accepted by Wageningen University.

Murage, E. W., N. K. Karanja, P. C. Smithson and P. L. Woomer (2000). Diagnostic indicators of soil quality in productive and non-productive smallholders' fields of Kenya's Central Highlands. *Agric. Ecosyst. Environ.* 79(1):1–8.

Naudin, K., E. Scopel, L.H. Andriamandroso, M. Rakotosolofo, L.H. Andriamarosoa, M. Ratsimbazafy, J. N. Rakotozandriny, P. Salgado and K. E. Giller (2011). Trade- offs between biomass use and soil cover. The case of rice-based cropping systems in the lake Alaotra region of Madagascar. *Exp. Agric.*, doi:10.1017 /S001447971100113X.

Nkonya, E., C. Kaizzi and J. Pender (2005). Determinants of nutrient balances in a maize farming system in eastern Uganda. *Agric. Syst.* 85(2):155–182.

Nocita, M., A. Stevens, B. van Wesemael, D. J. Brown, K. D. Shepherd, E. Towett, R. Vargase and L. Montanarella (2014). Soil spectroscopy: An opportunity to be seized. *Glob. Chang. Biol.* June 21, 2014, doi: 10.1111/gcb.12632.

Oldeman, L.R. (1994). Global extent of soil degradation. In *Soil Resilience and Sustainable Land Use*, D. J. Greenland and I. Szabolcs (eds.), pp. 99–118. CAB International, Wallingford, UK.

Pieri C (1989). Fertilité des terres des savanes. Ministère de la Coopération CIRAD, 444 pp.

Prudencio, C. Y. (1993). Ring management of soils and crops in the west African semi-arid tropics: The case of the mossi farming system in Burkina Faso. *Agric. Ecosyst. Environ.* 47(3):237–264.

Ramisch, J. (1999). In the balance. Evaluating soil nutrient budgets for an agro-pastoral village of southern Mali. Managing Africa's Soils 9.

Ramisch, J. J. (2005). Inequality, agro-pastoral exchanges, and soil fertility gradients in southern Mali. *Agric. Ecosyst. Environ.* 105(1–2):353–372.

Rushemuka, N. P., R. A. Bizoza, J. G. Mowo and L. Bock (2014). Farmers' soil knowledge for effective participatory integrated watershed management in Rwanda: Toward soil-specific fertility management and farmers' judgmental fertilizer use. *Agric. Ecosyst. Environ.* 183(0):145–159.

Rusinamhodzi, L., M. Corbeels, J. Nyamangara and K.E. Giller (2012). Maize–grain legume intercropping is an attractive option for ecological intensification that reduces climatic risk for smallholder farmers in central Mozambique. *Field Crops Research* 136:12–22.

Salako, F. K., G. Tian, G. Kirchhof and G. E. Akinbola (2006). Soil particles in agricultural landscapes of a derived savanna in southwestern Nigeria and implications for selected soil properties. *Geoderma* 137(1–2):90–99.

Samaké, O., E. M. A. Smaling, M. J. Kropff, T. J. Stomph and A. Kodio (2005). Effects of cultivation practices on spatial variation of soil fertility and millet yields in the Sahel of Mali. *Agric. Ecosyst. Environ.* 109(3–4):335–345.

Sanchez, P. A., C. A. Palm and S. W. Buo (2003). Fertility capability soil classification: A tool to help assess soil quality in the tropics. *Geoderma* 114:157–185.

Sanchez, P. A., S. Ahamed, F. Carré et al. (2009). Digital soil map of the world. *Science* 325:680–681.

Scopel, E., F. Da Silva, M. Corbeels, F. Affholder and F. Maraux (2004). Modelling crop residue mulching effects on water use and production of maize under semi-arid and humid tropical conditions. *Agronomie* 24:383–395.

Shepherd, K. D. and M. G. Walsh (2007). Infrared spectroscopy—Enabling an evidence-based diagnostic surveillance approach to agricultural and environmental management in developing countries. *J. Near Infrared Spec.* 15·1–19.

Shepherd, K. D. and M. G. Walsh (2002). Development of reflectance spectral libraries for characterization of soil properties. *Soil Sci. Soc. Am. J.* 66:988–998.

Shepherd, K. D. (2015). Plant Environment Facility of the ICRAF Soil-Plant Spectral Diagnostics Laboratory. http://wle.cgiar.org/blog/2014/02/01/addressing-nutrient-deficiencies-african-soils/.

Shepherd, K. D., G. Shepherd and M. G. Walsh (2015). Land health surveillance and response: A framework for evidence-informed land management. *Agric. Sys.* 132:93–106.

Smaling, E. M. A. and A. R. Braun (1996). Soil fertility research in sub-Saharan Africa: New dimensions, new challenges. *Commun. Soil Sci. Plant Anal.* 27(3–4):365–386.

Soil Cares (2014). The Soil Cares Initiative. http://www.soilcares.com.

Sommer, R., C. Thierfelder, P. Tittonell, L. Hove, J. Mureithi and S. Mkomwa (2014). Fertilizer use is not required as a fourth principle to define conservation agriculture—Response to the opinion paper of Vanlauwe et al. (2014) 'A fourth principal is required to define conservation agriculture in sub-Saharan Africa: The appropriate use of fertilizer to enhance crop productivity.' *Field Crop Res.* 169:145–148.

Soriano-Disla, J. M., L. J. Janik, R. A. Viscarra Rossel, L. M. Macdonald and M. J. McLaughlin (2014). The performance of visible, near-, and mid-infrared reflectance spectroscopy for prediction of soil physical, chemical, and biological properties. *Appl. Spectrosc. Rev.* 49:139–186.

Steiner, K. G. (1998). Using farmers' knowledge of soils in making research results more relevant to field practice: Experiences from Rwanda. *Agric. Ecosyst. Environ.* 69(3):191–200.

Stoop, W. A. (1987). Variations in soil properties along three toposequences in Burkina Faso and implications for the development of improved cropping systems. *Agric. Ecosyst. Environ.* 19(3):241–264.

Terhoeven-Urselmans, T., T.-G. Vagen, O. Spaargaren and K. D. Shepherd (2010). Prediction of soil fertility properties from a globally distributed soil mid-infrared spectral library. *Soil Sci. Soc. Am. J.* 74: 1792–1799.

Tesfahunegn, G. B., L. Tamene and P. L. Vlek (2011). A participatory soil quality assessment in Northern Ethiopia's Mai-Negus catchment. *CATENA* 86:1–13.

Tilahun, G. (2007). "Soil fertility status as influenced by different land soil fertility status as influenced by different land." MSc Thesis. Faculty of the Department of Plant Sciences, School of Graduate Studies Haramaya University. Ethiopia.

Tittonell, P. (2008). Msimu wa Kupanda: Targeting resources within diverse, heterogeneous and dynamic farming systems of East Africa, PhD thesis, Wageningen, The Netherlands: Wageningen Universtiy.

Tittonell, P. and K. E. Giller (2013). When yield gaps are poverty traps: The paradigm of ecological intensification in African smallholder agriculture. *Field Crops Res.* 143(0):76–90.

Tittonell, P., E. Scopel, N. Andrieu, H. Posthumus, P. Mapfumo, M. Corbeels, G.E. van Halsema, R. Lahmar, S. Lugandu, J. Rakotoarisoa, F. Mtambanengwe, B. Pound, R. Chikowo, K. Naudin, B. Triomphe and S. Mkomwa (2012). Agroecology-based aggradation-conservation agriculture (ABACO): Targeting innovations to combat soil degradation and food insecurity in semi-arid Africa. *Field Crop Res.* 132, 168–174.

Tittonell, P., A. Muriuki, K. D. Shepherd, D. Mugendi, K. C. Kaizzi, J. Okeyo, L. Verchot, R. Coe and B. Vanlauwe (2010). The diversity of rural livelihoods and their influence on soil fertility in agricultural systems of East Africa—A typology of smallholder farms. *Agric. Syst.* 103(2):83–97.

Tittonell, P., A. Muriuki, C.J. Klapwijk, K.D. Shepherd, R. Coe and B. Vanlauwe (2013). Soil heterogeneity and soil fertility gradients in smallholder agricultural systems of the East African highlands. *Soil Sci. Soc. Amer. J.* 77:525–538.

Tittonell, P., B. Vanlauwe, P. Leffelaar, E. Rowe and K. Giller (2005a). Exploring diversity in soil fertility management of smallholder farms in western Kenya: I. Heterogeneity at region and farm scale. *Agric. Ecosyst. Environ.* 110(3):149–165.

Tittonell, P., B. Vanlauwe, P. Leffelaar, K. D. Shepherd and K. E. Giller (2005b). Exploring diversity in soil fertility management of smallholder farms in western Kenya: II. Within-farm variability in resource allocation, nutrient flows and soil fertility status. *Agric. Ecosyst. Environ.* 110(3):166–184.

Tittonell, P., B. Vanlauwe, N. de Ridder and K. E. Giller (2007). Heterogeneity of crop productivity and resource use efficiency within smallholder Kenyan farms: Soil fertility gradients or management intensity gradients? *Agric. Syst.* 94:376–390.

Tittonell, P., B. Vanlauwe, M. Corbeels and K. E. Giller (2008). Yield gaps, nutrient use efficiencies and responses to fertilisers by maize across heterogeneous smallholder farms in western Kenya. *Plant Soil* 313:19–37.

UNEP (2012). *Land Health Surveillance: An Evidence-Based Approach to Land Ecosystem Management. Illustrated with a Case Study in the West Africa Sahel.* Nairobi, Kenya: United Nations Environment Programme. http://www.unep.org/dewa/Portals/67/pdf/LHS_Report_lowres.pdf.

Vågen, T.- G., K. D. Shepherd and M. G. Walsh (2006). Sensing landscape level change in soil quality following deforestation and conversion in the highlands of Madagascar using Vis-NIR spectroscopy. *Geoderma* 133:281–294.

Vanlauwe, B., K. Aihou, B.K. Tossah, J. Diels, N. Sanginga and R. Merckx (2005). Senna siamea trees recycle Ca from a Ca-rich subsoil and increase the topsoil pH in agroforestry systems in the West African derived savanna zone. *Plant Soil* 269:285–296.

Vanlauwe, B., P. Tittonell and J. Mukalama (2006). Within-farm soil fertility gradients affect response of maize to fertilizer application in western Kenya. *Nut. Cycl. Agroecosyst.* 76:171–182.

Vanlauwe, B. (n.d.). Status and challenges of soil management in Africa. International Institute of Tropical Agriculture.

Vanlauwe, B., J. Wendt, K. E. Giller, M. Corbeels, B. Gerard and C. Nolte (2014). A fourth principle is required to define Conservation Agriculture in sub-Saharan Africa: The appropriate use of fertilizer to enhance crop productivity. *Field Crops Res.* 155:10–13.

Vlek, P. L., Q. B. Le and L. Tamene (2008). Land decline in land-rich Africa. Science Council, Consultative Group on International Agricultural Research, London, Montpellier.

Wopereis, M., A. Tamélokpo, K. Ezui, D. Gnakpénou, B. Fofana and H. Breman (2006). Mineral fertilizer management of maize on farmer fields differing in organic inputs in the West African savanna. *Field Crops Res.* 96(2):355–362.

Zingore, S., H. K. Murwira, R. J. Delve and K. E. Giller (2007a). Soil type, management history and current resource allocation: Three dimensions regulating variability in crop productivity on African smallholder farms. *Field Crops Res.* 101(3):296–305.

Zingore, S., H. Murwira, R. Delve and K. Giller (2007b). Influence of nutrient management strategies on variability of soil fertility, crop yields and nutrient balances on smallholder farms in Zimbabwe. *Agric. Ecosyst. Environ.* 119(1):112–126.

Zingore, S., P. Tittonell, M. Corbeels, M. T. Wijk and K. E. Giller (2011). Managing soil fertility diversity to enhance resource use efficiencies in smallholder farming systems: A case from Murewa District, Zimbabwe. *Nut. Cycl. Agroecosyst.* 90:87–103.

9 Precision Farming for Coastal and Island Ecoregions
A Case Study of Andaman and Nicobar Islands

A. Velmurugan, T.P. Swarnam, Rattan Lal,
S.K. Ambast, and N. Ravisankar

CONTENTS

9.1   INTRODUCTION

The land surface of the earth is highly variable. The highly variable properties of different surface features are considered as spatial variation. This variation exists from global to field scale, and within the field variations exist in soil physical, chemical, and biological properties at the level of the microenvironment (Foster 1988; Mzuku et al. 2005; Papiernik et al. 2005; Ensign et al. 2006). Humankind has been engaged in understanding and managing these spatial variations throughout history. The degree of understanding has depended on the socioeconomic needs and the techno-logical advancement of any given time. Today, aerial photographs and satellite remote sensing have helped to understand these spatial variations in land use/cover more decisively than before at a regional to global scale (Moran et al. 1997). Simultaneously, scientific advances have also estab-lished the relationship between the distribution of natural resources and the food production system, which has paved the way for efficient resource management (Navalgund et al. 1991). Soil and water are the most important spatially varying natural elements and are critical to agricultural production, but the degree of variations is highly influenced by the scale at which these properties are described. Such variations can manifest in different soil types with features of waterlogging or inundation, dryness, erosion, and heterogeneity of soil physical, chemical, and biological properties. Thus, soils differ in their productive capacity to support human and animal life because of these variations.

In a quest to meet the world food requirement, indiscriminate application of agricultural inputs has increased the risk of environmental degradation (Lal 2004). As a result, human activities are increasingly concerned with judicious land use and conservation of soil and water resources while addressing the demand for world food. At the same time, the pressure of an increasing global popu-lation, urbanization, and demand for other essential items from the agricultural sector are even-tually passed on to the land. Land resources are affected by the consequences of global climate change and degradation also triggered by anthropogenic activities. Thus, it is prudent to use the land according to its production potential with the required level of inputs while minimizing risks of land degradation and restoring its productivity by appropriate management practices.

It is well known that the productivity of crops depends on their genetic features, environmental conditions, and management. Different inputs are used in an agricultural production system with the aim of producing more from the given piece of land. Such activities, implemented while ignoring the field level variations, lead to wastage of inputs, increase production cost, negatively impact the environment, and exacerbate global concerns (Mandal and Maity 2013). The intrafield variability is significantly too large to ignore because it varies strongly from place to place and from country to country. Thus, precise management of these variations is advantageous to agricultural production sys-tems (Pierce and Nowak 1999). Managing the production capacity of the land within the field is called precision farming. These within-field variations are also influenced by the field size. The field size and land holdings are relatively large in the United States and other developed countries where there is more available land for a given population. In contrast, the field size and land holdings are relatively small in most of the developing countries to quantify the within-field variations (Sahoo et al. 2007).

In several instances, the magnitude of variations prompts farmers to seek solutions for low yield in certain pockets and ways to reduce the input cost. The management challenge is to collect reli-able data and analyze and optimally manage the areas within the field that have different production capacities. Precision farming offers the potential to automate and simplify the collection, analysis, and use of information for variable rate management (McLoud et al. 2007). In the same context, there are circumstances or compelling conditions that necessitate the application of the concept of precision farming for larger areas or at the agroecosystem level. This is more relevant for coastal and island ecosystems where the area available for cultivation is finite and major determinants of agricultural production are more than just the input. In addition, the capacity of farmers to invest in precision agri-culture differs widely. Over the years, and often by trial and error, farmers have also learned how to use inputs and land judiciously. As a consequence, the starting point for developing a precision agricul-ture plan in such areas may differ depending on the goals (results) a farmer is trying to achieve within a specific environmental condition. Despite differences in its adoption, there should be no ambiguity in the purpose of precision farming, which aims to achieve food security with an optimal input use.

9.2 CONCEPT OF PRECISION FARMING

Precision farming is a site-specific management system based on information and technology. It also requires one or more reliable data points to implement at the field level such as soils, crops, nutrients, pests, moisture, or yield, for optimum profitability, sustainability, and protection of the environment (Precision Ag. 2003). Precision farming involves two aspects: (1) assessing the need and quantifying the input as well as (2) management effort to produce a particular quantity from a specific site. This strategy determines the required quantity that accommodates the spatial variability. Another aspect is knowing the site-specific requirements and how to apply them, which are the technology and information parts of precision farming. This involves collection and analysis of data to support the first part of the concept. The above understanding is in accord with Hugh Hammond Bennett's admonition to "… use every acre within its capability and treat them according to its needs" (McLoud et al. 2007).

In the same context, the concept of precision farming also accommodates the inherent capability of the soil to produce within the established equilibrium of its surroundings. This asserts the use of soil within its ecological limitations without affecting the surrounding environment. Therefore, there can be no compromise in the first part of the concept, but the second part of the concept should offer some kind of flexibility to accommodate different resource situations. Over the years, the concept of precision farming has evolved to include the entire production function of the farm and goes beyond yield mapping and variable rate fertilizer application. The conventional definition of precision farming is suitable when the land holdings are large and enough variability exists within the fields (Blackmore 1994). In some countries, the average land holdings are very small even with large and progressive farmers but other management efforts should also be precise to improve input efficiency. In this context, tillage, soil amendments, land configuration (land shaping), and drainage, in addition to input management, are all relevant and included in precision farming. Some of the components of precision farming and the concepts are given in Table 9.1.

TABLE 9.1
Components and Concepts of Precision Farming

Components	Concept	Reference
Precision farming	Precise application of inputs according to the within-field-level variations that later evolved to imply precise application of agricultural inputs based on soil, weather, and crop requirements and their management to maximize sustainable productivity, quality, and profitability.	Blackmore (1994); Robert et al. (1996); Bishop and McBratney (2002)
Precision tillage	Tilling the land only wherever required or hard pan/surfaces are encountered. Also includes control of traffic on fields to limit compaction to defined wheel tracks.	McLoud et al. (2007)
Variable rate technology	Application of different inputs such as seeds, fertilizers, weedicides, and pesticides at varying rates within the field according to their requirements with the help of a computer-controlled applicator. However, most of the studies focused on a single-factor response.	Robert et al. (1996); Lowenberg-DeBoer and Swinton (1997)
Variable irrigation application	Application of irrigation water within the field at different rate and amount according to the quantity of water required in different areas.	Pocknee et al. (1996); Evans and Harting (1999)
Variable rate harvest of biomass	Harvest based on the ability of the soil to withstand the loss of residues for erosion protection and sustainable soil quality.	Bunter (2007)

The central concept of precision farming involves a precise application of inputs according to the within-field-level variations. It has evolved along with the developments in data collection and variable application technology. Precision tillage is more applicable in high rainfall and arid zones where frequent tillage can lead to erosion and rapid oxidation of organic material. Similarly, variability in a cropped field results in different amounts of water being needed in different areas. However, at the field level installation/maintenance of equipment required for precision farming is rather expensive and it is essential that the increase in marginal return should justify the cost involved.

9.3 SCOPE/RATIONALE

Modern agriculture has helped to alleviate hunger from the world, although the world population more than doubled during the last half of the twentieth century (Lal 2000). Famines and scarcities have been known in India and other tropical countries from the earliest times (Randhawa 1983; Swaminathan 1996), while there was no major scarcity of food after the severe droughts of 1972 and 1987 (FAI 2004) due to modern agriculture. Nevertheless, an overuse of pesticides, fertilizers, and irrigation with a motive to produce more food grains and driven by economic benefits results in residues much above the safety levels (Carson 1963), as well as waterlogging and salinity in the surface. The indiscriminate use of inputs have caused nitrate enrichment of groundwaters, river waters, and estuaries and the release of ammonia (NH_3) and nitrous oxide (N_2O) to the atmosphere. The former added to the problem of acid rain, while the latter led to the reduction of the ozone layer (Laegreid et al. 1999). In addition, the factor productivity for many of the modern agricultural inputs has been decreasing (Rattan and Singh 1997).

In some regions across the world the land has several inherent constraints such as waterlogging, aridity, acidity, salinity, and erosion in addition to human-induced changes. Despite the constraint such lands are pressed into production activities to meet food, fodder, and fiber requirements that require variable treatments. The pressure on land becomes very high particularly in small islands, coastal regions, and island nations due to geographical and environmental limitations, which need more attention to sustain production and conserve resources (UN Secretariat 2010). In other words, those areas cannot afford resource degradation triggered by unsustainable practices following agricultural intensification or spend more on inputs that are uneconomical. In these cases, large variations within the field, ecosystem, and region occur resulting in the formation of hotspot areas that demand precision farming.

In general, precision agriculture brings to mind complex, intensely managed production systems using Global Positioning System (GPS) technology to spatially reference soil, water, yield, and other data for the variable rate application of agricultural inputs within a field. However, at the basic level, precision agriculture could be simple practices such as field scouting and the spot application of inputs such as seeds, fertilizers, pesticides, and irrigation water. Precision agriculture should also help farmers to recognize areas with productivity and environmental constraints and to select the best possible solution for each location.

9.4 ISLAND ECOSYSTEM OF ANDAMAN AND NICOBAR ISLANDS: PRELUDE

The humid tropical islands of Andaman and Nicobar are situated 1200 km off the east coast of India in the Bay of Bengal (Figure 9.1). They comprise about 556 small and large islands covering an area of 8249 km^2 with a coastline of 1962 km between 92°–94° E longitude and 6°–14° N latitude. The northern group of islands is called the Andaman Islands while the southern group of islands constitutes the Nicobar Islands. Andaman Islands are continental in origin whereas the Nicobar group consist mostly of coralline and sedimentary deposits. These two group of islands are separated by 10° channel.

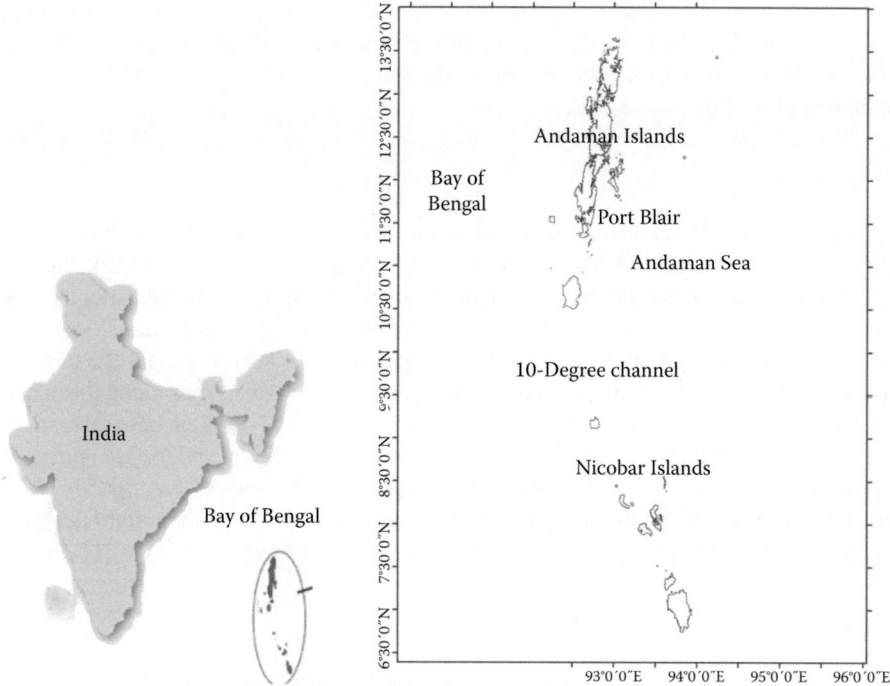

FIGURE 9.1 Location map of the Andaman and Nicobar Islands, India.

9.4.1 AGROCLIMATE

The climate of the Andaman and Nicobar Islands is typified by tropical conditions with little difference between mean summer and mean winter temperatures. The annual rainfall varies from 2900 to 3100 mm representing a perhumid climate. Evapotranspiration is very high due to intensive solar radiation especially during drier months (February–April), which far exceeds the rainfall resulting in a water deficit condition (Figure 9.2). The relative humidity varies from 68% to 86% and the mean

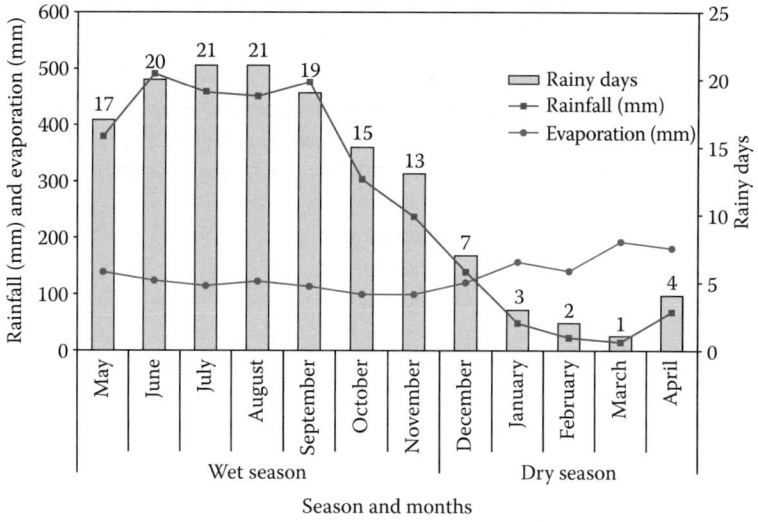

FIGURE 9.2 Climatic parameters of the Andaman and Nicobar Islands.

maximum and minimum temperatures is 32°C and 22°C, respectively. The length of the growing period is more than 210 days, which is long enough to support double cropping and plantation crops. These islands experience *udic* soil moisture and *isohyperthermic* soil temperature regime (Velayutham et al. 1999).

9.4.2 Soils

The topography of the islands is rolling with low-range hills to narrow valleys at the foothills forming undulating terrain ranging from steep slopes (>45°) to plains (<5°). The soils of the Andaman and Nicobar Islands are formed by the dominant influence of climate and vegetation. The uplands are under forest cover but are intensively leached due to very high runoff and slope. The valley floors are developed from the outwash of parent material from the surrounding hills. In general, the soils are medium to deep red loamy including marine-alluvium-derived soils along the coast. These are slightly to strongly acidic in nature and are moderate to low (40%–70%) in base saturation. These soils, under the great groups of *Hapludalfs*, *Dystropepts*, *Eutropepts*, and *Sulfaquents* (along the coast), have low to medium available water-holding capacity. The soils of Pahargaon, Dhanikhari, and Garacharma series typify the dominant soils observed in the Andaman and Nicobar Islands (Ganeshamurthy et al. 2002).

9.4.3 Land Cover/Use

The land cover is dominated by tropical rain forest in the longitudinal hills and mangroves in the sea front while agriculture is confined to specific areas around habitations. Out of the total geographical area of 8249 km², agricultural activities are confined to only 6%, which are dominated by plantation crops in the hill slopes followed by rice in the valley and coastal plains. About 4206 ha of agriculture land was permanently submerged due to the Indian Ocean tsunami in 2004 that decreased the area available for agriculture (DES 2011). There are no large, perennial streams and the physical conditions do not favor formation of large reservoir for water storage. In certain areas, coral sheets are exposed under shallow waters, which provides a habitat for large marine biodiversity.

9.5 RESOURCE BASE AND ITS DISTRIBUTION

9.5.1 Area under Major Crops and Productivity

The total area under major plantation crops in the Andaman and Nicobar Islands is around 69% of a gross cultivated area of 55,598 ha. Coconut (*Cocos nucifera*) and areca nut (*Areca catechu*) alone accounts for 53% of area followed by oil palm (*Elaeis guineensis*) and rubber (*Hevea brasiliensis*). Among the food grains, rice (*Oryza sativa*) accounts for 22% of the area while vegetables and fruits together accounts for 15% (Figure 9.3).

In the Andaman and Nicobar Islands, the area under coconut in 1979–80 was 20,787 ha with a production of 67.29 million nuts. During the last two decades, the area has increased by 4.3% with the production of 81.90 million nuts and productivity of 3749 nuts/ha. Although there is an increase of 20.21 million nuts, the productivity remains stagnant as the increase in nuts mostly comes from expansion in the area. These islands have 4147 ha under areca nut plantation and produce 5721 million tons (Mt) of arecanut. The cashew nut (*Anacardium occidentale*) production is reported to be 362 Mt from 800 ha. In addition, the favorable climate of these islands offers great opportunity for the cultivation of different types of vegetables in 5150 ha with the total production of 0.313 Mt (DES 2011).

The production statistics indicate that additional agricultural land is needed to meet the growing demand for food grains, vegetables, and fruits. The pressure for safer sites on plantation areas created by increasing population and the tourism sector is rising alarmingly. However, there is no scope

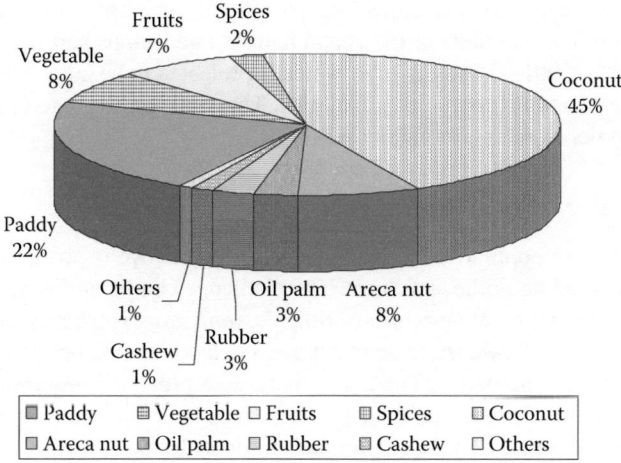

FIGURE 9.3 Area under major crops in the Andaman and Nicobar Islands. Area includes intercrop as well. (From DES, *Statistical Hand Book of Andaman and Nicobar Islands*, Directorate of Economic and Statistics, Andaman and Nicobar Administration, Port Blair, India, 2011.)

to increase the area under these crops either by converting forest land or diverting areas to other land uses. In addition, there is a problem of soil acidity and salinity. The high rainfall received in these islands leaches out the soluble salts from soils of upland and sloping areas resulting in soil acidity where mostly plantation crops are grown. In saline and waterlogged coastal areas, traditional rice varieties are grown with limited management practices resulting mostly in low productivity, while lack of technological implementation is hampering fruit and vegetable production. Therefore, marginal and degraded lands have to be reclaimed to explore their suitability for plantation and other crops in addition to phased conversion of existing areas into high-density plantations and increasing the cropping intensity through intercropping and crop rotations.

9.5.2 POPULATION GROWTH

The total population of these islands was 0.36 million in 2001, which increased to 0.46 million in 2011 (DES 2011) with a growth rate of about 24%. Assuming the population growth rate as in the last decade, it is projected to increase to 0.58 million in 2021 (Figure 9.4) with nearly 55% of the population living in rural areas. Thereafter, very high exponential growth in population is projected.

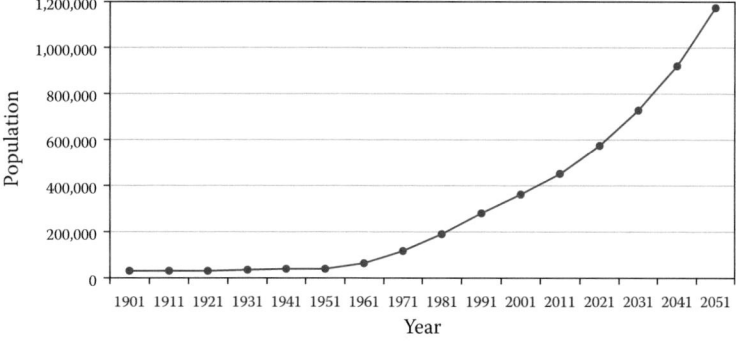

FIGURE 9.4 Trend of population growth and projected population in the Andaman and Nicobar Islands.

Apart from this, the tourism sector is expanding very rapidly, bringing in more floating population into these islands. Similarly, in most of the island nations and islands across the globe the decadal population growth rate (2001–11) was at 26%, which is projected to be 20%–30% in the next decade due to improved health and infrastructural facilities. The total fertility rate (TFR) in these regions is about 2.5 children per woman (Anonymous 2014a).

9.5.3 Agriculture and Food Requirement

The emerging situation in population explosion as described above is posing a major challenge in terms of demand for food and other products. The challenge can be addressed only by increasing the productivity and production of agriculture while the remaining gap between demand and supply has to be met from imports. However, the current trend in land use and production does not provide much scope to address the challenge. The commodity-wise projected requirement of food articles for the projected population is given in Table 9.2. Demands for cereals and vegetables will increase by one-third and that for pulses, milk, and animal products by 60% within the next two decades. Presently, two-thirds of rice comes from mainland India to meet the demand. While the scope for area expansion by land conversion is limited, there is a possibility of increasing crop production by introducing suitable high-yielding rice varieties, increasing the cropping intensity, and adopting improved crop and soil management practices. It is precisely in the context of improved soil and crop management that precision agriculture is of significant importance.

9.5.4 Agriculture and Water Resources

It is estimated that the present average food ingestion of 2800 kcal/person/day requires 1000 m^3 of water per annum. Therefore, water needed to produce the required food for the 0.58 million population in 2021, excluding water losses due to the irrigation system, is 0.61 km^3 (Table 9.3). Srivastava and Ambast (2009) have estimated that most of the water required for crop production is provided by rainfall stored in the soil profile, and only 1% is provided through ponds created under minor irrigation, which cover less than 5% of the cropped land. Therefore, irrigation needs 0.4 km^3 of water per annum for food crops alone. It is estimated that development of the available island water

TABLE 9.2
Projected Requirements of Food Items (as per Indian Council of Medical Research [ICMR] Standards) in the Andaman and Nicobar Islands

Commodity	Projected Food Requirement (MT yr^{-1})	
	2021	2031
Cereals	96,746	119,682
Pulses	5832	7290
Roots and tuber	29,616	37,020
Vegetables	34,399	42,998
Fruits	20,312	25,390
Fat and oils	8033	10,042
Milk	34,603	43,254
Meat/fish	17,177	21,471

Source: Modified and adopted from Srivastava, R.C. and Ambast, S.K., *Water Policy for Andaman and Nicobar Islands: A Scientific Perspective*, CARI, Port Blair, India, p. 18, 2009.

TABLE 9.3

Projected Water Requirement to Produce Different Commodities for the Andaman and Nicobar Islands

Commodity	Crop Water Requirement (m³/t)	Total Water Requirement (billion cubic meter)	
		2021	2031
Cereals (rice)	4254	0.412	0.509
Pulses	1000	0.006	0.007
Roots and tubers	1000	0.029	0.037
Vegetables	1000	0.034	0.043
Fruits	1000	0.020	0.025
Fat and oils	2000	0.016	0.020
Milk	1369	0.047	0.059
Meat	5187	0.089	0.111
Coconut and areca nut	–	0.048	0.048

resources by 7%–10% by 2021 and 15%–20% by 2031 could meet these requirements. However, the physiography does not favor a large-scale reservoir or irrigation network. Similar to this, the availability of freshwater is a major limiting factor for economic and social development for small island developing states. Many of these islands rely entirely on rainwater harvesting as a single source of water supply (Anonymous 2014b).

Consequently, a water resource development strategy for these islands should be based on utilization of rainwater either through surface storage or enhanced groundwater recharges. Ring wells of more than 5–6 m depth may result in saline water intrusion, and bore wells are not feasible in marine sedimentary formations (CGWB 2010). Considering these constraints, surface storage by *in situ* rainwater harvesting at different suitable locations instead of a centralized location should be preferred. This strategy necessitates mapping of physiography, crop coverage, and drainage lines of the Island to precisely find suitable locations. In all probabilities, the situation is similar for most small islands and island nations, and precision management of water resources can enhance use efficiency and advance sustainability.

9.5.5 CLIMATE CHANGE AND ITS EFFECT ON ISLAND ECOSYSTEM

Since the beginning of the twentieth century, there has been a marked change in surface temperature, rainfall, evaporation, and extreme events that pose serious threats to small islands and island nations. The atmospheric concentration of carbon dioxide (CO_2) has increased from about 280 parts per million by volume (ppmv) to about 400 ppmv and the global temperature of the earth has increased by about 0.6°C. The global mean sea level has risen by 10 to 20 cm, affecting mostly coastal and island regions. Due to global warming, the associated rise in global mean sea level is projected between 9 and 88 cm in the next 25 years (IPCC 2007). These changes in climate will affect the soil moisture, groundwater recharge, frequency of flood or drought episodes, and finally groundwater level in small islands such as the Andaman and Nicobar Islands. Additionally, the increasing frequency and intensity of dry spells and lesser number of rainy days may affect vegetable production particularly in the hilly areas where groundwater potential is low. Increase in cyclonic storms may result in waterlogging and crop lodging, particularly rice grown in the coastal areas. Nearly 20 million people are affected by tropical cyclones and floods in the tropical islands (Anonymous 2014b). Therefore, water resources are one of the most critical natural resources

TABLE 9.4
Status of Land Degradation in the Andaman Islands

Degradation Classes	Area (ha)	Severity Class
Coastal erosion	1650	Moderate (e2)
Acid sulfate soil	5600	Severe (A3)
Waterlogging and salinity	9854	Moderate (w2, s1)
Seasonal water logging	17,800	Severe (w3)
Erosion	21,460	Slight (e1)
Waterlogging and marshy	69,125	Slight (w1)
Erosion and acidity	449,079	Moderate (e1, m.a)

Note: A, acid sulfate soil; e, erosion; m.a, moderately acidic; s, salinity; w, waterlogging.

vulnerable to the perceived climate changes, which justifies the call for its precise management and judicious use through the principles of precision farming, particularly under the island conditions.

9.5.6 LAND DEGRADATION STATUS

Based on the interpretation of remote sensing data and field survey, Velmurugan et al. (2014) reported that erosion, waterlogging, acidity, salinity, and acid-saline conditions are the major types of land degradation observed in these islands (Table 9.4). More important, different types of land degradation categories exist together based on the dominating processes such as salinity and water-logging, acidity and soil erosion, and salinity and acidity. Soil erosion and acidity are commonly observed in forested areas on 78.7% of the total geographical area, followed by waterlogging and erosion and water logging and salinity. The coastal acid-saline soils generally extend up to 1–2 km inland, which coupled with waterlogging, critically affect the productivity of rice and other food crops. In addition, coastal erosion is observed on 0.23% of the total geographical area and acid sulfate soil occurs in 5600 ha with varying intensity and distribution. The Indian Ocean tsunami of 2004 also left a deep mark on these islands with varying severity of salinity and waterlogging in 14% of the cultivated area, mainly under rice (Velmurugan et al. 2006).

Thus, it is essential to address the issue of land degradation in a cost-effective manner at both the field and regional level depending on the degree and extent of problems. Therefore, it is imperative to adopt precision farming techniques for managing the land resources and increasing the produc-tion, which also provides an opportunity to effectively deal with the challenges posed by climate change on island ecosystems.

9.6 PRECISION FARMING TECHNIQUES FOR ISLAND CONDITIONS

For the purpose of variable management, specific areas within a field that respond to management practices in a similar way have to be demarcated into different management zones. In other words, being able to identify variability across fields should be the first step in defining management zones (Leon et al. 2003). However, it is difficult to demarcate the boundaries of management zones for a single crop type on a single field over time (Whelan and McBratney 2003). In practice, these zones are similar to smaller fields with bunds all around that exist in island conditions within which the standard deviation around the mean for any property does not vary significantly. This is justified by the smaller geographical extent and similar agroclimatic conditions within the island. In such situations, physiographic grouping as in the case of soil survey will be reasonable for deciding dif-ferent management options rather than continuous mapping of surface soil for different properties.

Some of the precision management techniques suitable for the island ecoregion of the Andaman and Nicobar Islands are given in Sections 9.6.2, 9.6.3, and 9.6.4.

9.6.1 Data Sources

The ability to micromanage fields will require a combination of technologies that include hardware, software, and the best management practices (BMPs). They are GPS, Geographic Information System (GIS), satellite remote sensing, large-scale soil map (1:8000), physiographic map, digital elevation model (DEM), and climatic parameters. The required information on surface soil properties such as electrical conductivity (EC), land use, waterlogging, soil type, and elevation could be prepared separately as thematic layers and integrated into a GIS framework (Figure 9.5). As these layers are georeferenced, they correspond to a particular location on the ground and represent the true nature of the surface. The intersection of these layers produces different polygons within which homogeneity and intrapolygon differences are observed in one or more properties. By this method, the field-level variations are represented as different polygons that facilitate variable treatments between the polygons.

There is a scarcity of land for agricultural purposes under island conditions, and the size of the field is relatively small. Hence, management zones can be delineated by finding the common areas of various thematic maps. The problems and potential of each zone are considered before deciding on the management options, and each selected option is the same for similar conditions elsewhere in the Island. In this method, the large field is subdivided into smaller uniform management zones in accordance with the variations in the field (Balkwill 2012). Variable rate of fertilizer application with precision equipment is not a viable option because the area is prone to waterlogging and the return from such an investment is likely to be uneconomical. In this context, topographical information is used to identify the potential management units that serve as a basis for directed soil sampling (Franzen et al. 1997) and/or the differential application of chemicals (Nolan et al. 1998).

9.6.2 Crop Selection

Physiographically the island can be divided into hills, upland, foot slope, and lowland, which end in the sea (Figure 9.6). The central part of the Island is elevated with longitudinal hill ranges that are densely covered with evergreen forest. The hilly upland and foot slopes are undulating where most of the plantation crops are grown. This is followed by the narrow coastal valley, which faces the seafront. Flat valleys are also found in between longitudinal hills where the washed-out materials from the surrounding hills are being deposited.

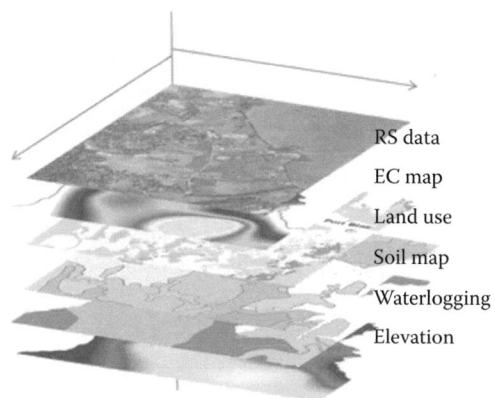

RS data

EC map

Land use

Soil map

Waterlogging

Elevation

FIGURE 9.5 Different thematic layers are used for delineation of management zones.

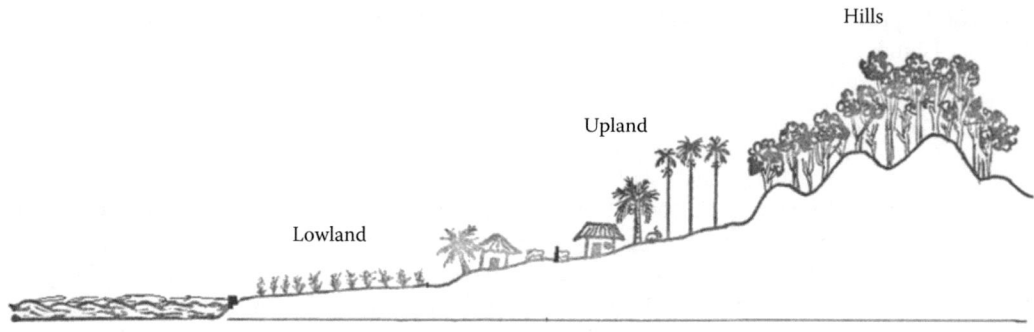

FIGURE 9.6 Cross-sectional view of typical physiography of the Andaman and Nicobar Islands.

Rice is the major staple crop grown in this island and the conditions such as high humidity, salinity, and waterlogging do not favor the growth of any other annual crop, especially during the monsoon season. Even the productivity of rice is low (2.6–3.0 t/ha) due to improper choice of varieties for different physiographic and field conditions and poor management, especially the unbalanced and inadequate use of fertilizers. The yield performance of rice varieties under different situations is evaluated to select suitable varieties for different management zones. The results show higher yield in normal soil with wet conditions while the yield decreases in waterlogged conditions irrespective of the varieties (Table 9.5). Similarly, the performance of rice varieties in saline soil conditions also varies significantly in which CSR-36 produces higher yield. The C-14-8 is a long-duration rice variety able to withstand waterlogging and matures at the end of the monsoon season, thereby escaping the lodging. In view of high demand for rice and prevailing climatic conditions of these islands, the model rice varieties should fit into the growing conditions and have resistance/tolerance to biotic and abiotic stresses, especially salinity and submergence.

Adoption of site-specific farming will be successful only when the effects of biotic and abiotic relationships on plant growth and yield are understood and managed accordingly (Machado et al. 2002). Therefore, selection of suitable rice varieties based on the physiographic location, salinity, and waterlogging will help to realize maximum yield potential under island conditions. Delineating management zones or grouping of areas of similar conditions into fields of convenient size as

TABLE 9.5

Performance of Salt-Tolerant Rice Varieties in the Andaman and Nicobar Islands

Situation	Varieties	Duration (Days)	Average Yield (t/ha)	
			Wet Condition[a]	Waterlogged
Normal soils (EC 0.1–1.0 dS/m	CSR 36	135–140	3.8	3.2
and pH 5.9–6.6)	Ranjeet	150–155	4.1	3.7
	CARI Dhan 5	140–145	3.3	2.9
	Jaya	135–145	2.9	2.6
	C-14-8	170–180	2.6	2.5
Saline soils (EC 3.7–9.4 dS/m	CSR 36	135–140	3.5	2.9
and pH 5.4–6.6)	Ranjeet	150–155	2.9	2.8
	CARI Dhan 5	140–145	3.0	2.8
	Jaya	135–145	2.7	2.6
	C-14-8	170–180	2.4	2.3

[a] Saturation to 0–5 cm standing water.

FIGURE 9.7 Crop management based on ground variations. (a) Bunding of fields based on depth of water-logging; (b) short (far end) and medium duration (front) rice varieties; (c) plastic mulch in the midland to save water; (d) vegetable cultivation in the raised bed in low-lying area.

discussed in Section 9.6.1 is preferable for selecting rice varieties for each zone, which will ensure optimum resource use and higher yield (Figure 9.7a). Short-duration rice varieties should be grown in areas where waterlogging is expected late in the monsoon season. In contrast, photosensitive long-duration rice variety (C-14-8) should be selected in areas where lodging is a problem during the early part of the monsoon season. Another viable option for the upper portion of the landscape (elevated land) is high-yielding, medium-duration varieties (Figure 9.7b). Similarly, in upland areas, mulching or conservation tillage is useful to conserve soil moisture (Figure 9.7c) while raised beds are highly beneficial for growing vegetables in the low-lying areas (Figure 9.7d). The yield obtained in each management zone and the yield determining conditions should be recorded for precision crop selection in the subsequent season. A similar approach can be practiced elsewhere in an island ecosystem based on the local conditions, which is a more viable solution than blanket recommendations of crops and varieties.

9.6.3 Land Configuration Technique (Broad Bed and Furrow System)

Paddy is the predominant crop cultivated in coastal plains and mountain valleys during monsoon season. After the Indian Ocean tsunami of 2004, several areas in the coastal plains became water-logged and saline due to drainage congestion and seawater intrusion during high tides. The depth and duration of submergence varies based on the distance from sea and the physiographic position.

In these areas, a land configuration technique called broad bed and furrow system (BBF) is a viable option to reclaim the degraded coastal areas and use them for agriculture.

9.6.3.1 Suitable Sites

The BBF system is not a panacea and cannot be recommended for every physiographic situation; note that the depth of furrows in a BBF should also increase while moving away from the coastal areas. Therefore, before installing BBF with particular dimensions, the entire area has to be sub-divided into smaller units with similar characteristics (management zone). The advantages of this system are the draining off of excess water from the field and improvement in soil aeration, which provides an opportunity for crop diversification in otherwise monocropped areas; *in situ* rainwater harvesting, and use of agricultural machineries that is otherwise not possible in a waterlogged area. The BBF system of different configurations and suitable rice varieties for different situations are shown in the satellite image in Figure 9.8. This is the outcome of implementation of the land configuration techniques in the coastal degraded areas of the islands under the National Agricultural Innovation Project sponsored by Global Environmental Facility. It also shows the delineated saline and waterlogged areas for which different rice varieties should be selected.

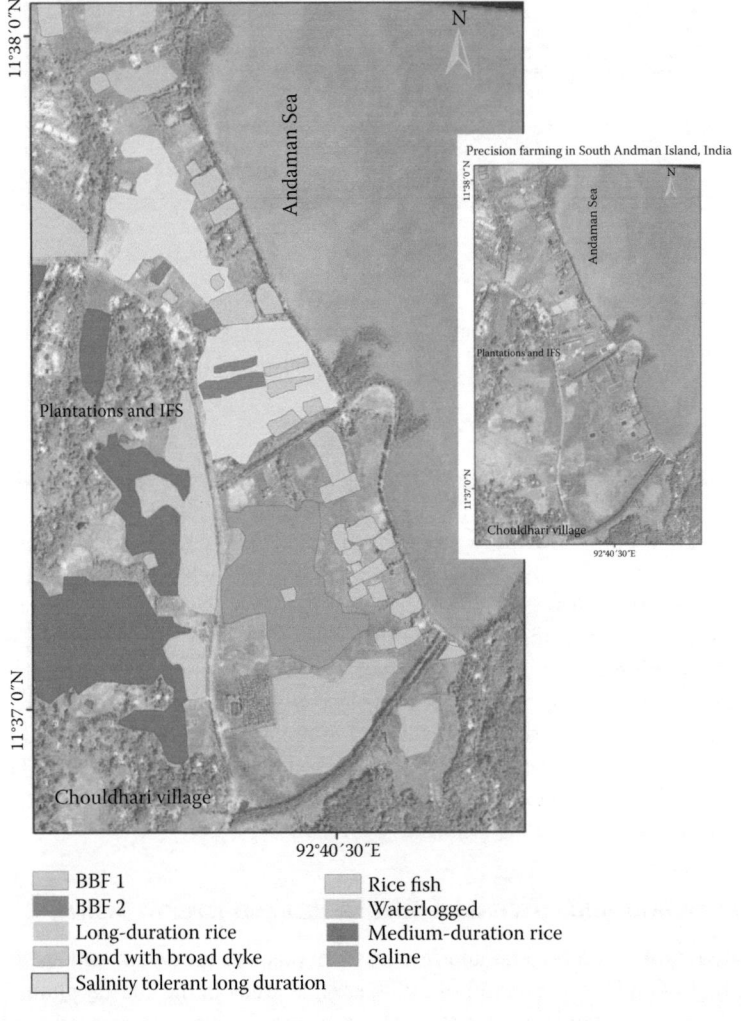

	BBF 1		Rice fish
	BBF 2		Waterlogged
	Long-duration rice		Medium-duration rice
	Pond with broad dyke		Saline
	Salinity tolerant long duration		

FIGURE 9.8 Delineation of different management zones based on its problems and potential.

9.6.3.2   Data Requirement

The BBF system is normally installed in low-lying areas. The purpose of this land configuration technique is to assess and utilize the saline and waterlogged coastal areas precisely for crop diversification. The management zones can be precisely delineated by overlaying of digital elevation models prepared from the contour lines (Survey of India toposheets), wetness index derived from satellite data (IRS P6), and soil map in a GIS framework (Arc GIS v. 9.2). These delineated zones can be precisely managed based on their potentials and limitations.

9.6.3.3   Design

The design of BBF involves making raised beds alternated with furrows by excavating the soil in waterlogged areas. Broad beds are made in the shape of an inverted trapezium by digging soil from either sides of the broad bed and putting it in the bed area layerwise by the cut-and-fill method (Figure 9.9). The design of a BBF system has been evaluated and standardized for different situations such as intensity of rainfall, physical, and chemical properties of soil, and drainage requirement under island conditions (Ravisankar et al. 2008).

The dry period from February to April is the appropriate time for making a BBF system because the soil is dry and can be easily manipulated. The beds of 4–5 m width (sandy soil lesser width) and furrows of 5–6 m width with minimum of 1.0–1.5 m depth are suitable for the island conditions of high rainfall intensity. The length and breadth of beds and furrows can be adjusted depending on the availability of land, soil type, and nature of the management zone. In sandy soil, the bed width should not exceed 4 m and should have at least 1:1 side slopes. In clayey soil, beds of 5 m with 5 m furrows are more stable and can easily be made. Close to the coast (up to 200 m from coastline), the furrow depth should be 1 m or less only up to the marine sediments (BBF 1). Away from the coast, furrows up to 1.5 m depth (BBF 2) can be made to store more rainwater and drain the seawater down below due to the pressure head gradient.

9.6.3.4   Cultivation

The degraded costal land can be configured into BBFs of different dimensions (Figure 9.10a–d). During the monsoon season, the furrow (excavated area) is used for rice + fish cultivation while vegetables are cultivated on the raised beds. During the postmonsoon season, vegetables and pulses are grown on the raised beds using the rainwater harvested in the furrows. Fodder/grasses are most suitable in the side slopes of raised beds and are grown in this physiography. The beds alternating with furrows in the BBF system reduce the average annual runoff to one-half and the soil loss to

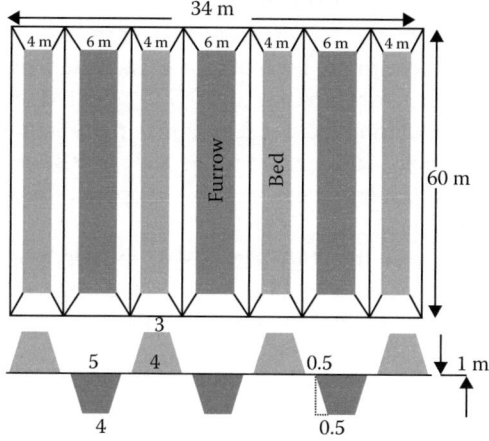

FIGURE 9.9   Schematic design of the broad bed and furrow system.

(a)   (b)

(c)   (d)

FIGURE 9.10 Land management based on field variations and distance from sea. (a) Degraded coastal undulating terrain; (b) BBF with 1.0 m deep furrow and 4 m width bed (vegetable + fish); (c) BBF with 1.5 m deep furrow and 3 m width bed (vegetable + fodder + fish); (d) BBF with 1.2 m deep furrow and 3 m width bed (banana + rice + fish).

one-fourth compared with that of the traditional method of cultivation. By adopting this system, nearly 67% of the rainfall is used by the crops while 14% and 19% is lost by evaporation and deep percolation, respectively. This system is also useful in decreasing water runoff and increasing infiltration, especially during the dry season.

9.6.3.5 Effect of Land Configuration Technique

The temporal variations in gravimetric soil moisture content in the top 15-cm soil layer has been observed continuously for 52 weeks in the beds of BBFs. The data indicates 5%–7% higher moisture content during the dry season between the 0 to 16th Standard Metrological Week (SMW) and 8%–12% lower moisture content during the wet season than the adjoining flatland. Therefore, irrigation up to 10%–15% of available soil moisture in the beds of BBF is sufficient to meet the crop demand during the dry season. Similar results have been reported for some high rainfall areas of Himachal Pradesh, India (Sharma 2003).

During the rainy season between the 20th–45th SMW, the depth of submergence ranges from 15 to 25 cm in the foot slope compared with 25 to 85 cm in the plains. Moreover, the plains adjoining the coastline are also subjected to waterlogging up to 15 cm even during the dry season due to tidal water intrusion (Figure 9.11). It has also been observed that soil in the plains is inundated and waterlogged until mid-February, which forces the farmers to leave the land fallow for the remaining 5 months in a year after the harvest of rice in December–January. However, the relationship between weekly rainfall and depth of submergence indicates a higher correlation for submergence of foot slope ($r^2 = 0.834$) than plain ($r^2 = 0.787$). This trend occurs because the plains are mainly influenced

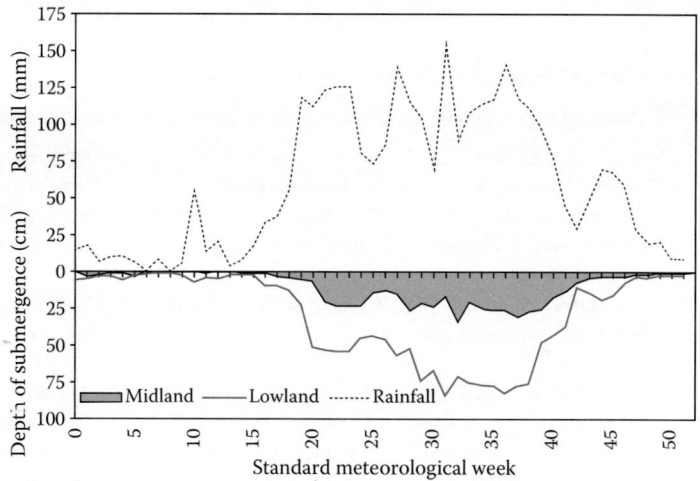

FIGURE 9.11 Relationship between rainfall and depth of submergence.

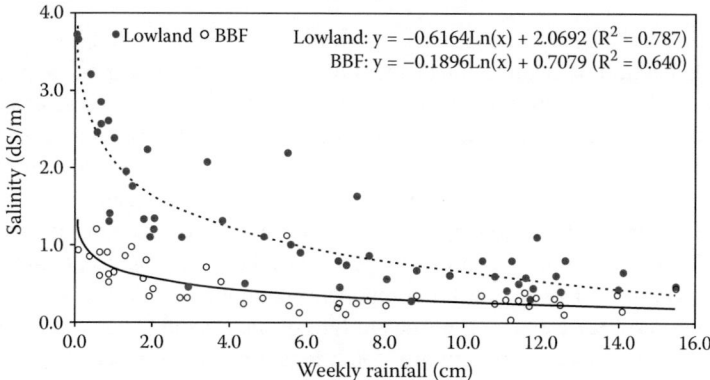

FIGURE 9.12 Relationship between salinity and weekly rainfall.

by seawater inundation during high tides in addition to rainfall. The depth of submergence becomes higher when it coincides with high rainfall events, leading to drainage congestion. During the same period, the raised beds of 1 to 1.5 m in a BBF system remain below the saturation level due to improved drainage. The latter enables growing of food crops (e.g., vegetables) in the beds throughout the year. Along with BBF, installing a one-way sluice gate in the drainage channel also helps to reduce seawater inundation and duration of submergence, while improving the drainage network reduces submergence in the foot slope.

Significant variations in soil salinity are also observed between beds of BBF and its surrounding areas. In general, salinity is higher in areas close to the coast because of seawater inundation during high tide than in the foot slopes, which are relatively far from the coastline. A close relationship also exists between soil salinity and the weekly rainfall amount (Figure 9.12) for lowland ($R^2 = 0.787$) compared to beds of BBF ($R^2 = 0.64$). Such a trend is mainly due to leaching of soluble salts from the raised beds of BBF resulting in a low salinity level than the surrounding water logged areas.

9.6.4 Integrated Farming System

In the Andaman and Nicobar Islands, farmers mostly grow traditional photosensitive and low-yielding cultivars of rice having a long vegetative period. The condition is not any better in coconut

TABLE 9.6
Suitable IFS Models for Different Physiographic Locations

Parameters	Flat Land/Coastal Plain	Hill Slope	Hilly Upland
Soil type	Sandy loam–clay loam	Sandy–loamy sand	Gravelly sand
Soil depth	Moderate to deep	Moderate–shallow	Very shallow
Vegetation	Field crops, grasses (*Echinocloa* sp., *Cynodon* sp., *Imperata* sp.)	Plantation, grass, Glyricidia shrub, *Mikania* sp.	Plantation, indigenous fruit trees, forest (evergreen type)
Crops	Paddy, pulses (red gram, green gram), vegetables and sweet potato	Coconut, areca nut, banana	Coconut, areca nut, banana, jack fruit
Major problems	Stem borer, Gundhi bug, lead folder, shoot borer; low level of diversification, subject to flooding and salinity, unstable system	Rhinoceros beetle, bud rot; low diversification, low income, erosion and soil acidity	Rhinoceros beetle, bud rot; low diversification, low income, dry during summer, erosion and soil acidity
Suitable IFS model	MFS III	MFS II	MFS I

gardens, which generate low income for farmers due to insufficient nutrient application and poor management. In addition, the unpredictable nature of rainfed farming leads to unstable production and financial risks. In order to meet these challenges, the farming system approach, based on the resources in a rainfed environment, is essential. In this approach, the small-size land holding of each farmer is considered as one management unit. Integrated farming systems (IFS) have the potential to enhance the food production, nutritional value, livelihood, and income of farmers while also reducing the risks of total failure in monocropped areas due to strong biotic and abiotic stresses. However, similar IFS models with similar components cannot be used in all situations, which may result in inefficient resource use or sometimes failure of the system. Therefore, precision resource use under specific natural and socioeconomic conditions is essential. In this approach, site-specific IFS models are required based on different components.

An accurate assessment of the agroecosystem situation of these islands has led to recognition of three distinct microfarming situations (MFSs) mainly distinguished on the basis of physiographic location and crop composition (Table 9.6). On-farm research on IFSs involving all MFSs has been carried out by selecting farmers representing specific scenarios (Ravisankar et al. 2007). The strategy involves integration of different enterprises (e.g., introduction of new technologies, integration of crop, dairy, goat, fishery, and backyard poultry) based on the resource endowment and their constraints in each location. The components and technologies have been intervened in different MFSs according to the client's need and resource availability. The details are given in the next sections.

9.6.4.1 Hilly Uplands (MFS I)

The hills, which have more than 25% slope, are suited for plantation crops (e.g., coconut and areca nut) along with growing some fruit trees (e.g., jackfruit [*Artocarpus heterophyllus*], guava [*Psidium guajava*], and sapota [*Manilkara sapota*]) in the homestead gardens. The soils are low in fertility and prone to erosion. Under such conditions, plantation-based farming systems involving crop + dairy + backyard poultry can increase both the production and productivity of the farm. One 15 m × 10 m lined pond at suitable sites for rainwater harvesting is provided. Growing of spices (e.g., black pepper [*Piper nigrum*], clove [*Eugenia caryophyllata*]) as an intercrop in a plantation is the best option for such a situation. The system generated additional employment of 163 mandays/ha/year. Cropping contributed toward the maximum net returns (84%) followed by dairy (13%) and poultry (3%). Among the cropping sequences, areca nut plantations with black pepper yielded the maximum return of U.S.$2,510 (approximately U.S.$1 = 60 Indian rupees). Cattle and poultry have a net return of U.S.$850 and $225, respectively.

9.6.4.2 Mid-Hills/Hill Slopes (MFS II)

In the mid-hill slopes, plantations are intercropped with fruit crops of banana (*Musa paradisiaca*) and pineapple (*Ananas comosus*). These hills have a medium slope of 10%–15%, soils of coarse texture with medium depth and low nutrient and water retention capacity. Accelerated soil erosion and low nutrient status are the major constraints. In these areas, upland rice is grown, followed by pulses, vegetables, flowers, and sugarcane (*Saccharum officinarum*). Sometimes, unmanaged fish-ponds are also installed at the end of the slope, which requires reshaping and rectification. Backyard poultry, cattle, and goat rearing are also common. Little modification of the existing system and integration of components like dairy + poultry + fish or dairy + goat + poultry + fish can provide an additional net income of about U.S.$1670 and employment generation of 270 mandays/ha/yr. Cropping contributes a net return of 71% to the total return followed by cattle (13%), goat (10%), poultry + fish + duck (5%), and backyard poultry (1%).

9.6.4.3 Low-Lying Valley (MFS III)

In low-lying valley areas, only rice cultivation is possible because of waterlogging for more than 6 months. Thus, long-duration and photosensitive paddy is the main crop. Soils are deep and fertile, and waterlogging during the monsoon season is the main constraint on these lands. Backyard poultry and cattle rearing are common. Modifications of existing cropping systems with short-duration rice followed by pulses or vegetables and integration with cattle, poultry, and fish can result in an additional net farm income of U.S.$1750 and employment generation of 201 mandays/ha/year. In low-lying and coastal areas, household income can be increased by land configuration (BBF), especially where vegetables are grown in beds even during the rainy season and rice + fish can be practiced in the furrows.

Thus, site-specific farming system can be implemented for efficient resource use and enhanced farm return under island conditions where the size of the holding is small and is limited because of varying physiography. These models exhibit greater stability because they facilitate rainwater harvesting, efficient organic waste recycling, product diversification, and optimal resource use. Even during the extreme weather events encountered in island conditions, the degree of damage is minimal than that under other specialized systems. The productivity potential can be realized by interlinking of different components with its physiographic conditions. In such situations, each farm holding can be treated as one unit and all other fields nearby with similar conditions constitute one management unit for practicing precision farming.

9.7 CONCLUSIONS AND FUTURE RESEARCH NEEDS

Precision farming is largely practiced and evaluated as a complex, intensely managed production system using GPS to spatially reference soil, water, yield, and other data for the variable rate application of agricultural inputs within a field. However, the success stories pertaining to precision farming are primarily from developed countries, wherein agriculture is a professionally managed enterprise. However, the goal of precision farming in practice is to optimize inputs for agricultural production according to the land capability and soil-specific conditions. Taking into account the predominance of fragmented land holdings, the heterogeneity of crops and livestock, and the concept of farm families in rural conditions, the model of precision farming representing the typical island conditions needs to be redefined within the larger concept of optimum resource use based on the land capabilities that exist within the field. In addition, farmers from most of the island nations across the tropical world are primarily focused on economic return from the limited resources while their governments are concerned with meeting the challenges of food security and climate change. Therefore, the concept of precision farming should be suitably evolved to accommodate the nature and specific needs of island ecosystems.

In the island ecosystem of Andaman and Nicobar, the spatiotemporal distribution of water-logging along with the varying landscape and soil degradation do not favor adoption of identical agricultural practices in all physiography, which results in low agronomic yields. The present and projected socioeconomic and climatic conditions necessitate the use of precision agriculture techniques to effectively manage the land resources. For that reason, different management zones should be delineated for adopting suitable and precise farming technologies using modern precision farming tools in a GIS framework. Such a grouping makes it possible to select suitable crop and variety, BBF design and locations, and different farming system models based on the MFSs prevailing in the Andaman and Nicobar Islands. Precision agriculture methods can help farmers to recognize areas that have productivity and environmental problems, and also help select the BMP for each situation.

Under island conditions, the potential of precision farming for economic and environmental benefits can be realized through reduced use of water, fertilizer, pesticides, seeds, and farm equipment. Instead of managing the entire field based on average conditions, the precision farming approach recognizes site-specific differences and adjusts management actions accordingly. The precision farming techniques as discussed in this chapter have wider applicability and exhibit stability. The analysis of the performance supports its application to similar situations prevailing in other small islands in the tropical regions. However, in all probability, the variable rate application technique with all its sophistication and high cost is applicable only for progressive farmers with large farm sizes and sufficient risk-bearing capacity. At the same time, several researchable priorities in precision farming have emerged that are applicable to island conditions:

1. Identification of resource constraints for which precision agriculture techniques can have a positive impact and are suitable for that location. Indiscriminate initiation and application of precision farming techniques may not provide the desired result. Therefore, a thorough analysis of feasibility and return is essential.
2. Undertaking of research to focus on efficient but simple ways of collection and analysis of the required background data of an area, and development of farm-specific plans to address the resource concern using precision agriculture.
3. Monitoring of the performance of precision farming techniques in the light of changing climate pattern and evolving socioeconomic needs of the local communities is required under site-specific conditions.
4. Dissemination of proven and the most appropriate precision farming techniques over a large area, including data collection and management options suitable for a particular agroclimatic condition.

These suggested precision farming methods serve the purpose of both large and small farms to optimize yields, minimize input, and reduce environmental pollution.

ACKNOWLEDGMENTS

The authors are grateful to the Director, Central Island Agricultural Research Institute, Port Blair for his support and the help rendered by the staff of the Division of Natural Resource Management is gratefully acknowledged. We sincerely acknowledge the financial assistance received from the Global Environmental Facility, National Agricultural Innovation Project of Indian Council of Agricultural Research, India, and the AP Cess fund of the Government of India.

REFERENCES

Anonymous. 2014a. Population and developments in SIDS. Department of Economic and Social Affairs, United Nations. http://www.unpopulation.org. Accessed on July 20, 2014.
Anonymous. 2014b. Small island developing states. Department of Economic and Social Affairs, United Nations. http://sustainable development.un.org/index.php?menu=203. Accessed on June 21, 2014.

Balkwill, T. 2012. Turning precision technology into agronomic solutions. *Crops Soils* 45(3):10–11.

Bishop, T.F.A., and A.B. McBratney. 2002. Creating field extent digital elevation models for precision agriculture. *Precis. Agric.* 3:37–46.

Blackmore, S. 1994. Precision farming: An introduction. *Outlook Agric.* 23:275–280.

Bunter, W. 2007. *Crop Residue Removal for Biomass Energy Production: Effects on Soils and Recommendations*. Soil Quality–Agronomy Technical Note No. 19, Natural Resources Conservation Service, Davis, CA. http://directives.sc.egov.usda.gov/.

Carson, R. 1963. *Silent Springs*. Hamish Hamilton, London.

CGWB. 2010. *Approach paper on ground water quality issues in Islands*. Central Ground Water Board, Ministry of Water Resources, GOI, New Delhi, India, p. 69.

DES. 2011. *Statistical Hand Book of Andaman and Nicobar Islands*. Directorate of Economic and Statistics, Andaman and Nicobar Administration, Port Blair, India.

Ensign, K.L., A. Elizabeth, W. Frederick, and J. Longstaffe. 2006. Micro-environmental and seasonal variations in soil water content of the unsaturated zone of a sand dune system at Pinery Provincial Park, Ontario, Canada. *Geoderma* 136:788–802.

Evans, G.W., and G.B. Harting. 1999. Precision irrigation with center pivot systems on potatoes. In: *Proceedings of the ASCE 1999 International Water Resources Engineering Conference*, Seattle, WA.

FAI. 2004. *Fertilizer Statistics 2003–04*. The Fertilizer Association of India, New Delhi, India.

Foster, R.C. 1988. Microenvironments of soil microorganisms. *Biol. Fertil. Soils* 6(3):189–203.

Franzen, D.W., V.L. Hofman, L.J. Cihacek, and L.J. Swenson. 1997. Soil nutrient relationships with topography as influenced by crop. *Precis. Agric.* 1:167–183.

Ganeshamurthy, A.N., R. Dinesh, N. Ravisankar, A. Nair, and S.P.S. Ahlawat. 2002. *Land Resources of Andaman and Nicobar Islands*. Central Agricultural Research Institute, Port Blair, India.

IPCC. 2007. *Climate Change Synthesis Report*, an Assessment of the Intergovernmental Panel on Climate Change (IPCC), November 12–17, Valencia, Spain, p. 73.

Laegreid, M., O.C. Beckman, and O. Kaarstad. 1999. *Agriculture, Fertilizers and the Environment*. CABI Publishing, Norsk Hydro ASA, Oxon, UK, p. 294.

Lal, R. 2000. *Controlling green house gases and feeding the globe through soil management*. University Distinguished Lecture, February 17, 2000, Ohio State University, Columbus, OH.

Lal, R. 2004. Soil carbon sequestration impacts on global climate change and food security. *Science* 304:1623–1627.

Leon, C.T., D.R. Shaw, M.S. Cox, M.J. Abshire, B. Ward, M.C. Wardlaw III, and C. Watson. 2003. Utility of remote sensing in predicting crop and soil characteristics. *Precis. Agric.* 4:359–384.

Lowenberg-DeBoer, J., and S.M. Swinton. 1997. Economics of site-specific management in agronomic crops. In: F.J. Pierce and E.J. Sadler (eds.), *The State of Site-Specific Management for Agriculture*, ASA, CSSA, SSSA, Madison, WI, pp. 369–396.

Machado, S., E.D. Bynum, Jr., R.J. Lascano, L.T. Wilson, and E. Segarra. 2002. Spatial and temporal variability of sorghum grain yield: Influence of soil, water, pests, and diseases relationships. *Precis. Agric.* 3:389–406.

Mandal, S.K., and A. Maity. 2013. Precision farming for small agricultural farm: Indian scenario. *Am. J. Exp. Agric.* 3(1):200–217.

McLoud, P.R., R. Gronwald, and H. Kuykendall. 2007. *Precision Agriculture: NRCS Support for Emerging Technologies*, Agronomy Technical Note No. 1. East National Technology Support Center, Natural Resources Conservation Service, Greensboro, NC, p. 9.

Moran, M.S., Y. Inoue, and E.M. Barnes. 1997. Opportunities and limitations for image based remote sensing in precision crop management. *Remote Sens. Environ.* 61:319–346.

Mzuku, M., R. Khosla, R. Reich, D. Inman, F. Smith, and L. MacDonald. 2005. Spatial variability of measured soil properties across site-specific management zones. *Soil Sci. Soc. Am. J.* 69:1572–1579.

Navalgund, R.R., J.S. Parihar, S. Ajai, and P.P. Nageshwar Rao. 1991. Crop inventory using remotely sensed data. *Curr. Sci.* 61(3–4):162–171.

Nolan, S.C., T.W. Goddard, D.C. Penney, and F.M. Green. 1998. Yield response to nitrogen within landscape classes. In: P.C. Robert, R.H. Rust, and W.E. Larson (eds.), *Precision Agriculture: Proceedings of the Fourth International Conference*, ASA/CSSA/SSSA, Madison, WI, pp. 479–485.

Papiernik, S.K., M.J. Lindstrom, J.A. Schumacher, A. Farenhorst, K.D. Stephens, T.E. Schumacher, and D.A. Lobb. 2005. Variation in soil properties and crop yield across an eroded prairie landscape, *J. Soil Water Conserv.* 60(6):388–395.

Pierce, F.J., and P. Nowak. 1999. Aspects of precision agriculture. *Adv. Agron.* 67:1–85.

Pocknee, S., B.H. Boydell, D. Green, and C. Kvien. 1996. Directed soil sampling. In: P.C. Robert, R.H. Rust, and W.E. Larson. (eds.), *Proceedings of the Third International Conference on Precision Agriculture*. American Society of Agronomy, Crops Science Society of America, Soil Science Society of America. Madison, WI.

Precision Agriculture. 2003. http://www.precisionag.org/html/introduction.html. Accessed June 28, 2014.

Randhawa, M.S. 1983. *A History of Agriculture in India*. Vol. III (1757–1947). Indian Council of Agricultural Research, New Delhi, India, p. 422.

Rattan, R.K., and A.K. Singh. 1997. Role of balanced fertilization in rice-wheat cropping system. *Fert. News* 42(4):79–97.

Ravisankar, N., S.C. Pramanik, S. Jayakumar, D.R. Singh, N. Bibi, S. Nawaz, and T.K. Biswas. 2007. Study on IFS under different resource conditions of island ecosystem. *J. Farming System Research and Development* 13(1):1–9.

Ravisankar, N., S.K. Ambast, and R.C. Srivastava. 2008. *Crop Diversification through Broad Bed and Furrow System in Coastal Regions*. Central Agricultural Research Institute, Port Blair, India, p. 156.

Robert, P. C., R.H. Rust, and W.E. Larson. 1996. *Precision Agriculture: Proceedings of the Third International Conference on Precision Agriculture, June 23–26*. ASA, CSSA, SSSA, Madison, WI.

Sahoo, R.N., R.K. Tomar, S. Pandey, P.M. Sahoo, D. Chakaborty, and N. Kalra. 2007. Precision farming: Concept and application in Indian context. *Indian J. Crop Sci.* 2(1):25–27.

Sharma, R.K. 2003. Raised-sunken bed system for increasing productivity of rice based cropping system in high rainfall areas of Himachal Pradesh. *J. Indian Soc. Soil Sci.* 51:10–16.

Srivastava, R.C., and S.K. Ambast. 2009. *Water Policy for Andaman and Nicobar Islands: A Scientific Perspective*. CARI, Port Blair, India, p. 18.

Swaminathan, M.S. 1996. *Sustainable Agriculture: Towards an Evergreen Revolution*. Konark Publishers Pvt. Ltd., New Delhi, India, p. 219.

UN Secretariat. 2010. *Trends in Sustainable Development: Small Island Developing States (SIDS)*. United Nations Secretariat, Department of Economic and Social Affairs, Division for Sustainable Development, New York, p. 46.

Velayutham, M., D.K. Mandal, C. Mandal, and J. Sehgal. 1999. Agro-ecological subregions of India for planning and development. NBSS and LUP, Publication No. 35.

Velmurugan, A., T.P. Swarnam, and N. Ravisankar. 2006. Assessment of tsunami impact in South Andaman using remote sensing and GIS. *J. Indian Soc. Remote Sensing* 34(2):193–202.

Velmurugan, A., T.P. Swarnam, S.K. Ambast, N. Ravisankar, and T. Subrmani. 2014. Land degradation and its spatio-temporal changes induced by natural events in Andaman Islands. *J. Andaman Sci. Assoc.* 19(1):65–74.

Whelan, B.M., and A.B. McBratney. 2003. Definition and interpretation of potential management zones in Australia, In: *Proceedings of the 11th Australian Agronomy Conference*, Geelong, Victoria, Australia, February 2–6, 2003.

10 Innovations in Soil and Water Management/Conservation Research through Integrated Approaches of Nuclear and Isotopic Techniques and Precision Agriculture

F. Zapata, M. Zaman, M.L. Nguyen, L.K. Heng,
K. Sakadevan, G. Dercon, and L. Mabit

CONTENTS

10.1 INTRODUCTION

The world is facing an unprecedented dual challenge of enhancing food security while ensuring environmental sustainability, in particular the conservation of a natural resources base (soil and water) and genetic (plant and animal) resources. The present world population of 7 billion will exceed 9 billion by 2050 (UN-DESA 2013). The majority of this population increase will occur in underdeveloped and developing countries that already face food shortages. Simplistically, a 60% increase in the current agricultural productivity will be required from existing available resources (land and water) to feed the growing human population. Worldwide soil degradation is currently estimated at 1.9 billion hectares and is increasing at a rate of 5 to 7 million hectares each year (Lal 2006). This soil and land degradation causes not only a productivity decline and biodiversity loss but also affects vital soil/water ecosystem services, all of which are intricately linked with long-term social, economic, and environmental impacts (Bruinsma 2003; UNEP 2010). Moreover, several environmental drivers also affect land and water resources. Among these, major impacts are related to climate change and variability (Nguyen et al. 2011). All of these are likely to have negative impacts and induce changes on agroecosystems, thus placing increased pressures on dwindling land and water resources to produce sufficient food for present and future generations (Lal 2004; Verchot and Cooper 2008).

Sustainable agricultural development would require the combined use of soil, nutrient, and water management strategies that enhance crop productivity and at the same time promote environmental sustainability. In this context, there is a strong need for high-quality innovative research and soil-water-specific technologies that will address the most strategically important issues of soil/land and water management and conservation in agroecosystems (Nguyen et al. 2011).

The objectives of this chapter are to (1) provide an overview of the development and application of nuclear and isotopic techniques (NITs) in soil and water management-conservation in agroecosystems and (2) identify potential areas where the utilization of precision agriculture (PA) technologies can enhance further the effectiveness of NITs in soil and water management, leading to innovative research and soil-water-specific technologies. However, NIT applications in soil and water research can also enhance PA. This analysis is based on the main project activities of the Soil and Water Management and Crop Nutrition (SWMCN) subprogram of the Joint Food and Agriculture Organization (FAO) and International Atomic Energy Agency (IAEA) Division of Nuclear Applications in Food and Agriculture. This chapter is not an exhaustive review of PA and only focuses on NITs that act as building blocks of precise information for PA to address present and emerging issues related to soil (land) and water management/conservation in agricultural research. For detailed information on the publications originating from the research projects of the SWMCN subprogram, readers are referred to their website (http://www-naweb.iaea.org/nafa/swmn).

10.2 DEVELOPMENT AND APPLICATION OF NITs IN AGRICULTURE

10.2.1 USE OF NITs

NITs, which are also called nuclear-based techniques, comprise the use of stable (natural abundance and enrichment by artificial labeling) and radioactive (radiation-emitting) isotopes as well as radiation sources such as neutron and gamma density probes. For example, the soil moisture neutron probe (SMNP) is portable nuclear equipment used to monitor soil water content changes for constructing field water balance and defining irrigation scheduling. Gamma density probes are used to measure soil bulk density changes resulting from farm management practices such as tillage systems and animal stocking rate.

Isotopes are utilized as tracers that provide unique, precise, and quantitative data on nutrient and water pools and fluxes in the soil-plant-water systems and to assess the relative value of selected soil-water management technologies tailored to specific agroecosystems for improving soil fertility, crop productivity, and water use efficiency in crop and livestock production systems. Specific chemical sources and pollutants from these systems can also be traced using NITs. For example, ^{15}N-stable isotopic techniques can be used to measure rates of nitrogen (N) processes such as N mineralization-immobilization, nitrification and denitrification, biological N fixation, N use efficiency, and sources of N pollution in ground- and surface waters. For details on the principles and applications of these NITs in soil, water, and plant nutrient studies in agroecosystems, readers are referred to the IAEA Training Manuals (IAEA 1990, 2001) and a review paper (Nguyen et al. 2011).

Nuclear-based techniques are a complement and not a substitute to non-nuclear conventional techniques. They are applied in the context of agricultural research under field and greenhouse conditions when they offer comparative advantages over conventional techniques. However, they demand skilled and trained personnel and adequate laboratory facilities, in particular measurement equipment/techniques or alternatively financial resources for analytical services. In the case of radioactive isotopes, strict compliance with safety regulations and radiation protection procedures are required.

Nuclear-based techniques like any other techniques have advantages and limitations (Nguyen et al. 2011). It is, therefore, the task of a research team leader to assess the usefulness and effectiveness of the nuclear-based techniques to meet specific research objectives taking into account the team's available resources.

10.2.2 FAO/IAEA Programme of Nuclear Techniques in Food and Agriculture

In 1964, two United Nations organizations, the Food and Agriculture Organization (FAO) and the International Atomic Energy Agency (IAEA) established the Joint FAO/IAEA Division of Nuclear Techniques in Food and Agriculture at the IAEA headquarters in Vienna, Austria, to create and strengthen capacities for using nuclear-based methods to develop technologies for sustainable food security and to disseminate these through international cooperation in research, training, and outreach activities in Member Countries of the FAO and IAEA (IAEA 2014a; FAO 2014a). To achieve this mission, the Division has five discipline-oriented Sections, namely the SWMCN, Plant Breeding and Genetics, Animal Production and Health, Insect and Pest Control, and Food and Environment Protection. Each Section is linked to a corresponding laboratory located at the Agriculture and Biotechnology Laboratories (ABL) in Seibersdorf, near Vienna, Austria (FAO/IAEA 2014a).

This Joint FAO/IAEA Division implements its regular program through medium-term networked research projects called Coordinated Research Projects (CRPs) involving scientists from agricultural institutes in developing countries as well as Consultative Group for International Agricultural Research (CGIAR) institutions and advanced research organizations from industrialized countries (IAEA 2014b). Since its creation, the projects implemented by this Division have addressed priority issues of the agricultural research agenda and developed new technologies using nuclear-based methods for sustainable food security. Furthermore, transfer of the generated technologies and scaling of supportive services are offered by the IAEA and FAO Technical Cooperation Programmes (TCPs) (IAEA 2014c).

In its initial period (1964–1990), the Soils subprogram was named Soil Fertility, Irrigation and Crop Production reflecting the discipline-orientation of the projects and its focus on developing management practices for increasing the efficient use of agricultural inputs (soil, water and nutrients for crop production). Early projects dating back to the 1960s and 1970s were related to maximize the nutrient use efficiency of the applied chemical N and phosphorus (P) fertilizers by major grain crops (cereals) using ^{15}N- and ^{32}P-labelled fertilizers. After the 1974 oil (energy) crisis, the scarcity of chemical fertilizers and their high prices led to studies on enhancing the inputs of biological nitrogen fixation (BNF) as a source of N in agricultural systems. In 1995, the subprogram became SWMCN and its objective was to develop and promote the adoption of NITs technologies to diagnose constraints and pilot test suitable interventions/management practices through the integrated management of soil, water, and nutrient resources in agroecosystems (FAO/IAEA 2014b). Since then, the SWMCN subprogram has developed a wide range of NITs technologies and applied them successfully to cropping systems for sustainable soil-water and nutrient management, arresting land degradation, and climate change adaptation and mitigation (Chalk et al. 2002; Nguyen and Zapata 2006; Nguyen et al. 2010).

10.3 RECENT DEVELOPMENTS IN SOIL-WATER MANAGEMENT AND CONSERVATION RESEARCH IN AGROECOSYSTEMS

The continuing need to enhance food security and reduce climate change impacts demands an activity program that leads to sustainable soil and water management/conservation. This section provides a brief account of the investigations using NITs grouped into three main project areas: (1) sustainable intensification of crop production, (2) development of conservation agriculture systems, and (3) sustainable land (soil)/water management/conservation. Recent developments in these topics were reported at the international symposium Managing Soils for Enhancing Food Security and Climate Change Mitigation/Adaptation (FAO/IAEA 2014b).

10.3.1 SUSTAINABLE INTENSIFICATION OF CROP PRODUCTION

Intensification of agricultural production on prime agricultural land demands more refined management of external inputs of water and nutrients and thus an increased need for both nuclear and nonnuclear methods to develop better water and nutrient management practices in both rainfed and irrigated agricultural systems.

Several CRPs were conducted adopting an integrated approach to soil, water, and nutrient management in selected cropping systems of the main agroecological zones of the world. The investigations included judicious management of external inputs (nutrients and water) to improve their use efficiency and enhance soil productivity in cropping systems (Chalk et al. 2002; Nguyen and Zapata 2006; Nguyen et al. 2010). NITs such as ^{15}N, ^{32}P and ^{35}S isotopes were employed as tracers to develop fertilizer management practices (e.g., sources, timing, and placement) tailored to local conditions and specific cropping systems that improve nutrient use efficiency and enhance soil fertility (IAEA 2005, 2006, 2009). Moreover, the use of crop genotypes best adapted to the local soil/climate conditions was found to be a key requirement for ensuring the productivity and sustainability of the cropping systems. These studies using NITs, which initially focused on the search of crop genotypes with superior nutrient use efficiency, have demonstrated great potential to assist in the breeding and selection of suitable germplasm with tolerance to particular abiotic stresses (drought, flooding, low nutrient status, salinity, aluminum [Al] toxicity, etc.) (Nguyen et al. 2011).

Recent trends highlight an agroecological management through nutrient recycling from on-farm organic resources (animal manure and crop residues) and inputs of BNF in integrated crop-livestock production systems of smallholder farmers (FAO 2014b).

10.3.1.1 Integrated Nutrient Management

Extensive tracts of land worldwide, particularly those in the tropical and subtropical regions of Asia, Africa, and Latin America, contain acid fragile soils with inherent poor soil fertility, where nutrient imbalances, especially low N and P status, are the main factors limiting crop yields in agroecosystems. Moreover, intensive cultivation and inadequate application of nutrient sources (both mineral and organic) can lead to severe land degradation due to rapid decline in soil organic carbon (SOC) and then ultimately to low crop production, food insecurity, and extreme rural poverty. Lal (2001) reported that the depletion of SOC occurs at a rate of 2%–12% per year and the cumulative loss can be as high as 50%–70% of the original carbon (C) pool over a cultivation period of 10 years. Such negative impacts of SOC loss can be reversed through the development and implementation of an integrated nutrient management approach.

The development of an integrated nutrient management package involves not only manufactured fertilizers but also natural sources of nutrients such as phosphate rocks, BNF, and animal and green manures, along with the recycling of crop residues to provide a range of nutrients for plant growth and soil microbial activity. These play an important role in enhancing soil fertility–land productivity and improving soil organic matter and biodiversity (Nguyen 2014).

10.3.1.1.1 Nitrogen

Among the essential plant nutrients, N is the key component of all agricultural production systems because it is required by plants in large amounts for protein synthesis and virtually all aspects of plant growth. Thus, N plays a major role in crop productivity and profitability. N inputs under most farming systems come predominantly from the application of chemical (manufactured) fertilizers, which are expensive and imported commodities in the majority of developing countries. This is followed by BNF, with excreta of grazing animals and animal manure being the third most common source.

In developed countries, modern agriculture is dependent on the use of high-energy-intensive inputs (mechanization, irrigation water, and chemical fertilizers) for achieving and maintaining the high yields of new crop cultivars. It is reported that adequate food production, in particular cereals, for present and future populations will not be achieved without external inputs of N fertilizers (FAO 2012). There has been a steady increase in the use of chemical N fertilizers with a forecast demand of 194 M ton, of N worldwide in 2016.

The amount of fertilizer N to be applied (optimum application rate) can be obtained through a response curve by measuring crop yield and total N uptake to increasing application rates of fertilizer N. However to achieve a high fertilizer N use efficiency (FNUE), additional information such as timing, placement, and sources is required. These should be properly evaluated to select the best fertilizer management practices tailored to specific cropping systems and local agroecological conditions. To achieve this, both direct and indirect (reverse) dilution methods using ^{15}N-labeled fertilizers can be deployed in field experiments (Hauck and Bremner 1976; Van Cleemput et al. 2008).

The findings from the FAO/IAEA networked research projects worldwide and those of many other researchers using NITs under a wide range of environments suggest that a large percentage of the N fertilizer input is essentially being wasted because the average N recovery by the crop was reported to be less than 50% of the applied N (IAEA 1980, 1984; Zapata and Hera 1996; Mosier et al. 2004). Such low FNUE has been attributed to a number of factors and processes influencing its plant availability and losses from the soil-plant system. Moreover, this low FNUE represents not only an environmental degradation (N pollution) but also a substantial economic loss for the farmer and the country (Keerthisinghe et al. 2003).

Quantitative information about the fate of applied fertilizer N in soil-plant-water-atmosphere is therefore critical for enhancing FNUE and identifying appropriate soil and fertilizer management practices that optimize crop production and protect the environment (water bodies and atmosphere). This can be obtained by conducting sequential experiments using NITs to (1) determine the recovery of the applied fertilizer N by the crop as influenced by management practices (FAO 1980), (2) ascertain the fate of the applied fertilizer N in the soil-plant system and obtain an estimate of unaccounted losses, and (3) measure direct losses by gas production processes (e.g., ammonia volatilization and denitrification) and/or leaching to reduce negative impacts on the environment (IAEA 1980, 1984; Follett et al. 1991; Freibauer et al. 2001; Van Cleemput et al. 2008).

Most organic residues, particularly animal manures and plant residues, contain valuable plant nutrients that can be recycled back onto farmland to improve soil quality and health and reduce the need for chemical fertilizers (Nguyen 2014). However, the extremely variable physical and chemical characteristics of the organic residues influence to a great extent the rate of release of plant-available nutrients. Therefore, research attention has focused on the standardization of methods for the characterization of selected quality parameters and the creation of databases for the application of simulation modeling (Cadisch and Giller 1997; Palm 2001; Vanlauwe et al. 2002; ORD 2004).

NITs based on ^{15}N are employed for the study of N released of organic residues and its uptake by crops (Hood-Novotny et al. 2008). Nitrogen transformations such as mineralization (i.e., the conversion of organic N to mineral N) and immobilization (N release from organic residues) are complex processes that are influenced by a variety of microbial and enzymatic activities, soil type, organic and inorganic amendments, and environmental/management conditions (Zaman et al. 1999a,b, 2002; IAEA 2003b; van Kessel 2008). Both these processes can occur concurrently and hence nonisotopic techniques cannot precisely measure N release from soil organic N. The use of ^{15}N stable isotope can provide precise quantitative information on the rate of N being released over time and its subsequent consumption processes including plant uptake, soil retention, and losses from the soil-plant systems (Barraclough 1995; Di et al. 2000). In addition, the production of greenhouse gas (nitrous oxide [N_2O]) and nongreenhouse gas (N_2) and their precise source in the soil can only be determined by using ^{15}N (Mosier and Klemedtsson 1994; Zaman et al. 2008; Müller et al. 2014).

Although it is well recognized that the application of chemical fertilizers plays an important role in the intensification of crop production, the lack of affordable and adequate supplies of chemical fertilizers in the developing world remains the major constraint to crop production. Under these circumstances, it is essential to consider the use of alternative N sources such as BNF and organic residues to provide N for plant growth and to enhance soil quality and health.

Among the BNF systems, the symbiotic relationship between *Rhizobium* (bacteria found in the root nodules of the legume) and legume plants can provide significant N inputs to the plants and it is considered the most effective system for enhancing the productivity of agroecosystems. Numerous nonisotopic and isotopic methods have been tested and applied under field conditions to assess the symbiotic N fixation in legumes (Giller 2001; Hardarson and Atkins 2003; Jensen et al. 2008). The ^{15}N isotopic methods both at enrichment and natural abundance levels are described by Jensen et al. (2008). Extensive research using NITs to assess and enhance BNF in agroecosystems has been done under several FAO/IAEA projects (Hardarson and Atkins 2003).

Additional studies using NITs have been also conducted to assess the N inputs from BNF to the agroecosystem (Jensen et al. 2008). Some studies using ^{15}N showed that the below-ground N contribution from legumes can be substantial. For instance, Poth et al. (1986) found about 53% to 71% of the N fixed by pigeon pea over a period of 225–252 days (equivalent to 150 to 180 kg N ha^{-1}) was recovered in the soil after removal of the coarse roots. Chalk (1998) reviewed the dynamics of biologically fixed N in legume-cereal rotations and highlighted the need for wider use of ^{15}N-based methodologies to estimate additions of legume N to the soil and its effects on subsequent crops so that more accurate N balances in soil-plant systems can be made.

In spite of all these investigations, there is a need for more research targeting specific aspects of BNF oriented to the development and introduction of viable and cost-effective technologies to farming systems considering the particular needs and constraints of small-scale resource-poor farmers in the developing world (Hardarson and Broughton 2003).

10.3.1.1.2 Phosphorous

Most acid soils of the tropics and subtropics often have an inherent low plant-available soil P status and a high P sorption capacity, and thus P deficiency becomes the main constraint to agricultural productivity. Moreover, P depletion by continuous cultivation without replenishment is severely affecting agricultural production in sub-Saharan Africa. Approximately 75 kg P ha^{-1} has been lost over the last 30 years from 200 million ha of cultivated land in 37 African countries (Smaling et al. 1997). Therefore, P application is needed to improve soil P status and to ensure normal plant growth and adequate crop yields. The application of manufactured water-soluble phosphate (WSP) fertilizers such as superphosphate is usually recommended to correct P deficiencies. However, most developing countries import WSP fertilizers, which are often in limited supply and represent a major outlay for resource-poor farmers. Thus, it is imperative to explore alternative P inputs such as reactive phosphate rocks (RPRs). RPRs have been shown to be as effective as WSP fertilizers for arable crops and pasture systems grown under suitable conditions of soil pH and rainfall for RPR dissolution (Bolan et al. 1990; Rajan et al. 1996; Sale et al. 1997; Zapata and Roy 2004; Quin and Zaman 2012).

Phosphorus deficiency is also a major constraint for crop-livestock production in temperate grasslands where the low concentration of solution P available for plant uptake limits the continuous production of animal products and crops. Legume-based pastures rely on a regular supply of adequate available P so that the *Rhizobium* in the legume root nodules is able to fix atmospheric N (Nguyen et al. 1989).

In agricultural areas under intensive livestock production, the continuous use of external P inputs can lead to its accumulation in the topsoil and increased nonpoint (diffuse) pollution risks and high potential for eutrophication of water bodies if such accumulated P finds its way to streams and rivers (Sharpley et al. 2001; Chardon and Withers 2003; Hart et al. 2004). Isotopic techniques using radioactive ^{32}P with its short half-life (14.3 days) can provide significant insights on P dynamics under laboratory conditions (Nguyen 2000; Nemery et al. 2005; Stroia et al. 2007) which in turn provide quantitative information on the potential adsorption-desorption of P forms under different simulated field conditions (water depth, P concentration in solution, etc.).

Several key issues relating to soil and fertilizer P management for crop production in tropical agroecosystems have been investigated using ^{32}P isotopic techniques under a CRP (IAEA 2002b). The ^{32}P isotope exchange, a kinetics technique, that provides a comprehensive description of soil P status (i.e., intensity [Cp], quantity [El], and capacity [Q] factors were found to be a very valuable tool to assess P dynamics in the soil with or without the addition of P fertilizers such as WSP fertilizers and phosphate rock [PR]) (Morel and Fardeau 1991; Fardeau et al. 1995; Fardeau 1996). Several soil P testing methods were compared using the ^{32}P isotope technique as a reference. It was concluded that there is no single soil chemical P test that can be universally used to estimate available P in soils amended with PR and WSP fertilizers (IAEA 2002b). Investigations were also conducted to develop standardized protocols to characterize PR sources and evaluate their relative agronomic effectiveness and where necessary to find ways/means of enhancing their effectiveness (IAEA 2002b). ^{32}P and ^{33}P isotopic techniques have been extensively used in both laboratory and glasshouse experiments to measure P uptake and utilization from the applied P fertilizers, in particular RPR (Zapata and Axmann 1995; Zaharah and Zapata 2003). Adequate soil-plant-fertilizer management practices that can be put in place to enhance the efficient use of soil P and added external P inputs, in particular the application of RPR to build up the soil P capital in tropical acid soils, were identified and pilot-tested in field experiments (Zapata 1995, 2002b; IAEA 2002b; Zapata and Roy 2004). All this information was used to develop databases and a decision support system (DSS) to provide better informed recommendations to soil/fertilizer practitioners, land managers, and policy makers (FAO/IAEA 2014c).

10.3.1.1.3   Sulfur

Sulfur (S) is an essential nutrient for plant growth and animal production. S deficiencies in crops and pastures can occur when replacement by fertilizers does not meet demands by cropping, such as following intensification of farming systems or change in crop types (e.g., legumes and brassica plants remove more S than cereals) (Nguyen and Goh 1994a).

Several studies have used both ^{35}S (radioactive) and ^{34}S (stable) as isotopic tracers to follow the pathways of S in soil-plant or soil-plant-animal systems and to construct S budgets in grazing systems (Till 1981; Chen et al. 1999). A direct method involving the use of ^{35}S-labeled materials has been applied in studies to determine the S uptake and recovery of ^{35}S-labeled fertilizer and animal excreta in flooded rice and pastoral systems (Samosir et al. 1993; Nguyen and Goh 1994b), the availability of subsoil sulfate using ^{35}S-labeled gypsum to crops (Bole and Pittman 1984; Nguyen and Goh 1994a), and the relative performance of different S fertilizers (Goh and Gregg 1982; Nguyen and Goh 1990).

When ^{35}S labeling of fertilizers, animal manure, or crop residues is not appropriate, the ^{35}S reverse dilution technique has been employed to determine (1) the ability of plants to acquire S from the atmosphere (Hoeft et al. 1972), (2) the release of S from elemental S sources, sulfate sources, and its uptake by plants (Shedley et al. 1979), (3) the sources of S taken up by ryegrass and measured by chemical extraction (Chinoim et al. 1997), and (4) the time course of S uptake from the S-coated urea by crops (Yasmin 2003).

The ^{34}S natural abundance has been used extensively to identify sources and to trace the fate of S in the environment. This approach requires that the S isotopic signatures of the different sources are known so that the contribution of S from the sources can be apportioned (Krouse 1977; Mayer et al. 1995; Alewell et al. 1999; Novak et al. 2001). This technique has been applied to study long-term changes in S deposition in the Broadbalk experiment in the United Kingdom (Zhao et al. 2003). Ion exchange membranes have been successfully used to collect soil water sulfate for investigating its isotopic (sulfur and oxygen isotope fractionation) composition (Kwon et al. 2008). For detailed information, readers are referred to the IAEA publication *Guidelines for the Use of Isotopes of Sulfur in Soil-Plant Studies* (IAEA 2003a).

10.3.1.1.4   Micronutrients

The seven micronutrients, including boron (B), copper (Cu), chlorine (Cl), iron (Fe), manganese (Mn), molybdenum (Mo), and zinc (Zn), are equally important to plant growth such as macronutrients in soils; however, they are required in very small quantities of only a few mg/kg in plant tissues. Micronutrients play many key roles in plant nutrition, enzyme systems, oxidation-reduction reactions, photosynthesis, and plant production. Increasing micronutrient deficiency is becoming a matter of concern due to their impact on agricultural production, in particular in intensively cultivated systems in Southeast Asia. Among the micronutrients, Zn deficiency is the most acute followed by B. According to Sillanpää (1990), Zn deficiency is the most commonly occurring micronutrient deficiency problem limiting crop growth in many parts of the world. In India, application of Zn resulted in spectacular yield increases in wheat-growing areas (Takkar et al. 1989). Studies using NITs such as ^{65}Zn reported improved Zn nutrition of flooded rice in Southeast Asia (IAEA 1983). In view of the widespread soil degradation and impacts of climate change on nutrient imbalances (deficiencies and toxicities) and their subsequent effects on human malnutrition (also called hidden hunger), especially in infants and children, there is renewed attention on micronutrient studies in soil-plant-man systems. Recent research focuses on the use of isotope and related techniques in micronutrient studies at several levels due to the urgent need to develop cost-effective interventions to control/mitigate these nutrient imbalances in the developing world (IAEA 2014d).

10.3.1.2   Agricultural Water Management

Water is the most precious resource that supports life on the planet and connects the various components of the ecosystems. Agriculture is the largest user of freshwater, accounting for about 75% of the global freshwater use. Cropland under irrigated agriculture contributes approximately 40% of world food production while the remaining 60% comes from the cropland under rainfed agriculture. Rising demand for food and livestock feed together with the use of biofuels and impacts from global climate change are placing a tremendous pressure on freshwater resources (Molden 2007).

Managing agricultural water to enhance crop water productivity (more crops per drop) and water use efficiency in crop production systems is therefore of paramount importance for both rainfed and irrigated agriculture. This can be accomplished by (1) increasing the marketable yield of crop for each unit of water transpired, (2) reducing all outflows (e.g., soil evaporation, drainage, seepage, and deep percolation), and (3) increasing the effective use of rainfall, stored water, and water of marginal quality (Molden 2007). In agroecosystems, due attention should be paid to the soil-water interactions and their influence not only on food production but also on the provision of essential ecosystem services (UNEP 2011). Therefore, it is envisaged that soil and water management/conservation for sustainable crop production will need to be smarter and adopt elements of PA. A paradigm shift will be necessary to change from supply-driven to a more demand-driven water management to conserve water and improve its use efficiency on-farm to produce more crops per drop of water in water-limited environments.

Under rainfed conditions, the storage of rainwater in the root zone is the critical factor for plant growth, which requires integrated land-water management practices (Rockström et al. 2007). Studies conducted by IAEA using NITs and related techniques have demonstrated that there is considerable scope to improve water use efficiency and crop productivity in rainfed agriculture. This can be achieved through the appropriate integration of soil-water-plant technological options such as conservation agriculture practices (e.g., zero or reduced tillage, mulching, crop residue retention, crop rotation, intercropping), water harvesting techniques, and improvement of the soil fertility status. Rainfed agriculture systems can be further improved by appropriate crop management such as use of crop varieties adapted to drought and saline conditions, use of deep-rooted crops to access soil water storage at deeper soil depths, or using early flowering and shorter season crop varieties, and changing sowing dates or planting density to suit the local conditions (IAEA 2005).

Conservation agricultural practices such as no-till, minimum tillage, tied ridge, and mulching can reduce soil or water losses under rainfed agriculture. Developing on-farm water storage facilities or having water conservation zones to provide supplementary irrigation to crops whenever possible can significantly minimize the risks of rainfed agriculture (Rockström et al. 2007). These risks can be further reduced if the preseason rainfall forecasting can be predicted with anticipated outcomes of different management decisions (Cooper et al. 2008).

SMNP, which measures slow neutrons after the collision of fast neutrons (emitted by the neutron radioactive source in the SMNP) with hydrogen atoms in the soil water, is an instrument that is well suited to precisely determine field-scale soil water content as well as to evaluate the impacts of different tillage systems in conserving soil moisture for crop production (IAEA 2008b). Isotopic techniques (^{18}O and ^{2}H) that quantify soil evaporation (E) and crop transpiration (T) fluxes are important research tools to determine their relative magnitudes and design better strategies for water management under different rainfed conditions so that such losses can be eliminated completely without affecting the rate of T (Williams et al. 2004; Heng et al. 2014).

Under irrigated agriculture, the combined use of PA and NITs may play a key role in improving water conservation and environmental protection (Nguyen 2014). Precision irrigation (PI) is defined here as the efficient, timely, and correct amount of water delivered to fields to maximize crop yield and quality, and to minimize environmental impacts, including the application of variable amounts of water over a field in response to spatial crop and soil heterogeneities (Fereres and Heng 2014). Three important prerequisites are necessary for successful application of irrigation: (1) the use of modern technologies (i.e., changing the method of irrigation to increase the efficiency of application), (2) knowledge of crop water requirements (evapotranspiration) with a certain degree of accuracy, and (3) the ability to effectively monitor the water status of the root zone so that precise irrigation frequency and the depth of application can be determined.

Improving irrigation scheduling to precisely determine the date and amount of irrigation is the first step toward optimizing on-farm water management and improving PI.

Accounting for soil variability across a field as well as nonuniform crop growing conditions can be met through the option of applying variable amounts of irrigation water within the field to match the requirements of every zone of the field with minimal environmental consequences compared with what a uniform irrigation rate applied over a variable field would have. Changing the method of irrigation from gravity and open channels to pipe network and/or pressurized systems (sprinkler or drip) can significantly reduce irrigation water use. Drip irrigation supplies water directly to the plant rooting zone and can thus cut water use by up to 50% while maintaining or even increasing crop yields (IAEA 2002a). Applying fertilizers and irrigation water together (fertigation) through drip irrigation system has the potential to further maximize irrigation and fertilizer use efficiencies (IAEA 2002a). In recent years, low-cost, small-scale drip-irrigation systems are encouraging farmers to adopt this technology to improve their water use efficiency on farms. Drip irrigation systems have the greatest potential to increase farm incomes, especially those of smallholders in developing countries through reduced water costs and increased crop yields (FAO 2001; Postel et al. 2001).

In a recently completed CRP project, integrated soil-water-plant approaches and recent advances in isotope techniques were utilized to better manage irrigation water for enhancing crop productivity under water-limiting conditions. Stable isotopes of water (^{18}O and ^{2}H), SMNP, and related conventional techniques (e.g., microlysimetry, sap flow) were used to quantify soil evaporation and transpiration fluxes at different stages of crop development. Data generated in the project is being used to validate the FAO's Aquacrop model (Raes et al. 2009; Steduto et al. 2009) for developing improved irrigation water, soil, and crop agronomic practices to achieve water productivity improvements (Heng et al. 2014).

Accurate knowledge of water losses via E and T—collectively known as evapotranspiration (ET)—is also necessary for effective water management on-farm. Isotopic techniques (using ^{18}O and ^{2}H) that quantify E and T are important research tools to determine the relative magnitudes of E and T in different situations (Williams et al. 2004; Heng et al. 2014). Quantifying E and T under surface and subsurface drip irrigation systems can lead to developing guidelines that enable farmers to realize water saving by minimizing soil evaporation.

Shifting from crops with high irrigation requirements, such as cotton or rice, to those that have much lower water needs, such as high-value vegetable crops, can cut water use, particularly in water-scarce areas where growing of traditional field crops under irrigation is no longer economically viable (Molden 2007).

The carbon isotopic discrimination (CID) technique is based on discrimination of the heavy isotope of carbon (^{13}C) in favor of the lighter and more abundant isotope (^{12}C) during the physical diffusion of CO_2 through the leaf stomata and subsequent enzymatic decarboxylation. Thus, CID is a time-integrated index of photosynthetic activity and is related to plant water use or transpiration efficiency that can be used as a surrogate marker for crop water use efficiency (Condon et al. 2002, 2004; IAEA 2005). This technique has been employed in a CRP on the selection of wheat genotypes tolerant to drought and irrigated rice genotypes tolerant to salinity for greater agronomic water use efficiency under a wide range of environments (IAEA 2012).

Simulation models are utilized to evaluate actual irrigation management and to identify new approaches to improve water use efficiency under both irrigated and rainfed conditions. Simulation models also provide insights to investigate the complexity of different climate change scenarios in terms of altered water and temperature regimes and elevated carbon dioxide concentration in the atmosphere (Steduto et al. 2012).

A major challenge is to assess the average soil water status over large areas because the use of point observations of soil water or plant water status is not feasible to reduce variability problems. The Cosmic-Ray Neutron Probe (CRNP) is a new instrument based on the collision of fast neutrons from the upper atmosphere with hydrogen atoms in the soil that offers potential in assessing area-wide soil moisture status (Zreda et al. 2008; Shuttleworth et al. 2010). This CRNP does not contain a radioactive neutron source so it can be transported easily and left in the field unattended, providing measurements of area-wide scale over a circle of about 600m diameter (~30 ha in area) and over depths varying between 15 and 70 cm (Zreda et al. 2008, 2012; Franz et al. 2012). Monitoring soil water status at these scales allows the integration of variations caused by differences in soil-crop properties and distribution of irrigation water. Several other uses could be implemented (e.g., early warning for flood events in mountainous areas). A sequence of observations over time also permits the computation of the components of the field water balance if the appropriate inputs and outputs are recorded in databases. This technique could also be used to calibrate remote sensing data but this aspect is still under investigation. A very recent development of a cosmic-ray rover (Desilets et al. 2010) with a principle similar to that of the stationary CRNP allows the assessment of average soil moisture with high spatial resolution along the path of the probe to be determined (Chrisman and Zreda 2013).

10.3.2 Development of Conservation Agriculture Systems

Conservation agriculture (CA) refers to the sustainable intensification of agricultural production systems with an integrated approach to improve crop productivity for food security as well as to restore soil quality and enhance its resilience against degradation and risks associated with climate change impacts. This is achieved following best management practices such as (1) the strategic application of the required amount and ratio of essential plant nutrients at the right timings to meet crop nutrient demand and minimize losses, (2) minimum mechanical soil disturbance by tillage/cultivation, (3) retention of crop residues (mulching), and (4) the use of crop rotations/plant associations, including cover/green manure crops (FAO 2014c). Such conservation systems are currently practiced in over 100 million ha to enhance the food security of small holders in the developing world (Derpsch and Friedrich 2009).

10.3.2.1 Improving Soil Organic Matter and Enhancing Soil Quality

Increasing and preserving soil organic matter (SOM) plays a key role in improving soil fertility and quality and increasing crop production while ensuring long-term sustainability of agricultural ecosystems. Increased SOM also plays a key role in the global C cycle by acting as a sink for atmospheric CO_2 (Paustian et al. 1997). This combination of CA practices mentioned above, when applied on a continuous basis over a period of time, provides a means of enhancing soil C sequestration on agricultural lands and thus, sustaining and increasing SOM levels and providing ecosystem services for enhancing crop production and contributing to environmental sustainability (Lal and Kimble 1997; Lal 2007).

In an FAO/IAEA CRP entitled "Integrated soil, water and nutrient management in conservation agriculture," the influence of soil, water, and crop management practices on SOM accumulation and its subsequent impacts on soil water, nutrient, and C dynamics were investigated using NITs in various cropping systems worldwide. The results from this global project demonstrated that CA can bring benefits such as increased soil moisture retention, BNF, N retention, and soil C sequestration. However, these effects were highly variable and site-specific, in some cases the benefits being negated by the influence of crop residues on plant diseases that could reduce crop yields and quality. One of the major lessons, with great implications for adoption strategies, is that CA can only be sustainable and successfully implemented if specific local constraints such as soil compaction, low soil fertility, and lack of SOM are first removed (Dercon et al. 2010).

Further research is therefore needed to develop and pilot-test specific packages of integrated technologies and practices tailored to targeted agroecological zones and local agronomic management. Furthermore, this information is required for a comprehensive assessment of socioeconomic and environmental benefits and the development of appropriate policies to facilitate and encourage the adoption of CA by farmers. With this in mind, another CRP entitled "Soil quality and nutrient management for sustainable food production in mulch-based cropping systems in sub-Saharan Africa" was initiated in 2011 to investigate the potential of mulch-based cropping systems to enhance soil resilience against degradation and climate change risks and to increase soil fertility for sustainable food production in sub-Saharan Africa (SSA) (Nguyen et al. 2011). Stable isotope techniques (^{15}N and ^{13}C) at enriched and/or natural abundance levels will enable an in-depth analysis and understanding of basic soil biological-physical processes, including soil C and nutrient cycling in mulch-based cropping systems. The selection and characterization of benchmark sites in the moist and dry savannahs will provide a platform for extrapolation of results to other relevant agroecological zones in SSA (FAO/IAEA 2014b).

10.3.2.2 Greenhouse Gas Emissions and Mitigation Options Including C Sequestration

Among the three greenhouse gases (GHGs)—N_2O, carbon dioxide (CO_2) and methane (CH_4)—emission of N_2O from agricultural soils has received particular attention. Increased and inefficient use of chemical N fertilizers, dairy manures and irrigation water, high stocking rate (number of grazing livestock per hectare), and intensive cultivation are the major influencing factors for increased N_2O emissions into the atmosphere. Soil can be either a source or sink of CO_2 emission depending on land use while livestock and flooded rice systems can be a major source of CH_4 emissions.

Mitigation options to reduce both N_2O and CO_2 emissions from farmlands need a holistic approach by integrating farm management practices that include enhancing nutrients and water use efficiencies on farm, sequestering more C, managing animal grazing, and improving soil fertility (physical, chemical, and biological).

Nitrogen cycling processes occurring in the soil-plant-water interface including N use efficiency and uptake by plants, N retention in soil, and losses to groundwater and atmosphere are very complex and therefore cannot be precisely measured by nonisotopic techniques. NITs such as ^{15}N can help to quantify and identify the relative importance of these processes. For example, strategic use of chemical fertilizers and organic materials (i.e., applying the right amount of N at the right time in the right way) could lead to improved nutrient use efficiency (Zaman et al. 2013). Enhancing biological N fixation through legumes minimizes the use and dependence on chemical fertilizers (Ledgard et al. 1999).

N transformations such as nitrification (both autotrophic and heterotrophic), denitrification, dissimilatory nitrate reduction to ammonium (DNRA), and conversion of organic N to mineral N are considered the major microbial processes that produce greenhouse N_2O and nongreenhouse gas (N_2) and they can occur concurrently in a given soil system. Measuring the contribution of N_2O production from each of these microbial processes is a key element for recommending specific mitigation options. Moreover, the International Panel on Climate Change (IPCC) uses a default global emission factor of 1% for fertilizer-induced emission (FIE). However, a number of studies have shown that the amount of N lost as N_2O (1% to 20% of applied N) varies with soil type, N inputs, and soil and crop management practices, and is often greater than the default value (1%) of the IPCC. The available conventional techniques such as acetylene (C_2H_2) inhibition (Zaman and Nguyen 2010) and the closed chamber methods cannot accurately measure both N_2O and N_2 and their sources in a given soil. The use of ^{15}N offers the best option to quantify both N_2O and N_2 concurrently and to identify their precise sources in soil (Mosier and Klemedtsson 1994; Zaman et al. 2008; Müller et al. 2014).

Soil C sequestration is one of the strategies to offset anthropogenic CO_2 emissions by capturing atmospheric C through photosynthesis and storing it in soil. Carbon sequestration is also a key factor in enhancing soil fertility, nutrients, and water retention, and improving soil quality and health. Appropriate farm management practices and land uses can enable agricultural soils to be a net sink for sequestering atmospheric CO_2 and other GHG (Lal and Kimble 1997; Paustian et al. 1997; West and Post 2002). For example, CA practices such as mulching, cover crops, zero or minimum tillage, strategic use of manures, and biochar add more C and nutrients into soil, which could lead to improved soil structure and its drainage and hence reduce both N_2O and CH_4 emissions. With the help of the stable isotope ^{13}C in soil C sequestration studies, one can assess the relative contribution of C from different plant species (C_3 and C_4) (Nguyen et al. 2011).

Enhancing water use efficiency on-farm is another strategy that could potentially reduce GHG emissions from soils. The differences in soil types, fertility, texture, and land use management practices result in significant variability in soil water content after irrigation/rainfall (Zaman et al. 2012). This patchy distribution of applied irrigation water can enhance localized high emission potential for GHG. It is well known that peak emissions of GHG are associated with areas of high fertility (Parkin 1987), so-called hot spots, or temporary areas of high GHG production potential (Groffman et al. 2009; Müller and Clough 2013). Thus it is important to understand the conditions under which these critical source areas of high emission potential develop. Therefore, to optimize agricultural management practices to reduce GHG emissions, a holistic approach is needed. The type of irrigation system is crucial to avoid conditions of high GHG emission potential. Generally, irrigation systems that provide evenly distributed soil water, such as drip irrigation systems, seem to have a lower potential for GHG emissions than systems that are irrigated with a traveling gun (Bruckler et al. 2000; Scheer et al. 2008; Lv et al. 2014). This was confirmed by a recent study where both conventional N treatments compared with a fertigation system (drip irrigation with N fertilizer solution) showed very low emission factors of N_2O emissions (<0.1% of N applied). Net methane (CH_4) oxidation in this system was reduced when ammonium (NH_4^+) was present, which is related to stimulation of N dynamics (Vallejo et al. 2014).

10.3.3 Sustainable Land (Soil) and Water Management/Conservation

The continuing need to enhance food security and reduce the impacts of climate change demands an action plan that leads to sustainable land (soil) and water management/conservation in area-wide (landscape and watershed scale) studies. Natural resources base land (soil) and water must be managed in an integrated way to ensure the provision of ecosystem services and environmental sustainability.

10.3.3.1 Soil Erosion Assessment and Control

Soil losses by water and wind erosion are one of the most serious threats affecting world food production. Each year, about 10 million hectares of cropland are lost due to soil erosion (Pimentel and Burgess 2013). As a result of climate change and variability, the soil losses by erosion are likely to increase further (Yang et al. 2003; Nearing et al. 2004).

Soil erosion is the main mechanism of land degradation worldwide, having major impacts on water and biogeochemical cycles, plant primary productivity, and biodiversity. It is therefore important to deploy effective soil conservation measures to counteract soil erosion losses. Measuring soil erosion and identifying the sources of erosion are the key elements for designing effective soil conservation measures. Reliable data on the extent and actual rates of soil redistribution (erosion and associated sedimentation) is needed to provide a comprehensive assessment of the magnitude of the erosion problems and to underpin soil conservation measures, including the assessment of their economic and environmental impacts. The quest for techniques of soil erosion assessment to complement existing classical methods and to meet new requirements has led to the development of NITs based on the use of fallout radionuclides (FRNs), also called environmental radionuclides (Nguyen et al. 2011).

Initial IAEA investigations focused on the refinement and standardization of the Cesium-137 (^{137}Cs) technique for its worldwide application in agricultural landscapes under a range of environmental conditions (Zapata 2002a, 2003). Several conversion models for deriving soil erosion/sedimentation rates from FRN measurements were developed, tested, and refined (Walling et al. 2002, 2011). The results from this research paved the way for extending not only the ^{137}Cs technique but also other FRNs as tracers for soil erosion/sedimentation investigations. In a follow-up project, combined FRN techniques involving beryllium-7 (^{7}Be), ^{137}Cs, and lead-210 (^{210}Pb) as soil/sediment tracers were further developed to document short-term (<30 days), medium-term (~40 years), and long-term (~100 years) soil redistribution (erosion and deposition) rates and spatial patterns in the landscape, respectively, under different conditions (climate, soil, topography, and land uses). This combined application has shown that they are powerful tools to assess the relative effectiveness of soil conservation measures on soil erosion and land degradation (IAEA 2011, 2014f; Dercon et al. 2012). Since then, FRNs have been used worldwide to obtain rates and patterns of soil erosion and deposition at several temporal and spatial scales (Zapata 2002a; Zapata and Nguyen 2009; Mabit et al. 2008, 2013; Taylor et al. 2013).

The IAEA TCP through national and regional TCPs is assisting developing Member States in the establishment and strengthening of their human and institutional capacities as these were essential requirements for the successful application of the FRN techniques to enhance sustainable agriculture and minimize land degradation. Under regional TC project RAS5045 "Sustainable land use and management strategies for controlling soil erosion and improving soil and water quality" involving 14 Member States in the East Asia and the Pacific region, FRN methodologies have been successfully used to assess soil erosion and evaluate soil conservation measures as well as better understand the link between soil redistribution and soil quality (e.g., soil organic matter) in the landscape (IAEA/RCA 2010). Another regional TC project, RLA5051, "Using environmental radionuclides as indicators of land degradation in latin american, caribbean and antarctic ecosystems," was implemented to enhance soil conservation and environmental protection in the ecosystems of the region in order to ensure sustainable agricultural production and reduce the impacts of land degradation (Dercon et al. 2012).

10.3.3.2   Areawide (Watershed) Land and Water Conservation

The resulting soil redistribution across the landscape involves not only on-site effects of soil erosion but also off-site sedimentation problems. Hence, the impacts of soil erosion in agricultural lands need to be investigated at the watershed (areawide) scale. These studies involve links (connectivity) between upstream-downstream and upland-lowland environments. Connectivity issues between agricultural and natural ecosystems should also be taken into account. Therefore, there was a need to develop comprehensive areawide sediment budgets for better understanding of sediments mobilization and their source-sink transfer and storage in watersheds. It should be noted that most studies using FRNs have been conducted at the field/landscape scale, although some work has been undertaken in small basins ranging from a few hectares to a few km^2 (e.g., Mabit et al. 2007).

In a recently completed CRP, "Integrated isotopic approaches for area-wide precision conservation to control the impacts of agricultural practices on land degradation and soil erosion," integrated isotopic and conventional approaches were studied and developed to support the implementation of precision conservation at the watershed scale (Nguyen et al. 2011; Dercon et al. 2012).

FRN techniques were successfully applied to develop sediment budgets on an areawide basis (catchment) over different time scales. Furthermore, compound-specific stable isotope (^{13}C and ^{15}N) techniques (CSSI) (Gibbs 2008) have been utilized not only to identify the sources of soils in sediments (fingerprints) but also to apportion their relative contribution from different land uses in catchments worldwide (Nguyen et al. 2011). Protocols for using CSSI of carbon as well as other fingerprints associated with plants, animal manure, and soil samples have been developed to identify the sources of soil loss/sediment production (Blake et al. 2012; Gibbs and Mabit 2012; Gibbs 2014).

Such integrated application will help identify critical source areas (hot spots) of soil loss/ sediment production and assist land managers/practitioners/farmers to target appropriate soil conservation measures in agroecosystems, and thus provide effective guidelines for areawide sustainable management of land and water resources (Arbuckle 2013; Gibbs 2014). The integrated use of FRNs and CSSI is being explored in Africa, Asia, and Latin America through IAEA-funded regional TC projects RAF5063, RAS5055, and RLA5051, respectively (FAO/IAEA 2014b).

10.3.3.3 Areawide Soil-Water Conservation for Pollution Control

In many regions across the world, water conservation zones (specific areas within landscapes that are used to capture and store water and nutrients) are considered as water resource management strategies that have been developed to prevent downstream water quality degradation, improve soil water status, promote flood mitigation, and enhance groundwater recharge (Greenway 2004). They include (1) farm ponds, (2) constructed and natural wetlands, and (3) riparian buffer zones that can be used collectively or individually in agricultural catchments (Burns and Nguyen 2002; Rutherford and Nguyen 2004; Aye et al. 2006). Farm ponds add value to other farming activities such as serving domestic and livestock water supplies as well as irrigation for high-value crops and vegetables (Sakadevan et al. 2014).

NITs play an important role in identifying the sources of water captured and stored in water conservation zones. Studies conducted in Tunisia using ^{18}O and ^{2}H isotopic signatures showed that water conservation zones are the major source of water for groundwater recharge (Sakadevan et al. 2014). In Iran, isotopic signatures of water (^{18}O and ^{2}H) in runoff, rainwater, and stream water showed that more than 90% of water captured by water conservation zones is by surface runoff during rainy periods (Sakadevan et al. 2014).

Installation and restoration of wetlands in agricultural catchments is an important strategy to help sustainable agricultural development while reducing the impacts of nonpoint source pollution from agricultural activities to aquatic systems (Matheson et al. 2002, 2003; Zedler 2003). Wetlands have been shown to retain widely variable amounts of nitrogen (30%–99%) and phosphorus (0%–99%) and reduced their input to aquatic systems (Moreno-Mateos and Comin 2010).

Riparian buffer zones are widely accepted for reducing nonpoint source pollution from agricultural landscapes. Studies carried out in Estonia have shown that riparian buffer zones with alder trees removed between 170 and 350 kg N ha^{-1} from the incoming water. The use of ^{15}N natural abundance in these studies showed that more than 60% of N was removed by denitrification mainly as N_2 gas, thus reducing both nutrient pollution to downstream water and GHG emissions (Sakadevan et al. 2014). The effectiveness of riparian buffer zones for removing N is controlled by specific factors that include water flow, N removal by denitrification, and hydrogeology of the site (Noij et al. 2012). In the last few years the design and placement criteria of the water conservation zones has been continuously improving to optimize nutrient removal (Diebel et al. 2008; Sakadevan et al. 2014).

10.3.3.4 Areawide Management of Salinization

There is a need to control salinity buildup under irrigation in the long run, especially in areas where annual rainfall is insufficient. Soil salinization is one of the major causes of soil degradation threatening productive lands (FAO 2011). Recent data showed that about 11% of the irrigated land has been salt-affected globally, and up to 50% in some irrigated regions in Central Asia, North Africa, South Asia, and the Arabian Peninsula. Major impacts of soil and water salinization include (1) reduced crop yields, (2) flooding, runoff, soil erosion and desertification, (3) reduced suitability of surface and groundwaters for domestic and agricultural uses, (4) increased cost of water treatments, (5) replacement freshwater plants with halophytes, and (6) reducing recreational and commercial fisheries (Sakadevan and Nguyen 2010). Efficient investment in salinity mitigation requires an understanding of how salinity responds to alternative land and water use/management options at the field and landscape scales (Sakadevan and Nguyen 2010).

Investigations are being currently conducted under a CRP titled "Landscape salinity and water management for improving agricultural productivity" to address soil and water salinity in agricultural landscapes and to optimize the use of salt-affected soils and saline water through innovative soil, water, and crop management technologies and practices. A suite of NITs such as ^{13}C, ^{2}H, ^{18}O, SMNP, and CRNP is employed to quantify E and T, improve irrigation scheduling, minimize water losses under saline conditions, and assess salt tolerance of crops.

10.4 FURTHER DEVELOPMENTS ON SOIL AND WATER MANAGEMENT AND CONSERVATION RESEARCH IN AGROECOSYSTEMS

10.4.1 BASIC CONSIDERATIONS

Small-scale farmers who grow crops and rear animals are the main source of currently produced food supplies and the mainstay of economic development in the developing world. Fostering smallholder agriculture is therefore vital to reduce poverty and enhance food security as well as to ensure environmental sustainability. Recent studies highlight the recycling of locally available resources for an integrated agroecological management of smallholder crop-livestock production systems (FAO 2014b).

NITs are employed as tracers in mechanistic studies to achieve a better understanding of relevant processes and their main influencing factors occurring in the soil-water-plant/animal systems so that a knowledge-based management can be implemented. NITs can be also utilized to assess the relative value of selected interventions (management practices) of agricultural inputs (right product at the right time and place) suitable for achieving high-use efficiency and enhancing crop yields and/or controlling potentially harmful environmental impacts. Thus, NITs are mostly utilized in combination with the conventional (classical) techniques (not in isolation) when they offer comparative advantages and adequate human and institutional capacities are available. As such, the main counterpart institutions of the FAO/IAEA program are the National Agricultural Research Systems (NARS) with active research programs. These are commonly located in the organizational chart of the Ministry of Agriculture or Rural Development of their countries. Table 10.1 displays selected applications of NITs in FAO/IAEA CRPs relating to soil and water management/conservation. The references contain the contributions from the participating NARS in the network and a summary of results and main conclusions of the CRPs.

Moreover, appropriate coordination should be established with relevant organizations of the agricultural/rural sector for the transfer and application of the generated technologies by the small-scale farmers. Outreach activities should also be planned and implemented to deliver the generated information to policy and decision makers (IAEA 2014b,c).

With regard to the management/conservation of land (soil) and water (natural resources), the mandate lies with the Department of Natural Resources and Environment of FAO (FAO 2014c). Established national research programs can be found in the United States and Canada where a number of land and water investigations in landscapes and watersheds relating to environmental issues have been implemented. In view of the need to develop more efficient and cost-effective technologies for soil and water conservation, significant advances in precision (target) conservation are reported (Delgado and Berry 2008; Delgado et al. 2011). In the European Union (EU) countries, attention has recently focused on land and water use/management and conservation for monitoring and control of Good Agricultural Practices (GAP) under the Common Agriculture Policy (CAP) reform 2014–2020 (EU 2014).

However, most countries in the developing world do not have established land (soil) and water conservation programs. Sometimes short-term monitoring studies are implemented under specific watershed development or soil conservation projects in a given area when financial support from international aid is available (Sheng 2007). It is also important to consider that the ministry of environment is often responsible for land and water management/conservation issues. Some developing countries where water issues are vital for their socioeconomic development have an independent ministry of water to ensure the sustainable use/management of water resources and their adequate governance with participation of all stakeholders.

The United Nations Office for Outer Space Affairs (UNOOSA) is implementing a Program on Space Applications to support developing countries in incorporating space-based technologies in Natural Resources Management and Environmental Monitoring (i.e., inventory, survey, and monitoring studies in agriculture, hydrology, geology, mineralogy, land use and cover, and ecology/environmental sciences). This program also includes activities addressing climate change issues (UNOOSA 2014).

TABLE 10.1

Selected NITs Applications under FAO/IAEA Projects

NITs	Investigations Topics/Issues	Key References
^{15}N, ^{32}P	Influence of fertilizer management practices such as sources, timing, and placement, on N and P use efficiency for major grain crops	FAO 1980
^{15}N	i. Influence of plant legume genotypes, *Rhizobium* strains, agronomic practices, and environmental factors on biological nitrogen fixation (BNF) in *Rhizobium*-legume systems: grain legumes, pastures/forages, and trees	Hardarson 1994; Hardarson and Atkins 2003
	ii. *Azolla* and blue green algae (BGA)/N fertilization of rice systems	Kumarasinghe and Eskew 1993
	iii. Fate of fertilizer N in the soil-plant system and N cycling in cropping systems	IAEA 1980, 1984; Zapata and Hera 1996
	iv. ^{15}N aided studies on fertilizer N, BNF, and organic residues	IAEA 2008a
	v. N supply from crop residues in crop rotation	IAEA 2003b
^{32}P	i. Dynamics of P in soils, available soil P, and P fertilizer management including the use of phosphate rocks (PRs)	IAEA 2002b
	ii. Agronomic effectiveness of PR sources	Zapata 1995, 2002b
	iii. PR utilization for sustainable agriculture	Zapata and Roy 2004
	iv. Decision support system for direct application PR (DAPR)	FAO/IAEA 2014c
$^{34}S/^{35}S$	S fertilization/S cycling studies	IAEA 2003a
$^{15}N/^{32}P$	Selection food crop genotypes adapted to low N and P soils	IAEA 2014e
^{13}C	i. SOC cycling/sequestration studies in agroecosystems	Nguyen et al. 2011
^{13}C	ii. Enhancing agronomic water use efficiency in wheat and rice	IAEA 2012
SMNP	i. Irrigation scheduling for field crops	IAEA 1996
	ii. Deficit irrigation practices	Kirda et al. 1999; FAO 2002
SMNP and ^{15}N	i. Optimizing fertilizer N application to irrigated wheat	IAEA 2000
	ii. Water balance and fertigation in West Asia	IAEA 2002a
^{18}O, ^{2}H	Partitioning ET in field crops	Williams et al. 2004; Heng et al. 2014
NITs suite	i. SWNM for crop production in rainfed agriculture in arid/semiarid areas	IAEA 2005
	ii. SWNM for crop production in tropical and subtropical acid savannah soils	IAEA 2006
	iii. Resource use efficiency and crop productivity in agroforestry systems	IAEA 2009
	iv. Crop production in irrigated agriculture under water-limited conditions	FAO 2012; Heng et al. 2014
FRN ^{137}Cs	i. Harmonization ^{137}Cs methodology for assessment of soil erosion and sedimentation in agricultural landscapes	Zapata 2002a
	ii. Field application of ^{137}Cs technique in soil erosion and sedimentation studies	Zapata 2003
FRN ^{137}Cs, $^{210}Pb_{ex}$, ^{7}Be	Combined use of FRNs for assessing the impacts of soil conservation measures on erosion control and soil quality	IAEA 2011; Dercon et al. 2012
FRNs	i. Harmonized protocols for using FRN to assess soil erosion and effectiveness of soil conservation strategies	Mabit et al. 2008; IAEA 2014f
	ii. Soil erosion/sedimentation studies conducted in regional TC projects: Asia and Latin America	Dercon et al. 2012; IAEA/RCA 2010
Integration of FRN and CSSI	Area-wide (watershed) sediment budgets, FRNs for soil redistribution and CSSI for precision conservation: determination of critical source areas (soil/sediment)	Nguyen et al. 2011; IAEA 2014f; Gibbs 2008, 2014; Blake et al. 2012; Hancock and Revill 2013

PA encompass a wide range of scientific and technological disciplines, such as soil science, water sciences, natural sciences, crop sciences, farm mechanization, meteorology, agronomy, food sciences, agroindustry, marketing, and economics. Other areas of PA include spatial sciences and technologies, earth sciences and geoinformation, geomatics and related technologies such as remote sensing (RS), Global Positioning Systems (GPSs), Geographic Information System (GIS), and geospatial tools; sensor technologies; information and communication technology (ICT): computing and processing digital information; use of databases: time series data collection, analysis, and interpretation; advanced modeling and PC-based software for implementation; and development of DSS including socioeconomic evaluation. A great number of techniques ranging from sophisticated high tech to simple low tech are available for their integrated use and application in research on natural resources land and water and environmental sustainability as well as in management of farm resources (Stafford 2013).

PA, also known as site-specific crop management, has the potential to identify spatial and temporal variability in soil fertility status, plant nutrients, and water status on-farm using a wide range of technologies and tools such as RS, GPS, GIS, soil maps, drone technology, and a suite of sensors (US-NASA 2014; US-USDA 2014; Canada, Alberta Government 2014a). PA thus helps to guide land users, farmers, consultants, and researchers on how to use farm resources (nutrients, water and chemicals, etc.) in the right amount, at the right place, at the right time, and in the right way for optimum crop productivity and profitability, sustainability, and environmental protection (Bongiovanni and Lowenberg-Deboer 2004).

In several developed countries, PA technologies and tools are utilized to control farming operations following standard regulations and policies established by the local governments. To facilitate its application by farmers (precision farming), all field operations are mechanized and controlled by a DSS. Recent PA applications include agrobusiness and marketing issues (Canada, Alberta Government 2014b).

Significant developments in electronics, communications, and other disciplines, and sensor technology tools for the collection of large amounts of real-time geospatial data (crop, soil, water, and nutrient databases) as well as computer modeling advances allow the incorporation of several layers of information and better analysis of the spatial variability. The real-time data and computer modeling potentially lead to a better informed DSS at targeting appropriate soil and water management practices (precision conservation) at critical areas across a watershed (Berry et al. 2005; Delgado and Berry 2008). With time, a large amount of real-time data (databases) that has been collected should be analyzed and properly interpreted by a team of experts for further fine-tuning of the system. PA systems are, therefore, dynamic systems and should be continuously updated for achieving more accurate site-specific information in time and in space (Delgado and Berry 2008).

The development and application of these emerging new technologies and tools allows for a more holistic approach to the intensification of agricultural production through better management of soil, water, fertilizers, and other agricultural inputs in agroecosystems. More important, these technologies and tools offer the possibility of integrating the management of land and water (natural resources) across the landscape and their connectivity with natural areas in agricultural watersheds, which will contribute to environmental sustainability (Berry et al. 2003; US-NRCS 2014). Advances on the use of PA technologies to improve soil and water management/conservation on farming systems and agroecosystems have been reported by Delgado and Berry (2008). Precision (target) conservation agriculture is considered a key tool to achieve food security in the twenty-first century (Delgado et al. 2011).

10.4.2 DEVELOPMENT OF INTEGRATED APPROACHES FOR NIT AND PA APPLICATIONS IN SOIL AND WATER MANAGEMENT/CONSERVATION

By developing and implementing integrated approaches using NITs and selected PA technologies and tools, innovations in soil, water, and nutrient management (SWNM) in agroecosystems can be further achieved, thus reducing much of the guesswork and developing a more efficient knowledge- and information-based management.

In practice, this integration will involve (1) the use of NITs to develop SWNM practices followed by the application of a suitable PA technology to target the data generated to a specific geographic position (site, location, zone, region), and/or (2) the use of PA technologies to generate recommendations on SWNM based on existing agroexpert databases followed by the calibration/validation of the recommended management practices in field experiments using NITs. In both cases, it is envisaged the use of models for further transfer of the technologies to the beneficiaries and end users. Case studies based on selected networked research projects using NITs (Table 10.1) have been prepared to illustrate the potential for developing innovative soil and water management practices in agroecosystems through the combined use of NITs and PA technologies.

10.4.2.1 Innovations of the Soil (Nutrient) and Water Management in Agroecosystems

For the sustainable intensification of crop production systems, the development and pilot-testing of SWNM practices has focused on predominant cropping systems located on main agroecological zones/geographical regions (i.e., tropical and subtropical savanna acid soils of Africa and Latin America, arid and semiarid regions worldwide, and rice-wheat in subtropical Asia). However, it was found that the studied agroecological zones encompass a complex pattern of climatic conditions, soils, vegetation, terrain/landscapes, and land use while the farmer's demand was for location-specific SWNM recommendations for intensification of their plant and animal production systems in their farms.

In general, currently available PA technologies offer great potential for strengthening the development and application of NITs by collecting reliable data and providing site-specific information (soil, climate, crop data) for better selection and characterization of the experimental field locations, thus facilitating the use of models for better prediction of the crop yields. The use of the variable rate technology (VRT) taking into account the spatial variability in soil-crop properties (by zone management) also enable the provision of site-specific recommendations of the agricultural inputs of nutrients and water best suited to the farmer's fields. Overall, this increased precision will contribute to the formulation of more efficient and cost-effective management practices of agricultural (energy-intensive) inputs such as chemical fertilizers and irrigation water for productivity and profitability as well as environmental sustainability. Moreover, the potential economic benefits that can be achieved at the country level are enormous considering the costs of the inputs and their limited supplies to small-scale farmers. An evaluation of cost estimates relating to the use of fertilizer N and the BNF in the developing world can be found in Hardarson and Broughton (2003).

Table 10.2 summarizes the main strategies, approaches, and potential integration of NITs and PA technologies for developing innovative SWNM practices in agroecosystems. The following selected case studies highlight such developments.

10.4.2.1.1   Case Study 1: Optimizing Fertilizer N Management of Irrigated Wheat

In intensively cultivated areas of the world, high rates of fertilizer N are applied to cereals such as wheat and rice grown under irrigation to achieve high grain yields with increasing concerns about their environmental and economic impacts. A CRP was carried out to optimize fertilizer N management of irrigated wheat with the following objectives: (1) to increase the efficient use of irrigation water and N fertilizer and consequently reduce environmental pollution through the use of NITs (SMNP and ^{15}N) and related conventional techniques, and (2) build up a database for the application of the Ceres-wheat growth simulation model of the Decision Support System for Agrotechnology Transfer (DSSAT) and to assist in the formulation of a fertilizer N recommendation by relating specific fertilizer N management to expected yields (IAEA 2000; Pathak et al. 2009).

Field experiments were carried out to investigate the best timing for the split application of fertilizer N applied to irrigated wheat systems in 10 countries over a range of soil and environments. The recovery of the first split application (1/3 N applied at planting time) was usually less than the second split application (2/3 N applied at wheat tillering stage). These results were obtained in 7 out of 10 countries over the period 1995–1998, thus indicating that the fraction of total fertilizer N

TABLE 10.2

Integration of NITs and PA Technologies for Developing Innovative Soil, Water, and Nutrient Management

Main Strategy	Approach	Studies Using NITs and Conventional Techniques	PA Application
Sustainable intensification and crop production systems Integrated crop-livestock production systems	Managing soils/ nutrients for crop production and ecosystem services	• Field experiments in representative locations • Recovery efficiency and losses fertilizer N, fertilizer P, and other nutrient sources	• Site-specific information • Avoid blanket application • Define best fertilizer and other nutrient management practices • Site selection and characterization • Application VRT/zone management • Databases for model application
	Managing agricultural water for climate change adaptation	• Field experiments in representative locations • Irrigation methods/ fertigation • Irrigation scheduling • Deficit irrigation • Monitoring soil water status/field water balances	• Site-specific information • Control irrigation water amount/ application method/timing • Site selection and characterization • Application VRT/zone management • Databases for model application
Conservation agriculture systems	Improve/maintain soil quality through enhanced SOM (net carbon sequestration) in targeted agroecosystems	• Field experiments in selected locations • SOC assessment • Measuring GHG emissions	• Location-specific information • Define interventions to enhance SOC and mitigate GHG emissions • Site selection and characterization • Databases for model application
Plant breeding programs	Support implementation of targeted plant breeding programs	• Breed plant genotypes resilient to specific soil and climate conditions • Selection of best-fit plant genotypes to abiotic stresses such as low N and P fertility, acidity, drought, and salinity	• Target specific agroecologies • Environment-specific information • Site selection and characterization • Databases for model application

applied as second split application could be increased. Losses of irrigation water (surface irrigation) and N fertilizer were observed in a few countries under some specific conditions (Egypt, China, and India). In one country, the use of well-scheduled sprinkler irrigation improved fertilizer N recovery. With regard to the best timing, a chlorophyll meter was tested to monitor real-time crop N status at selected crop development stages. Harmonized protocols for the appropriate use of this device were developed and standardized readings allowed comparisons of crop N status across growth stages, locations, cultivars, and years.

A minimum data set Irrigated Wheat Database XLS was created for testing the Ceres-wheat growth simulation model. Good agreement between observed and simulated results was obtained for most growth parameters (dry matter production, leaf area index, seasonal ET, crop development stages, and others plant attributes).

The use of PA technology and tools can assist these studies to produce better databases over time and locations for further calibration and testing of the model, and thus can be utilized for providing recommendations related to fertilizer N and irrigation scheduling of wheat-cropping systems. Special computer programs (e.g., WeatherMan) within DSSAT can be used to generate long-term weather data to assess temporal variations of the production system under study. The combined use of remote sensing and sensor technologies can be utilized to monitor crop N status for setting the best timing for the second split application. GIS can be also used to analyze variability within field site-specific farming systems or to interpolate the simulated results to the regional scale.

10.4.2.1.2 Case Study 2: Evaluating the Agronomic Effectiveness of P Fertilizers, in Particular PRs

A CRP entitled "The use of nuclear and related techniques for evaluating the agronomic effectiveness of phosphate fertilizers, in particular rock phosphates" (1993–1998) generated a wealth of data from laboratory and greenhouse studies using the radioisotope ^{32}P as a tracer in soil-plant systems. Field experiments and complementary studies were also carried out to evaluate the agronomic effectiveness of PR sources with regard to WSP fertilizers such as superphosphates. In addition, the standard characterization of soils and rock phosphates utilized in the project was made allowing direct comparison of the information and better interpretation of the results obtained to formulate P fertilizer recommendations (IAEA 2002b; Zapata 2002b). A database was created from the experimental data of the above CRP for testing a P simulation submodel of the Ceres-DSSAT. This model, which was jointly developed and promoted by International Fertilizer Development Center (IFDC) and the University of Michigan, was found unsuitable because it demanded a large number of parameters that are often unavailable for most agronomic experiments, thus significantly limiting its application (Heng 2000).

In a follow-up CRP (2000–2004), PR application studies were pilot-tested in field experiments to build up the soil P capital for improving sustainable crop production in tropical acid soils (IAEA 2006). FAO/IAEA produced a PR publication on the use of rock phosphate for sustainable agriculture (Zapata and Roy 2004). A joint venture FAO/IAEA and IFDC was established to develop a global decision support system for direct application of PR (DSS-DAPR). This DSS is an effective approach to integrate both biophysical and economic factors to assess the feasibility of using either PR or WSP as a P source to crops. A conceptual framework for a simple expert system was developed and standardized databases for soil and PR were created (Heng 2004; Singh et al. 2006). Field validation of the DSS was carried out in a small network of benchmark sites (Smallberger et al. 2006). A web version of the DSS was developed and posted in the DAPR website of the Joint FAO/IAEA Program (FAO/IAEA 2014c).

The application of PA technology and tools would greatly enhance the capabilities of the DSS-DAPR in several ways: (1) generate better records of weather (time series) and soil for creating new databases, (2) expand and upgrade the available PR database including more PR sources and defining the specific geographical location (and mapping) of the geological deposits, (3) refine and update the model by creating a new version, and (4) field validation of the model in cropping sequences to assess immediate and residual effects in a network of benchmark sites distributed worldwide. The improved PR-DSS will result in better informed decisions by researchers, farm managers, extension workers, progressive farmers, fertilizer companies, and policy makers. This information can be used for socioeconomic analysis to identify and map potential areas for DAPR and to provide site-specific recommendations for DAPR including possible modifications/alterations if a particular natural PR source is not suitable.

10.4.2.1.3 Case Study 3: Improving Nutrient and Water Management of Rainfed Cropping Systems

A networked research project was implemented with a main objective to optimize and sustain the productivity of rainfed farming systems by investigating management strategies that increase the efficiency of utilization of nutrients and water. Various options for utilizing organic manures and fertilizers, recycling of crop residues, and inclusion of legumes in rotations and soil-water conservation practices that are sustainable and economically attractive to farmers were examined using NITs (SMNP and ^{15}N) and conventional techniques. Field experiments were carried out by the participants in a wide range of arid and semiarid regions and cropping systems such as wheat-maize on the Loess plateau of China, sorghum-castor rotations in Andra Pradesh, India, maize-based crop systems in Machakos, Kenya, maize-peanut rotation in Senegal peanut basin, and wheat-vetch rotations in Safi-Abda, Morocco (IAEA 2005).

Another objective was to collect minimum data sets from all experiments for developing a database and testing the performance of the crop simulation model Agricultural Production Systems Simulator Model (APSIM). This model was tested using available data from two locations in the rainfed environments of the West Asia and North Africa (WANA) region, Morocco, and Jordan. APSIM-N wheat was successfully used to simulate wheat grain yield and grain N content in these locations. It was subsequently used to simulate the long-term effect of soil type, rate, and timing of N fertilizer, initial stored soil water, different cultivars (early and late), and supplemental irrigation on increasing wheat production in Morocco, where historical weather records were available. The simulation results indicated that yields were mainly limited by the amount and timing of rainfall. While fertilizer N improved grain yields in wet years, its effect was minimal or detrimental in drier years. Strategic management of fertilizer N to improve and sustain yields can be achieved through early sowing, good storage of initial soil moisture at the start of the season, and a small amount of supplemental irrigation at sowing (IAEA 2005).

Crop production in these rainfed arid and semiarid areas is dependent on a number of environmental (climatic), soil, and agronomic management factors and interactions. To formulate specific recommendations of fertilizer N management by conducting field experiments alone is expensive and a time-consuming and labor-intensive task. Using a simulation model such as APSIM, especially when it is combined with long-term climate and soil databases, has proven very effective to integrate key factors such as rainfall (timing and amount), soil type/conditions, agronomic management, and crop development to identify the best SWNM practices to increase crop production in arid and semiarid areas. These databases can be readily achieved through the use of PA technologies and tools, thus allowing the formulation of more reliable fertilizer N recommendations. The use of GIS can further assist in the interpolation of the results over the study area.

10.4.2.1.4 Case Study 4: Improving Agricultural Water Management of Irrigated Crops

In view of the increasing scarcity of freshwater resources, there is an urgent need for improving agricultural water management of irrigated crops. The three main strategies for developing precision irrigation (Fereres and Heng 2014) have been described above under Section 10.3.1.2, "Agricultural Water Management." There is significant scope to maximize water uptake by crops through demand-based irrigation scheduling that takes into account their specific water requirements according to their growth and development stage and prevailing soil and environmental conditions. Networked research field experiments and basic studies on irrigation schedules and deficit irrigation practices have been carried out in crops and regions (Kirda et al. 1999; FAO 2002).

Irrigation water management can be further improved by using PA technologies and tools. RS allows the assessment of spatial variability of soil water status and identification of areas that need further irrigation improvement. Significant advances in RS techniques have been made in recent years through the development of sensor technologies for detecting a number of vegetation properties such as water stress or drought indicator with very high resolution (Zarco-Tejada et al. 2009). RS can also be linked to crop ET in testing crop growth simulation models to better understand the soil–water balance in the root zone and this combined approach has the potential to be used as an operational tool to predict on-farm irrigation requirements.

Precision farming that takes into account the spatial variability of soil properties in relationship to moisture level and crop productivity has a high potential to minimize GHG emissions. Applying variable rate of water through PI and measuring water use efficiency using NITs (^{18}O and ^{2}H) could lead to control and minimize the high spatial variability in GHG emissions from agroecosystems (Li et al. 2013).

PI scheduling may also lead to reduced GHG emissions. Soil moisture distribution and potential hot spots in arable fields can be identified by sensor techniques. In New Zealand, commercially available PA tool(s) allows farmers to control all sprinklers on a center-pivot or lateral-move irrigator. This technology allows farmers to apply the right amount of water to specific areas under irrigation, significantly saving water from being lost, and maximize yields and profitability (Precision VRI 2014).

Precision (targeted) SWNM practices not only have the potential to reduce the spatial N_2O emissions through the use of controlled release fertilizer and nitrification inhibitors (Delgado and Mosier 1996) but also nitrate leaching in irrigated crops (Delgado and Bausch 2005).

10.4.2.1.5 Case Study 5: Developing Soil, Water and Nutrient Management for Conservation Agriculture Systems

In an FAO/IAEA CRP entitled "Integrated soil, water and nutrient management in conservation agriculture (CA)," the influence of soil, water, and crop management practices on SOM accumulation and its subsequent impacts on soil water, nutrient, and C dynamics were investigated using NITs in various cropping systems worldwide. The results from this global project demonstrated that CA can bring benefits such as increased soil moisture retention, BNF, N retention, and soil C sequestration. However, these effects were highly variable and site-specific, in some cases the benefits being negated by the influence of crop residues on plant diseases that could reduce crop yields and quality (Dercon et al. 2010).

Based on the analysis of these results, changes were made in the formulation of subsequent CA projects. Location-specific SWNM practices should be identified and selected for the provision of targeted soil and water ecosystem services. A follow-up CRP entitled "Soil quality and nutrient management for sustainable food production in mulch-based cropping systems in sub-Saharan Africa (SSA)" was initiated in 2011 to investigate the potential of mulch-based cropping systems to enhance soil resilience against degradation and climate change risks and to increase soil fertility for sustainable food production in SSA (Nguyen et al. 2011).

PA technologies and tools can be used in the selection and characterization of benchmark sites in the moist and dry savannahs of Africa to provide a platform for extrapolation of results to other relevant agroecological zones in SSA (FAO/IAEA 2014b). Because of the nature of the sustainability issues, long-term field experiments with treatments of a continuous nature are needed to develop and pilot-test adequate technologies over time, thus demanding time series data that can be readily achieved using PA technologies and tools for testing simulation models.

10.4.2.1.6 Case Study 6: Supporting Plant-Breeding Programs for Crop Adaptation to Harsh Environments

In developing an integrated SWNM approach, a major finding was that the use of the best-adapted crop genotypes to the local soil/climate conditions is a key requirement for ensuring the productivity and sustainability of the cropping systems. These studies using NITs, which initially focused on the search of crop genotypes with superior nutrient use efficiency, have demonstrated great potential to assist plant-breeding programs in the identification of suitable germplasm resilient to particular harsh environments such as drought, flooding, NP infertility, salinity, and Al toxicity. These interdisciplinary projects call for a close interaction of plant breeders, crop scientists, and soil and water specialists working across these topics of research (Nguyen et al. 2011; Nguyen 2014).

Traditional approaches in plant-breeding programs for crop improvement utilize yield as the ultimate criterion of success or failure and they are expensive, labor-intensive, and time-consuming. Thus, any technique that can predict yield well in advance of harvest has the potential to be used as a selection tool with specific advantages.

The CID technique has been employed in a CRP on the selection of wheat genotypes tolerant to drought and irrigated rice genotypes tolerant to salinity for greater agronomic water use efficiency under a wide range of environments. Relationships between yield and CID were studied across a range of soils having different water-holding capacities under a range of natural or imposed water regimes (IAEA 2012).

Although significant progress has been achieved in the selection of crop genotypes for greater agronomic water use efficiency, some constraints were identified in these studies:

1. Experiments were conducted under controlled (greenhouse/phytotron) or field conditions. Multiple field sites were used over several seasons to cover natural variations in the soils' water-holding capacities and rainfall. Partial, full, or no irrigation was used to artificially manipulate the water regimes. Thus, there is a need to provide detailed site-specific information of the targeted environment under which the selection is made.
2. Many variables are known to affect CID within a given genotype, including phenotype, plant organ, age of plant organ, environmental conditions (especially water regime), and edaphic factors (e.g., salinity) while very little is known about other soils factors (e.g., plant nutrition). Thus, it is necessary to develop harmonized protocols for the application of CID in these projects.

PA technologies and tools would greatly assist in establishing a solid platform to support targeted plant/animal breeding programs through (1) the creation of databases of climate and soil conditions for the detailed characterization of the environment under which the selection is made, and (2) the creation of a database of the CID data and related key measurement conditions/parameters to make comparisons over locations and seasons.

10.4.2.2 Innovations in Areawide (Watersheds) Land (Soil) and Water Conservation Studies

Much of the initial work using NITs in the soil erosion research in agricultural watersheds has been based on the adoption of approaches commonly utilized in catchment hydrology and soil survey. The selected FRN involving ^7Be, ^{137}Cs, and ^{210}Pb as soil/sediment tracers have the main advantages of being able to document short-term (<30 days), medium-term (~40 years), and long-term (~100 years) average soil redistribution rates and spatial patterns in the landscape, respectively, under different local conditions. These time frames are relevant to the average duration of land use in agroecosystems (IAEA 2014f).

10.4.2.2.1   Case Study 7: Soil Erosion Assessment and Control in Agricultural Watersheds

The development and application of NITs, such as selected of FRNs ^{137}Cs, ^{210}Pb, and ^7Be in these land (soil) conservation studies, have been made for over 20 years through the implementation of several CRPs of the Joint FAO/IAEA Program. Initial CRPs (1995–2000) focused on the development of harmonized protocols for the application ^{137}Cs technique and testing approaches for the application of ^{210}Pb and ^7Be on the assessment of soil redistribution (erosion/deposition) in agricultural landscapes (Zapata 2002a, 2003). The results from this research paved the way for extending not only the ^{137}Cs technique but also other FRNs as tracers for soil erosion/sedimentation investigations in natural and agricultural landscapes. In a follow-up CRP, "Assessing the effectiveness of soil conservation measures for sustainable watershed management using fallout radionuclides" (2002–2007), combined FRN techniques involving ^7Be, ^{137}Cs, and ^{210}Pb as soil/sediment tracers were further developed to document short-term (<30 days), medium-term (~40 years), and long-term (~100 years) average soil redistribution rates and patterns in the landscape, respectively, under different local conditions (climate, soil, topography, and land uses). This combined application has shown that they are powerful tools to assess the relative impacts of soil conservation measures on soil erosion and land degradation (IAEA 2011; Dercon et al 2012). The information generated from these CRPs has led to the production of guidelines for using FRN to assess erosion and effectiveness of soil conservation measures (IAEA 2014f).

The IAEA TCP through national and regional TCPs is assisting developing Member States in the establishment and strengthening of their human and institutional capacities because these were essential requirements for the successful application of the FRN techniques to enhance sustainable agriculture and minimize land degradation. Laboratory quality control assurance services and relevant expert services were also provided to support the national capacities. Currently there are 37 Member States using FRN to address issues relating to sustainable land management (Nguyen 2014).

In a recently terminated CRP, "Integrated isotopic approaches for area-wide precision conservation to control impacts of agricultural practices on land degradation and soil erosion" (2009–2013), a sediment budget approach and a suite of NITs (combined use of FRNs and CSSI techniques), and nonnuclear techniques, including advanced models, have been applied and tested (Nguyen et al. 2011). The participating teams are currently involved in processing and analysis of the data related to the areawide (watershed) studies. Moreover, it is envisaged to produce guidelines for the combined application of FRNs and CSSI techniques in precise conservation studies.

There are a number of opportunities for using PA technologies and related tools in these studies. They can be grouped into two main categories: methodological and applications. The FRN methodologies can be significantly improved as follows: (1) the production of harmonized database and mapping of FRN inventories worldwide for the application of these techniques, (2) developing an improved field sampling design (land segmentation, selection sampling strategy, spatial analysis to identify reference sites, erosional and deposition zones), (3) providing more precise assessment of FRN inventories and improving the conversion models, and (4) implementing detailed studies (causes and extent) of the spatial variability across the landscape for scaling-up information. To date, most studies using FRNs have been conducted at the field/landscape scale, although some work has been undertaken in small basins ranging from a few hectares to a few km^2 (e.g., Mabit et al. 2007). Because of the large spatial variability over short distances across the landscape, innovative areawide (catchment/watershed) approaches combining hydrological (sediment budgets, fingerprints) and geoinformation systems and related techniques (GIS and geostatistical tools) are commonly required to make a spatial analysis to scale up the data in watersheds (IAEA 2014f).

With regard to the applications, such combined approaches have been successfully applied by some investigators for the implementation of precision conservation (e.g., for the identification of spatial patterns of erosion through various approaches [Schumacher et al. 2005] and for the comanagement of C and N in the agricultural landscapes of the Canadian prairies [Pennock 2005]). Further developments of precision conservation through the integration of technologies and tools relating to surface modeling, spatial data mining, and map analysis are needed to better assess land and water connectivity issues within agroecosystems and also between neighboring natural and agroecosystems (Berry et al. 2005; Delgado and Berry 2008). More important, the combined use of NITs and PA technologies can be instrumental for the identification of hot spots (critical source areas) to target specific soil/water management practices in the landscape. They can provide data for the creation of databases to calibrate and validate available models to detect hot spots for implementation of precision conservation (Renscher and Lee 2005) as well as for the application of new advanced models in watershed studies (Delgado and Berry 2008; IAEA 2014f).

Table 10.3 provides an overview of the integration of NITs and PA technologies in areawide (watershed) studies for developing innovative land (soil) and water management/conservation.

Studies on water conservation zones using NITs (isotopic signatures ^{18}O and ^{2}H) have shown that farm ponds are an important source of recharge for the underlying groundwater during the rainfall/runoff period whereas during the dry season the groundwater discharges water into the farm pond. It should be noted that this recharge/discharge process occurs only on some farm ponds depending on their location in the landscape, suggesting that NITs are useful to identify the ideal location for constructing farm ponds and in general water conservation zones (Sakadevan et al. 2014). Recent advances of precision conservation applied at the watershed scale indicate that the strategic location/management of the water conservation zones can be used not only to collect runoff water but also for nutrient (N and P) trapping by minimizing transport of nutrients downstream (Delgado and Berry 2008). Thus, there is great potential to improve water quality and reduce losses and impacts of nutrients to rivers using integrated PA and NIT technologies to manage the location of the water conservation zones in the watershed.

In these areawide soil and water conservation studies, PA technologies can be a powerful tool because they offer a number of comparative advantages to develop and implement precision

TABLE 10.3

Integration of NITs and PA Technologies for Developing Innovative Land (Soil) and Water Management/Conservation

Main Strategy	Approach	Studies Using NITs and Conventional Techniques	PA Application
Control soil erosion	Measuring soil redistribution (erosion/deposition) across landscape	• Use FRN of ^{137}Cs in cultivated landscapes • Use/validation of erosion models	• Location-specific information • Improving protocol application FRN technique/avoid guess-work • Field sampling design • Scaling up FRN inventories • Developing conversion models • Control spatial variability • Databases for model • Application/validation
	Assessing the impact of soil conservation measures to control erosion and improve soil quality	• Combined use FRNs, ^{137}Cs, ^{7}Be, and ^{210}Pb$_{excess}$ • Use/validation of erosion models	• Location-specific information • Improving protocols application FRN technique • Field sampling design • Scaling up FRN inventories • Control spatial variability • Databases for model application/validation
Precision soil and water conservation at an area-wide scale (catchment level)	Targeted soil and water management/conservation measures	• Combined use of FRNs and CSSI and other fingerprints • Sediment budget studies • Advanced models	• Location and site-specific information within watershed (sinks and sources) • Field sampling design • Scaling up FRN inventories • Detailed spatial analysis • Control spatial variability • Databases for model application/validation
Water conservation zones at an area-wide scale (catchment level)	Targeting placement sites for water conservation and water pollution control	• Stable isotopic ^{18}O and ^{2}H signatures of water sources • Water balance studies • Water pollution control studies (denitrification reactors and P trapping zones)	• Location-specific information • Strategic placement of the water conservation zones • Construction of nutrient-trapping elements across landscapes
Land and water salinization management at an area-wide scale (catchment level)	Spatial analysis of crop response to land and water management in salt-affected areas	• SMNP and CNRP for soil water monitoring • Isotopic studies for soil, water, and plant studies • Monitoring changes to soil and water salinity, water balance studies	• Location-specific information (saline soil/water and climatic conditions) • Better understanding of the soil/water salinity changes and crop response to land and water management at the landscape level

conservation in areawide studies. For potential applications of PA in these studies, the reader is referred to the reviews of Delgado and Berry (2008) and Delgado et al. (2011).

Further developments on natural resource (land and water) management and environmental monitoring at a large scale (national, regional, and global) include a range of studies such as (1) land use development and implementing regional land use planning and strategies in the long term (Foley et al. 2005), (2) land use vulnerabilities assessment based on the ecosystem services and their adaptive capacity (Metzger et al. 2006), (3) digital soil mapping and modelling at continental scales (Grunwald et al. 2011), and (4) RS of soil salinization: impact on land management (Metternicht and Zinck 2008).

10.5 FUTURE PROSPECTS: KEY CHALLENGING ISSUES

The development of innovative soil and water management in agroecosystems through integrated approaches includes the use of not only a suite of nuclear-based and conventional (nonnuclear) techniques, but also a platform of technology and tools for precision agriculture. The latter demand the establishment of adequate infrastructures to support the generated technologies, data processing/ modeling, and transfer of the results to the beneficiaries.

Novel PA technologies (targeting location/site and time and collection of spatial and time-series data) will greatly improve the selection and characterization of the experimental sites, in particular soil, climatic, and crop data; thus enabling the use of models and the extrapolation of the information to similar areas. Their application in areawide (watershed) studies for developing land (soil) and water conservation calls for the collection of geospatial data, creation of time series databases, spatial mapping and analysis, and application of advanced models to integrate large and complex sets of data obtained under a range of specific natural and agroecosystems (Berry et al. 2005; Delgado and Berry 2008). The experimental data can be also used to validate and refine existing models as well as to develop decision support systems to make better-informed decisions on the technologies for sustainable land and water conservation (Delgado et al. 2011).

The adoption of these integrated approaches need the formation of multidisciplinary and often interinstitutional teams with the required skills and expertise and the collaboration of all the stakeholders involved in both PA and soil/water management/conservation. However, these measures will demand more strategic networking, close coordination, the use of suitable PA technologies and tools and up-to-date ICT, and will increase the overall cost and complexity of the research efforts. It is therefore of utmost importance to establish coordination and close collaboration with existing national space agencies and regional centers for space science and technology operating under the program of UNOOSA (UNOOSA 2014).

Human and institutional capacity building is the key essential element required for the successful application of these technologies for developing innovative soil and water precision (target) management/conservation in agroecosystems and contributing to sustainable agricultural development for food security and environmental sustainability.

The planning and use of PA technologies and tools will require analysis of (1) a definition of the PA objectives and suitable types and levels of technology, (2) assessment of accessibility and affordability to the desired PA technology in the study area, and (3) availability of resources—human resources with an ideal combination/blend of skills and expertise and financial resources, among others. It is also envisaged the use of specialized experts to assist in the planning and application of PA technologies and tools and support further education including training modules for all the stakeholders through the FAO and IAEA TCPs.

10.6 CONCLUSIONS

This chapter provided an overview of the use of NITs for developing innovative SWNM and conservation to achieve food security through improved productivity while ensuring environmental sustainability in the developing world. In particular, it highlighted opportunities for integrated approaches of utilizing NITs and currently available PA technologies for targeting best SWNM/conservation practices in support of sustainable agriculture and development under climate change adaptation and mitigation. Both NITs and PA technologies have comparative advantages and limitations. Their full potential in terms of efficiency and cost-effectiveness can only be achieved through the use of appropriate approaches and correct application of the techniques. Moreover, the integrated mixes of these technologies must be properly inserted into national and/or regional research project plans.

It is foreseen that the implementation of integrated approaches by research teams will be made stepwise in several approximations to further support the development of innovative soil and water management/conservation for sustainable agriculture using both conventional and NIT approaches. In this context, a continuous dynamic process will be established by setting an entry point and a baseline that will be updated and expanded over time by including more PA elements to achieve value-added information services. The final step would be the development of a DSS, including a socioeconomic evaluation to facilitate the provision of information to policy makers for formulation of appropriate policies and regulations for adoption by farmers and other stakeholders involved in the land and water management/conservation. Close coordination between several related disciplines and cooperation partnerships with regional, national, and international organizations (e.g., FAO Global Soil Partnership) can greatly facilitate the provision of such information to land/water use and management/conservation practitioners in agroecosystems for sustainable land and water management/conservation.

REFERENCES

Alewell, C., M.J. Mitchell, G.E. Likens, and H.R. Krouse. 1999. Sources of stream sulphate at the Hubbard Brook Exp. Forest: Long term analysis using stable isotopes. *Biogeochem.* 44: 281–299.

Arbuckle, J.G. 2013. Farmer attitudes toward proactive targeting of agricultural conservation programs. *Soc. Nat. Resour.* 26: 625–641.

Aye, T.M., M.L. Nguyen, N.S. Bolan, and M.J. Hedley. 2006. Phosphorus in soils of riparian and non-riparian wetland and buffer strips in the Waikato area, New Zealand. *N. Z. J. Agr. Res.* 29: 349–358.

Barraclough, D. 1995. ^{15}N dilution techniques to study soil nitrogen transformations and plant uptake. *Fert. Res.* 42: 185–192.

Berry, J.K., J.A. Delgado, R. Khosla, and F.J. Pierce. 2003. Precision conservation for environmental sustainability. *J. Soil Water Conserv.* 58: 332–339.

Berry, J.K., J.A. Delgado, F.J. Pierce, and R. Khosla. 2005. Applying spatial analysis for precision conservation across the landscape. *J. Soil Water Conserv.* 60: 363–370.

Blake, W.H., K. Ficken, P. Taylor, M. Russell, and D.E. Walling. 2012. Tracing crop-specific sediment sources in agricultural catchments. *Geomorphology* 139–140: 322–329.

Bolan, N.S., R.E. White, and M.J. Hedley. 1990. A review of the use of phosphate rocks as fertilisers for direct application in Australia and New Zealand. *Aust. J. Exp. Agric.* 30: 297–313.

Bole, J.B., and U.J. Pittmann. 1984. Availability of soil sulphates to barley and rapeseed. *Can. J. Soil Sci.* 64: 301–312.

Bongiovanni, R., and J. Lowenberg-Deboer. 2004. Precision agriculture and sustainability. *Prec. Agric.* 5: 359–387.

Bruckler, L., F. Lafolie, S. Ruy, J. Granier, and D. Baudequin. 2000. Modelling the agricultural and environmental consequences of non-uniform irrigation on a maize crop. 1. Water balance and yield. *Agronomie* 20: 609–624.

Bruinsma, J. 2003. *World Agriculture: Towards 2015/2030: An FAO Perspective*. London: Earthscan Publications.

Burns, D.A., and M.L. Nguyen. 2002. Nitrate movement and removal along a shallow groundwater flow path in a riparian wetland within a sheep-grazed pastoral catchment: Results of a tracer study. *N. Z. J. Mar. Fresh. Res.* 36: 371–385.

Cadisch, G., and K.E. Giller (eds). 1997. *Driven by Nature: Plant Litter Quality and Decomposition*. Wallingford, UK: CAB International.

Canada, Alberta Government, Agriculture and Rural Development. 2014a. Decision making tools (available at http://www.agriculture.alberta.ca/app21/ldcalc).

Canada, Alberta Government 2014b. What is precision farming? (available at http://www1.agric.gov.ab .ca/$department/deptdocs.nsf/all/sag1951).

Chalk, P.M. 1998. Dynamics of biologically fixed N in legume-cereal rotations: A review. *Aust. J. Agric. Res.* 49: 303–316.

Chalk, P.M., F. Zapata, and G. Keerthisinghe. 2002. Towards integrated soil, water and nutrient management in cropping systems: The role of nuclear techniques. In: *Transactions 17th World Congress of Soil Science*, Symposium 59, Paper 2164, August 14–20, 2002, Bangkok, Thailand: WCSS.

Chardon, W., and P. Withers. 2003. Introduction to papers from EU COST Action 832, Quantifying the agricultural contribution to eutrophication. *J. Plant Nutr. Soil Sci.* 166: 401.

Chen, W., G.J. Blair, J. Scott, and R. Lefroy. 1999. Nitrogen and sulphur dynamics of contrasting grazed pastures. *Aust. J. Agric. Res.* 50: 1381–1392.

Chinoim, N., R. Lefroy, and G.J. Blair. 1997. The effect of crop duration and soil type on the ability of soil sulphur tests to predict plant response to sulphur. *Aust. J. Agric. Res.* 35: 1131–1141.

Chrisman, B., and M. Zreda. 2013. Quantifying mesoscale soil moisture with the cosmic-ray rover. *Hyd. Earth Sys. Sci. Discuss.* 10: 7127–7160.

Condon, A.G., R.A. Richards, G.J. Rebetzke, and G.D. Farquhar. 2002. Improving intrinsic water-use efficiency and crop yield. *Crop Sci.* 42: 122–131.

Condon, A.G., R.A. Richards, G.J. Rebetzke, and G.D. Farquhar. 2004. Breeding for high water-use efficiency. *J. Exp. Bot.* 55: 2447–2460.

Cooper, P.J.M., J. Dimes, K.P.C. Rao, B. Shapiro, B. Shiferaw, and S. Twomlow. 2008. Coping better with current climatic variability in the rain-fed farming systems of sub-Saharan Africa: An essential first step in adapting to future climate change? *Agric. Ecosyst. Environ.* 126: 24–35.

Delgado, J.A., and J. K. Berry. 2008. Advances in precision conservation. *Adv. Agron.* 98: 1–44.

Delgado, J.A., and W. Bausch. 2005. Potential use of precision conservation techniques to reduce nitrate leaching in irrigated crops. *J. Soil Water Conserv.* 60: 379–387.

Delgado, J.A., and A.R. Mosier. 1996. Mitigation alternatives to decrease nitrous oxides emissions and urea-nitrogen loss and their effect on methane flux. *J. Environ. Qual.* 25: 1105–1111.

Delgado, J.A., R. Khosla, and T. Mueller. 2011. Recent advances in precision (target) conservation. *J. Soil Water Conserv.* 66: 167A–170A.

Dercon, G., M.L. Nguyen, M. Aulakh et al. 2010. Soil, water and nutrient management under conservation agriculture across agro-ecosystems worldwide. In: *Agro-2010, Proceedings 11th ESA Congress*, Montpellier, France.

Dercon, G., L. Mabit, G. Hancok et al. 2012. Fallout radionuclide based techniques for assessing the impact of soil conservation measures on erosion control and soil quality: An overview of the main lessons learnt under an FAO/IAEA Coordinated Research Project. *J. Environ. Radioact.* 107: 78–85.

Derpsch, R., and T. Friedrich. 2009. Global overview of conservation agriculture adoption. In: *4th Congress Conservation Agriculture*, February 2009, New Delhi, India.

Desilets, D., M. Zreda, and T.P.A. Ferre. 2010. Nature's neutron probe: Land-surface hydrology at an elusive scale with cosmic rays. *Water Resour. Res.* 46: W011505, doi:10.1029/2009WR008726.

Di, H.J., K.C. Cameron, and R.G. McLaren. 2000. Isotopic dilution methods to determine the gross transformation rates of nitrogen, phosphorus, and sulfur in soil: A review of the theory, methodologies, and limitations. *Aust. J. Soil Res.* 38: 213–230.

Diebel, M.W., J.T. Maxted, P.J. Nowak, and M. Jake Vander Zanden. 2008. Landscape planning for agricultural non-point source pollution reduction. I: A geographical allocation framework. *J. Environ. Manage.* 42:789–802.

EU 2014. *The Common Agricultural Policy after 2013*. Agriculture and rural development (available at http://ec.europa.eu/agriculture/cap-post-2013).

FAO. 1980. *Maximizing the Efficiency of Fertilizer Use by Grain Crops*, FAO Fertilizer and Plant Nutrition Bulletin No. 3. Rome: FAO.

FAO. 2001. *Smallholder Irrigation Technology: Prospects for Sub-Saharan Africa*. International Programme for Technology and Research in Irrigation and Drainage.

FAO. 2002. *Deficit Irrigation Practices*. FAO Water Report 22. Rome: FAO and IAEA.

FAO. 2011. *Proceedings Global Forum on Salinization and Climate Change*. World Soils Report No. 105, R.P. Thomas (ed.). Rome: FAO.

FAO. 2012. *Current World Fertilizer Trends and Outlook to 2016* (available at ftp://ftp.fao.org/ag/agp/docs/cwfto16.pdf).

FAO. 2014a. *Home Portal* (available at http://www.fao.org).

FAO. 2014b. *Sustainable Crop Production Intensification* (available at http://www.fao.org/agriculture/crops/thematic-sitemap/theme/spi/en).

FAO. 2014c. *Conservation Agriculture* (available at http://www.fao.org/ag/ca).

FAO/IAEA. 2014a. *The Joint FAO/IAEA Program* (available at http://www-naweb.iaea.org/nafa/index.html).

FAO/IAEA. 2014b. *The Soil and Water Management and Crop Nutrition Subprogram* (available at http://www-naweb.iaea.org/nafa/swmn/index.html).

FAO/IAEA. 2014c. *Direct Application of Phosphate Rock* (available at http://www-iswam.iaea.org/dapr/srv/en).

Fardeau, J.C., G. Guiraud, and C. Marol. 1995. Bioavailability soil P as a key to sustainable agriculture. Functional model determined by isotopic tracers. In *Nuclear and Related Techniques in Soil-Plant Studies on Sustainable Agriculture and Environmental Protection*, Proc. Int. Symp. STI/PUB/947, ed. IAEA, 131–144. Vienna, Austria: IAEA.

Fardeau, J.C. 1996. Dynamics of phosphate in soils: An isotopic outlook. *Fert. Res.* 45: 91–100.

Fereres, E., and L.K. Heng. 2014. Enhancing the contribution of isotopic techniques to the expansion of precision irrigation. In *Managing Soils for Food Security and Climate Change Adaptation and Mitigation*, eds. L.K. Heng, K. Sakadevan, G. Dercon, and M.L. Nguyen, Proceedings FAO/IAEA Int. Symposium, July 23–27, 2012, Rome: FAO.

Foley, J., R. DeFries, G.P. Asner et al. 2005. Global consequences of land use. *Science* 309: 570–575.

Follett, R.F., D.R. Keeney, and R.M. Cruze (eds.). 1991. *Managing Nitrogen for Groundwater Quality and Farm Profitability*, Madison, WI: SSSA.

Franz, T.E., M. Zreda, T.P.A. Ferre, R. Rosolem, C. Zweck, S. Stillman, X. Zeng, and W.J. Shuttleworth. 2012. Measurement depth of the cosmic-ray soil moisture probe affected by hydrogen from various sources. *Water Resour. Res.* 48: W08515.

Freibauer, A., A. Mosier, J. Clemens et al. (eds.). 2001. Biogenic emissions of greenhouse gases caused by arable and animal agriculture, Special Issue *Nutr. Cycling Agroecosyst.* 60: 1–326.

Gibbs, M.M. 2008. Identifying source soils in contemporary estuarine sediments: A new compound-specific isotope method. *Estuar. Coasts* 31: 344–359.

Gibbs, M.M., and L. Mabit. 2012. Test of compound-specific stable isotope (CSSI) technique to investigate sediment provenance in a small Austrian watershed. *Soils Newsletter* 34(2): 29.

Gibbs, M.M. 2014. Protocols on the use of the CSSI technique to identify and apportion soil sources from land use. NIWA report HAM2013-106 to Joint FAO/IAEA Division of Nuclear Techniques in Food and Agriculture (Contract Number 15491/R1) (available at http://www-naweb.iaea.org/nafa/swmn/public/CSSI-technique-protocols-revised-2013.pdf).

Giller, K.E. 2001. *Nitrogen Fixation in Tropical Cropping Systems*, 2nd Edition. Wallingford, Oxon, UK: CAB International.

Goh, K.M., and P.E.H. Gregg. 1982. Field studies on the fate of radioactive sulphur fertilizer applied to pastures. *Fert. Res.* 3: 337–351.

Greenway, M. 2004. Constructed wetlands for water pollution control–processes, parameters and performance. *Asia-Pacific J. Chem. Eng.* 12(5–6): 491–504.

Groffman, P., K. Butterbach-Bahl, R.W. Fulweiler, A.J. Gold, J.L. Morse, E.K. Stander, C. Tague, C. Tonitto, and P. Vidon. 2009. Challenges to incorporating spatially and temporally explicit phenomena (hotspots and hot moments) in denitrification models. *Biogeochem.* 93: 49–77.

Grunwald, S., J.A. Thompson, and J.L. Boetlinger. 2011. Digital soil mapping and modelling at continental scales: Finding solutions for global issues. *Soil Sci. Soc. Am. J.* 75: 1201–1213.

Hancock, G.J., and A.T. Revill. 2013. Erosion source discrimination in a rural Australian catchment using compound-specific isotope analysis (CSIA). *Hydrol. Process.* 27(6): 923–932.

Hardarson, G. 1994. International FAO/IAEA programmes on biological nitrogen fixation. In *Symbiotic Nitrogen Fixation*, eds. P.H. Graham, M.J. Sadowsky, and C.P. Vance, 189–202, Dordrecht, The Netherlands: Kluwer.

Hardarson, G., and C. Atkins. 2003. Optimising biological nitrogen fixation by legumes in farming systems. *Plant Soil* 252: 41–54.

Hardarson, G., and W.J. Broughton (eds.). 2003. *Maximising the Use of Biological Nitrogen Fixation in Agriculture*. Special Issue *Plant Soil* 252, Dordrecht, The Netherlands: Kluwer, and Rome: FAO.

Hart, M.R., B.F. Quin, and M.L. Nguyen. 2004. Phosphorus runoff from agricultural land and direct fertilizer effects: A review. *J. Environ. Qual.* 33: 1954–1972.

Hauck, R.D., and J.M. Bremner. 1976. Use of tracers for soil and fertilizer nitrogen research, *Adv. Agron.* 28: 219–266.

Heng, L.K. 2000. Modelling, database and the P submodel. In *Management and Conservation of Tropical Acid Soils for Sustainable Crop Production*. IAEA TECDOC 1159, IAEA ed., 101–108. Vienna, Austria: IAEA.

Heng, L.K. 2004. Towards developing a decision support system for phosphate rock direct application in agriculture. In *Direct Application of Phosphate Rock and Related Technology: Latest Developments and Practical Experiences*. Special Publications IFDC-SP-37, eds. S.S.S. Rajan, and S.H. Chien, 225–235. Muscle Shoals, AL: IFDC.

Heng, L.K., T.C. Hsiao, D. Williams et al. 2014. Managing irrigation water to enhance crop productivity under water-limiting conditions: A role for isotopic techniques. In: *Managing Soils for Food Security and Climate Change Adaptation and Mitigation*, eds. L.K. Heng, K. Sakadevan, G. Dercon, and M.L. Nguyen, Proceedings FAO/IAEA Int. Symposium, July 23–27, 2012. Rome: FAO.

Hoeft, R.G., D.R. Keeney, and L.M. Walsh. 1972. Nitrogen and sulphur precipitation and sulfur dioxide in the atmosphere in Wisconsin. *J. Environ. Qual.* 1: 203–208.

Hood-Novotny, R., C. Van Kessel, and B. Vanlauwe. 2008. Use of tracer technology for the management of organic sources. In: *Guidelines on Nitrogen Management in Agricultural Soils*, IAEA-TCS-CD 29, ed. IAEA, 181–234. Vienna, Austria: IAEA.

IAEA. 1980. *Soil Nitrogen as Fertilizer or Pollutant*, Panel Proc. Series STI/PUB/535. Vienna, Austria: IAEA.

IAEA. 1983. *Zn Fertilization of Flooded Rice*. IAEA-TECDOC-245. Vienna, Austria: IAEA.

IAEA. 1984. *Soil and Fertilizer Nitrogen*. Technical Report Series No. 244. Vienna, Austria: IAEA.

IAEA. 1990. *Use of Nuclear Techniques in Studies of Soil-Plant Relationships*. IAEA Training Course Series No. 2. Vienna, Austria: IAEA.

IAEA. 1996. *Nuclear Techniques to Assess Irrigation Schedules for Field Crops*. IAEA-TECDOC-888. Vienna, Austria: IAEA.

IAEA. 2000. *Optimizing Nitrogen Fertilizer Application to Irrigated Wheat*. IAEA-TECDOC-1164. Vienna, Austria: IAEA.

IAEA. 2001. *Use of Isotope and Radiation Methods in Soil and Water Management and Crop Nutrition*. IAEA Training Course Series No. 14. Vienna, Austria: IAEA.

IAEA. 2002a. *Water Balance and Fertigation for Crop Improvement in West Asia*. IAEA-TECDOC-1266. Vienna, Austria: IAEA.

IAEA. 2002b. *Assessment of Soil Phosphorus Status and Management of Phosphatic Fertilizers to Optimize Crop Production*. IAEA-TECDOC-1272 (also available on CD-ROM). Vienna, Austria: IAEA.

IAEA. 2003a. *Guidelines for the Use of Isotopes of Sulphur in Soil-Plant Studies*. IAEA Training Course Series No. 20. Vienna, Austria: IAEA.

IAEA. 2003b. *Management of Crop Residues for Sustainable Crop Production*. IAEA-TECDOC-1354. Vienna, Austria: IAEA.

IAEA. 2005. *Nutrients and Water Management Practices for Increasing Crop Production in Rainfed Arid/Semi-Arid Areas*. IAEA-TECDOC-1468. Vienna, Austria: IAEA.

IAEA. 2006. *Management Practices for Improving Sustainable Crop Production in Tropical Acid Soils*. IAEA Proceedings Series, STI/PUB/1285. Vienna, Austria: IAEA.

IAEA. 2008a. *Guidelines on Nitrogen Management in Agricultural Soils*. IAEA-TCS-CD 29, Vienna, Austria: IAEA.

IAEA. 2008b. *Field Estimation of Soil Water Content: A Practical Guide to Methods, Instrumentation and Sensor Technology*. IAEA-Training Course CD Series No. 30. Vienna, Austria: IAEA.

IAEA. 2009. *Management of Agroforestry Systems for Enhancing Resource Use Efficiency and Crop Productivity*. IAEA-TECDOC/CD-1606. Vienna, Austria: IAEA.

IAEA. 2011. *Impact of Soil Conservation Measures on Erosion Control and Soil Quality*. IAEA-TECDOC-1665. Vienna, Austria: IAEA.

IAEA. 2012. *Greater Agronomic Water Use Efficiency in Wheat and Rice Using Carbon Isotope Discrimination*. IAEA-TECDOC-1671. Vienna, Austria: IAEA.

IAEA. 2014a. *Home portal/work* (available at http://www-naweb.iaea.org/na/index.html).

IAEA. 2014b. *Co-Ordinated Research Activities. NACA Welcome* (available at http://cra.iaea.org/cra/index.html).

IAEA. 2014c. *Technical Co-Operation Programme* (available at http://iaea.org/technicalcooperation/Home/index.html).

IAEA. 2014d. *Nutritional and Human Related Environmental Studies* (available at http://www-naweb.iaea.org/nahu/NAHRES).

IAEA. 2014e. *Optimizing Productivity of Food Crop Genotypes in Low Nutrient Soils*. IAEA-TECDOC-1741. Vienna, Austria: IAEA.

IAEA. 2014f. *Guidelines for Using Fallout Radionuclides to Assess Erosion and Effectiveness of Soil Conservation Strategies*. IAEA-TECDOC-1741. Vienna, Austria: IAEA.

IAEA/RCA (Regional Co-operative Agreement for Research, Development and Training Related to Nuclear Science and Technology for Asia and the Pacific). 2010. *Combating Soil Erosion-Caused Land Degradation in the Asia and the Pacific Region.* RCA success story in 2010. Daejeon Metropolitan City, Rep. of Korea: RCA Regional Office (available at http:///wwwrcaro.org/news/articles/view/tableid/news /page/23/id/633).

Jensen, E.S., G. Hardarson, and M.B. Peoples. 2008. Use of tracer technology in biological nitrogen fixation research. In *Guidelines on Nitrogen Management in Agricultural Soils*, IAEA-TCS-29CD, ed. IAEA, 127–180. Vienna, Austria: IAEA.

Keerthisinghe, G., F. Zapata, and P.M. Chalk. 2003. Plant nutrition: Challenges and tasks ahead. In *Global Food Security and the Role of Sustainable Fertilization*, Proceedings IFA-FAO Agricultural Conference, March 26–28, 2003, Rome, Italy (available at http://www.fertilizer.org/ifa/publicat/PDF/2003_rome _keerthisinghe.pdf and at http://www-naweb.iaea.org/nafa/swmn/public/soil-nl-jun03.pdf).

Kirda, C., P. Moutonnet, C. Hera, and D.R. Nielsen (eds.). 1999. *Crop Yield Response to Deficit Irrigation.* Dordrecht, The Netherlands: Kluwer.

Kumarasinghe, K., and D.L. Eskew (eds.). 1993. *Isotopic Studies of Azolla and Nitrogen Fertilization for Rice.* Development in Plant and Soil Sciences 51. Dordrecht, The Netherlands: Kluwer.

Krouse, H.R. 1977. Sulphur isotope abundance elucidates uptake of atmospheric sulphur emissions by vegetation. *Nature* 265: 45–46.

Kwon, Jang-Soon, B. Mayer, Seong-Taek Yun, and M. Nightingale. 2008. The use of ion exchange membranes for isotope analysis on soil water sulfate: Laboratory experiments. *J. Environ. Qual.* 37: 501–508.

Lal, R. (ed.). 2001. *Soil Carbon Sequestration and the Greenhouse Effect.* SSSA Special Publication 57, Madison, WI: SSSA.

Lal, R. 2004. Soil carbon sequestration impacts on global climate change and food security. *Science* 304: 1623–1627.

Lal, R. 2006. *Encyclopaedia of Soil Science.* Second Edition, Boca Raton FL: CRC Press.

Lal, R. 2007. Farming carbon. *Soil Till. Res.* 96: 1–5.

Lal, R., and J.M. Kimble. 1997. Conservation tillage for carbon sequestration. *Nutr. Cycling Agroecosyst.* 49: 243–253.

Ledgard, S.F., J.W. Penno, and M.S. Spronsen. 1999. Nitrogen inputs and losses from clover/ryegrass pastures grazed by dairy cows, as affected by nitrogen fertilizer application. *J. Agric. Sci. Camb.* 132: 215–225.

Li, Y., X. Fu, X. Liu, J. Shen, Q. Luo, R. Xiao, Y. Li, C. Tong, and J. Wu. 2013. Spatial variability and distribution of N_2O emissions from a tea field during the dry season in subtropical central China. *Geoderma* 193–194: 1–12.

Lv, J., X. Liu, X. Wang, K. Li, C. Tian, and P. Christie. 2014. Greenhouse gas intensity and net annual global warming potential of cotton cropping systems in an extremely arid region. *Nutr. Cycling Agroecosyst.* 98: 15–26.

Mabit, L., C. Bernard, and M.R. Laverdière. 2007. Assessment of erosion in the Boyer River watershed (Canada) using a GIS oriented sampling strategy and ^{137}Cs measurements. *Catena* 71(2): 242–249.

Mabit, L., M. Benmansour, and D.E. Walling. 2008. Comparative advantages and limitations of Fallout radionuclides (^{137}Cs, ^{210}Pb, and [7]Be) to assess soil erosion and sedimentation. *J. Environ. Radioact.* 99(12): 1799–1807.

Mabit, L., K. Meusburger, E. Fulajtar, and C. Alewell. 2013. The usefulness of ^{137}Cs as a tracer for soil erosion assessment: A critical reply to Parsons and Foster (2011). *Earth-Sci. Rev.* 127: 300–307.

Matheson, F.E., M.L. Nguyen, A.B. Cooper, T. Burt, and D. Bull. 2002. Fate of N15-nitrate in unplanted, planted and harvested riparian wetland soil microcosms. *Ecol. Eng.* 19: 249–264.

Matheson, F.E., M.L. Nguyen, A.B. Cooper, and T.P. Burt. 2003. Short-term nitrogen transformation rates in riparian wetland soil determined with nitrogen-15. *Biol. Fertil. Soils* 38: 129–136.

Mayer, B., K.H. Feger, A. Giesemann, and H.J. Jager. 1995. Interpretation of sulfur cycling in two catchments in the Black Forest (Germany) using stable sulfur and oxygen-isotope data. *Biogeochemistry* 30: 31–58.

Metternich, G., and A. Zinck (eds.). 2008. *Remote Sensing of Soil Salinization: Impact on Land Management.* Boca Raton, FL: CRC Press.

Metzger, M.J., M.D.A. Rounsewell, L. Acosta-Michlik, R. Leemans, and D. Schroter. 2006. The vulnerability of ecosystem services to land use change. *Agric. Ecosyst. Environ.* 114: 69–85.

Molden, D. (ed.) 2007. *Water for Food, Water for Life: A Comprehensive Assessment of Water Management in Agriculture.* London: Earthscan, and Colombo: International Water Management Institute.

Morel, C., and J.C. Fardeau. 1991. Phosphorus availability of fertilizers: A predictive laboratory method for its evaluation. *Fert. Res.* 28: 1–9.

Moreno-Mateos, D., and F.A. Comin. 2010. Integration objectives and scales for planning and implementing wetland restoration and creation in agricultural landscapes. *J. Environ. Manage.* 91: 2087–2095.

Mosier, A.R., and L. Klemedtsson. 1994. Measuring denitrification in the field. In: *Methods of Soil Analysis*, Part 2, SSSA Book Series, ed. R.W. Weaver, 1047–1065. Madison, WI: SSSA.

Mosier, A.R., K. Syers, and J.R. Freeney (eds.). 2004. *Agriculture and the Nitrogen Cycle: Assessing the Impacts of Fertilizer Use on Food Production and the Environment.* Washington, DC: Island Press.

Müller, C., and T.J. Clough. 2013. Advances in understanding nitrogen flows and transformations: Gaps and research pathways. *J. Agric. Sci.* FirstView Article 1–11.

Müller, C., R.J. Laughlin, O. Spott, and T. Rütting. 2014. Quantification of N_2O emission pathways via a ^{15}N tracing model. *Soil Biol. Biochem.* 72: 44–54.

Nearing, M.A., F.F. Pruski, and M.R. O'Neal. 2004. Expected climate change impacts on soil erosion rates: A review. *J. Soil Water Conserv.* 59: 43–50.

Nemery, J., J. Garnier, and C. Morel. 2005. Phosphorus budget in the Marne watershed (France): Urban vs. diffuse sources, dissolved vs. particulate forms. *Biogeochem.* 72: 35–66.

Nguyen, M.L. 2000. Phosphate incorporation and transformation in surface sediments of a sewage-impacted wetland as influenced by sediment sites, sediment pH and added phosphate concentration. *Ecol. Eng.* 14: 139–155.

Nguyen, M.L., and F. Zapata. 2006. Use of nuclear techniques in addressing soil-water-nutrient issues for sustainable agricultural production. In *Transactions 18th World Congress of Soil Science*, Abstracts-Session No.1-1, July 2006, Philadelphia: WCSS.

Nguyen, M.L. 2014. Towards sustainable land management for enhancing food security while mitigating climate change impacts: The role of nuclear and isotopic techniques. In: *Managing Soils for Food Security and Climate Change Adaptation and Mitigation*, eds. L.K. Heng, K. Sakadevan, G. Dercon, and M.L. Nguyen, Proceedings FAO/IAEA Int. Symposium, July 23–27, 2012, Rome: FAO.

Nguyen, M.L., and K.M. Goh. 1990. Effects of grazing animals on the plant availability of sulphur fertilisers in grazed pastures. *Proc. NZ Grassland Assoc.* 52: 181–185.

Nguyen, M.L., and K.M. Goh. 1994a. Sulphur cycling and its implications on sulphur fertilizer requirements of grazed grassland ecosystems. *Agric. Ecosyst. Environ.* 49: 173–206.

Nguyen, M.L., and K.M. Goh. 1994b. Distribution, transformations and recovery of urinary sulphur and sources of plant-available soil sulphur in irrigated pasture soil-plant systems treated with ^{35}sulphur-labelled urine. *J. Agric. Sci. (Camb.)* 122: 91–105.

Nguyen, M.L., D.S. Rickard, and S.D. McBride. 1989. Pasture production and changes in phosphorus and sulphur status in irrigated pastures receiving long-term applications of superphosphate fertilizer. *N Z J. Agric. Res.* 32: 245–262.

Nguyen, M.L., F. Zapata, and G. Dercon. 2010. Zero tolerance on land degradation for sustainable intensification of crop production. In: *Land Degradation and Desertification: Assessment, Mitigation and Remediation*, eds. P. Zdruli, M. Pagliai, S. Kapur, and F. Faz Cano, 37–47, Dordrecht, The Netherlands: Springer.

Nguyen, M.L., F. Zapata, R. Lal, and G. Dercon. 2011. Role of isotopic and nuclear techniques in sustainable land management: Achieving food security and mitigating impacts of climate change. In: *World Soil Resources and Food Security, Advances in Soil Science*, Vol. 18; eds. R. Lal, and B.A. Stewart, 345–418, Boca Raton, FL: CRC Press.

Noij, I.G.A.M., M. Heinen, H.I.M. Heesmans, J.T.N.M. Thissen, and P. Groenendijk. 2012. Effectiveness of unfertilized buffer strips for reducing nitrogen loads from agricultural lowland to surface waters. *J. Environ. Qual.* 41(2): 322–333.

Novak, M, I. Jackova, and E. Prechova. 2001. Temporal trends in the isotope signature of air-borne sulphur in Central Europe. *Environ. Sci. Techn.* 35: 255–260.

Organic Resources Database (ORD). 2004. (available at ftp://iserver.ciat.cgiar.org/webciat/ORD).

Palm, C.A. 2001. Organic inputs for soil fertility management in tropical agroecosystems: Application of an organic resource database. *Agric. Ecosyst. Environ.* 83: 27–42.

Parkin, T.B. 1987. Soil microsite as a source of denitrification variabilitiy. *Soil Sci. Soc. Am. J.* 51: 1194–1199.

Pathak, H., J.K. Ladha, A. Yadvinder-Singh, A. Hussain, F. Hussain, R. Munankarmy, M.K. Gathala, S. Verma, U.K. Singh, and M.L. Nguyen. 2009. Resource-conserving technologies in the rice-wheat system of South Asia: Field evaluation and simulation analysis. In: *Integrated Crop and Resource Management in the Rice-Wheat Systems of South East Asia*, eds. J.K. Ladha, A. Yavinder-Singh, O. Erenstein, and B. Hardy, 297–318, Metro Manila, Philippines: IRRI.

Paustian, K., O. Andren, H.H. Janzen, R. Lal, P. Smith, G. Tian, H. Tiessen, M. Van Noordwijk, and P.L. Woomer. 1997. Agricultural soils as a sink to mitigate CO_2 emissions. *Soil Use Manag.* 13: 230–244.

Pennock, D. 2005. Precision conservation for co-management of carbon and nitrogen on the Canadian prairies. *J. Soil Water Conserv.* 60: 396–401.

Pimentel, D., and M. Burgess. 2013. Soil erosion threatens food production. *Agriculture* 3(3): 443–463.

Postel, S., P. Polak, F. Gonzales, and J. Keller. 2001. Drip irrigation for small farmers: A new initiative to alleviate hunger and poverty. *Water Int.* 26(1): 3–13.

Poth, M., J.S. La Favre, and D.D. Focht. 1986. Quantification by direct ^{15}N dilution of fixed N_2 incorporation into soil by *Cajanus cajan* (pigeon pea). *Soil Biol. Biochem.* 18: 125–127.

Precision Variable Rate Irrigation. 2014. *Overview* (available at http://www.precisionirrigation.co.nz/en/pages /product/).

Quin, B.F., and M. Zaman. 2012. RPR revisited (1): Research, recommendations, promotion and use in New Zealand. *Proc. NZ Grassland Assoc.* 74: 255–268.

Raes, D., P. Steduto, T.C. Hsiao, and E. Fereres. 2009. AquaCrop–the FAO crop model to simulate yield response to water: II. Main algorithms and software description. *Agron. J.* 101: 438–447.

Rajan, S.S.S., J.H. Watkinson, and A.G. Sinclair. 1996. Phosphate rocks for direct application to soils. *Adv. Agron.* 57: 77–159.

Renscher, C.S., and T. Lee. 2005. Spatially distributed assessment of short term and long term impacts of multiple best management practices in agricultural watersheds. *J. Soil Water Conserv.* 62: 446–456.

Rockström, J., N. Hatibu, T. Oweis, and S.P. Wani. 2007. Managing water in rainfed agriculture. In: *Water for Food, Water for Life: A Comprehensive Assessment of Water Management in Agriculture*, ed. D. Molden, 315–348. London: Earthscan, and Colombo: International Water Management Institute.

Rutherford, J.C., and M.L. Nguyen. 2004. Nitrate removal in riparian wetlands: Interactions between surface flow and soils. *J. Environ. Qual.* 33: 1133–1143.

Sakadevan, K., and M.L. Nguyen. 2010. Global response to soil and water salinization in agricultural landscapes. In: FAO World Soils Report No. 105, ed. R.P. Thomas, 35, *Proc. Global Forum Salinization and Climate Change*, October 25–29, 2010, Valencia, Spain.

Sakadevan, K., L.K. Heng, and M.L. Nguyen. 2014. Water conservation zones in agricultural catchments for biomass production, food security and environmental protection. In: *Managing Soils for Food Security and Climate Change Adaptation and Mitigation*, eds. L.K. Heng, K. Sakadevan, G. Dercon, and M.L. Nguyen, Proceedings FAO/IAEA Int. Symposium, July 23–27, 2012. Rome: FAO.

Sale, P.W.G., P.G. Simpson, C.A. Anderson, and L.L. Muir. 1997. The role of reactive phosphate rock fertilizers for pastures in Australia. *Aust. J. Exp. Agric.* 37: 845–1023.

Samosir, S.S.R., G.J. Blair, and R.D.B. Lefroy. 1993. Effects of placement of elemental S and sulfate on the growth of two rice varieties under flooded conditions. *Aust. J. Agric. Res.* 44: 1775–1788.

Scheer, C., R. Wassmann, K. Kienzler, N. Ibragimov, and R. Eschanov. 2008. Nitrous oxide emissions from fertilized, irrigated cotton (*Gossypium hirsutum* L.) in the Aral Sea Basin, Uzbekistan: Influence of nitrogen applications and irrigation practices. *Soil Biol. Biochem.* 40: 290–301.

Schumacher, J.A., T.C. Kaspar, J.C. Ritchie et al. 2005. Identifying spatial patterns for use in precision conservation. *J. Soil Water Conserv.* 60: 355–362.

Sharpley, A.N., P. Kleinman, and R. McDowell. 2001. Innovative management of agricultural phosphorus to protect soil and water resources. *Commun. Soil Sci. Plant Anal.* 32: 1071–1100.

Shedley, C.D., A.R. Till, and G.J. Blair. 1979. A radiotracer technique for studying the nutrient release from different fertilizer materials and its uptake by plants. *Commun. Soil Sci. Plant Anal.* 10: 737–745.

Sheng, T.C. 2007. Principles and practices of monitoring and evaluation for watershed and conservation projects. In: *Monitoring and Evaluation of Watershed and Soil Conservation Projects*, eds. J. de Graaf, J. Cameron, S. Sombatpanit, C. Pieri, and J. Woodhill, 13–21, Enfield, NH: Science Publishers.

Shuttleworth, W.J., M. Zreda, X. Zeng, C. Zweck, and T.P.A. Ferre. 2010. The Cosmic-Ray Soil Moisture Observing System (COSMOS): A non-invasive, intermediate scale soil moisture measurement network. In *Role of Hydrology in Managing Consequences of a Changing Global Environment*, Proceedings British Hydrological Society's Third International Symposium, Newcastle University, July 19–23, 2010.

Sillanpää, M. 1990. *Micronutrient Assessment at the Country Level: An International Study*. FAO Soil Bulletin No. 63. Rome, Italy: FAO.

Singh, U., S.A. Smallberger, S.H. Chien, and P.W. Wilkens. 2006. Development and testing of a phosphate rock decision support system (PRDSS) for direct application. In *Management Practices for Improving Sustainable Crop Production in Tropical Acid Soils*. IAEA Proceedings Series, STI/PUB/1285. IAEA ed., 219-2 Vienna, Austria: IAEA.

Smaling, E.M.A., S.M. Nandwa, and B.H. Janssen. 1997. Soil fertility in Africa is at stake. In *Replenishing Soil Fertility in Africa*, SSSA Special Publication 51, eds. R.J. Buresh, P.A. Sanchez, and F. Calhoun, 47–61. Madison, WI: SSSA.

Smallberger, S.A., U. Singh, S. Chien, J. Henao, and P.W. Wilkens. 2006. Development and validation of a phosphate rock decision support system. *Agron. J.* 98: 471–483.

Stafford, J.V. (ed.). 2013. *Precision Agriculture '13.* Proceedings 9th European Conference on Precision Agriculture, Lleida, Spain. Wageningen, The Netherlands: Wageningen Academic Publishers.

Steduto, P., T.C. Hsiao, D. Raes, and E. Fereres. 2009. AquaCrop–the FAO crop model to simulate yield response to water: I. Concepts and underlying principles. *Agron. J.* 101: 426–437.

Steduto, P., T.C. Hsiao, E. Fereres, and D. Raes. 2012. *Crop Yield Response to Water.* Food and Agricultural Organization of the United Nations.

Stroia, C., C. Morel, and C. Jouany. 2007. Dynamics of diffusive soil phosphorus in two grassland experiments determined both in the field and laboratory conditions. *Agric. Ecosyst. Environ.* 119: 60–74.

Takkar, P.N., I.M. Chibba, and S.K. Mehta. 1989. *Twenty Years of Co-ordinated Research of Micronutrients in Soil and Plants 1967–1987.* Bull. Indian Institute of Soil Science. Bhopal, India: IISS.

Taylor, A., W.H. Blake, H.G. Smith, L. Mabit, and M.J. Keith-Roach. 2013. Assumptions and challenges in the use of fallout beryllium-7 as a soil and sediment tracer in river basins. *Earth-Sci. Rev.* 126: 85–95.

Till, A.R. 1981. The study of sulphur recycling processes in a grazing system using S. In *Animal World Science, B1, Grazing Animals.* Chapter 3, 33–54. Amsterdam: Elsevier.

UN-DESA (Department of Economic and Social Affairs) 2013. *World Population Prospects: The 2012 Revision* (available at http://esa.un.org/undp/wpp/index.html).

UNEP. 2010. *The United Nations System-wide Earthwatch* (available at http://earthwatch.unep.ch/emerging issues/desertification/landdegradation.php).

UNEP. 2011. *An Ecosystem Services Approach to Water and Food Security.* Nairobi: UNEP and Colombo: IWMI.

UN Office for Outer Space Affairs (UNOOSA). 2014. *UN Programme on Space Applications: Natural Resource Management and Environmental Monitoring* (available at http://ww.oosa.unvienna.org/oosa /SAP/rs/index.html).

US-NASA. 2014. *Precision farming.* Feature article (available at http://earthobservatory.nasa.gov/Features /PrecisionFarming/precision_farming5.php).

US-NRCS. 2014. *Conservation Practices that Save: Precision Agriculture* (available at http://www.nrcs.usda .gov/wps/portal/nrcs/site/national/home).

US-USDA, NIFA. 2014. *Program Precision, Geospatial and Sensor Technologies* (available at http://www .csrees.usda.gov/precisiongeospatialsensortechnologies.cfm).

Vallejo, A., A. Meijide, P. Boeckx, A. Arce, L. García-Torres, P.L. Aguado, and L. Sanchez-Martin. 2014. Nitrous oxide and methane emissions from a surface drip-irrigated system combined with fertilizer management. *Eur. J. Soil Sci.* 65: 386–395.

Van Cleemput, O., F. Zapata, and B. Vanlauwe. 2008. Use of tracer technology in mineral fertilizer management. In: *Guidelines on Nitrogen Management in Agricultural Soils,* IAEA-TCS-CD 29, ed. IAEA, 19–126. Vienna. Austria: IAEA.

Van Kessel, C. 2008. Nitrogen transformation and turnover in soils amended with organic sources. In: *Guidelines on Nitrogen Management in Agricultural Soils,* IAEA-TCS-29CD, ed. IAEA, 204–219. Vienna, Austria: IAEA.

Vanlauwe, B., C.A. Palm, H.K. Murwira, and R. Merckx. 2002. Organic resource management in sub-Saharan Africa: Validation of a residue quality-driven decision support system. *Agron.* 22: 839–846.

Verchot, L.V., and P. Cooper. 2008. International agricultural research and climate change: A focus on tropical systems. *Agric. Ecosyst. Environ.* 126: 1–3.

Walling, D.E., Q. He, and P.G. Appleby. 2002. Conversion models for use in soil-erosion, soil-redistribution and sedimentation investigations. In *Handbook for the Assessment of Soil Erosion and Sedimentation Using Environmental Radionuclides,* ed. F. Zapata, 111–164. Dordrecht, The Netherlands: Kluwer.

Walling, D.E., Y. Zhang, and Q. He. 2011. Models for deriving estimates of erosion and deposition rates from fallout radionuclide (caesium-137, excess lead-210, and beryllium-7) measurements and the development of user friendly software for model implementation. In *Impact of Soil Conservation Measures on Erosion Control and Soil Quality,* IAEA-TECDOC-1665, ed. IAEA, 11–33. Vienna, Austria: IAEA.

West, T.O., and W.M. Post. 2002. Soil organic carbon sequestration rates by tillage and crop rotation: A global data analysis. *Soil Sci. Soc. Am. J.* 66: 1930–1946.

Williams, D.G., W. Cable, K. Hultine et al. 2004. Evapotranspiration components determined by stable isotopes, sap flow and eddy covariance techniques. *Agric. Forest. Meteor.* 125: 241–258.

Yang, D., S. Kanae, T. Oki, T. Koike, and K. Musiake. 2003. Global potential soil erosion with reference to land use and climate changes. *Hydrol. Process.* 17(14): 2913–2928.

Yasmin, N. 2003. Availability of fertilizer S and suitable time of application for rice, mustard and cotton. In *Guidelines for the Use of Isotopes of Sulphur in Soil-Plant Studies*, IAEA TCS-20, ed. IAEA, 97–98. Vienna, Austria: IAEA.

Zaharah, A.R., and F. Zapata. 2003. The use of P-32 isotope techniques to study soil P dynamics and to evaluate the agronomic effectiveness of phosphate fertilizers. In *Direct Application of Phosphate Rock and Related Technology: Latest Developments and Practical Experiences*, Special Publications IFDC-SP-37, eds. S.S.S. Rajan, and S.H. Chien, 225–235. Muscle Shoals, AL: IFDC.

Zaman, M., H.J. Di, K.C. Cameron, and C.M. Frampton. 1999a. Gross N mineralization and nitrification rates and their relationships to enzyme activities and soil microbial biomass in soils treated with dairy shed effluent and ammonium fertilizer at different water potentials. *Biol. Fert. Soils* 29(2): 178–186.

Zaman, M., H.J. Di, and K.C. Cameron. 1999b. Gross N-mineralization and nitrification rates and their relationships to enzyme activities and soil microbial biomass in soils treated with dairy shed effluent and ammonium fertilizer in the field. *Soil Use Manag.* 15(3): 188–194.

Zaman, M., K.C. Cameron, H.J. Di, and K. Inubushi. 2002. Changes in mineral N, microbial biomass and enzyme activities in different soil depths after surface applications of dairy shed effluent and ammonium fertilizer. *Nutr. Cycling Agroecosyst.* 63: 275–290.

Zaman, M., M.L. Nguyen, A.J. Gold, P.M. Groffman, D.Q. Kellogg, and R.J. Wilcock. 2008. Nitrous oxide generation, denitrification and nitrate removal in a seepage wetland intercepting surface and subsurface flows from a grazed dairy catchment. *Aust. J. Soil Res.* 46: 565–577.

Zaman, M., and M.L. Nguyen. 2010. Effect of lime or zeolite on N_2O and N_2 emissions from a pastoral soil treated with urine or nitrate-N fertiliser under field conditions. *Agric. Ecosyst. Environ.* 136: 254–261.

Zaman, M., M.L. Nguyen, M. Šimek, S. Nawaz, M.J. Khan, M.N. Babar, and S. Zaman. 2012. Emissions of nitrous oxide (N_2O) and di-nitrogen (N_2) from agricultural landscape, sources, sinks, and factors affecting N_2O and N_2 ratios. In: *Greenhouse Gases–Emission, Measurement and Management*, ed. G. Liu, 1–32. Croatia: Intech.

Zaman, M., S. Zaman, C. Adhinarayanan, M.L. Nguyen, and S. Nawaz. 2013. Effects of urease and nitrification inhibitors on the efficient use of urea for pastoral systems. *Soil Sci. Plant Nutr.* 59: 649–659.

Zapata, F. (ed.). 1995. Evaluation of the agronomic effectiveness of phosphate fertilizers through the use of nuclear and related techniques. Special issue *Fert. Res.* 41: 167–242.

Zapata, F. (ed.). 2002a. *Handbook for the Assessment of Soil Erosion and Sedimentation Using Environment Radionuclides*. Dordrecht, The Netherlands: Kluwer.

Zapata, F. (ed.). 2002b. Utilisation of phosphate rocks to improve soil P status for sustainable crop production in acid soils. Special issue, *Nutr. Cycl. Agroecosyst.* 63(1): 1–98.

Zapata, F. (ed.). 2003. Field application of the caesium-137 technique in soil erosion and sedimentation studies. Special issue, *Soil Till. Res.* 69: 1–162.

Zapata, F., and H. Axmann. 1995. P-32 isotopic techniques for evaluating the agronomic effectiveness of rock phosphate materials. *Fert. Res.* 41: 189–195.

Zapata, F., and C. Hera. 1996. FAO/IAEA International networking in soil/plant nitrogen research. In *Nitrogen Economy in Tropical Soils*, ed. N. Ahmad, 415–425. Dordrecht, The Netherlands: Kluwer.

Zapata, F., and M.L. Nguyen. 2009. Soil erosion and sedimentation studies using environmental radionuclides. Chapter 7. In *Environmental Radionuclides: Tracers and Timers of Terrestrial Processes*, ed. K. Froehlich, 295–322. Amsterdam: Elsevier B.V.

Zapata, F., and R.N. Roy (eds.). 2004. *Use of Phosphate Rocks for Sustainable Agriculture*. FAO Fertilizer and Plant Nutrition Bulletin No. 13. Rome, Italy: FAO.

Zarco-Tejada, P.J., J.A.J. Berni, L. Suárez, G. Sepulcre-Cantó, F. Morales, and J.R. Miller. 2009. Imaging chlorophyll fluorescence from an airborne narrow-band multi-spectral camera for vegetation stress detection. *Remote Sensing Environ.* 113: 1262–1275.

Zedler, J.B. 2003. Wetlands at your service: Reducing impacts of agriculture at the watershed scale. *Front. Ecol. Environ.* 1: 65–72.

Zhao, F.J., J.S. Knights, Z.Y. Hu, and S.P. McGrath. 2003. Stable sulphur isotope ratio indicates long term changes in sulphur deposition in the Broadbalk Experiment since 1845. *J. Environ. Qual.* 32: 33–39.

Zreda, M., D. Desilets, T.P.A. Ferre, and R.L. Scott. 2008. Measuring soil moisture content non-invasively at intermediate spatial scale using cosmic-ray neutrons. *Geophys. Res. Lett.* 35: 5.

Zreda, M., W.J. Shuttleworth, X. Zeng, C. Zweck, D. Desilets, T.E. Franz, and R. Rosolem. 2012. COSMOS: The Cosmic-Ray Soil Moisture Observing System. *Hydrol. Earth Syst. Sci.* 16: 4079–4099.

11 Precision Agriculture for Improving Water Quality under Changing Climate

Sami Khanal and Rattan Lal

CONTENTS

11.1 INTRODUCTION

Declining water quality has become a global concern due to worldwide expansion of agricultural land and intensification of agricultural practices, industrialization, and climate change. Every day, 2 million tons of agricultural and industrial waste and sewage are discharged into the world's water. The amount of wastewater produced annually is about 1500 km^3, six times more water than exists in all the rivers of the world (UN WWAP 2003). Waterborne infections account for 80% of all infectious diseases worldwide and 90% in developing countries. In the United States, almost one-third of the nation's stream miles and 20% of the lakes contain high total phosphorus (P) and total nitrogen (N) concentrations (US EPA 2010), and waterborne diseases account for 900,000 infections and about 900 deaths each year (Pimentel et al. 2007). Globally, 24% of mammals and 12% of birds connected to inland waters are considered threatened due to both degrading water quality and excessive water harvesting (UN WWAP 2003).

Agricultural nonpoint source pollution (NPSP) is the leading source of water impairment in the streams and wetlands and is a major contributor to groundwater contamination around the world (Bennett et al. 2001). Sources for NPSP include agricultural activities including improper, excessive, or poorly timed application of fertilizer, pesticides, and irrigation water, plowing too often or at the wrong time, poorly located or managed animal feeding operations, or excessive use of tile drainage (Kling et al. 2014). High concentration of N and P in water bodies from agricultural runoff causes algal blooms that promotes hypoxic zones and impairs water supplies. The number of hypoxic zones due to agricultural runoffs is increasing globally in both number and severity (Diaz and Rosenberg 2008). They are now observed annually in the Gulf of Mexico, the Great Lakes (Zhou et al. 2013), Chesapeake Bay, Long Island Sound, the Baltic Sea, and the Black Sea (Rabalais et al. 2002). As a result of contaminated drinking water due to algal bloom in Lake Erie, from which the city of Toledo draws its water, around 0.5 million people in Toledo were without access to safe drinking water for two days in August 2014 (Fitzsimmons 2014). The annual cost of NPSP is high when replacement cost for nutrients, lost water, and soil depth are included (Pimentel et al. 1995). In the United States, P-related water quality problems alone cost over $2.2 billion/year including losses in recreational water usage, waterfront real estate, in recovery of threatened and endangered species, and in cleaning of drinking water (Dodds et al. 2009). It costs around $4.8 billion per year to remove nitrate from sources of drinking water, and $1.7 billion to remove nitrate resulting from agriculture alone (Ribaudo et al. 2011).

Weather patterns including precipitation, wind, and temperature are also major causes of nutrient transport from agricultural watersheds. Increased climatic variability in combination with agricultural intensification and expansion in the past few decades have already degraded soil quality and altered regional runoff regimes (IPCC 2013). Projected future increase in temperature and variability in precipitation under current agricultural practices is anticipated to further degrade the soil quality, increase the susceptibility to runoff and erosion (Lal 2004), deplete water resources (Schewe et al. 2014), and exacerbate the current water quality problems (Park et al. 2010). Preventing soil degradation and erosion, improving water quality, and achieving sustainable food security thus have been major challenges facing the world. Improving the resilience of the agricultural sector to climate change and mitigating agricultural-driven water and other environmental problems requires development and implementation of soil and water conservation practices and programs (Lal et al. 2011).

Water quality problems contributed by agriculture can be solved through (1) input management of water and nutrient applications, (2) controlling and trapping water and nutrient runoff at the edge-of-field, and (3) controlling and trapping sediment and nutrients in the aquatic system adjacent to an agricultural area (Kroger et al. 2012). Input management of water and nutrient applications is the first area where improvement in water quality can occur. First, if farmers apply nutrients according to recommendations and correctly manage applications, the amount of nutrients exported can be minimized. Site- and soil-specific farming practice such as precision agriculture (PA) contribute to the assessment of optimal requirement of inputs. Sustained yields with optimized fertilizer and water applications can result in reduced runoff and increased economic benefits. Second, the implementation of conservation structures in areas contributing to higher nutrient pollution in the watershed (also called critical source areas [CSAs]) help maximize environmental benefits. Conservation structures include edge-of-field manipulation (e.g., buffer strips, riparian corridors, terraces) in the agricultural watershed that slow runoff, lower sedimentation loss, and increase nutrient reductions by controlling and trapping water and nutrients (Mayer et al. 2007; Schoonover et al. 2005; Tomer et al. 2007). Third, aquatic systems such as drainage ditches, grassed waterways, and creeks adjacent to agricultural fields can be manipulated and managed to control and trap nutrients. These systems reduce nutrient delivery downstream through settling and enhanced biogeochemical processes (Kroger et al. 2009; Moore et al. 2010).

In this chapter, we review and synthesize relevant information on the impacts of agricultural practices and climate change on water quality, the role of PA on improving soil, crop, and water quality, and how the use of precision technologies helps guide effective farm management practices for water quality improvement. We also highlight some of the knowledge gaps and suggestions for policy makers relating to PA.

11.2 LAND USE AND MANAGEMENT PRACTICES AND WATER QUALITY

Land use and management practices are the important drivers of regional and global hydrologic cycles determining water quality (Kaushal et al. 2014). Changes in vegetation type alter leaf area index (LAI), surface cover, and soil quality that significantly affect hydrologic responses, including the timing and magnitude of evaporative losses to the atmosphere, water infiltration, soil moisture, runoff, and base flow. In forests and other areas with good vegetation cover and little human disturbance, overall soil health is good with enhanced soil infiltration and water buffering capacity. Consequently, most rainfall infiltrates through the soil rather than running off the ground; stream flows are fairly steady, and water quality is good. Agricultural and urban watersheds on the other hand provide less surface cover and have relatively poor soil quality (Guo and Gifford 2002) and thus generate flashier hydrographs across storm sizes, and most nutrient exports occur during high-flow events (Vidon et al. 2009).

Hydrologic connectivity (i.e., transfer of nutrients within or between elements of the hydrologic cycle) in an agricultural watershed is often enhanced by tile drains and ditches. Tile drainage has been an important component of agricultural practices in the Midwest region of the United States for over 150 years. More than 50% of the cropland in the Midwest has subsurface drainage system (Sugg 2007). In agricultural regions with seasonally high water tables, drainage facilitates seedbed preparation and planting, thereby minimizing plant stress and yield reduction resulting from poor soil aeration due to waterlogging. However, a dense network of tile drainage leads to the transfer of infiltrated precipitation to streams and facilitates rapid delivery of runoff, chemicals, and sediments to streams. For example, in small watersheds where up to 50% of cropland is tile-drained, water can move from uplands to tile drains and to streams in as little as 15 minutes to a few hours during storms (Vidon and Cuadra 2010). Baker et al. (2007) observed that tile-drain flow contributes between 56%–99% of streamflow depending on storm characteristics, demonstrating that tile drains effectively move both water and nutrients at the watershed scale. Soil and nutrient runoffs from agricultural fields adversely impact soil quality and crop yield. Efficient use of tile-drainage system, soil surface cover through residue management, and effective use of fertilization help control nutrient runoff and leaching from the agricultural fields.

11.3 CLIMATE CHANGE AND AGRICULTURAL NUTRIENT RUNOFFS

Many parts of the world are already experiencing long-term trends in climate change with significant implications on agricultural productivity, water supply and quality, and ecosystem services. There have been substantial variations in the spatial and temporal distribution of extreme weather around the world. For example, while the amount of heavy precipitation in the most recent 30-year period (1983–2012) is greater than in the previous 30 years (1953–1982) and the largest changes of greater than 20% are in the Midwest and Northeast United States (Peterson et al. 2014), west and southwest regions of the United States have experienced record-breaking droughts in current and past years. These changes in weather pattern have significant implications on agricultural crop production, water availability, and quality. In mid-August 2012, the extent of the drought conditions was significant: over 80% of the land area of the contiguous United States was affected by abnormally dry and drought conditions. The widespread nature of drought and excessive heat conditions in the Midwest and Great Plains in the summer of 2012 contributed to sharply lower yields for major crops, including corn (*Zea mays*) and soybeans (*Glycine max*) (Lal et al. 2012). More than one-third of agricultural land areas in California's fertile Central Valley are likely to be fallowed due to the 2014 drought (Carlton and Lazo 2014). Exports of nutrients from agricultural watersheds vary significantly in response to record drought and extreme wet years (Rabalais et al. 2001). Nitrate and P concentrations, respectively, increased by seven- and 10-fold in an agricultural watershed of Florida during Hurricane Katrina (Delgado et al. 2011; Zhang et al. 2009). Although nutrient runoff from farmland can be much lower during drought periods (Endale et al. 2011), low water levels can contribute to deterioration in water quality due to the lack of dilution of wastewater discharges and existing nutrients.

Future climate projections from various global circulation models (GCMs) based on various greenhouse gas (GHG) emission scenarios provided by the Intergovernmental Panel on Climate Change (IPCC) indicate that temperature and the intensity and frequency of large storm events will increase. Some areas will likely experience a low precipitation, while others (such as the tropics and high latitudes) are expected to see high precipitation (IPCC 2007, 2013). Projected changes in precipitation and temperature are likely to affect soil and nutrient runoffs through changes in hydrologic and biophysical processes, such as growing season, evapotranspiration, soil moisture, nutrient mineralization, and nutrient availability. Warmer temperature during the growing season increases evapotranspiration, limits soil moisture, and demands more water. Similarly, it promotes faster nutrient mineralization, which depletes soil organic matter (SOM) (Lal 2004). Crop yield will reduce if warming exceeds a crop's optimum temperature or if sufficient water and nutrients are not available. In regions where rainfall would decrease, system feedbacks related to decreased biomass production could lead to greater susceptibility of the soil to erode (Nearing et al. 2004). Historical temperature, precipitation, and discharge records indicate that global runoff increases by 4% for each 1°C rise in temperature (Labat et al. 2004). Applying this projection to changes in evapotranspiration and precipitation, it is likely that runoff will increase to 7.8% globally by the end of the century (Oki and Kanae 2006).

More precipitation will increase a region's susceptibility to a variety of factors, including soil moisture availability, flooding, rate of soil erosion, and nutrient and pesticide runoff. In locations where both precipitation and soil moisture decrease, land surface drying is magnified and areas are left increasingly susceptible to reduced water supplies. Water availability is likely to be further exacerbated by poor land use management, overuse from increasing populations, and an increase in water demand primarily from increased agricultural production (IPCC 2007). Further, drylands have higher risks of reduced crop yields and increased nutrient runoff not only because rainfall is irregular or insufficient, but also because up to 40% of the rainfall may be lost as runoff. This poor use of rainfall is partly the result of natural phenomena (relief, slope, and rainfall intensity), but is also caused by improper and inadequate land management practices (removal of crop residues, excessive tillage, eliminating hedges, etc.) that destroy soil structure, reduce organic matter levels, eliminate beneficial soil fauna, and lower water infiltration. A key challenge is how to manage limited rainfall so that avoidable surface runoff does not occur. Crops depend not only on precipitation but also on the ability of the soil to absorb and store water. Inadequate or improper agricultural practices may worsen the natural main causes of soil moisture scarcity and then add water stress in crops (FAO 2004).

11.4   PA AND WATER QUALITY MANAGEMENT

Agriculture is a part of the problem leading to water pollution, but through the adoption of sustainable land management options, it can be a part of the solution for mitigating water quality concerns. To mitigate agriculturally driven water quality concerns and improve the resilience of agriculture sector to climate change, the key challenges are how to better manage the soil–water interaction to reduce water runoff and soil loss to erosion, improve nutrient cycling, increase water infiltration and storage and availability of moisture, increase resilience under variable and extreme precipitation conditions (i.e., drought, and inundation), and ensure storage capacity to buffer seasonal fluctuations in river flows (Morton 2014).

PA is a farming practice based on observing, measuring, and responding to inter- and intrafield variability within the field, and therefore offers promise in addressing agriculturally related environmental issues while maximizing farmers' profitability (Berry et al. 2003; Delgado et al. 2011). By realizing that there is a large variation within a parcel of land (Figure 11.1), PA focuses on getting the right practices in the right place at the right rate and manner for nutrient and water stewardship. By applying precision technologies, site-specific data, such as soil characteristic, fertility and nutrient data, topographic and drainage characteristics, and yield data are collected, stored, and analyzed

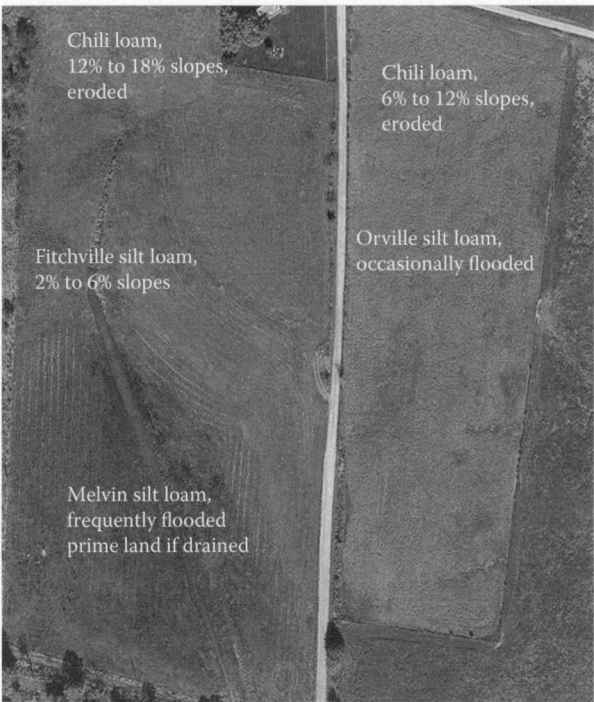

Chili loam,
12% to 18% slopes,
eroded

Chili loam,
6% to 12% slopes,
eroded

Fitchville silt loam,
2% to 6% slopes

Orville silt loam,
occasionally flooded

Melvin silt loam,
frequently flooded
prime land if drained

FIGURE 11.1 Map of fields showing variation at the field level (NRCS Soils) near Coshocton County, OH.

to identify patterns in the field (e.g., identification of areas with greater or lesser yield are in need of irrigation and nutrient, are infected with weed, or have compact soil, and the correlation between yield and topography and soil quality). Once the patterns and correlation are identified, management practices can be modified to optimize farm input needs, including nutrient and pesticide application, tillage and irrigation, and drainage throughout an individual field (Table 11.1). Some benefits of PA

TABLE 11.1
Application Areas for Precision Agriculture and Technologies

Application Areas	Role of Precision Technologies
Soil quality management	Identification of surface roughness, soil compaction, and soil organic matter for residue management, and scheduling of site-specific tillage practice (i.e., depth and type) (Daughtry et al. 2006; Liu et al. 2008; Vaudour et al. 2013)
Nutrient management	Assessment of soil nutrients (i.e., nitrogen, phosphorus, potassium, other nutrients, and pH) (Delgado et al. 2011; Ge et al. 2011; Mallarino and Wittry 2004; Zhang et al. 2010), and crop nutrient stress (Eitel et al. 2007; Goel et al. 2003; Seelan et al. 2003) for optimization of nutrient application
Soil moisture management	Monitoring of crop water stress, soil moisture, and water table depth for irrigation scheduling (Hedley et al. 2013) and drainage management (Ghane et al. 2012; Raine et al. 2007)
Weed management	Identification of weed-infected fields (Lopez-Granados 2011) and site-specific weed management (Christensen et al. 2009)
Seeding	Management of seeding space and depth (Yazgi and Degirmencioglu 2007)
Crop yield monitoring	Crop yield monitoring for strategic planning of inputs (Doraiswamy et al. 2004; Yang et al. 2006) and selective harvesting (Bramley et al. 2011)
Conservation structures	Siting of new conservation structures (Goetz et al. 2003; Williams et al. 2011)

include improvement in crop yield, reduction in soil compaction by limiting multiple trips to field, reduced production cost through reduced implementation overlap, more accurate farming records, improved carbon sequestration, and lower adverse environmental impacts. This also helps target places on the landscape where best management practices (BMPs) will be most effective. Thirty-year (1975–2004) observation data sets on nutrient loads from tributaries of Lake Erie in the United States show some successes of BMPs in reducing concentration of suspended sediment and P loads (Bosch et al. 2014; Richards et al. 2009).

11.5 PRECISION TECHNOLOGIES FOR ENHANCING THE EFFECTIVENESS OF PA

PA relies on massive field data and information collection and processing in time and space to make more efficient use of farm inputs, leading to improved crop production and environmental quality. This is possible due to a combination of technologies, including Global Positioning System (GPS)-driven farm equipment, Geographic Information Systems (GISs) and remote sensing tools, and decision support tools (Figure 11.2). While GPS provides accurate time and space information on the earth surface, remote sensing technologies gather information about the earth surface in the form of electromagnetic radiation from a distant platform, usually a satellite or airborne sensor or sensor mounted on a track, and GIS is used for data management, processing, and analysis. At present, GPS technology is used for various farm operations including soil preparation, seeding, fertilizer application, and harvesting (Figure 11.3). GPSs record the position of the field and help locate and navigate agricultural vehicles within the field. Remote sensing technologies, on the other hand, offer a diagnostic approach for site-specific management of crop and land management. Various remote sensing platforms, ranging from remote (airborne-, satellite-, and unmanned-based platforms) to proximal (ground-based sensors and cameras) sensors, collect data from a distance to evaluate soil and crop health (moisture, nutrients, compactions, crop disease, and crop yield) and help plan for variable rate applications of fertilizer, pesticides, and irrigation. GIS offers analytical capabilities in the areas

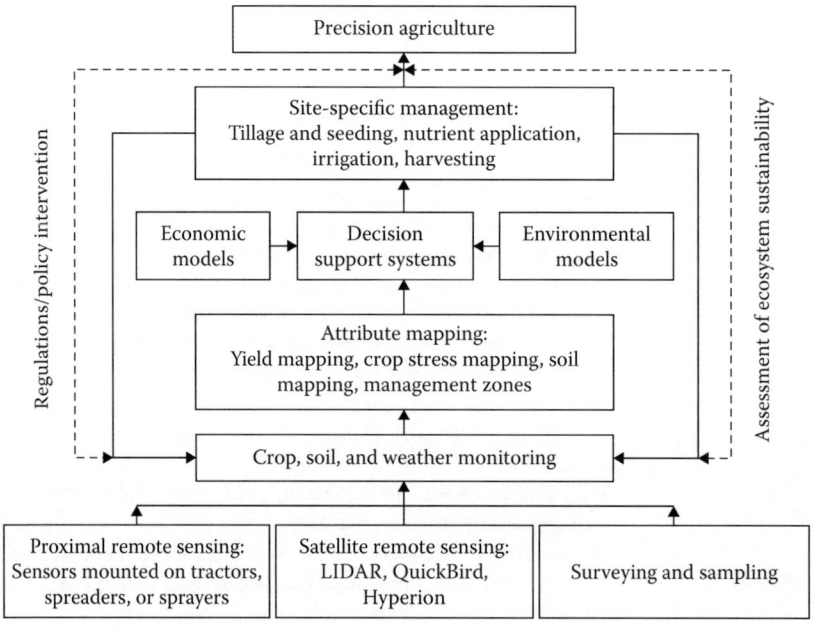

FIGURE 11.2 Precision agriculture framework.

FIGURE 11.3 Precision farm machinery. (From http://www.deere.com.)

of surface modeling, spatial data mining, and map analysis, and help find the spatial relationships within and among mapped field data. The ability to analyze spatial relationships within and among mapped data using various crop and hydrologic models provides new insight into conservation applications and contributes to new evaluation of application of conservation management practices (Berry et al. 2003).

Some of the application areas of PA for water quality improvement include soil moisture management through tillage, drainage, and irrigation, soil nutrient management, crop management through yield monitoring and weed management, and proper installment of edge-of-field structures (e.g., buffer strips, contouring, and terraces) (Figure 11.4). Any solutions related to land management for water quality improvement also enhances carbon sequestration and offers ways to mitigate climate change as water and carbon cycles are tightly interconnected.

11.5.1 REMOTE SENSING: A DIAGNOSTIC APPROACH FOR PA

Typically, remote sensing involves the measurement of reflected radiation (also called spectral signature) from soil and plants. Plant reflectance is high in the near-infrared (NIR) (700–1300 nm) region as a result of leaf density and canopy structure. The sharp contrast in reflectance signal between the red and NIR portion of the spectrum has been the motivation for development of spectral indices based on ratios of reflectance values in the visible and NIR regions (Sripada et al. 2005). Several satellite and remote sensors have emerged over the years with varying spectral, temporal, and spatial coverages (Table 11.2). Multispectral broadband sensors usually collect data for 3–7 bands of around 100 nm width, whereas hyperspectral sensors detect many narrow and contiguous wavelengths, usually <10 nm width. With narrower bandwidths in hyperspectral scanner systems, small variations in reflectivity can be detected that might otherwise be masked within the broader bands of multispectral scanner systems. Repetitive coverage and synoptic view from remotely sensed multi- and hyperspectral sensors are the two important advantages compared to ground observations and hyperspectral airborne data. These multi- and hyperspectral sensors have successfully been able to quantify a large array of agriculturally important soil properties (texture,

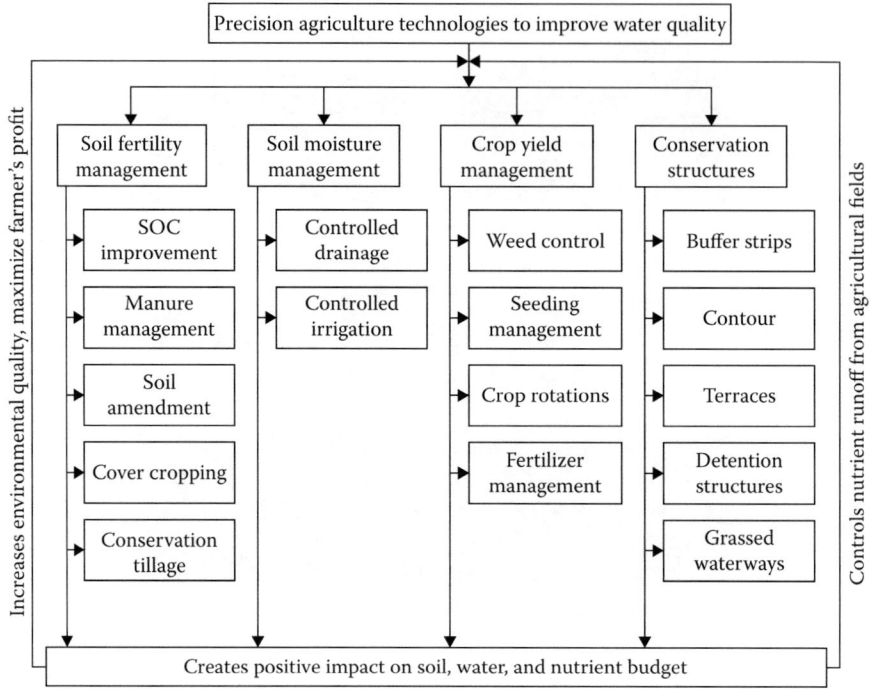

FIGURE 11.4 Areas for water quality improvement.

organic and inorganic carbon content, micro- and macronutrients, moisture content), crop properties (yield, water, and nutrient stress), and weed problems (Table 11.3).

Remotely sensed images encourage the investigation of long-term management practices by providing a visual approach for understanding the effects of management inputs including fertilizer, tillage, and drainage, and their impact on soil quality, agricultural productivity, and other ecosystem services. Images collected several times throughout the growing season also allow timely input to management decisions for correcting problems or deficiencies in the current crop. Further, these maps are useful in developing scouting plans for direct examination of the detected soil and plant condition and in developing site-specific treatment plans (Casady and Palm 2002). Some of the areas where remote sensing information will be useful to manage nutrients and limit water and nutrient runoffs are listed next.

Crop nutrient mapping. Crop nutrients such as chemical fertilizers and manure help achieve optimum crop yields when applied in proper quantities and at appropriate times. However, improper application of nutrients can cause water quality problems locally and downstream. Remotely sensed images can be used primarily for early diagnosis of crop nutrient stress and in developing fertilizer prescription to avoid yield or quality losses (Bausch and Khosla 2010; Kyveryga et al. 2011) or limit excessive fertilizer application (Baush and Diker 2001). Bausch and Delgado (2005) and Delgado and Bausch (2005) demonstrated that use of remote sensing in PA techniques can reduce nitrogen applications by close to 50% in sprinkler-irrigated systems of northeastern Colorado, increasing nitrogen use efficiency and significantly reducing nitrate leaching losses without reducing yields. By integrating remotely sensed information, the effectiveness of PA can be improved in lowering the adverse impact of agriculture on water quality and environment. Management zones can be identified for variable and site-specific nutrient applications to increase nitrogen use efficiency and reduce nitrate leaching losses without decreasing yields (Delgado et al. 2011; Inman et al. 2008; Khosla et al. 2008).

Vegetation and yield mapping. Healthy vegetation can tap nutrients and water from soil through their root systems and is usually associated with high crop yield. Spatial and temporal mapping of

TABLE 11.2
Some of the Remote Sensing Platforms with Their Suitability for Precision Agriculture

Remote Sensing Platforms	Spatial Resolution (m)	# of Bands	Return Frequency (Day)	Suitability for PA
Multispectral Satellite				
Landsat TM (4 and 5), and ETM+ (7)	30	7, 8	16	Medium
SPOT 1–4	20	4	3–5	Medium
LiDAR	VIS (vertical RMSE 10 cm)			High
IKONOS	1–4	4	3–5	High
QuickBird	0.61–2.4	4	1–3.5	High
ASTER	15–90	15	16	Medium
RapidEye	6.5	4	1–2	High
GeoEye-1	1.6 m	4	2–8	High
WorldView-2	0.5 m	8	1.1	High
Hyperspectral Satellite				
EO-1 Hyperion	30	220	16	High
Hyperspectral Airborne				
AVIRIS	4	224		High
CASI	0.25–1.5	288		High
HyMap	2–10	128		High
Airborne				
Unmanned aerial vehicle	Variable	Variable		High

Source: Modified from Mulla, D.J., *Biosyst Eng* 114(4):358–371, 2013; Lopez-Granados, F., *Weed Res* 51(1):1–11, 2011; de Paul Obade, V. et al., *Water Air Soil Poll* 224(9):1–27, 2013.

Note: AVIRIS, airborne visible and near-infrared; CASI, compact airborne spectrographic imager; SAR, synthetic aperture radar.

vegetation yield guide farmers in selecting resource efficient approaches for achieving maximum yield. Remote sensing provides an alternative to the more resource-consuming field measurements and surveys typically used to estimate crop yields at regional to national scales (Sibley et al. 2014). Several studies have demonstrated accurate corn yield prediction with statistical models by relating measured yield to vegetation indices (VIs) derived from remotely sensed reflectance measurements (Cicek et al. 2010; Panda et al. 2010; Shanahan et al. 2001). Using vegetation indices from Landsat-5 and SPOT-4 satellite imageries, Cicek et al. (2010) found corn/soybean grain yield from controlled tile drainage to be higher than those from uncontrolled tile drainage. Mapping of vegetation yield also offers farmers in assessing the site suitability for crop, crop rotation, and optimization of farm inputs.

Soil carbon mapping. The amount and composition of soil organic carbon (SOC) are important indicators of soil quality and ecosystem services (Lal 2013). SOC stores and supplies nutrient and herbicides, increases soil water-holding capacity (Liu et al. 2008), and is also positively correlated with high plant yield. It is an important parameter in developing appropriate site-specific nutrient (Lal et al. 2011) and herbicide application recommendations (Locke and Bryson 1997). SOC, however, varies spatially within a field and sampling its distribution is time-consuming and costly. A remote sensing approach has been suggested as more accurate and less costly than the grid-sampling method (Chen et al. 2008). Reflectance in various spectral bands has been correlated with

TABLE 11.3
Remote Sensing of Soil and Crop Properties for Precision Agriculture

Previous Studies	SOM	SOC	Soil Texture	Soil Tillage	Soil Nutrient	Drainage	Crop Nutrient and Water Stress	Weed Detection	Crop Yield Monitoring	Remote Sensing Platforms
Coleman et al. (1993)	X									Landsat 5TM
Agbu et al. (1990)	X									SPOT
Lopez-Granados (2011)	X									HyMap
Vaudour et al. (2013)		X								SPOT
Stevens et al. (2006)		X								CASI
Sullivan et al. (2005)				X						IKONOS
Bricklemyer et al. (2006)				X						Landsat 5TM
Daughtry et al. (2006)				X						EO-1 Hyperion
Clark et al. (2003)					X					AVIRIS
Sridhar et al. (2009)					X					Landsat 5TM
Weng et al. (2010)					X					EO-1 Hyperion
Seelan et al. (2003)						X	X		X	IKONOS
Griffin et al. (2011)						X				Landsat 5TM
Wan et al. (2014)						X				LIDAR
Goel et al. (2003)							X			CASI
Eitel et al. (2007)							X			RapidEye
Berni et al. (2009)							X			Unmanned aerial vehicle
Chen et al. (2010)							X			CASI
Lopez-Granados (2011)								X		IKONOS, QuickBird, GeoEye-1, AVIRIS, CASI
Andujar et al. (2013)								X		LIDAR
Doraiswamy et al. (2004)									X	Landsat 5TM
Yang et al. (2006)									X	QuickBird

soil properties such as SOM, and the relationships developed between reflectance and SOM are used for mapping concentration of SOM (Chen et al. 2008, Gelder et al. 2011; Sullivan et al. 2005). Chen et al. (2008) used aerial photography and soil sample to produce continuous estimates of SOC concentrations at the field scale, and advanced this technique by using similarity analysis to predict SOC in surrounding fields without additional sample collection. Effective nutrient management at field scale can be facilitated through the monitoring of spatial and temporal distribution of SOC using remote sensing.

Soil moisture mapping. Soil moisture is a key player in land–atmosphere interactions (Pielke 2005), and it influences hydrologic processes in agricultural and forest landscapes. Assessment of soil moisture at various temporal and spatial scales helps in scheduling of differential irrigation treatment as opposed to uniform treatment. Several remote sensing sensors (Lakshmi et al. 2011; Minet et al. 2012) have been used to provide soil moisture estimates at field and watershed scales and also have been used widely in hydrologic studies (Lakshmi et al. 1997). Precision irrigation based on remotely acquired soil moisture information allows farmers not only to boost crop yield while using less water but also help them reduce nutrient runoff and leaching by avoiding excess water application (Raine et al. 2007). For example, Cardenas-Lailhacar and Dukes (2012) showed that use of soil moisture sensor system on turfgrass plots in Florida reduced irrigation frequency from 7 days per week (d week^{-1}) to 1 and 2 d week^{-1}. Similarly, information on soil moisture and soil compaction is useful for farmers in selection of precision tillage practice. Usually, no-till (NT) conservation tillage, a practice that leaves crop residue on fields before and after planting the crop, reduces soil erosion and runoff and has other important advantages such as maintenance of soil moisture level during drought and enhancement of SOC. NT in wet soil, however, slows soil warming in the spring and delays germination and growth. Residue in the surface can physically impair planting by causing plugging within the planting unit and can influence crop yield. In a situation where soil is wet, the tillage practice known as strip-till offers the soil conservation benefits of NT and captures the production advantages of full-width tillage (Wolkowski et al. 2009).

Drainage mapping. Artificial drainage systems are necessary to remove the excess water from the root zone of the crop and prevent flooding of field. These drainage systems increase crop yield by enhancing soil oxygen and lower losses of gaseous nitrogen and microbial activity. Remote sensing can be used to locate the subsurface drainage tile lines (Naz et al. 2009) installed decades ago for maintenance of old tile drain lines or for the installation of new additional drainage systems. With the installation of controlled drainage structures that control drainage during dry seasons, farmers may benefit over the conventional system (Ghane et al. 2012). Information on elevation combined with yield data obtained through remote sensing can help locate the areas that are waterlogged or prone to erosion risk and need drainage improvements (Liu et al. 2008; Seelan et al. 2003).

Mapping of conservation structures. Management of conservation structures such as riparian buffers, grassed waterways, contour, and terraces adjacent to agricultural fields helps to control nutrient and sediment runoff from agricultural fields to streams. Remote sensing can assist in monitoring and management of areas in the watershed that are prone to erosion and flooding. Williams et al. (2011) used digital elevation data and terrain analysis to develop precise contour planting maps for managing winter wheat and to help capture runoff and minimize erosion. Goetz et al. (2003) used remote sensing imagery for mapping of impervious surface areas and riparian buffer zones in relation to stream health ratings. Remote sensing enables site suitability assessment for placement of various control structures in the fields that farmers can use to lower agricultural runoffs from their fields.

11.5.2 Hydrologic Model: A Simulation Tool to Evaluate the Effectiveness of PA

While remote sensing helps to locate the areas within the field that need to be managed for a variety of reasons, including poor crop yield, soil quality (Pinter et al. 2003), or susceptibility to runoff

(Williams et al. 2011), hydrologic models provide unique opportunities in assessing the implications of land use and agricultural management practices as well as climate change on soil, water, and crop quality (Arnold and Fohrer 2005; Flanagan et al. 2001; Gassman et al. 2007, 2009; Powers et al. 2011; Williams and Singh 1995) by combining complex and known ecological relations and hydrologic processes. Information on land use cover, management practices, topographic variability, and soil properties obtained through remote sensing and GIS analysis strengthens the effectiveness of the models used for simulating soil erosion, water quality, and quantity at temporal and spatial scales (Stisen et al. 2008). This allows hydrologists to deal with large-scale, complex, and spatially distributed hydrologic processes (Berry et al. 2003).

It is often difficult to quantify the role of agricultural practices on water quality given the high monitoring cost and variable weather patterns. Further, the effectiveness of BMPs varies from site to site and with the type of BMPs applied. Changing climate further introduces uncertainty in the role of current BMPs (Giri et al. 2014). For example, current practices that are effective at present conditions may not be effective under future climates to address future water quality issues. Calibrated and validated hydrologic models that integrate modules for erosion and crop growth thus have often been used to identify and prioritize the critical areas within the watershed needing improvement in agricultural practices as well as in examining the effectiveness of various BMPs for water quality management (Table 11.4). Examining the effectiveness of BMPs across various geographic scales under different climate scenarios helps to develop NPSP mitigation strategies under both current and future climates (Bosch et al. 2014; Giri et al. 2014).

TABLE 11.4
Summary of Prior Studies on Conservation Practices Using Hydrologic Modeling

Location	Model	BMPs Studied	Reductions (%)				Reference
			Sediment	TN	Nitrate	TP	
West Fork Watershed, TX	SWAT	Nutrient management	85–97	77–93		53–78	Santhi et al. (2006)
West Fork Watershed, TX	SWAT	Forage harvest management	21–76	4–23		1–11	Santhi et al. (2006)
Iowa	SWAT	Land set-asides, terraces, grassed waterways, contouring, conservation tillage, nutrient reduction	6–65		6–20	28–59	Secchi et al. (2007)
Maumee Watershed, OH	SWAT	No-till, cover crops, filter	42	28	26	30	Bosch et al. (2013)
Riesel, TX	EPIC	No-till	89		52		King et al. (1996)
Illinois NRI Cropland Sites	EPIC	No-till	80				Phillips et al. (1993)
Mill Creek Watershed, TX	APEX	Terrace, contour farming, conservation cropping (i.e., with less tillage), and nutrient management	96	87–89		73–82	Tuppad et al. (2010a)
Coshocton, OH	WEPP	5 m wide field border	12				Renschler and Lee (2005)

Among the various NPSP evaluation models, models such as the Soil & Water Assessment Tool (SWAT) (Bosch et al. 2013, 2014; Giri et al. 2014; Jha et al. 2006), Agricultural Policy/ Environmental eXtender Model (APEX) (Williams and Izaurralde 2000), Water Erosion Prediction Project (WEPP) (Foster and Lane 1987; Shen et al. 2009), and Environmental Policy Integrated Climate Model (EPIC) (Izaurralde et al. 2006) have been widely used to evaluate the current and future soil, crop, water quality, and hydrologic impacts of various agricultural and land use management practices. These models have the ability to simulate various crop rotations, tillage, and BMP practices (e.g., contours, terraces, grazing, filter strips, cover crops, and bio-fuel crops), allowing one to test alternative management scenarios aimed at reducing sediment and nutrient transport from the watershed. Each of these models has its own strength and weakness over other models. For example, SWAT is a basin-scale, continuous-time model (Gassman et al. 2007) whereas APEX, EPIC (Gassman et al. 2009), and WEPP (Defersha and Melesse 2012; Laflen et al. 1991) are farm/field scale continuous time models. Similarly, hillslope or landscape profile application of process-based WEPP in simulating runoff, rill, and interrill erosion offers major advantages (Flanagan and Nearing 1995) over SWAT that uses the empirical Modified Universal Loss Equation (MUSLE) for estimating runoff (Shen et al. 2009). SWAT, on the other hand, is a comprehensive hydrologic model and is widely used around the world (Gassman et al. 2007). Models, in combination with observation data from historical and current monitoring programs, help provide information for total maximum daily loads (TMDLs) waste/load allocation and implementation strategies (Santhi et al. 2006) as well as evaluation measures for various BMPs. Understanding the spatial and temporal variability in BMPs performance using models allows water resource managers to incorporate uncertainty analysis into mitigation strategies that aim to reduce negative impacts of management practices and climate change on water resources (Woznicki and Nejadhashemi 2014).

Prioritization of CSAs for BMP implementation. Hydrologic models help to identify/prioritize CSAs generating a disproportionate amount of pollutants and sediments in the watershed that need to be managed for NPSP reduction (Tuppad et al. 2010b). Critical source areas can be determined based on the type of water-quality mitigation targets (i.e., whether the goal is to establish wastewater treatment by focusing on total pollutant load from subbasin reach, identify hot spots of excessive loads that threaten aquatic health by considering total pollutant load from each subbasin, or identify local concerns within an area by considering an average pollutant load per unit area from each subbasin) (Tuppad and Srinivasan 2008). When BMPs are strategically placed in CSAs rather than implemented randomly in the watershed or field, the percent of land required to reduce sediment/ nutrient from a given watershed will be less (Bosch et al. 2013; Tuppad et al. 2010b). For example, in the Smoky Hill River watershed in Kansas, for a 10% reduction in sediment yield, a strategic targeting approach would require implementation of vegetative filter strips (VFSs) (6 m) on only 8% of the cropland in the watershed compared to 25% of the watershed using random implementation (Tuppad et al. 2010b). More than double the land area is estimated to be needed to achieve a 10% reduction in total N and total P yields for a random approach compared to a targeted approach. Implementing BMPs in only high- and medium-priority areas can achieve similar results as applying BMPs on all areas (Giri et al. 2014). Prioritizing areas for BMPs implementation through hydrologic modeling thereby offers economic advantage to farmers and provides an environmental advantage at the same time.

Assessment of spatiotemporal variability of CSAs and BMPs. Variability in weather patterns often modifies the effect of BMPs on water quality improvement. Additionally, over the BMP implementation period, geographic distribution of CSAs may change (i.e., the high-priority CSAs in a given period may change to medium- to low-priority and vice versa). Giri et al. (2014) observed faster changes in priority areas for BMP implementation for both sedimentation and P reduction over the implementation years. Assessment of spatiotemporal variability of CSAs and the effectiveness of BMPs leads to informed implementation for maximizing positive impact while assuming minimal risk.

Although a BMP may be effective in reducing nutrient and sediment yield when assessed from a long-term perspective, its performance varies at certain times of the year when examined temporally. For example, based on their hydrologic modeling Woznicki and Nejadhashemi (2014) showed that although native grass results in less pollutant loads for most of the time throughout the year compared to NT, pollutant loads from native grass and NT are about equal in the winter season. The effectiveness of native grass in limiting sediment and nutrient is highly uncertain and correlates with climatic variables. As watershed managers and stakeholders begin to plan adaptive water management strategies, understanding the temporal effectiveness and uncertainty of BMPs will be an important decision-making factor.

Evaluation of BMP effectiveness under current and future climate scenarios. Various factors including targeting method, type of BMPs, and the priority area on which BMP is applied determine the changes in sediment and nutrient pollutants (Giri et al. 2014). Hydrologic models help to assess the role of BMPs in lowering the nutrient and runoff from the fields (Table 11.4). Using SWAT modeling, Tuppad et al. (2010b) found VFSs and terraces (TR) to be more effective in reducing pollutants than reduced tillage in the Smoky Hill River watershed in Kansas. Bosch et al. (2013) found cover crops to be the most effective in reducing flow, sediments, and nutrients compared to NT and filter strips in the watersheds around Lake Erie. Giri et al. (2014) showed that BMPs including native grass, TR, recharge structures, and residue management at the levels of 1000 and 2000 kg/ha reduced significant amounts of total sediment and N. While BMPs like native grass and contour farming demonstrated the highest pollutant reduction, conservation tillage and NT showed the least impact on nutrient reduction.

Hydrologic models can also be used to evaluate the performance of various BMPs in water quality under future climate change scenarios. For example, under an A2 climate scenario (i.e., a future scenario dominated by fossil fuel usage and greater carbon dioxide (CO_2) and temperature increases), Woznicki and Nejadhashemi (2014) showed higher uncertainty of current BMPs in reduction of sediment, total N, and total P compared to a B1 scenario (i.e., a future scenario with alternative energy and sustainability, lesser CO_2 and temperature increases). This is likely due to greater ranges in monthly temperature increases and more variable precipitation changes under a B1 scenario from the current climate. They also showed that native grass and residue management were highly effective compared to other BMPs including NT, grazing management, contour farming, TR, and conservation tillage. Use of hydrologic models under various climate change scenarios helps farmers or policy makers in adapting/promoting BMPs strategically.

11.6   CONSTRAINTS TO THE ADOPTION OF PA

The water quality benefits of PA can be realized at local, watershed, and regional scales only when it is adopted at a larger scale. Despite numerous benefits of PA practices and technological advancement, its adoption is slow and only a minority of farmers in developed countries (e.g., the United States, Canada, Germany, Australia, and Chile) are practicing it. Several factors including cultural perception, lack of local technical expertise, infrastructure and institutional constraints, knowledge and technical gaps, high start-up costs, and in some cases, a risk of insufficient return on the investment, are some of the major obstacles to the adoption of PA (Zarco-Tejada et al. 2014).

PA is often misinterpreted by developing countries as a complex technological intervention to agriculture meant for large crop fields in developed countries. In developing countries such as those in Southeast Asia or sub-Saharan Africa, farms are small and most of the agricultural practices are based on subsistence farming, and the use of sophisticated precision technologies is far beyond the reach of small farm holders. However, PA in these countries can take various forms without large technological inputs. For example, the use of low-cost technological tools such as the chlorophyll meter (e.g., SPAD), leaf color chart (LCC), and portable diagnostic tools for *in situ* measurements of the crop and soil nutrient status combined with experience and intuition can help farmers to adopt PA. Analyses based on 28 projects in 57 developing countries demonstrated that use of various forms of conservation practices (such as reduced tillage, water harvesting, and integration of livestock into farming systems) improved food crop productivity, increased water use efficiency (WUE) and C sequestration, and reduced pesticide use (Pretty et al. 2006).

Lack of willingness and technical expertise of traditional farmers to undertake computer analysis and decision making may be another major constraint in the adoption of PA. While most farmers recognize that site-specific farming offers both economic and environmental benefits, not every farmer is interested in collecting data or has the expertise to make use of it. Many farmers choose agriculture for the active outdoor lifestyle and are reluctant to spend time in front of a computer (Lowenberg-DeBoer 2003). Most farmers in developing countries not only lack basic information about their soil productivity, they also lack good infrastructures for soil testing and yield monitoring. These farmers may benefit from a network of agricultural extension programs or local agrodealers that help them understand their soil fertility or crop productivity through soil sampling or field sensors and provide them with accurate soil fertility/crop growth maps and recommendations for fertilizers. As well, the availability of subsidies can help economically constrained farmers in developing countries to cover the initial start-up cost or lower the risk of insufficient return on the investment of PA. In subsistence farming regions, major advantages of PA come from the reduction in fertilizer application, reduction in loss of soil nutrients, and marginal improvement in grain yield (Khosla 2010).

Similarly, there are gaps between gathering information via precision technologies and the use of gathered information to make management decisions. Although farmers in developed countries collect data using GPS or vehicle-driven sensors, they lack the technical expertise in making use of the massive amount of production data they collect. Thus, in parallel with the technical advancement in precision technologies, there is a need for training and extension support to leading growers and consultants in data acquisition, management, and analysis. The growth of local service providers dealing with various aspects of PA should also be encouraged (Bramley 2009). Attempts to address the knowledge gap areas detailed below will help to increase the wide adoption of PA.

11.6.1 Utility of Remote Sensing for Precision Agriculture

As crop response to management practices varies significantly across years in response to changing climate, PA has begun to emphasize the importance of spatial-temporal data analysis of management practices rather than spatial data analysis alone (Miao et al. 2009; Varvel et al. 2007). However, this approach requires massive data collection and analysis involving stationary or mobile sensors that can measure the characteristics of individual plants in real time. Sensors in the future could range from satellites (Bausch and Khosla 2010) to airplanes (Goel et al. 2003), from unmanned aerial vehicles (Berni et al. 2009) to tractors, or attached to mobile robots to record weed densities, crop height, leaf reflectance, soil moisture, and other properties necessary for making decisions about irrigation, fertilizer, and pest management. Along with the emergence of several new technologies, there is a need for the evaluation of the most appropriate and cost-effective spatial resolution and the optimal time of image acquisition. Benefits of airborne compared with satellite-based remote sensing platforms need to be assessed (Mulla 2013).

There is also a need to consider the limitations in the applications of remote sensing in soil, water, and crop quality assessment that arise from (1) spectral confusions (i.e., impure pixels) from simultaneous detection of reflectance signals at the sensors, (2) difficulty in distinguishing plant species, (3) scaling issues, (4) low accuracy due to soil and water heterogeneity, and (5) absence of long-term satellite data. In situations with heterogeneous landscapes and missing or limited spatial and temporal data, data fusion, which combines data from multiple sources (i.e., field and multiple sensors with different spatial resolutions) acquired at the same time, may offer a solution (de Paul Obade et al. 2013).

11.6.2 Integration of Historical Archives of Satellite Remote Sensing Data

Historical archives of satellite remote sensing data at moderate to high spatial resolutions are available at many locations for satellites including Landsat, SPOT, IRS, IKONOS, and QuickBird. For example, Landsat satellite data is available from 1972. There is a need to integrate these historical archives of satellite data with real-time remote sensing data at high spatial and spectral resolution for improved decision making in PA. Satellite images at a fixed location could be analyzed across various crop growth stages, seasons, and years in order to identify relatively homogeneous subregions of fields that differ from one another in LAI, normalized difference vegetation index (NDVI; an indicator for density of green vegetation), and potential productivity. Use of auxiliary data, including crop yield maps, digital elevation models, and soil series maps along with historical remote sensing helps to identify potential management zones where PA input operations can be customized. Real-time remote sensing with high spatial and spectral resolution satellites, such as the EO-1 Hyperion or the upcoming (2016) National Aeronautics and Space Administration (NASA) Hyperspectral Infrared Imager (HyspIRI) satellite, or comparable data collected with aerial platforms could then be used to refine the location of management zones for real-time PA decision making (Mulla 2013).

11.6.3 Development of Decision Support Systems

Mapping of various soil, crop, and environmental factors within a field produces large quantities of data that are not processed into a form suitable for the crop manager to use as a quantified data source. There are not many formal decision support systems (DSSs) and well-designed strategies that are flexible enough to incorporate all the field-related information to guide a farmer's decision. Thus the data overload for the manager needs to be addressed through the development of data integration tools, expert systems, and DSSs (McBratney et al. 2005).

11.6.4 Documentation of Economic and Environmental Benefits from PA

Although the potential use of PA in improving environmental quality and reducing pollution is appealing, the economic and environmental benefits of PA have not been well assessed and documented with quantified figures. There is also little on-site monitoring and verification of a BMP's effectiveness once it is implemented, which would help improve the accuracy of estimation methodologies. The benefit of PA is not only limited at farm level but also at adjacent landscapes (e.g., stream, wetlands, and vegetation) and takes years to gain. Many BMPs provide environmental cobenefits that are not easily quantified—such as wildlife benefits, flood control, and decreases in stream quality (Guiling and St. John 2007). Thus, there is a need for increased site-specific research on estimating the environmental outcomes and environmental cobenefits of PA, and these studies need to add benefits to those from the broader scale and those obtained at the farm level (Bramley 2009).

11.6.5 Continuous Improvement of Hydrologic Models

There is a need to refine empirical equations underlying many of the ecological and hydrological models that were developed decades ago based on little observed data. For example, in the SWAT model, surface runoff due to daily rainfall is estimated using the empirical soil conservation service (SCS) curve number method (soil conservation service 1973), which is the function of soil permeability, land use, and antecedent soil water conditions (Woznicki and Nejashashemi 2014). Similarly, SWAT does not account for increases in LAI under increased CO_2 concentration, which may result in overestimation of stomatal conductance and decreased evapotranspiration (Eckhardt and Ulbrich 2003; Ficklin et al. 2009). Empirical equations that were developed decades ago may not be sufficient to understand the ecological processes under a changing climate. Data collected from PA constitutes a useful databank on spatial and temporal variability of crop and soil performance, and thus contributes to a better understanding of the impacts of soil properties, fertilizer/pesticides efficiency, topography, climate, and other factors. This information provides a better understanding of the ecological processes and helps fine-tune some of the processes within the existing ecological and hydrological models that are currently being used for simulating climate change effects.

11.6.6 Assessment of Model Uncertainty in BMP Effectiveness

Despite the progress in understanding BMP effectiveness through field and modeling studies, there is a lack of quantitative information on model uncertainty in the spatiotemporal performance of BMPs in future climates. A knowledge gap still exists regarding the uncertainty bounds for BMPs in changing climates. The use of empirical equations and values relating to soil and management (such as the SCS curve number) adds uncertainty in the simulated output from most of the watershed models. There is thus a need for accounting uncertainty in the processes, model specification, and parameters governing the model used in evaluating BMP performance. This provides an understanding of the risk and reliability of BMP implementation for mitigation and adaptation under changing future climates (Woznicki and Nejadhashemi 2014). Using this information, watershed managers and stakeholders will be able to select BMPs that will be more effective and reliable in future climates, allowing for a greater return on the implementation investment.

11.7 SUGGESTIONS FOR POLICY MAKERS

Concerns for the quality of water resources and their access and availability are forcing a meaningful dialog among competing and conflicting urban and rural interests, and among research, education, industry, technical, and regulatory entities (Morton 2014). Precision conservation management at a field and/or watershed level is suggested as one of the mitigation strategies for some of these concerns. Despite various technological advancements in precision technologies, one of the main reasons for the slow adoption of PA is in the inability to convince the general public and farmers that precision techniques are easy to handle, economically feasible, and even profitable. There is a need for regulation, leadership, and proper infrastructure to highlight the advantages of PA in terms of its sophistication, crop yield improvement, fertilizer use efficiency, and environmental benefits.

11.7.1 Leadership

There is a need for leadership at local, state, and national levels to create core groups of people with skills and a common vision centered on (1) the growing recognition that water security is a systems problem and healthy agroecosystems are necessary to sustain a well-functioning hydrologic cycle, (2) strategies to incorporate soil performance/health improvement into crop production planning and business plans, and (3) building social, political, and economic infrastructures necessary to enable widespread adoption of soil–water management. Investments in time and energy, infrastructure, and natural and economic resources are needed for maintaining and repairing the hydrological cycle necessary for a fresh, safe, and abundant water supply (Morton 2014). Investments should be focused on greater public participation in connecting scientific facts with social values and increasing the willingness to adapt water and land use decisions. There is also a need for networks of agricultural retailers and service providers, training opportunities, industry backing, and government support.

11.7.2 Payment for Environmental Service

Implementation of agricultural conservation programs that provide farmers an incentive to conserve agricultural and water resources is an effective way to promote the use of PA around the world. For example, the United States Department of Agriculture (USDA) offers various voluntary conservation initiatives to encourage farmers to implement BMPs to reduce nutrient and soil loss on farms. These programs provide financial and technical assistance, in the form of subsidies or tax credits, to agricultural producers to implement conservation practices on agricultural land to manage risk and address natural resource issues such as conserving surface and groundwater quality and improving water quality. The effectiveness of incentive payments improves when performance-based approaches are used. Performance-based approaches use incentive payments based on actual environmental outcomes rather than paying for actions and implementations of practices (Greenhalgh et al. 2007).

11.7.3 ENVIRONMENTAL LEGISLATION/TRACEABILITY

Up to now, private sector suppliers have been the clear driver in the development of precision technologies and their adoption. The support from governments and other public institutions can play an important role in a wider adoption of PA (Zarco-Tejada et al. 2014). Environmental fees, service charges, and taxes can be used to create incentives for PA adoption for reducing pollution from agricultural sources, increasing effectiveness of pollution control, and promoting innovation in pollution control strategies (Hoffman and Boyd 2006). For example, Netherlands directly taxes nutrient effluent as a way of creating a strong incentive for intensive nutrient management.

Good traceability systems such as ecolabeling from the perspective of PA throughout the food chain (i.e., from production to distribution) also helps to minimize the potential for environmental pollution. Ecolabeling is a voluntary method of certifying products that are produced in a way that is environmentally sustainable and preferable to other products in the same product category based on life-cycle considerations. Ecolabeling of agricultural products can provide incentives for farmers who wish to certify their products and adopt sustainable agricultural systems (Selman and Greenhalgh 2009).

11.7.4 NUTRIENT TRADING

Although there are several programs aimed at reducing sediment and nutrient runoff in an effort to protect and preserve water resources (Shortle et al. 2012), these efforts are targeted mainly at point sources, and NPSP remains largely uncontrolled (Brown and Froemke 2012). Limited success in NPSP control is primarily due to the difficulty of identifying specific problem areas that are significant sources of pollution and lack of related regulation and enforcement (EPA 2005). Policy makers should focus on adoption and utilization of a watershed general permit and market-based incentives to help achieve NPSP. Programs such as water-quality trading, which allows point sources (such as wastewater treatment plants) to meet nutrient reduction requirement by acquiring nutrient reduction credits from NPS (such as farming operation), should be encouraged. To generate reduction credits, an NPS must first ensure that it is meeting baseline compliance and then meeting minimum requirements referred to as threshold. For example, the installation of BMPs that are above baseline and threshold requirement can generate credits that are eligible for sale. With private sources paying for reduction in farm-generated water quality impairments, there is motivation for farmers to adopt BMPs that are equally beneficial to them (Ribaudo and Gottlieb 2011).

11.8 CONCLUSION

Water-quality problems contributed by agriculture are likely to exacerbate under changing climates if immediate actions for improving soil, water, and agricultural management practices are not taken. It is important that we develop and implement resource-efficient approaches for soil conservation and enhancement of agricultural productivity for the recovery of degraded soils and water. This can only be possible through PA that contributes to (1) improving the efficiency of agricultural inputs, (2) conserving and improving soil quality, (3) recovering degraded soils by accounting and managing spatially degraded soils, (4) enhancing soil carbon sequestration, (5) increasing agricultural productivity, (6) reducing off-site transport of soil nutrients, chemicals, and sediments, and (7) enhancing environmental quality. We propose that PA technologies that use GIS and remote sensing information on soil, weather, crops, landscape, and other information with hydrologic and crop models offers the development and implementation of reliable site-specific practices for soil and water conservation and adaptation to climate change.

REFERENCES

Agbu, P.A., D.J. Fehrenbacher, and I.J. Jansen. 1990. Soil property relationships with SPOT satellite digital data in east central Illinois. *Soil Sci Soc Am J* 54(3):807–812.

Andujar, D., A. Escolà, J.R. Rosell-Polo, C. Fernández-Quintanilla, and J. Dorado. 2013. Potential of a terrestrial LiDAR-based system to characterise weed vegetation in maize crops. *Comput Electron Agr* (92):11–15.

Arnold, J.G., and N. Fohrer. 2005. SWAT2000: Current capabilities and research opportunities in applied watershed modelling. *Hydrol Process* 19(3):563–572.

Baker, N.T., M. Meyer, W. Stone, and J. Wilson. 2007. *Occurrence and Transport of Agricultural Chemicals in Leary Weber Ditch Basin, Hancock County, Indiana, 2003–04.* U.S. Geological Survey Scientific Investigation Report 2006, 5251:1–44.

Bausch, W.C., and J.A. Delgado. 2005. Impact of residual soil nitrate on in-season nitrogen applications to irrigated corn based on remotely sensed assessments of crop nitrogen status. *Precis Agric* 6(6):509–519.

Bausch, W.C., and K. Diker. 2001. Innovative remote sensing techniques to increase nitrogen use efficiency of corn. *Commun Soil Sci Plant Anal* 32(7–8):1371–1390.

Bausch, W.C., and R. Khosla. 2010. QuickBird satellite versus ground-based multi-spectral data for estimating nitrogen status of irrigated maize. *Precis Agric* 11(3):274–290.

Bennett, E.M., S.R. Carpenter, and N.F. Caraco. 2001. Human impact on erodable phosphorus and eutrophication: A global perspective increasing accumulation of phosphorus in soil threatens rivers, lakes, and coastal oceans with eutrophication. *BioScience* 51(3):227–234.

Berni, J., P.J. Zarco-Tejada, L. Suárez, and E. Fereres. 2009. Thermal and narrowband multispectral remote sensing for vegetation monitoring from an unmanned aerial vehicle. *Geosci Remote Sens* 47(3):722–738.

Berry, J.K., J.A. Delgado, R. Khosla, and F. Pierce. 2003. Precision conservation for environmental sustainability. *J Soil Water Conserv* 58(6):332–339.

Bosch, N.S., J.D. Allan, J.P. Selegean, and D. Scavia. 2013. Scenario-testing of agricultural best management practices in Lake Erie watersheds. *J Great Lakes Res* 39(3):429–36.

Bosch, N.S., M.A. Evans, D. Scavia, and J.D. Allan. 2014. Interacting effects of climate change and agricultural BMPs on nutrient runoff entering Lake Erie. *J Great Lakes Res* 40(3):581–589.

Bramley, R.G.V., J. Ouzman, and C. Thornton. 2011. Selective harvesting is a feasible and profitable strategy even when grape and wine production is geared towards large fermentation volumes. *Aust J Grape Wine Res* 17(3):298–305.

Bramley, R.G.V. 2009. Lessons from nearly 20 years of precision agriculture research, development, and adoption as a guide to its appropriate application. *Crop Pasture Sci* 60(3):197–217.

Bricklemyer, R.S., R.L. Lawrence, P.R. Miller, and N. Battogtokh. 2006. Predicting tillage practices and agricultural soil disturbance in north central Montana with Landsat imagery. *Agric Ecosyst Environ* 114(2):210–216.

Brown, T.C., and P. Froemke. 2012. Nationwide assessment of nonpoint source threats to water quality. *BioScience* 62(2):136–146.

Cardenas-Lailhacar, B., and M.D. Dukes. 2012. Soil moisture sensor landscape irrigation controllers: A review of multi-study results and future implications. *Trans ASABE* 55(2):581–590.

Carlton, J., and A. Lazo. 2014. California declares drought emergency. *Wall Street Journal*, January 18, p. A3.

Casady, W.W., and H.L. Palm. 2002. Precision agriculture: Remote sensing and ground truthing. University of Missouri Extension.

Chen, F., D.E. Kissel, L.T. West, W. Adkins, D. Rickman, and J.C. Luvall. 2008. Mapping soil organic carbon concentration for multiple fields with image similarity analysis. *Soil Sci Soc Am J* 72(1):186–193.

Chen, P., D. Haboudane, N. Tremblay, J. Wang, P. Vigneault, and B. Li. 2010. New spectral indicator assessing the efficiency of crop nitrogen treatment in corn and wheat. *Remote Sens Environ* 114(9):1987–1997.

Christensen, S., H.T. Søgaard, P. Kudsk, M. Nørremark, I. Lund, E.S. Nadimi, and R. Jørgensen. 2009. Site-specific weed control technologies. *Weed Res* 49(3):233–241.

Cicek, H., M. Sunohara, G. Wilkes, H. McNairn, F. Pick, E. Topp, and D.R. Lapen. 2010. Using vegetation indices from satellite remote sensing to assess corn and soybean response to controlled tile drainage. *Agric Water Manage* 98(2):261–270.

Clark, R.N., G.A. Swayze, K.E. Livo, R.F. Kokaly, S.J. Sutley, J.B. Dalton, R.R. McDougal, and C.A. Gent. 2003. Imaging spectroscopy: Earth and planetary remote sensing with the USGS Tetracorder and expert systems. *J Geophys Res-Planet* 108:1991–2012.

Coleman, T.L., P.A. Agbu, and O.L. Montgomery. 1993. Spectral differentiation of surface soils and soil properties: Is it possible from space platforms? *Soil Sci* 155:283–293.

Daughtry, C.S., P.C. Doraiswamy, E.R. Hunt, A.J. Stern, J.E. McMurtrey, and J.H. Prueger. 2006. Remote sensing of crop residue cover and soil tillage intensity. *Soil Tillage Res* 91(1):101–108.

Defersha, M.B., and A.M. Melesse. 2012. Field-scale investigation of the effect of land use on sediment yield and runoff using runoff plot data and models in the Mara River basin, Kenya. *Catena* 89(1):54–64.

Delgado, J.A., and W.C. Bausch. 2005. Potential use of precision conservation techniques to reduce nitrate leaching in irrigated crops. *J Soil Water Conserv* 60(6):379–387.

Delgado, J.A., P.M. Groffman, M.A. Nearing, T. Goddard, D. Reicosky, R. Lal, N.R. Kitchen, C.W. Rice, D. Towery, and P. Salon. 2011. Conservation practices to mitigate and adapt to climate change. *J Soil Water Conserv* 66(4):118A–129A.

de Paul Obade, V., R. Lal, and J. Chen. 2013. Remote sensing of soil and water quality in agroecosystems. *Water Air Soil Poll* 224(9):1–27.

Diaz, R.J., and R. Rosenberg. 2008. Spreading dead zones and consequences for marine ecosystems. *Science* 321: 926–929.

Dodds, W.K., W.W. Bouska, J.L. Eitzmann, T.J. Pilger, K.L. Pitts, A.J. Riley, J.T. Schloesser, and D.J. Thornbrugh. 2009. Eutrophication of U.S. freshwaters: Analysis of potential economic damages. *Environ Sci Technol* 43:12–19.

Doraiswamy, P.C., J.L. Hatfield, T.J. Jackson, B. Akhmedov, J. Prueger, and A. Stern. 2004. Crop condition and yield simulations using Landsat and MODIS. *Remote Sens Environ* 92(4):548–559.

Eckhardt, K., and U. Ulbrich. 2003. Potential impacts of climate change on groundwater recharge and streamflow in a central European low mountain range. *J Hydrol* 284(1–4):244–252.

Eitel, J.U.H., D.S. Long, P.E. Gessler, and A.M.S. Smith. 2007. Using in-situ measurements to evaluate the new RapidEye™ satellite series for prediction of wheat nitrogen status. *Int J Remote Sens* 28(18):4183–4190.

Endale, D.M., D.S. Fisher, L.B. Owens, M.B. Jenkins, H.H. Schomberg, C.L. Tebes-Stevens, and J.V. Bonta. 2011. Runoff water quality during drought in a zero-order Georgia Piedmont Pasture: Nitrogen and total organic carbon. *J Environ Qual* 40:969–979.

FAO. 2004. Food and Agricultural Organization of the United Nations. *Drought-resistant soils: Optimization of soil moisture for sustainable plant production*. Rome, Italy.

Ficklin, D.L., Y. Luo, E. Luedeling, and M. Zhang. 2009. Climate change sensitivity assessment of a highly agricultural watershed using SWAT. *J Hydrol* 374(1):16–29.

Fitzsimmons, E.G. 2014. Tap water ban for Toledo residents. *New York Times*, August 3.

Flanagan, D.C., and M.A. Nearing. 1995. *USDA-Water Erosion Prediction Project: Hillslope Profile and Watershed Model Documentation*. NSERL Report No. 10. West Lafayette, IN: USDA-ARS National Soil Erosion Research Laboratory.

Flanagan, D.C., J.C. Ascough II, M.A. Nearing, and J.M. Laflen. 2001. *The Water Erosion Prediction Project (WEPP) Model*. In: *Landscape Erosion and Evolution Modeling*, R.S. Harmon, and W.W. Doe, III (eds), 145–199. New York: Kluwer Academic/Plenum Publishers.

Foster G.R., and L.J. Lane. 1987. *User Requirements: USDA-Water Erosion Prediction Project (WEPP)*. NSERL Report No. 1. West Lafayette, IN: USDA-ARS National Soil Erosion Research Laboratory.

Gassman, P.W., M.R. Reyes, C.H. Green, and J.G. Arnold. 2007. The soil and water assessment tool: Historical development, applications, and future research directions. Ames, IA: Center for Agricultural and Rural Development, Iowa State University.

Gassman, P.W., J.R. Williams, X. Wang, A. Saleh, E. Osei, L. Hauck, R.C. Izaurralde, and J. Flowers. 2009. The Agricultural Policy Environmental Extender (APEX) model: An emerging tool for landscape and watershed environmental analyses. Ames, IA: Center for Agricultural and Rural Development (CARD) Publications.

Ge, Y., J.A. Thomasson, and R. Sui. 2011. Remote sensing of soil properties in precision agriculture: A review. *Front Earth Sci* 5(3):229–238.

Gelder, B.K., R.P. Anex, T.C. Kaspar, T.J. Sauer, and D.L. Karlen. 2011. Estimating soil organic carbon in Central Iowa using aerial imagery and soil surveys. *Soil Sci Soc Am J* 75(5):1821–1828.

Ghane, E., N.R. Fausey, V.S. Shedekar, H.P. Piepho, Y. Shang, and L.C. Brown. 2012. Crop yield evaluation under controlled drainage in Ohio, United States. *J Soil Water Conserv* 67(6):465–473.

Giri, S., A.P. Nejadhashemi, S. Woznicki, and Z. Zhang. 2014. Analysis of best management practice effectiveness and spatiotemporal variability based on different targeting strategies. *Hydrol Process* 28(3):431–445.

Goel, P.K., S.O. Prasher, J.A. Landry, R.M. Patel, R.B. Bonnell, A.A. Viau, and J.R. Miller. 2003. Potential of airborne hyperspectral remote sensing to detect nitrogen deficiency and weed infestation in corn. *Comput Electron Agr* 38(2):99–124.

Goetz, S.J., R.K. Wright, A.J. Smith, E. Zinecker, and E. Schaub. 2003. IKONOS imagery for resource management: Tree cover, impervious surfaces, and riparian buffer analyses in the mid-Atlantic region. *Remote Sens Environ* 88(1):195–208.

Greenhalgh, S., M. Selman, J. St. John, and J. Guiling. 2007. *Paying for Environmental Performance: Using Reverse Auctions to Allocate Money for Conservation.* Washington, DC: WRI Policy Note, Environmental Markets: Reverse Auctions (3).

Griffin, C.G., K.E. Frey, J. Rogan, and R.M. Holmes. 2011. Spatial and interannual variability of dissolved organic matter in the Kolyma River, East Siberia, observed using satellite imagery. *J Geophys Res* 116(G3), doi:10.1029/2010JG001634.

Guiling, J., and J. St. John. 2007. *Paying for Environmental Performance: Estimating the Environmental Outcomes of Agricultural Best Management Practices.* Washington, DC: WRI Policy Note, Environmental Markets: Estimating Environmental Outcomes (4).

Guo, L.B., and R.M. Gifford. 2002. Soil carbon stocks and land use change: A meta-analysis. *Glob Chang Biol* 8(4):345–360.

Hedley, C.B., P. Roudier, I.J. Yule, J. Ekanayake, and S. Bradbury. 2013. Soil water status and water table depth modelling using electromagnetic surveys for precision irrigation scheduling. *Geoderma* 199:22–29.

Hoffmann, S. and J. Boyd. 2006. *Environmental Fees: Can Incentives Help Solve the Chesapeake's Nutrient Pollution Problems.* Washington, DC: Resources for the Future, RFF DP 06-38.

Inman, D., R. Khosla, R. Reich, and D.G. Westfall. 2008. Normalized difference vegetation index and soil color-based management zones in irrigated maize. *Agron. J.* 100:60–66.

IPCC. 2007. *Climate Change 2007: The Physical Science Basis.* Contribution of Working Group I to the Fourth Assessment Report of the Intergovernmental Panel on Climate Change, Solomon, S., D. Qin, M. Manning, Z. Chen, M. Marquis, K.B. Averyt, M. Tignor and H.L. Miller (eds.). Cambridge, UK: Cambridge University Press.

IPCC. 2013. *Climate Change 2013: The Physical Science Basis.* Working Group I Contribution to the IPCC Fifth Assessment Report of the Intergovernmental Panel on Climate Change, T.F. Stocker, D. Qin, G.-K. Plattner, M. Tignor, S.K. Allen, J. Boschung, A. Nauels, Y. Xia, V. Bex, and P.M. Midgley (eds.). Cambridge, UK: Cambridge University Press.

Izaurralde, R.C., J.R. Williams, W.B. McGill, N.J. Rosenberg, and M.C. Jakas. 2006. Simulating soil C dynamics with EPIC: Model description and testing against long-term data. *Ecol Modell* 192(3):362–384.

Jha, M., J.G. Arnold, P.W. Gassman, F. Giorgi, and R.R. Gu. 2006. Climate change sensitivity assessment on Upper Mississippi River Basin Streamflows using SWAT. *J Am Water Resour Assoc* 42(4):997–1015.

Kaushal, S.S., P.M. Mayer, P.G. Vidon, R.M. Smith, M.J. Pennino, T.A. Newcomer, S. Duan, C.W. Director, and K.T. Belt. 2014. Land use and climate variability amplify carbon, nutrient, and contaminant pulses: A review with management implications. *J Am Water Resour Assoc* 50(3):585–614.

Khosla, R. 2010. *Precision Agriculture: Challenges and Opportunities in a Flat World.* Brisbane, Australia: 19th World Congress of Soil Science, Soil Solutions for a Changing World.

Khosla, R., D. Inman, D.G. Westfall, R. Riech, W.M. Frasier, M. Mzuku, B. Koch, and A. Hornung. 2008. A synthesis of multi-disciplinary research in precision agriculture: Site-specific management zones in the semi-arid western Great Plains of the USA. *J Precis Agric* 9(1–2):5–100.

King, K.W., C.W. Richardson, and J.R. Williams. 1996. Simulation of sediment and nitrate loss on a vertisol with conservation tillage practices. *Trans ASAE* 39(6):2139–2145.

Kling, C.L., Y. Panagopoulos, S.S. Rabotyagov et al. 2014. LUMINATE: Linking agricultural land use, local water quality and Gulf of Mexico hypoxia. *Eur Rev Agric Econ* 41(3):431–459.

Kroger, R., M.T. Moore, M.A. Locke, R.F. Cullum, R.W. Steinriede, Jr, S. Testa, C.T. Bryant, and C.M. Cooper. 2009. Evaluating the influence of wetland vegetation on chemical residence time in Mississippi Delta drainage ditches. *Agric Water Manage* 96:1175–1179.

Kroger, R., M.T. Moore, K.W. Thornton, J.L. Farris, J.D. Prevost, and S.C. Pierce. 2012. Tiered on-the-ground implementation projects for Gulf of Mexico water quality improvements. *J. Soil Water Conserv.* 67(4):94A–99A.

Kyveryga, P.M., T.M. Blackmer, R. Pearson, and T.F. Morris. 2011. Late-season digital aerial imagery and stalk nitrate testing to estimate the percentage of areas with different nitrogen status within fields. *J Soil Water Conserv* 66(6):373–385.

Labat, D., Y. Godderis, J. Probst, and J. Guyot. 2004. Evidence for global runoff increase related to climate warming. *Adv Water Resour* 27(6):631–642.

Laflen, J.M., L.J. Lane, and G.R. Foster. 1991. WEPP: A new generation of erosion prediction technology. *J Soil Water Conserv* 46(1):34–38.

Lakshmi, V., E.F. Wood, and B.J. Choudhury. 1997. Evaluation of special sensor microwave/imager satellite data for regional soil moisture estimation over the Red River Basin. *J Appl Meteorol* 36(10):1309–1328.

Lakshmi, V., S. Hong, E.E. Small, and F. Chen. 2011. The influence of the land surface on hydrometeorology and ecology: New advances from modeling and satellite remote sensing. *Hydrol Res* 42(2–3):95–112.

Lal, R. 2004. Soil carbon sequestration to mitigate climate change. *Geoderma* 123(1):1–22.

Lal, R., J.A. Delgado, P.M. Groffman, N. Millar, C. Dell, and A. Rotz. 2011. Management to mitigate and adapt to climate change. *J Soil Water Conserv* 66(4):276–285.

Lal, R., J.A. Delgado, J. Gulliford, D. Nielsen, C.W. Rice, and R.S. Van Pelt. 2012. Adapting agriculture to drought and extreme events. *J Soil Water Conserv* 67(6):162A–166A.

Lal, R. 2013. Enhancing ecosystem services with no-till. *Renew Agr Food Syst* 28(02):102–114.

Locke, M.A., and C.T. Bryson. 1997. Herbicide-soil interactions in reduced tillage and plant residue management systems. *Weed Sci* 45:307–320.

Lowenberg-DeBoer, J. 2003. Precision agriculture or convenience agriculture. *Proceedings of the 11th Australian Agronomy Conference*, Geelong, Victoria.

Liu, J., E. Pattey, M.C. Nolin, J.R. Miller, and O. Ka. 2008. Mapping within-field soil drainage using remote sensing, DEM and apparent soil electrical conductivity. *Geoderma* 143(3):261–272.

Lopez-Granados, F. 2011. Weed detection for site-specific weed management: Mapping and real-time approaches. *Weed Res* 51(1):1–11.

Mallarino, A.P., and D.J. Wittry. 2004. Efficacy of grid and zone soil sampling approaches for site-specific assessment of phosphorus, potassium, pH, and organic matter. *Precis Agric* 5(2):131–144.

Mayer, P.M., S.K. Reynolds, M.D. McCutchen, and T.J. Canfield. 2007. Meta-analysis of nitrogen removal in riparian buffers. *J Environ Qual* 36:1172–1180.

McBratney, A., B.Whelan, T. Ancev, and J. Bouma. 2005. Future directions of precision agriculture. *Precis Agric* 6(1):7–23.

Miao, Y., D.J. Mulla, G. Randall, J. Vetsch, and R. Vintila. 2009. Combining chlorophyll meter readings and high spatial resolution remote sensing images for in-season site-specific nitrogen management of corn. *Precis Agric* 10(1):45–62.

Minet, J., P. Bogaert, M. Vanclooster, and S. Lambot. 2012. Validation of ground penetrating radar full-waveform inversion for field scale soil moisture mapping. *J Hydrol* 424:112–123.

Moore, M.T., R. Kroger, M.A. Locke, R.F. Cullum, R.W. Steinriede Jr., S. Testa, J.R.E. Lizotte, C.T. Bryant, and C.M. Cooper. 2010. Nutrient mitigation capacity in Mississippi Delta, USA drainage ditches. *Environ Pollut* 158:175–184.

Morton, L.W. 2014. Achieving water security in agriculture: The human factor. *Agron J* 106:1–4. doi:10.2134/agronj14.0039.

Mulla, D.J. 2013. Twenty five years of remote sensing in precision agriculture: Key advances and remaining knowledge gaps. *Biosyst Eng* 114(4):358–371.

Naz, B.S., S. Ale, and L.C. Bowling. 2009. Detecting subsurface drainage systems and estimating drain spacing in intensively managed agricultural landscapes. *Agric Water Manage* 96(4):627–637.

Nearing, M.A., F.F. Pruski, and M.R. O'Neal. 2004. Expected climate change impacts on soil erosion rates: A review. *J Soil Water Conserv* 59(1):43–50.

Oki, T. and S. Kanae. 2006. Global hydrological cycles and world water resources. *Science* 313(5790):1068–1072.

Panda, S.S., D.P. Ames, and S. Panigrahi. 2010. Application of vegetation indices for agricultural crop yield prediction using neural network techniques. *Remote Sens* 2(3):673–696.

Park, J.H., L. Duan, B. Kim, M.J. Mitchell, and H. Shibata. 2010. Potential effects of climate change and variability on watershed biogeochemical processes and water quality in Northeast Asia. *Environ Int* 36(2):212–225.

Peterson, T.C., T.R. Karl, J.P. Kossin, K.E. Kunkel, J.H. Lawrimore, J.R. McMahon, R.S. Vose, and X. Yin. 2014. Changes in weather and climate extremes: State of knowledge relevant to air and water quality in the United States. *J Air Waste Manage Assoc* 64(2):184–197.

Phillips, D.L., P.D. Hardin, V.W. Benson, and J.V. Baglio. 1993. Nonpoint source pollution impacts of alternative agricultural management practices in Illinois: A simulation study. *J Soil Water Cons* 48(5):449–457.

Pielke, R.A., Sr. 2005. Land use and climate change. *Science* 310:1625–1626.

Pimentel, D., C. Harvey, P. Resosudarmo, K. Sinclair, D. Kurz, M. McNair, S. Crist, L. Shpritz, L. Fitton, R. Saffouri, and R. Blair. 1995. Environmental and economic costs of soil erosion and conservation benefits. *Science* 267(5201):1117–1123.

Pimentel, D., S. Cooperstein, H. Randell, D. Filiberto, S. Sorrentino, B. Kaye, C. Nicklin, J. Yagi, J. O'Hern, A. Habas, and C. Weinstein. 2007. Ecology of increasing diseases: Population growth and environmental degradation. *Hum Ecol* 35(6):653–668.

Pinter, P.J., Jr., J.L. Hatfield, J.S. Schepers, E.M. Barnes, M.S. Moran, C.S. Daughtry, and D.R. Upchurch. 2003. Remote sensing for crop management. *Photogramm Eng Rem S* 69(6):647–664.

Powers, S.E., J.C. Ascough II, R.G. Nelson, and G.R. Larocque. 2011. Modeling water and soil quality environmental impacts associated with bioenergy crop production and biomass removal in the Midwest USA. *Ecol Modell* 222(14):2430–2447.

Pretty, J.N., A.D. Noble, D. Bossio, J. Dixon, R.E. Hine, F.W. Penning de Vries, and J.I. Morison. 2006. Resource-conserving agriculture increases yields in developing countries. *Environ Sci Technol* 40(4):1114–1119.

Rabalais, N.N., R.E. Turner, and W.J. Wiseman. 2001. Hypoxia in the Gulf of Mexico. *J Environ Qual* 30(2):320–329.

Rabalais, N.N., R.E. Turner, and D. Scavia. 2002. Nutrient policy development for the Mississippi River watershed reflects the accumulated scientific evidence that the increase in nitrogen loading is the primary factor in the worsening of hypoxia in the northern Gulf of Mexico. *BioScience* 52(2):129–142.

Raine, S.R., W.S. Meyer, D.W. Rassam, J.L. Hutson, and F.J. Cook. 2007. Soil–water and solute movement under precision irrigation: Knowledge gaps for managing sustainable root zones. *Irrigation Sci* 26(1):91–100.

Renschler, C.S., and T. Lee. 2005. Spatially distributed assessment of short-and long-term impacts of multiple best management practices in agricultural watersheds. *J Soil Water Conserv* 60(6):446–456.

Ribaudo, M.O., and J. Gottlieb. 2011. Point-nonpoint trading—Can it work? *J Am Water Resour Assoc* 47(1):5–14.

Ribaudo, M., J.A. Delgado, L. Hansen, M. Livingston, R. Mosheim, and J. Williamson. 2011. *Nitrogen in Agricultural Systems: Implications for Conservation Policy*. Washington, DC: Economic Research Report No. (ERR-127).

Richards, R.P., D.B. Baker, and J.P. Crumrine. 2009. Improved water quality in Ohio tributaries to Lake Erie: A consequence of conservation practices. *J Soil Water Conserv* 64(3):200–211.

Santhi, C., R. Srinivasan, J.G. Arnold, and J.R. Williams. 2006. A modeling approach to evaluate the impacts of water quality management plans implemented in a watershed in Texas. *Environ Model and Soft* 21(8):1141–1157.

Schewe, J., J. Heinke, D. Gerten et al. 2014. Multimodel assessment of water scarcity under climate change. *PNAS* 111(9):3245–3250.

Schoonover, J.E., K.W.J. Williard, J.J. Zaczek, J.C. Mangun, and D. Carver. 2005. Nutrient attenuation in agricultural surface runoff by riparian buffer zones in southern Illinois, USA. *Agrofor Syst* 64:169–180.

Secchi, S., P.W. Gassman, M. Jha, L. Kurkalova, H.H. Feng, T. Campbell, and C.L. Kling. 2007. The cost of cleaner water: Assessing agricultural pollution reduction at the watershed scale. *J Soil Water Cons* 62(1):10–21.

Seelan, S.K., S. Laguette, G.M. Casady, and G.A. Seielstad. 2003. Remote sensing applications for precision agriculture: A learning community approach. *Remote Sens Environ* 88(1):157–169.

Selman, M., and S. Greenhalgh. 2009. *Europhication: Policies, Actions, and Strategies to Address Nutrient Pollution*. Washington, DC: WRI Policy Note, Water Quality: Eutrophication and Hypoxia (3).

Shanahan, J.F., J.S. Schepers, D.D. Francis, G.E. Varvel, W.W. Wilhelm, J.M. Tringe, M.R. Schlemmer, and D.J. Major. 2001. Use of remote-sensing imagery to estimate corn grain yield. *Agron J* 93(3):583–589.

Shen, Z.Y., Y.W. Gong, Y.H. Li, Q. Hong, L. Xu, and R.M. Liu. 2009. A comparison of WEPP and SWAT for modeling soil erosion of the Zhangjiachong Watershed in the Three Gorges Reservoir area. *Agric Water Manage* 96(10):1435–1442.

Shortle, J.S., M. Ribaudo, R.D. Horan, and D. Blandford. 2012. Reforming agricultural nonpoint pollution policy in an increasingly budget-constrained environment. *Environ Sci Technol* 46(3):1316–1325.

Sibley, A.M., P. Grassini, N.E. Thomas, and K.G. Cassman. 2014. Testing remote sensing approaches for assessing yield variability among maize fields. *Agron J* 106(1):24–32.

Sridhar, B.B., R.K. Vincent, J.D. Witter, and A.L. Spongberg. 2009. Mapping the total phosphorus concentration of biosolid amended surface soils using LANDSAT TM data. *Sci Total Environ* 407(8):2894–2899.

Sripada, R.P., R.W. Heiniger, J.G. White, and R. Weisz. 2005. Aerial color infrared photography for determining late-season nitrogen requirements in corn. *Agron J* 97(5):1443–1451.

Stevens, A., B. Van Wesemael, G. Vandenschrick, S. Touré, and B. Tychon. 2006. Detection of carbon stock change in agricultural soils using spectroscopic techniques. *Soil Sci Soc Am J* 70(3):844–850.

Stisen, S., K.H. Jensen, I. Sandholt, and D.I. Grimes. 2008. A remote sensing driven distributed hydrological model of the Senegal River basin. *J Hydrol* 354(1):131–148.

Sugg, Z. 2007. *Assessing US Farm Drainage: Can GIS Lead to Better Estimates of Subsurface Drainage Extent*. Washington, DC: World Resources Institute.

Sullivan, D.G., J.N. Shaw, and D. Rickman. 2005. IKONOS imagery to estimate surface soil property variability in two Alabama physiographies. *Soil Sci Soc Am J* 69(6):1789–1798.

Tomer, M.D., T.B. Moorman, J.L. Kovar, D.E. James, and M.R. Burkart. 2007. Spatial patterns of sediment and phosphorus in a riparian buffer in western Iowa. *J Soil Water Conserv* 62(5):329–338.

Tuppad, P., and R. Srinivasan. 2008. *Bosque River Environmental Infrastructure Improvement Plan: Phase II BMP Modeling Report*. Publication No. TR-313. College Station, TX: Texas A&M University, Texas AgriLife Research.

Tuppad, P., C. Santhi, X. Wang, J.R. Williams, R. Srinivasan, and P.H. Gowda. 2010a. Simulation of conservation practices using the APEX model. *Appl Eng Agric* 26(5):779–794.

Tuppad, P., K.R. Douglas-Mankin, and K.A. McVay. 2010b. Strategic targeting of cropland management using watershed modeling. *Agr Eng Int* 12(3–4):12–24.

UN WWAP. 2003. United Nations World Water Assessment Programme. *The World Water Development Report 1: Water for People, Water for Life*. Paris, France: UNESCO.

US EPA. 2010. US Environmental Protection Agency. *National Lakes Assessment Fact Sheet*. Washington, DC: USEPA.

US EPA. 2012. US Environmental Protection Agency. Nonpoint Source. Accessed at http://water.epa.gov/polwaste/nps/nonpoin1.cfm.

Varvel, G.E., W.W. Wilhelm, J.F. Shanahan, and J.S. Schepers. 2007. An algorithm for corn nitrogen recommendations using a chlorophyll meter based sufficiency index. *Agron J.* 99(3):701–706.

Vaudour, E., L. Bel, J.M. Gilliot, Y. Coquet, D. Hadjar, P. Cambier, J. Michelin, and S. Houot. 2013. Potential of SPOT multispectral satellite images for mapping topsoil organic carbon content over peri-urban croplands. *Soil Sci Soc Am J* 77(6):2122–2139.

Vidon, P., and P.E. Cuadra. 2010. Impact of precipitation characteristics on soil hydrology in tile-drained landscapes. *Hydrol Process* 24(13):1821–1833.

Vidon, P., L.E. Hubbard, and E. Soyeux. 2009. Seasonal solute dynamics across land uses during storms in glaciated landscape of the US Midwest. *J Hydrol* 376(1):34–47.

Wan, H., D.J. Mulla, and J.C. Galzki. 2014. Using LiDAR and geographic information system data to identify optimal sites in southern Minnesota for constructed wetlands to intercept nonpoint source nitrogen. *J Soil Water Conserv* 69(4):115A–120A.

Weng, Y.L., P. Gong, and Z. Zhu. 2010. A spectral index for estimating soil salinity in the Yellow River Delta region of China using EO-1 Hyperion data. *Pedosphere* 20(3):378–388.

Williams, J.R., and R.C. Izaurralde. 2000. *The APEX model*. Texas A&M Blackland Research Center Temple, BRC Report No. 00-06.

Williams, J.R., and V.P. Singh. 1995. The EPIC model. In *Computer Models of Watershed Hydrology*, V.P. Singh (ed), 909–1000. Highlands Ranch, CO: Water Resources Publication.

Williams, J.D., D.S. Long, and S.B. Wuest. 2011. Capture of plateau runoff by global positioning system–guided seed drill operation. *J. Soil Water Conserv* 66(6):355–361.

Wolkowski, R., T. Cox, and J. Leverich. 2009. *Strip-Tillage: A Conservation Option for Wisconsin Farmers*. UW-Extension Cooperative Extension A3883.

Woznicki, S.A., and A.P. Nejadhashemi. 2014. Assessing uncertainty in best management practice effectiveness under future climate scenarios. *Hydrol Process* 28(4):2550–2566.

Yang, C., J.H. Everitt, and J.M. Bradford. 2006. Comparison of QuickBird satellite imagery and airborne imagery for mapping grain sorghum yield patterns. *Precis Agric* 7(1):33–44.

Yazgi, A., and A. Degirmencioglu. 2007. Optimisation of the seed spacing uniformity performance of a vacuum-type precision seeder using response surface methodology. *Biosyst Eng* 97(3):347–356.

Zarco-Tejada P.J., N. Hubbard, P. Loudjani. 2014. *Precision Agriculture: An Opportunity for EU Farmers—Potential Support with the CAP 2014–2020*. Directorate-General for Internal Policies, Policy Department B: Structural and Cohesion Policies, European Parliament, Brussels, Belgium.

Zhang, J.Z., C.R. Kelble, C.J. Fischer, and L. Moore. 2009. Hurricane Katrina induced nutrient runoff from an agricultural area to coastal waters in Biscayne Bay, Florida. *Estuar Coast Shelf Scie* 84(2):209–218.

Zhang, X., L. Shi, X. Jia, G. Seielstad, and C. Helgason. 2010. Zone mapping application for precision-farming: A decision support tool for variable rate application. *Precis Agric* 11(2):103–114.

Zhou, Y., D.R. Obenour, D. Scavia, T.H. Johengen, and A.M. Michalak. 2013. Spatial and temporal trends in Lake Erie hypoxia, 1987–2007. *Environ Sci Technol* 47(2):899–905.

12 Total Carbon and Labile Fractions Inventory and Mapping by Soil Orders under Long-Term No-Till Farming to Promote Precision Agriculture

Daniel Ruiz Potma Gonçalves, João Carlos de Moraes Sá,
Allison José Fornari, Flávia Juliana Ferreira Furlan,
Lucimara Aparecida Ferreira, and Ademir de Oliveira Ferreira

CONTENTS

12.1 INTRODUCTION

Studies of the spatial distributions of carbon (C) and nitrogen (N) contents and stocks at a landscape scale have received increasing attention since the 1990s because of the development of Geographic Information System (GIS) software (Davidson and Lefebvre 1993; Levine et al. 1994; Poier and Richter 1992). The use of GIS software has led to significant advances in landscape-scale studies (Hartemink 2008; Lilburne et al. 2012) because GIS enables the grouping of a large number of variables, the creation of different scenarios, and a panoramic view of the landscape.

Research of landscape-scale variations in C inventories among soil orders has facilitated the creation of C stock maps (Batjes 1996; Bernoux et al. 2002; Sá et al. 2013a; Tornsquist et al. 2009) and has promoted substantial advances in the understanding of C dynamics. With the availability of C inventory data, it has become possible to predict the long-term accumulation of C using mathematical models (Easter et al. 2007; Tornquist et al. 2009). These predictions can be used as a baseline for identifying public policies to decrease the greenhouse effect, increase food safety, and diminish critical problems such as hunger (Cerri et al. 2007; Pan et al. 2010). Therefore, farm-scale total organic C (TOC) mapping is important because it provides a fundamental agronomic decision-making tool for developing more sustainable production systems. Precision agriculture advanced the understanding about C accumulation, identifying areas with high or low potential to accumulate C and can be used to determine specific management practices to increase crop yield.

In studies examining soil C, the total C is usually divided into different pools, and one such division is based on the lability of C compounds. Highly labile C is comprised of polysaccharides associated with microbial biomass; it varies in the short term because of the sensitivity of the microbial population to several factors (e.g., environment and the application of agrochemicals and fertilizers) and the addition of C, which is a substrate for microbial growth and activity (Ghani et al. 2003; Mazzarino et al. 1993). Labile C is mainly comprised of aliphatic organic compounds derived from the decomposition of crop residues that supply organic matter (OM) to the soil (Blair et al. 1995). Labile C influences the effect of the TOC on nutrient cycling from plants because it consists of C compounds that can be easily oxidized by microbes (Culman et al. 2012, 2013; Six et al. 2002). The TOC consists of the set of all types of C compounds and varies in the long term because it is mainly comprised of humic substances with a mean residence time (MRT) of >2000 years (Stevenson 1986).

Thus, the aims of this study were to (1) perform an inventory of the total C and labile C contents in various soil orders to a depth of 1 m at a landscape scale, (2) map these C pools on a farm under 30 years of no-till management, and (3) identify the chemical, physical, biological, topographic, and microclimatic variables that affect the spatial variation of soil C.

12.2 MATERIALS AND METHODS

12.2.1 LOCATION AND DESCRIPTION OF THE STUDY AREA

This study was conducted at the Paiquerê Farm, located in the municipality of Piraí do Sul/Arapoti, Paraná, Brazil, at 24°20′20″ S latitude and 50°07′31″ W longitude (Figure 12.1). This study site was selected because of the existence of a detailed database for its production components and soil attributes and because of the long-term (30 years) continuous no-till (NT, represent minimum disturbance in the row planting, soil covered all year long and adoption of crop rotation) farming at this site (Figure 12.2).

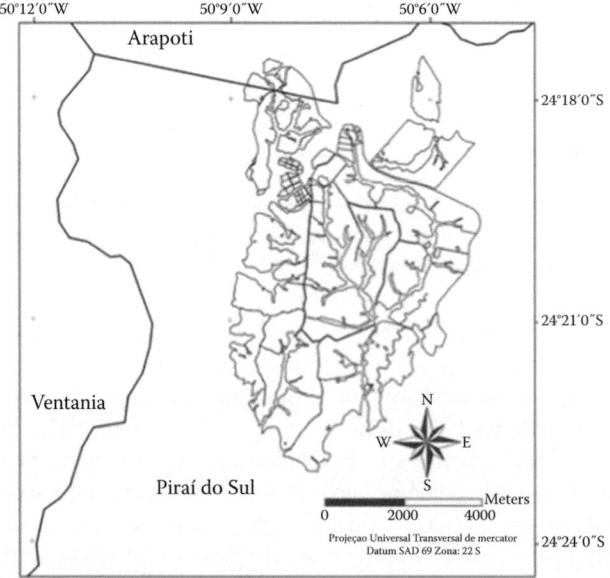

FIGURE 12.1   Location of the study area, Paiquerê Farm in Piraí do Sul, Paraná, Brazil.

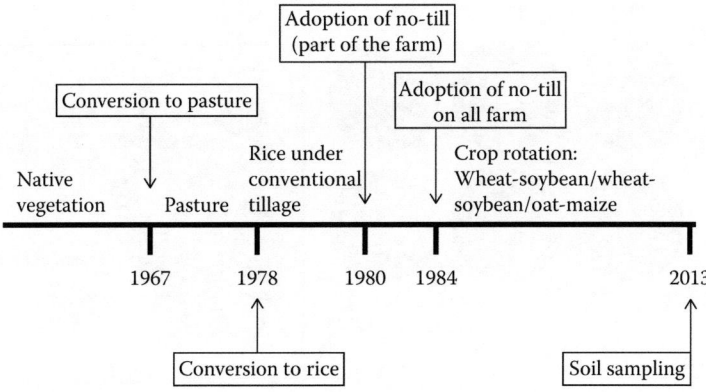

FIGURE 12.2 Chronology of soil use at Paiquerê Farm.

The mean altitude of the farm is 970 m a. s. l. The climate of the region is classified as Cfb (Köppen classification), which corresponds to a humid, subtropical climate with well-defined seasons and mean maximum and minimum temperatures of 25.9°C and 13.5°C, respectively (IAPAR 2013). The mean annual rainfall at the farm is approximately 1700 mm, and is well distributed throughout the year, with no prolonged periods of drought.

The original soil material in the region is, in general, clastic sediments from the Devonian period, arenites from the Furnas formation, and shales from the Ponta Grossa formation (MINEROPAR 2013). The typical vegetation of the region originally consisted of fields dominated by C4 species such as *Andropogon* sp., *Aristida* sp., *Paspalum* sp., and *Panicum* sp. (Behling 1997).

The farm's relief is predominantly wavy-smooth to wavy, with plot slopes varying between 3% and 10%. The main soil orders are Red Latosols (RLs), Red-Yellow Latosols (RYLs) (Brazilian classification, EMBRAPA 2006, and equivalent to Oxisols, USDA, Soil Survey Staff 2010), Humic Cambisols, Haplic Cambisols (Brazilian classification, and equivalent to Inceptisols, USDA, Soil Survey Staff 2010), Litholic Neosols (Brazilian classification, and equivalent to Entisols, USDA, Soil Survey Staff 2010), and Melanic Gleysols (Brazilian classification, and equivalent to Histosols, USDA, Soil Survey Staff 2010). The farm's soil use plan was developed based on a detailed soil survey performed at a 1:10,000 scale. The farm was divided into 24 plots, which were assembled into three groups. Each group represented one-third of the surface area of the farm and assigned a crop sequence during the agricultural year. In the winter, two-thirds of the cultivated area was planted with wheat (*Triticum aestivum* L.), and one-third was planted with oat (*Avena sativa* L.). In summer, soybean (*Glycine max* L.) was cultivated after the wheat was harvested, and maize (*Zea mays* L.) was cultivated after the oat was harvested. The distribution of the three successional sequences was as follows: one-third of the farm was under the wheat/soy sequence, henceforth designated as S1; another one-third was under the wheat/soy sequence, henceforth designated as S2; and one-third was under the oat/maize sequence, henceforth designated as S3 (Figure 12.3). Throughout each year, each plot was subsequently replaced by the next numeric sequence (e.g., S1 was replaced by S2). Every 2 years, soil sampling was performed in management areas within each plot to analyze the soil fertility components and to guide the crop fertilization strategies. The soil fertilization strategy on the farm has enhanced soil fertility and there has been an increase in the mean crop yield over the last 17 years (Table 12.1).

12.2.2 Soil Sampling and Sample Preparation

The soil samples were collected from predefined areas in each soil order, designated as reference areas (Figure 12.4), which represented each soil order on the landscape. The dimensions of the reference areas were 30 × 30 m. Soil samples were collected at depths of 0–10, 10–20, 20–40, 40–70, and 70–100 cm. At each sampling depth, five subsamples were collected (Figure 12.4) and composited.

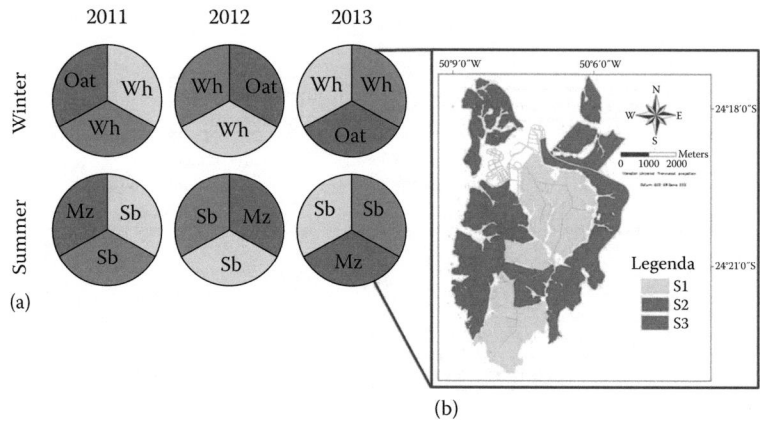

(a)

(b)

FIGURE 12.3 (a) Description of the crop successional process cultivated on the farm, and (b) sequence stages of crops in the study area during sampling in 2013. S1, wheat/soybean; S2, wheat/soybean; S3, oat/maize.

The criteria used to define the sampling sites (Figure 12.4) were based on the farm operation and the soil order. Collections from each soil order were divided into three categories (upper third, middle third, and bottom third) to encompass the textural gradient represented by the position on the landscape. The locations of the reference areas for each soil order and landscape position were defined by a hypsometric map that was generated based on the contour lines from the Paiquerê Farm database.

Soil samples were transported to the Laboratory of Soil Organic Matter (http://labmos.com .br/pt/) and dried in an oven at 40°C to a constant weight. Next, with the help of a wooden roller, samples were crushed and sieved through a 2-mm sieve. At this stage, samples were ready for analysis of the labile C (HWE-C and POX-C). A volume of 1 cm³ of soil was finely ground with a mortar and pestle to determine total C and N contents.

12.2.3 Determinations of Total C and Labile Fractions in the Soil

The determination of TOC and total nitrogen (TN) contents was performed using the dry combustion method at a temperature of 950° C and a C and N elemental analyzer (Truspec CN LECO® 2006, St. Joseph). Total soil C was considered to approximate the TOC because the inorganic C is <0.1% of the TOC in Oxisols (Oxisols, USDA classification) and <0.25% for Cambisols (Inceptisols, USDA classification) and the other soil orders are not derived from carbonaceous minerals (Sá et al. 2013a).

The hot-water extractable C (HWE-C) was extracted by the methodology described by Ghani et al. (2003) and a determination of the C content was performed using oxidation in an acidic medium and titration with ferrous sulfate based on Walkley and Black (1934). Briefly, 3 g of soil was weighed in an Erlenmeyer flask, and 9 mL of deionized water was added. The sample was incubated in an oven at 80°C for 16 hours, and, 6 mL of the supernatant was removed using a pipette for a measuring of C concentration. This step consisted of adding 10 mL of potassium dichromate, 10 mL of sulfuric acid, and 3 mL of phosphoric acid to the 6 mL sample and allowing it to react for 1 hour before performing a titration of the excess dichromate with a ferrous sulfate solution.

The permanganate oxidized C (POX-C) was determined according to the methods of Weil et al. (2003). Briefly, 3 g of soil was weighed into a tube and 6 mL of 0.2 M $KMnO_4$ was added. The sample was then stirred for 15 min at 200 oscillations per minute and centrifuged for 15 min at 4000 rpm (equivalent to 2851 × g). The sample was then rested for 10 min to allow for C oxidation to occur. Next, a 2.0-mL aliquot of the solution was pipetted into an Erlenmeyer flask and 50 mL of deionized water was added. The quantification of POX-C was then performed using a spectrophotometer with the absorbance adjusted to 565 nm.

TABLE 12.1

Historical Grain Yield of Biomass-C Inputs from the Above and Below Ground

Year	Crops	Grain Yield	Biomass-C Input[a]		
			Above Ground (A)	Below Ground (B)	Total (A + B)
			Mg ha^{-1}		
97/98	Soybean	3.2	1.42	0.28	1.38
	Wheat	3.7	1.73	0.26	1.81
	Corn	7.2	3.00	0.75	4.45
98/99	Soybean	3.3	1.44	0.29	1.40
	Wheat	3.2	1.50	0.23	1.57
	Corn	6.4	2.64	0.66	3.92
99/2000	Soybean	3.1	1.38	0.28	1.34
	Wheat	2.8	1.35	0.20	1.41
	Corn	6.9	2.87	0.72	4.26
2000/2001	Soybean	3.2	1.40	0.28	1.36
	Wheat	2.2	1.04	0.16	1.09
	Corn	8.3	3.44	0.86	5.11
2001/2002	Soybean	3.4	1.49	0.30	1.44
	Wheat	4.3	2.04	0.31	2.13
	Corn	9.5	3.95	0.99	5.86
2002/2003	Soybean	3.6	1.58	0.32	1.54
	Wheat	1.3	0.61	0.09	0.64
	Corn	8.9	3.66	0.92	5.44
2003/2004	Soybean	3.4	1.51	0.30	1.47
	Wheat	4.7	2.21	0.33	2.31
	Corn	9.5	3.93	0.98	5.83
2004/2005	Soybean	3.0	1.35	0.27	1.31
	Wheat	3.1	1.49	0.22	1.55
	Corn	8.6	3.55	0.89	5.27
2005/2006	Soybean	3.4	1.49	0.30	1.44
	Wheat	3.4	1.63	0.24	1.70
	Corn	9.6	3.98	1.00	5.91
2006/2007	Soybean	3.7	1.62	0.32	1.57
	Wheat	3.0	1.41	0.21	1.47
	Corn	8.6	3.55	0.89	5.27
2007/2008	Soybean	3.0	1.33	0.27	1.29
	Wheat	2.6	1.21	0.18	1.27
	Corn	8.0	3.30	0.83	4.90
2008/2009	Soybean	3.1	1.40	0.28	1.35
	Wheat	3.5	1.68	0.25	1.76
	Corn	8.8	3.63	0.91	5.39
2009/2010	Soybean	3.3	1.46	0.29	1.41
	Wheat	2.1	0.97	0.15	1.02
	Corn	10.3	4.25	1.06	6.31
2010/2011	Soybean	4.0	1.79	0.36	1.74
	Wheat	4.4	2.08	0.31	2.18
	Corn	10.3	4.26	1.07	6.33

(*Continued*)

TABLE 12.1 (CONTINUED)
Historical Grain Yield and of Biomass-C Inputs from the Above and Below Ground

			Biomass-C Input[a]		
			Above Ground (A)	Below Ground (B)	Total (A + B)
Year	Crops	Grain Yield	Mg ha⁻¹		
2011/2012	Soybean	3.5	1.55	0.31	1.51
	Wheat	3.1	1.47	0.22	1.54
	Corn	10.4	4.30	1.08	6.39
2012/2013	Soybean	4.0	1.78	0.36	1.72
	Wheat	3.6	1.73	0.26	1.80
	Corn	10.5	4.33	1.08	6.44
Total		246.9	106.8	23.1	135.6

[a] The values of C inputs from the roots, shoots, and crop totals were estimated based on Sá et al. (2014).

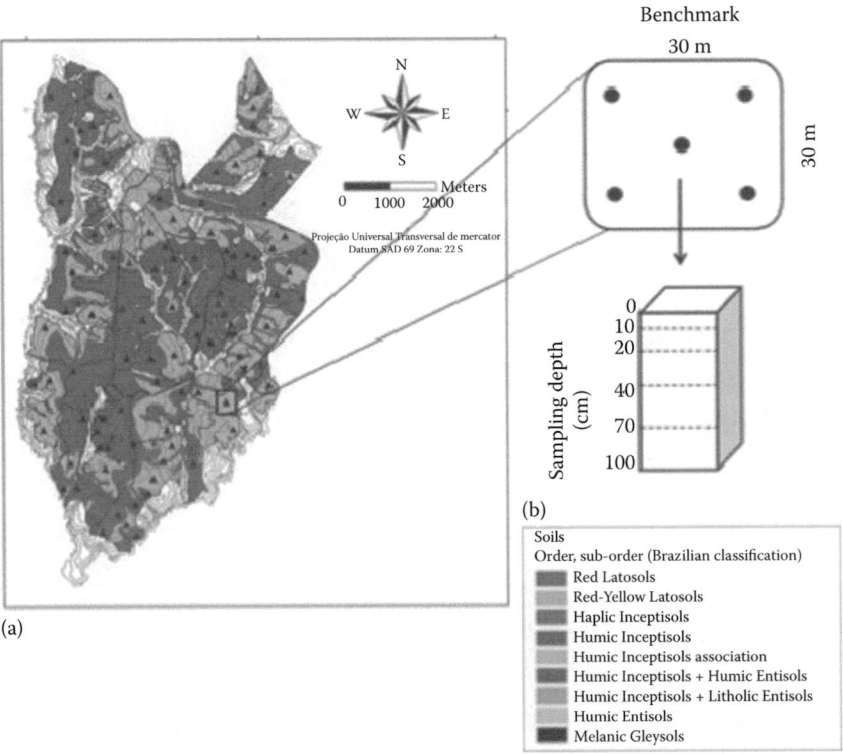

FIGURE 12.4 (a) General representation of sampling sites based on the contour lines and the main soil orders of the farm totaling 100 sites and (b) schematic representation of a reference area.

12.2.4 STATISTICAL ANALYSIS

The variations among the classes (soil order and landscape position) were assessed to determine the means and standard deviations of the soil TOC, TN, POX-C, HWE-C, and C/N ratio in the following categories: (1) the soil order, (2) the landscape position within each soil order, and (3) the crop sequence stage of the study area at the moment of sampling at all depths. The calculated means for

TABLE 12.2
Mean of Clay, Silt, and Sand Content at 0–20 cm Depth for Each Soil Order and Landscape Position

Solo	Clay	Silt	Sand	n
Red Latosol (Oxisol)	646.0 ± 16.1	111.3 ± 6.6	242.7 ± 15.4	15
Red-Yellow Latosol (Oxisol)	604.4 ± 16.1	127.2 ± 6.9	268.3 ± 17.9	18
Humic Cambisol (Inceptisol)	517.4 ± 15.1	190.0 ± 9.4	292.6 ± 16.6	23
Humic Cambisol (Inceptisol) + Neosol Lithic, NL (Entisol)	485.6 ± 14.7	192.2 ± 9.2	322.2 ± 17.7	9
Haplic Cambisol (Inceptisol)	526.7 ± 16.5	143.3 ± 10.1	330.0 ± 22.6	3
Landscape Position[a]				
Upper third	576.7 ± 13.9	312.7 ± 7.0	264.0 ± 12.8	15
Middle third	581.1 ± 15.6	295.2 ± 7.3	285.9 ± 17.0	27
Bottom third	545.6 ± 18.1	266.3 ± 10.4	279.3 ± 18.8	27

[a] Mean of all collected samples by soil orders at each landscape position.

each crop sequence stage were analyzed because the more labile TOC pools can be affected by soil management over short time scales (Culman et al. 2012).

The clay, silt, and sand contents were obtained from the Paiquerê Farm database for the same sampling sites (Table 12.2). The effect of soil texture on the TOC accumulation was analyzed by performing a simple linear regression between the clay content and the TOC concentration. The standard deviations of the clay, silt, and sand contents were also calculated based on their soil order and landscape position.

Visual representations of the spatial variations of the TOC, POX-C, and HWE-C were produced based on maps generated from spline models using the plot polygons as barriers. Spline models were chosen because of the lower requirement for homogeneity in the spacing among the sampling sites.

The visual representations of the spatial variation of maize and wheat yields were obtained from the Paiquerê Farm database for 2012. Therefore, it was possible to assess which soil orders were receiving the greatest plant biomass input by analyzing the yield maps together with the soil maps (Figure 12.4).

ArcGIS v. 10.2.1 software was used for the analysis of digital elevation models, map organization, georeferenced information, and to generate maps of the spatial variation of the analyzed variables. For the calculations of means and the regression analysis, R Development Core Program Team (2012) v. 2.15.2 software was used.

12.3 RESULTS AND DISCUSSION

12.3.1 C POOLS BY SOIL ORDER AND LANDSCAPE POSITION

The C pools varied according to the soil order. There was higher TOC content in the upper third of the Oxisols, ranging from 40.8 ± 6.1 g C kg^{-1} at a depth of 0–10 cm to 14.4 ± 6.2 g C kg^{-1} at a depth of 70–100 cm. In contrast, in the Inceptisols, the highest TOC contents were recorded in the bottom third, ranging from 44.9 ± 15.8 g C kg^{-1} at a depth of 0–10 cm to 16.3 ± 7.0 g C kg^{-1} at a depth of 70–100 cm (Table 12.3). Although a small textural gradient was observed between the upper and bottom third (Table 12.2), most likely because of the high clay content in the study site soils, no significant variations in the C content were observed (Table 12.3).

TABLE 12.3
Mean TOC Content, Labile C Content (POX-C and HWE-C), TN, and C/N Ratio by Soil Order and Landscape Location

Soil Order	Depth cm	Upper Third					Middle Third					Bottom Third				
		TOC	TN	POX-C	HWE-C	C/N	TOC	TN	POX-C	HWE-C	C/N	TOC	TN	POX-C	HWE-C	C/N
		g kg⁻¹					g kg⁻¹					g kg⁻¹				
Oxisol (RL)	0–10	40.8±6.1[a]	2.6±0.7	4.51±1	0.78±0.6	16.8±5.4	34.8±9.5	2.4±0.7	4.29±0.9	0.63±0.5	14.6±2.1	36.1±5.3	2.0±0.2	3.16±1	0.55±0.5	18.7±4.6
	10–20	35.7±7.4	2.1±0.7	3.95±1	0.59±0.6	19.3±7.5	30.8±8.6	2.0±0.6	3.95±0.8	0.61±0.6	15.3±1.6	32.4±4.6	1.5±0.3	3.07±1.2	0.72±0.8	22.8±8.3
	20–40	26.8±5.9	1.4±0.4	3.21±0.8	0.46±0.6	20.7±6.2	23.8±7.5	1.3±0.4	3.19±0.5	0.55±0.5	18.0±5.1	27.6±2.4	1.3±0.4	2.73±0.7	0.48±0.5	21.6±8.3
	40–70	18.3±6.0	1.0±0.3	2.58±0.5	0.44±0.6	19.5±3.6	15.6±6.0	0.9±0.4	2.70±0.4	0.33±0.3	19.7±9.9	18.8±0.2	0.8±0.5	1.94±1.0	0.46±0.5	30.1±18.9
	70–100	14.4±6.2	0.7±0.3	2.74±0.8	0.37±0.5	20.8±9.0	14.1±4.7	0.6±0.2	2.45±0.4	0.23±0.3	22.8±7.0	14.7±2.0	0.6±0.4	2.19±0.8	0.52±0.2	29.4±22.0
Oxisol (RYL)	0–10	36.7±7.6	2.4±0.4	4.59±0.5	0.70±0.6	15.4±3.7	36.4±8.4	2.4±0.8	4.56±0.6	0.59±0.5	15.8±3.2	39.7±5.3	2.7±0.5	3.73±1.2	0.59±0.3	15.0±1.6
	10–20	31.9±9.0	1.7±0.6	3.89±0.5	0.63±0.2	20.7±3.6	31.7±9.0	1.9±0.6	4.01±0.5	0.38±0.2	17.2±3.6	36.0±5.1	2.3±0.4	3.34±1.4	0.47±0.3	15.6±1.2
	20–40	26.7±7.0	1.6±0.6	3.15±0.5	0.57±0.6	18.4±7.4	25.4±7.5	1.4±0.5	3.30±0.6	0.27±0.2	20.7±9.0	28.8±3.3	1.7±0.3	3.09±1.1	0.39±0.3	16.8±1.8
	40–70	17.6±5.3	0.9±0.5	2.63±0.5	0.49±0.5	24.4±9.6	16.8±5.5	0.8±0.4	2.65±0.5	0.18±0.1	23.0±9.8	19.9±1.8	1.1±0.3	2.34±0.8	0.35±0.4	19.8±6.2
	70–100	13.2±3.0	0.6±0.2	2.33±0.4	0.46±0.5	23.5±5.2	12.7±5.7	0.6±0.3	2.51±0.4	0.18±0.3	25.4±12.1	20.8±9.7	1.2±0.7	2.34±0.9	0.28±0.3	17.8±5.0
Inceptisol (HC)	0–10	37.2±8.7	2.5±0.1	4.35±0.8	0.75±0.2	16.1±4.4	37.5±10.8	2.5±1	4.66±0.8	0.62±0.2	16.0±2.4	35.0±10.3	2.4±0.9	4.36±0.9	0.67±0.4	15.7±4.5
	10–20	32.6±6.8	2.0±0.7	3.83±0.5	0.64±0.2	16.8±3.6	33.7±8.6	1.9±0.6	4.05±0.7	0.47±0.2	18.1±2.5	26.4±10.6	1.6±0.9	3.62±0.8	0.34±0.3	17.5±12.8
	20–40	25.9±3.7	1.3±0.3	3.24±0.4	0.35±0.1	20.8±4.0	24.6±5.9	1.3±0.5	3.63±0.6	0.40±0.1	20.1±6.0	21.6±8.0	1.2±0.5	3.19±0.7	0.40±0.3	19.7±6.3
	40–70	19.4±4.9	0.9±0.5	2.94±0.5	0.23±0.1	30.1±22.5	18.3±9.7	1.1±0.5	2.85±0.6	0.26±0.1	16.6±0.8	17.7±8.3	0.8±0.5	2.85±0.7	0.36±0.3	26.3±11.9
	70–100	12.0±4.7	0.6±0.2	2.48±0.9	0.18±0.1	25.8±17.6	11.8±4.6	0.8±0.1	2.74±0.3	0.25±0.2	14.4±6.3	14.8±6.7	0.7±0.4	2.68±0.6	0.29±0.3	23.2±13.5

Sampling Location in the Landscape Scale at Each Soil Order

(Continued)

TABLE 12.3 (CONTINUED)

Mean TOC Content, Labile C Content (POX-C and HWE-C), TN, and C/N Ratio by Soil Order and Landscape Location

Soil Order	Depth	Upper Third					Middle Third					Bottom Third				
		Sampling Location in the Landscape Scale at Each Soil Order														
	cm	TOC	TN	POX-C	HWE-C	C/N	TOC	TN	POX-C	HWE-C	C/N	TOC	TN	POX-C	HWE-C	C/N
		g kg⁻¹							g kg⁻¹					g kg⁻¹		
Inceptisol + Entisol (HC+RL)	0–10	29.2 ± 0	2.3 ± 0	3.52 ± 0	0.70 ± 0	12.8 ± 0	38.9 ± 2.7	2.4 ± 0.1	4.03 ± 0.8	0.52 ± 0.3	18.4 ± 7.2	30.0 ± 9.3	2.0 ± 0.5	3.88 ± 0.2	0.43 ± 0.2	15.7 ± 4.3
	10–20	28.9 ± 0	2.8 ± 0	3.33 ± 0	0.50 ± 0	10.2 ± 0	37.4 ± 3.0	1.9 ± 0.8	3.65 ± 0.6	0.41 ± 0.3	21.4 ± 7.6	26.4 ± 9.0	1.6 ± 0.6	3.62 ± 0.3	0.34 ± 0.1	17.5 ± 5.4
	20–40	26.5 ± 0	1.9 ± 0	3.25 ± 0	0.50 ± 0	13.8 ± 0	27.0 ± 2.7	1.6 ± 0.5	2.95 ± 0.4	0.33 ± 0.2	18.4 ± 5.4	21.8 ± 9.5	1.2 ± 0.3	3.05 ± 0.3	0.23 ± 0.1	18.4 ± 5.5
	40–70	12.1 ± 0	1.4 ± 0	2.72 ± 0	0.39 ± 0	8.2 ± 0	17.3 ± 1.7	0.8 ± 0.1	2.35 ± 0.3	0.17 ± 0.1	21.1 ± 2.4	15.2 ± 5.0	0.7 ± 0.2	2.76 ± 0.1	0.20 ± 0.1	21.5 ± 9.0
	70–100	5.3 ± 0	2.0 ± 0	2.5 ± 0	0.5 ± 0	22.3 ± 0	13.6 ± 3.0	0.5 ± 0.1	2.27 ± 0.3	0.11 ± 0.8	25.6 ± 6.7	12.2 ± 4.3	0.5 ± 0.2	2.55 ± 0.3	0.14 ± 0.1	31.7 ± 14.2
Inceptisol (XC)	0–10	37.7 ± 1.5	2.8 ± 0.1	4.86 ± 1.2	0.60 ± 0.2	14.6 ± 4.3	40.3 ± 18.7	2.9 ± 0.2	5.76 ± 0.2	0.45 ± 0.2	16.0 ± 4.6	44.9 ± 15.8	3.0 ± 0.1	4.73 ± 1	0.62 ± 0.1	15.0 ± 1.1
	10–20	32.8 ± 12.3	2.0 ± 1.3	3.53 ± 0.1	0.45 ± 0.1	20.8 ± 11.3	33.7 ± 1.4	2.2 ± 1.8	4.76 ± 0.1	0.32 ± 0.2	19.1 ± 6.6	41.1 ± 20.8	2.6 ± 1.4	3.97 ± 0.4	0.57 ± 0.1	15.8 ± 1.7
	20–40	29.2 ± 9.8	1.7 ± 0.7	3.76 ± 0.8	0.52 ± 0.2	17.7 ± 2.7	24.3 ± 6.4	1.3 ± 0.7	3.79 ± 0.8	0.31 ± 0.1	19.8 ± 5.6	32.7 ± 16	1.9 ± 1.2	3.68 ± 0.6	0.36 ± 0.2	18.5 ± 6.1
	40–70	20.1 ± 6.1	1.2 ± 0.7	2.87 ± 0.8	0.33 ± 0.6	20.1 ± 7.3	16.4 ± 4.5	1.0 ± 0.3	3.48 ± 0.7	0.19 ± 0.2	17.9 ± 5.0	26.5 ± 18.1	1.3 ± 0.1	2.86 ± 0.3	0.11 ± 0.1	20.8 ± 1.5
	70–100	11.8 ± 3.4	0.5 ± 0.2	2.39 ± 1.0	0.21 ± 0.1	27.2 ± 8.1	8.6 ± 0.3	0.6 ± 0.1	2.87 ± 0.6	0.10 ± 0.8	14.3 ± 2.0	16.3 ± 7.0	0.7 ± 0.3	2.78 ± 0.2	0.23 ± 0.3	23.4 ± 7.5

Note: C/N, soil C/N ratio; HC, Humic Cambisol; HWE-C, carbon extracted by hot water; POX-C, carbon extracted by permanganate; RL, Red Latosol; RYL, Red-Yellow Latosol; TN, total nitrogen; TOC, total organic carbon; XC, Haplic Cambisol.

[a] Refers to the standard deviation in relation to the mean.

Soil-Specific Farming

The high TOC content for both the RL and RYL can be explained as follows:

1. The higher clay content in these soil orders compared to the others can partially explain their higher TOC content (Table 12.2). The effect of clay on C accumulation has been previously reported in the literature. Sá et al. (2013a) reported that this relationship was significant for Oxisols in the 0–20 cm and 20–70 cm layers and that the angular coefficient indicated increases of 0.047 and 0.026 g of C for each kilogram of clay, respectively. Sá and colleagues also observed similar ratios (0.049 and 0.030 g C kg^{-1} clay, respectively) in the same layers of Inceptisols. Wang et al. (2012) reported that the correlation coefficient between the soil clay content and the TOC content was 0.4 ($P < 0.001$), which is similar to the value obtained in this study, 0.35 ($P < 0.001$) for all of the soil orders combined. Different soil orders were analyzed, meaning that the variation in mineralogy may significantly affect the accumulation of TOC because the association of C with iron (Fe) and aluminum (Al) oxides is an important mechanism for C storage and is also one of the main factors that regulates the accumulation of C in acid soils in subtropical and tropical regions (Inda Junior et al. 2013; Kaiser and Guggenberger 2003; Kleber et al. 2005).
2. Greater input of OM caused by higher crop yields can partially explain the higher TOC contents (Figure 12.5). It is widely recognized that grain yield and plant biomass production are strongly correlated (Araujo and Teixeira 2012; Geraldo et al. 2000; Sá et al. 2014), which suggests that in areas where grain yields are high, the inputs of dry plant biomass via crop waste are also high. The RL and RYL soils were associated with the highest grain yields (Figure 12.5); therefore, they should also have produced the greatest plant biomass and its return to the soil compared to the other soil orders.

In the upper third of the soils, the Humic Cambisol + Litholic Neosol (HC+LN) soil orders contained 22% lower TOC than the mean for the other soil types. The lower effective depth of these soil types (EMBRAPA 2006) along with their lower clay content (Table 12.2) contributed to lower

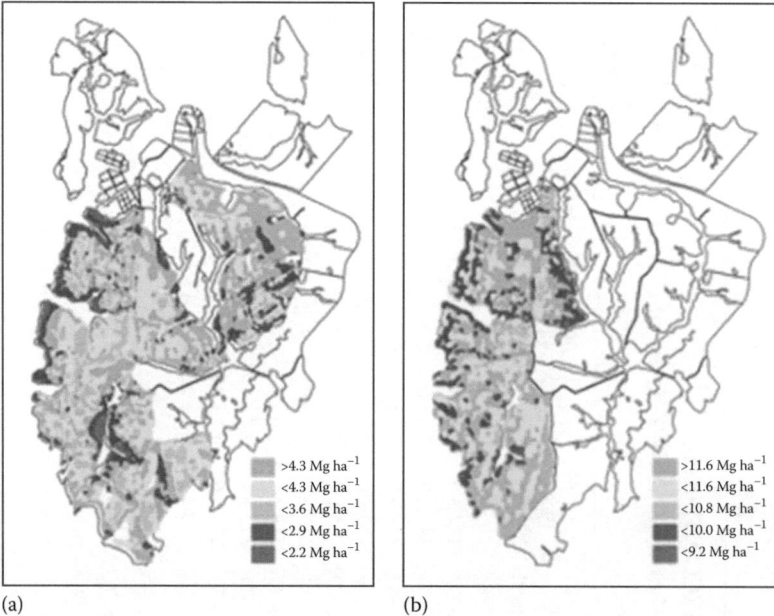

(a)   (b)

FIGURE 12.5 Maps of (a) wheat (2012) and (b) maize (2011/2012) harvests. The maximum and minimum wheat yields (2012) on the farm were 4355 kg ha^{-1} and 1939 kg ha^{-1}, respectively. The maximum and minimum maize yields (2011/2012) on the farm were 11,610 kg ha^{-1} and 9167 kg ha^{-1}, respectively.

water retention capacity in the soil, which was reflected by lower yields. The lower yields and the weaker chemical protection of C due to reduced clay content (Six et al. 2002) can explain the lower TOC content.

The high TOC content that was observed in the bottom third of the Inceptisols (Haplic Cambisols) can be explained by the transport of organic acids by superficial and subsuperficial flows because this soil type occurs at a lower elevation than the Oxisols (Figure 12.4). During the summer austral months, especially December, January, and February, rainfall is the highest. The increased rainfall, in combination with the lower effective depth of this soil order (EMBRAPA 2006), favors the formation of an anaerobic environment, inhibiting C oxidation and enhancing its accretion.

The spatial variation in TN was similar to that of the TOC (Table 12.3). The correlation between these variables was 0.8 ($P < 0.001$); the vast majority of N in the soil is present in organic forms that are part of the soil organic matter (Poirier et al. 2009; Rasmussen et al. 1998). Gil et al. (2008) reported a correlation of 0.89 ($P < 0.05$) between the OM and TN content in the soil, which was similar to the value obtained in the present study. The C/N ratio was the highest in the RL (16.7), then decreased as follows: HC (15.9) > RYL (15.6) > HC+LN (15.5) > XC (15.2); this decreasing ratio was similar for TOC (Table 12.3), except in the Haplic Cambisol (XC) where the high TN resulted in a small C/N ratio.

The spatial variation of POX-C was similar to that of the TOC (Table 12.3), whereas the variation in HWE-C did not display the same trend as the TOC, TN, and POX-C (Table 12.3). The occurrence of an anaerobic environment, even over a short period of time, contributed to a reduction in HWE-C because the XC contained a HWE-C content that was 18% lower than the mean in the other soils (Table 12.3). In contrast, both the HWE-C and the POX-C varied as a function of crop sequence, and the residence times of these more labile pools were approximately 0.5 and 2 years, respectively (Stevenson 1986).

Soil texture and water content, which are associated with the topography, are the strong determinants of C accumulation. Wang et al. (2012) reported that altitude, slope, clay content, and water content explained 70.3% of the spatial variation in the TOC and 67.1% of that in the TN. In the Flanders region of Belgium, Meersmans et al. (2008) evaluated draining and texture in a multiple regression analysis to assess the TOC stocks. Guo et al. (2006) considered topography to be the main control of TOC accumulations.

The C and N pools in the 10–20 cm layer followed the pattern observed in the 0–10-cm layer (Table 12.3), which was also the case for the 20–40-cm layer (Table 12.3). High HWE-C contents were observed in the 20–40-cm layer for all soil types, as well as in the 40–70- and 70–100-cm layers for Oxisols (Table 12.3). This trend represents the contribution from the crop root system, which releases exudates, thereby stimulating microbial communities in the soil zone closest to the plant roots (Jones et al. 2004). Similar results have been reported by Sá et al. (2013b, 2014).

The observed TOC content in the deepest layers, $14.4^{\pm6.2}$ and $13.2^{\pm3}$ g C kg^{-1} for RL and RYL, respectively, can be explained by the movement of C through the soil profile, which is distinct in NT farming systems (Chabbi et al. 2009; Rumpel et al. 2009). Underlying reasons of this trend involve the transport of soluble organic acids via preferential paths such as the pores formed by the roots. Therefore, as C saturation occurs in the superficial layers, newly introduced compounds descend through the soil profile. The contribution from the decomposition of the crop root system is also high.

12.3.2 Effect of Crop Sequence

Although the observed difference was small, the POX-C in S3 was 10% higher than in S2 and 16% higher than in S1 (Table 12.4) as a result of more biomass input from maize (Table 12.1). This variation in the most labile TOC compartment is in accordance with Blair et al. (1995), who reported that short-term alterations in labile C could reflect agroecosystem management (Culman et al. 2012).

TABLE 12.4

Mean TOC Content, Labile C Content (POX-C and HWE-C), TN, and C/N Ratio by Crop Sequence in 2013

Depth cm	Crop Sequence	TOC	TN	POX-C	HWE-C	C/N
		g kg^{-1}				
0–10	S1	37.4 ± 7.9	2.4 ± 0.1	4.11 ± 0.7	0.78 ± 0.5	16.7 ± 4.4
	S2	33.3 ± 9.6	2.3 ± 0.1	4.37 ± 0.6	0.33 ± 0.2	14.8 ± 3.1
	S3	39.2 ± 10.7	2.7 ± 0.1	4.87 ± 1.2	0.65 ± 0.3	15.4 ± 2.9
10–20	S1	33.3 ± 8.0	1.9 ± 0.1	3.59 ± 0.6	0.63 ± 0.4	19.6 ± 6.3
	S2	28.8 ± 9.5	1.9 ± 0.1	3.94 ± 0.6	0.26 ± 0.1	16.5 ± 6.2
	S3	34.3 ± 10.7	2.1 ± 0.1	4.12 ± 1.0	0.50 ± 0.3	17.6 ± 4.9
20–40	S1	26.2 ± 6.3	1.4 ± 0.01	3.09 ± 0.5	0.52 ± 0.4	20.6 ± 5.6
	S2	21.3 ± 7.7	1.3 ± 0.01	3.26 ± 0.6	0.19 ± 0.1	16.3 ± 4.2
	S3	27.1 ± 7.9	1.5 ± 0.1	3.56 ± 0.8	0.39 ± 0.2	20.4 ± 7.7
40–70	S1	17.6 ± 6.3	0.8 ± 0.01	2.56 ± 0.5	0.37 ± 0.3	24.5 ± 12.2
	S2	16.2 ± 7.4	1.0 ± 0.01	2.76 ± 0.4	0.13 ± 0.1	16.5 ± 4.2
	S3	19.7 ± 7.8	1.0 ± 0.01	2.95 ± 0.7	0.31 ± 0.2	22.8 ± 8.4
70–100	S1	13.2 ± 5.3	0.6 ± 0.01	2.41 ± 0.5	0.33 ± 0.4	23.7 ± 11.2
	S2	13.2 ± 7.6	0.8 ± 0.01	2.64 ± 0.4	0.10 ± 0.1	17.2 ± 6.4
	S3	15.0 ± 5.0	0.7 ± 0.01	2.73 ± 0.7	0.23 ± 0.2	26.1 ± 11.5

Note: C/N, soil C/N ratio; HWE-C, carbon extracted by hot water; POX-C, carbon extracted by permanganate; TN, total nitrogen; TOC, total organic carbon.

The HWE-C concentration indicated that S1 > S3 > S2 at all depths. For the 0–10-cm layer, this result can be explained by soybean straw additions having a low C/N ratio (13 to 16) in S1, which increased the availability of N for microbial biomass and stimulated the decomposition of maize and oat straw with higher C/N ratios (>50 and >25, respectively) in S3. In the 10–20-cm and deeper layers, this trend may be attributed to the decomposition of the root system and rhizodeposition because a large portion of the TOC input into the soil is derived from root decomposition. Santos et al. (2011) estimated the magnitude of root deposition, 2.36 Mg C ha^{-1} y^{-1} for an oat/maize/wheat/soybean sequence and 1.91 Mg C ha^{-1} y^{-1} for a wheat/soybean sequence. Sá et al. (2014) reported additions of 1.41 Mg C ha^{-1} y^{-1} in an oat/maize sequence and 0.45 Mg C ha^{-1} y^{-1} in a wheat/soybean sequence. The C/N ratio of the soil was the lowest in S2 after two wheat/soybean sequences due to the lower C/N ratios commonly observed in legumes compared to grasses.

12.3.3 Maps of Total C and Labile C Fractions

The maps of the variation in TOC and POX-C were very similar (Figure 12.6), illustrating the close relationship between these two variables and indicating that the TOC accumulation occurred as a function of greater additions of labile C, which was reflected by the POX-C. In a number of plots in the center of the farm, the TOC content did not vary as a function of the POX-C (Figure 12.6). The predominance of Inceptisols, which favor the formation of an anaerobic environment during the rainier periods of the year, alters the C dynamics and causes C accumulation because of the inhibition of oxidation, hence creating this deviation from the trend. As the HWE-C primarily varied in association with the crop sequence, the lighter areas of this map (Figure 12.6) occur in plots that were under S1 at the time of sampling (Figure 12.3).

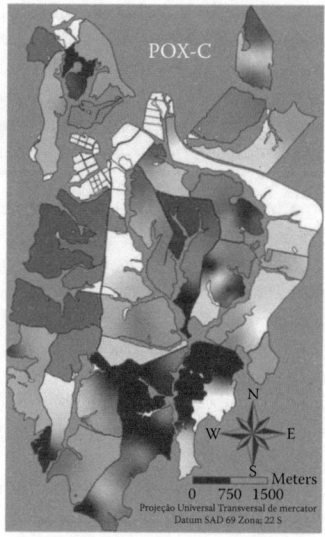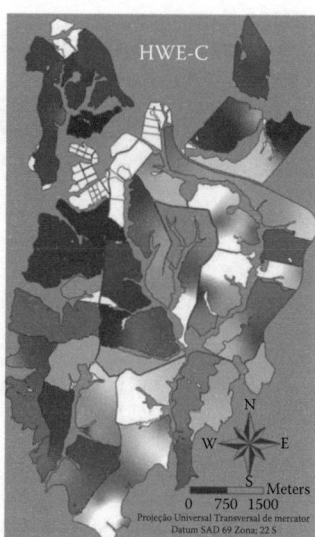

FIGURE 12.6 Spatial distribution of the analyzed variables fit by spline models. Lighter areas correspond to higher levels. White areas correspond to no data. HWE-C, carbon extracted by hot water; POX-C, carbon extracted by permanganate; TOC, total organic carbon.

12.4 CONCLUSION

The TOC content was the highest in the upper third of the Oxisols because of the greater addition of plant biomass associated with higher clay content. In the Inceptisols, the highest C content was observed in the bottom third of the profile because of the inhibition of C oxidation during the periods of anaerobiosis. In lower areas without inhibition of C oxidation was less efficient in accumulate C and would be managed in a site-specific strategies. Both the POX-C and HWE-C varied as a function of crop sequence because they are very sensitive to the effect of soil management. The highest HWE-C concentrations were observed in S1 after the addition of plant biomass derived from soybeans, which have a high N concentration that stimulates the microbial biomass.

REFERENCES

Araújo, A. P.; Teixeira, M. G. Variabilidade dos Índices de Colheita de Nutrientes em Genótipos de Feijoeiro e Suas Relações com a Produção de Grãos. *Revista Brasileira de Ciência do Solo*, Vol. 36, pp. 137–146, 2012.

Batjes, N. H. Total carbon and nitrogen in the soils of the world. *European Journal of Soil Science*, Vol. 47, pp. 151–163, 1996.

Behling, H. Late quaternary vegetation, climate and fire history of the Araucaria forest and Campos region from Serra Campos Gerais, Parana State (South Brazil). *Review of Palaeobotany and Palynology*, Vol. 97, pp. 109–121, 1997.

Bernoux, M.; Carvalho, M. C. S.; Volkoff, B.; Cerri, C. C. Brazil's soil carbon stocks. *Soil Science Society of America Journal*, Vol. 66, pp. 888–896, 2002.

Blair, G. J.; Lefroy, R. D. B.; Lisle, L. Soil carbon fractions based on their degree of oxidation, and the development of a carbon management index for agricultural systems. *Australian Journal of Agriculture Research*, Vol. 46, pp. 1459–1466, 1995.

Cerri, C. E. P.; Easter, M.; Paustian, K. et al. Simulating SOC changes in 11 land use change chronosequences from the Brazilian Amazon with RothC and Century models. *Agriculture, Ecosystems and Environment*, Vol. 122, pp. 46–57, 2007.

Chabbi, A.; Kögel-Knabner, I.; Rumpel C. Stabilised carbon in subsoil horizons is located in spatially distinct parts of the soil profile. *Soil Biology Biochemistry*, Vol. 41, pp. 256–271, 2009.

Culman, S. W.; Snapp, S. S.; Freeman, M. A. et al. Permanganate oxidizable carbon reflects a processed soil fraction that is sensitive to management. *Soil Science Society of America Journal*, Vol. 76, No. 2, 2012.

Culman, S. W.; Snapp, S. S.; Green, J. M.; Gentry, L. E. Short- and long-term labile soil carbon and nitrogen dynamics reflect management and predict corn agronomic performance. *Agronomy Journal*, Vol. 105, No. 2, 2013.

Davidson, E. A.; Lefebvre, P. A. Estimating regional carbon stocks and spatially covarying edaphic factors using soil maps at three scales. *Biogeochemistry*, Vol. 22, pp. 107–131, 1993.

Easter, M.; Paustian, K.; Killian, K. et al. The GEFSOC soil carbon modelling system: A tool for conducting regional-scale soil carbon inventories and assessing the impacts of land use change on soil carbon. *Agriculture, Ecosystems and Environment*, Vol. 122, pp. 13–25, 2007.

EMBRAPA. Empresa *Brasileira de Pesquisa Agropecuária. Sistema Brasileiro de Classificação de Solos* (in Portuguese.) 2nd ed., Centro Nacional de Pesquisa de Solos, Rio de Janeiro, Brazil, 2006.

Geraldo, J.; Rossielo, R. O. P.; Araújo, A. P.; Pimentel, C. Diferênças em Crescimento e Produção de Grãos Entre Quatro Cultivares de Milho Pérola. *Pesquisa Agropecuária Brasileira*, Brasília, Vol. 35, No. 7, pp. 1367–1376, 2000.

Ghani, A.; Dexter, M.; Perrott, K. W. Hot-water extractable carbon in soils: A sensitive measurement for determining impacts of fertilisation, grazing and cultivation. *Soil Biology & Biochemistry*, Vol. 35, pp. 1231–1243, 2003.

Gil, S. V.; Meriles, J.; Conforto, C.; Figoni, G.; Basanta, M.; Lovera, E.; March, G. J. Field assessment of soil biological and chemical quality in response to crop management practices. *World Journal of Microbiology and Biotechnology*, Vol. 25, pp. 439–448, 2008.

Guo, Y.; Amundson, R.; Gong, P.; Yu, Q. Quantity and spatial variability of soil carbon in the conterminous United States. *Soil Science Society of America Journal*, Vol. 70, pp. 590–600, 2006.

Hartemink, A. E. Soil map density and a nation's wealth and income. In A. E. Hartemink, A. McBratney, and M. L. Mendonça-Santos (eds.), *Digital Soil Mapping with Limited Data*. Springer, New York, pp. 53–66, 2008.

IAPAR. Istituto Agronômico do Paraná, available at http://www.iapar.br/. Accessed on December 10, 2013.

Inda Junior, A. V.; Torrent, J.; Barrón, V.; Bayer, C.; Fink, J. R. Iron oxides dynamics in a subtropical Brazilian Paleudult under long-term no-tillage management. *Scientia Agricola*, Vol. 70, No. 1, pp. 48–54, 2013.

Jones, D. L.; Hodge, A.; Kazyakov, Y. Plant and mycorrhizal regulation of rhizodeposition. *New Phytologist*, Vol. 163, pp. 459–480, 2004.

Kaiser, K.; Guggenberger, G. Mineral surfaces and soil organic matter. *European Journal of Soil Science*, Vol. 54, pp. 219–236, 2003.

Kleber, M.; Mikutta, R.; Torn, M. S.; Jahn, R. Poorly crystalline mineral phases protect organic matter in acid subsoil horizons. *European Journal of Soil Science*, Vol. 56, pp. 717–725, 2005.

Levine, E. R.; Knox, R. G.; Lawrence, W. T. Relationships between soil properties and vegetation at the Northern Experimental Forest, Howland, Maine. *Remote Sensing Environment*, Vol. 47, pp. 231–241, 1994.

Lilburne, L. R.; Hewitt, A. E.; Webb, T. W. Soil and informatics science combine to develop S-map: A new generation soil information system for New Zealand. *Geoderma*, Vol. 170, pp. 232–238, 2012.

Mazzarino, M. J.; Szott, L.; Jimenez, M. Dynamics of soil total C and N, microbial biomass, and water-soluble C in tropical agroecosystems. *Soil Biology Biochemestry*, Vol. 25, No. 2, pp. 205–214, 1993.

Meersmans, J.; Ridder, F. D.; Canters, F.; Baets, D. S.; Van Molle, M. A multiple regression approach to assess the spatial distribution of soil organic carbon (SOC) at the regional scale (Flanders, Belgium). *Geoderma*, Vol. 143, pp. 1–13, 2008.

MINEROPAR. Serviço Geológico do Paraná, available at http://www.mineropar.pr.gov.br/. Accessed on December 13, 2013.

Pan, G.; Xu, X.; Smith, P.; Pan, W.; Lal, R. An increase in topsoil SOC stock of China's croplands between 1985 and 2006 revealed by soil monitoring. *Agriculture, Ecosystems and Environment*, Vol. 136, pp. 133–138, 2010.

Poier, K. R.; Richter, J. Spatial distribution of earthworms and soil properties in an arable loess soil. *Soil Biological Biochemistry*, Vol. 24, No. 12, pp. 1601–1608, 1992.

Poirier, V.; Angers, D. A.; Rochette, P.; Chantigny, M. H.; Ziadi, N.; Tremblay, G.; Fortin, J. Interactive effects of tillage and mineral fertilization on soil carbon profiles. *Soil Science Society of America Journal*, Vol. 73, No. 1, 2009.

R Development Core Program Team. R: A language and environment for statistical computing. ISBN 3-900051-07-0, 2012.

Rasmussen, P. E.; Douglas, C. L.; Collins, H. P.; Albrecth, S. L. Long-term cropping system effects on mineralizable nitrogen in soil. *Soil Biology Biochemistry*, Vol. 30, No. 13, pp. 1829–1837, 1998.

Rumpel, C.; Ba, A.; Darboux, F.; Chaplot, V.; Planchon, O. Erosion budget and process selectivity of black carbon at meter scale. *Geoderma*. Vol. 154, pp. 131–137, 2009.

Sá, J. C. M.; Santos, J. B.; Lal, R. et al. Soil-specific inventories of landscape carbon and nitrogen stocks under no-till and native vegetation to estimate carbon offset in a subtropical ecosystem. *Soil Science Society of America Journal*, Vol. 77, pp. 2094–2110, 2013a.

Sá, J. C. M.; Séguy, L.; Tivet, F. et al. Carbon depletion by plowing and its restoration by no-till cropping systems in oxisols of sobtropicaland tropical agro-ecoregions in Brazil. *Land Degradation and Development* (Available at http://www.wileyonlinelibrary.com), 2013b, doi: 10.1002/ldr.2218.

Sá, J. C. M.; Tivet, F.; Lal, R.; Briedis, C.; Hartman, D. C.; Santos, J. Z.; Santos, J. B. Long-term tillage systems impacts on soil C dynamics, soil resilience and agronomic productivity of a Brazilian Oxisol. *Soil & Tillage Research*, Vol. 136, pp. 38–50, 2014.

Santos, N. Z.; Dieckow, J.; Bayer, C.; Molin, R.; Favaretto, N.; Pauletti, V.; Piva, J. T. Forages, cover crops and related shoot and root additions in no-till rotations to C sequestration in a subtropical Ferralsol. *Soil & Tillage Research*, Vol. 111, pp. 208–218, 2011.

Six, J.; Conant, R. T.; Paul, E. A.; Paustian, K. Stabilization mechanisms of soil organic matter: Implications for C-saturation of soils. *Plant and Soil*, Vol. 241, pp. 155–176, 2002.

Soil Survey Staff. *Keys to Soil Taxonomy*, 11th ed. United States Department of Agriculture-Natural Resources Conservation Service, Washington, DC, 2010.

Stevenson, F. J. *Cycles of Soil*. John Wiley & Sons, Inc. New York, 1986.

Tornquist, C. G.; Gassman, P. W.; Mielniczuk, J.; Giasson, E.; Campbell, T. Spatially explicit simulations of soil C dynamics in Southern Brazil: Integrating century and GIS with i_Century. *Geoderma*, Vol. 150, pp. 404–414, 2009.

Walkley, A.; Black, I. A. An examination of the Degtjareff method for determining organic carbon in soils: Effect of variations in digestion conditions and of inorganic soil constituents. *Soil Science*, Vol. 63, pp. 251–263, 1934.

Wang, S.; Wang, X.; Ouyang, Z. Effects of land use, climate, topography and soil properties on regional soil organic carbon and total nitrogen in the Upstream Watershed of Miyun Reservoir, North China. *Journal of Environmental Sciences*, Vol. 24, No. 3, pp. 387–395, 2012.

Weil, R. R.; Islam, K. R.; Stine, M. A.; Gruver, J. B.; Sanson-Liebig, S. E. Estimating active carbon for soil quality assessment: A simplified method for laboratory and field use. *American Journal of Alternative Agriculture*, Vol. 18, pp. 3–17, 2003.

13 Laser-Assisted Precision Land Leveling Impacts in Irrigated Intensive Production Systems of South Asia

M.L. Jat, Yadvinder Singh, Gerard Gill, H.S. Sidhu,
Jeetendra P. Aryal, Clare Stirling, and Bruno Gerard

CONTENTS

13.1 INTRODUCTION

South Asian countries, comprising India, Pakistan, Nepal, and Bangladesh and having a total geographical area of only 401.72 million hectares (m ha), hold nearly half the world's population of 3.1 billion (FAO 1999). Nearly half of the land area in South Asia is devoted to agriculture, which provides livelihood and food security for 59% of the world's population. For agricultural production in South Asia (SA), irrigation is the most crucial input because about 40% of cropland is irrigated agriculture that accounts for 60%–80% of food production (Yadvinder-Singh et al. 2014). Agricultural water use is estimated to consume almost 95% of the withdrawn water in South Asian countries, which is well above the global average of 70% (Babel and Wahid 2008). Groundwater in the northwestern Indo-Gangetic plains (IGP) is being depleted at 13 to 17 $km^3 yr^{-1}$ (Rodell et al. 2009), which is a major concern for the current growth rate and sustainability of South Asian agriculture (Hira 2009). Negative environmental effects related to irrigation are increasing as overexploitation of groundwater and poor water management lead to the steady rate of decline in the depth to the groundwater in much of the rice-wheat (RW) areas of northwestern India and parts of Pakistan (Qureshi et al. 2003; Ambast et al. 2006; Hira 2009; Rodell et al. 2009). The number of tube wells in Indian Punjab, for example, has increased from 0.30 million in 1975 to 1.232 million in 2007 (Anonymous 2007), which led to a groundwater table decline from 18 m to 27 m from 2000–2009. Similarly, in Pakistan, both groundwater and surface water resources are decreasing due to overexploitation of water resources and the adverse impact of climate change on water resources (Gill 2001).

Rice-rice, RW, cotton-wheat, rice-maize, maize-wheat, and sugarcane-wheat are the major cropping systems in irrigated ecologies of SA. Rice is the topmost, followed by wheat and maize in terms of area, production, and yield. The total water requirement for the RW system is estimated to vary between 138 to 184 cm in the IGP, accounting for more than 80% of the rice-growing season (Jat et al. 2006). Intensively cultivated RW is the most important cropping system for food security in SA with 12.4 m ha in the subtropical areas of the IGP providing food for more than 400 million people (Ladha et al. 2003; Yadvinder-Singh et al. 2014). This represents 32% of the total rice area and 42% of the wheat area (Hobbs and Morris 1996; Ladha et al. 2000) and contributes more than 80% of the total cereal production (Timsina and Connor 2001). The productivity and sustainability of the system are threatened because of the inefficiency of current production practices, shortage of resources such as water and labor, and socioeconomic changes (Ladha et al. 2003). The major obstacle to boosting crop yields is the availability and efficient use of water, since there is increasing stress on water resources used for agriculture in many parts of SA (Yadvinder-Singh et al. 2014). To meet the need of increasing food production in the face of the emerging challenges of climate change, the demand on water for irrigation is likely to increase in SA. For example, present total water requirement in India has been estimated at slightly more than 600 billion cubic meters (BCM). The requirement is likely to increase to 1093 BCM by 2025 and it will further increase to 1447 BCM by 2050 (Yadvinder-Singh et al. 2014).

Most of the agricultural land in the IGP are poorly leveled using traditional land leveling (TLL) practices (Jat et al. 2006; Ladha et al. 2009). The land slopes vary from transect I to transect V of the IGP. The average field slope in transect I and II (includes Pakistan Punjab, Indian Punjab, and Haryana and western Uttar Pradesh) ranges between 1 and 3 degrees. The average field slopes range between 3 and 5 degrees in transect IV and V (Jat et al. 2006). Uneven fields and unlined irrigation channels cause the loss of a huge amount of irrigation water (Gill 1994). Rickman (2002) found a strong correlation between the levelness of the land and crop yield. Hence, there is an urgent need to manage water efficiently using cost-effective and eco-friendly techniques. Water management by proper irrigation scheduling in combination with better crop management techniques (i.e., laser-assisted precision land leveling, conservation-agriculture-based management practices) are potential options for saving of water and increasing water productivity (WP).

The majority of growers in SA practice surface irrigation either through flood or check basin methods. These irrigation practices in unleveled or traditionally leveled bunded units normally result in overirrigation (Corey and Clyma 1973), causing high loss of irrigation water through deep percolation. Several researchers (Cook and Peikert 1960; Asif et al. 2003; Sattar et al. 2003; Jat et al. 2006; Kahlown et al. 2006; Bhatt and Sharma 2009) have reported that irrigation application efficiency on unleveled fields is reduced by up to 30% to 50% as compared to an attainable level of 60% to 80%. In surface irrigation, land leveling is essential for high application efficiency that ensures high water-use efficiency and crop yield. Land leveling is an important operation for good agronomic, soil, and crop management practices. For an efficient irrigation system, the level difference between high and low spots of a field should not exceed 20 mm, whereas under actual field conditions, a difference of 50 to 150 mm is very common. Traditional methods of leveling lands using animal-drawn or tractor-drawn levelers are not only more cumbersome and time-consuming but also more expensive. It is seen that even the best leveled fields using traditional land leveling practices are not precisely leveled and this leads to uneven distribution of irrigation water (Jat et al. 2006).

Laser-assisted precision land leveling (PLL) is the foremost step in the judicious use of irrigation water and enhancing water productivity. It is a laser-guided (light amplification by stimulated emission of radiation) precision leveling technique used for achieving very fine leveling with the desired grade on the field within ±2 cm of its average microelevation. Hill et al. (1991) rated the development of laser leveling technology as second only to the breeding of high-yielding crop varieties for meeting challenges of food security. PLL will increase the water application efficiency and consequently increase the yield of crops (Ahmed et al. 2001). It can help even the distribution of soluble salts in salt-affected soils (Khan 1986), increase cultivable land area due to reduction in bunds and channels in the field (Choudhary et al. 2002; Naresh et al. 2011), reduce weed intensity (Rickman 2002), increase fertilizer use efficiency (Jat et al. 2011), increase crop yields, and result in saving in irrigation through elimination of unnecessary depression and elevated contours (Khattak et al. 1981; El-Guindy et al. 1994; Mathankar et al. 2005; Abdullaev et al. 2007).

Conservation agriculture (CA)-based management practices such as zero tillage, raised bed planting, and dry seeding of rice are being advocated for increasing crop yields and input use efficiency, especially WP, in RW and other major cropping systems of SA (Hobbs and Gupta 2003; Gupta and Seth 2007; Humphreys et al. 2010). PLL is considered as a precursor technology for resource conservation technologies (RCTs) and has been reported to improve crop yields and input use efficiency, including water and nutrients (Jat et al. 2006). RCTs coupled with PLL facilitate uniform water application and reduce deep percolation losses of water. This chapter will focus on improving crop yields, input use efficiency (specifically of water, energy, and nutrients), and farmers' profits through PLL in SA.

13.2   LASER LAND LEVELING: TECHNIQUES

Land leveling is essential for enhancing the use efficiency of water, fertilizer, nutrients, and improving the crop stand and yields. Unevenness of fields leads to inefficient use of irrigation water and also delays tillage and the crop establishment process. Most of the agricultural lands in SA are unleveled or poorly leveled using traditional tools and practices. Land leveling is a bigger issue in the Eastern IGP (EIGP) (India and Bangladesh) compared to Western IGP (WIGP) where fields are uneven and smaller in size and irrigation water is lifted mainly using diesel-operated pumps. The average difference in height between the highest and lowest portions of rice fields in SA is 160 mm (Jat et al. 2006), which means that in an unleveled field an extra 80 to 100 mm of water must be applied to the field to provide complete water coverage to the entire field. Farmers normally level their fields using tractor-operated levelers or bullocks using wooden planks, wherein even after best leveling efforts, an 80–150-mm within-field elevation difference still exits (Jat et al. 2006). A larger quantity of irrigation water is therefore required because of poor water distribution in fields, and part of the area suffers due to water stress and part due to an excess of water that leads to uneven crop stand, poor crop performance, and low profits for the farmer (Rickman 2002). This problem is more pronounced in the case of rice fields than in other crops. Farmers recognize this and therefore devote considerable attention and resources in leveling their fields properly. Effective land leveling is meant to optimize water-use efficiency, improve crop establishment, and reduce irrigation time and effort required to manage the crop. Traditional methods of leveling land by eyesight are not only more cumbersome and time-consuming but are less accurate. Traditionally, farmers use scrapers or leveling boards drawn by draft animals or tractors to level their fields. The advanced technology used to level or grade the field is laser-guided land leveling. The laser guided land leveling for 4 wheel tractor in operation can be seen in Figure 13.1.

The laser-assisted PLL technique is well known for achieving a high level of accuracy in leveling (Khepar et al. 1982; Jat et al. 2006). The quality of land leveling is important since it affects all the other farming operations. PLL results in a much more level field—up to 50% better than leveling using other techniques. PLL improves irrigation efficiency and reduces the potential for nutrient loss through better irrigation and runoff control. It facilitates uniformity in the placement of seedlings by methods such as a rice transplanter or direct drilling, which helps in achieving higher yield levels. The precisely leveled surface leads to uniform soil moisture distribution, which results in good germination and enhanced input use efficiency and yields, whereas undulated land conditions lead to consumption of more energy and ultimately to a higher cost of production. Use of a soil matric potential approach for scheduling irrigation in rice (Kukal et al. 2005) assumes that the soil water deficit at the time of irrigation is uniform over the whole field. PLL can assist in reducing spatial variability in slopes and moisture conservation of fields. It has been shown to be very effective in enhancing the productivity of a RW system by increasing grain yields of both rice and in the efficient use of irrigation water and other inputs such as fertilizers and pesticides (Jat et al. 2006).

PLL is the process of smothering the land surface within ±2 cm from the average elevation of land using a laser-guided bucket that scrapes from higher places and spreads onto the low-lying areas. PLL involves altering the fields in such a way as to create a constant slope of 0% to 0.2%. Before running the laser leveler, the field is surveyed at a 3–5 m distance for recording the elevation. The elevation points are averaged to the desired elevation for leveling the field. The average elevation value is entered in to the control box for controlling the scrapper at this elevation point. Rudragouda et al. (2012) reported that slopes of the field before TLL in the X and Y directions were 0.29% and 0.47%, respectively, and after leveling slopes were reduced to 0.20% and 0.23%,

FIGURE 13.1   The most commonly used four-wheel, tractor-drawn, laser-assisted precision land leveler in operation.

respectively. Similarly, the slopes of the field before PLL in the X and Y directions were 0.32% and 0.19%, respectively, and after leveling, slopes were reduced to 0.021% and 0.016%, respectively. The % slope reductions were 31% and 51% in the X and Y directions, respectively, in TLL and the corresponding reductions in the case of PLL were 93% and 91%, respectively. The highest slope reduction in PLL is helpful to maintain uniform topography of the land surface, for conservation of scanty rains, and for water savings in the arid and semiarid regions of SA, and can prevent the erosion of soil because higher slopes contribute to higher erosion. The frequent microrelief that is a common characteristic of saline-alkaline soils at a study site was eliminated through laser leveling (Jat et al. 2011). The leveling index (LI) and land uniformity coefficient (LUC) are used to assess the smoothness of the land surface. The lower limit of LI is zero, which indicates to perfect leveling of the field. LUC represents the magnitude as well as the frequency of occurrence of successively larger undulations in the field. The highest value of LUC is 1.0, which reflects a perfectly level surface. Decreasing values of LUC represent successively poorer quality of land leveling. From another study, Anuraja et al. (2013) observed considerably higher accuracy of grading under PLL in comparison to using the TLL system. In their study, PLL reduced the standard deviation before and after leveling by 86% compared with 47% for TLL.

The introduction of laser leveling in the 1970s has raised the potential of surface irrigation efficiency to the levels of sprinkler and drip irrigation (Erie and Dedrick 1979). It is a recent introduction in SA. PLL was introduced at farm level in Pakistan during the 1980s, in India during 2001, and in Nepal and Bangladesh during 2010.

13.3 PLOT-SCALE IMPACTS OF PLL

13.3.1 Crop Productivity

A considerable increase in yield of rice, wheat, and sugarcane crops is possible due to PLL (Table 13.1). The absolute increase in rice yield under PLL compared with TLL ranged from 0.32 to 1.30 Mg ha^{-1} and that in wheat ranged from 0.30 to 0.62 Mg ha^{-1} (Table 13.1). The data suggests that

TABLE 13.1
Effect of Laser-Assisted PLL on Increase in Grain Yields of Rice, Wheat, and Sugarcane over TLL in South Asia

Crop	No. of Years/Sites/ Farmers	Yield Increase over TLL (Mg ha^{-1})		Yield Increase over TLL (%)		Reference
		Range	Mean	Range	Mean	
Rice	6 sites	0.42–1.08	0.74	5.4–14.6	10.8	Jat et al. (2009b)
Rice	2 sites, 54 farmers	0.20–1.10	0.62			Jat et al. (2009b)
Rice	3 years		0.38		7.1	Naresh et al. (2014)
Wheat	3 years		0.24		5.7	
Sugarcane	3 years		8.40		12.2	
Rice	2 years	0.30–0.33	0.32	4.4–6.1	5.4	Jat et al. (2009a)
Wheat	2 years	0.32–0.43	0.38	8.4–8.9	8.6	
Rice	71 farmers	0.26–0.40	0.33	4.8–11.4	7.3	Jat et al. (2006)
Rice	2 years		1.30		25.8	Jat et al. (2006)
Wheat	2 years		0.30		6.9	
Wheat	71 farmers	0.09–0.40	0.32	1.5–9.5	6.1	
Wheat	59 farmers		0.83		21	Latif et al. (2013)
Rice	25 farmers		0.32		4.3	Kaur et al. (2012)

rice showed greater response to PLL than wheat. The values for percent increase in productivity under PLL are, however, similar for rice (5%–26%) and wheat (6%–21%) due to higher productivity of rice than wheat in the region (Table 13.1). Using a structured survey, Kaur et al. (2012) estimated that PLL increased rice and wheat yields by 0.32 Mg ha^{-1} (4.3%) over TLL. The increase in cane yield due to PLL was 8.4 Mg ha^{-1} (12.2%) over TLL (Table 13.1). The higher crop yields in PLL are attributed to increased spike density and more grain weight due to improved weed control, and more efficient use of inputs and uniform availability of soil moisture in the effective root zone of the crop (Jat et al. 2004; Naresh et al. 2014). Jat et al. (2006) reported that PLL increased wheat yields by 15%, sugarcane yields by 42%, rice yields by 61%, and cotton yields by 66%. Their study showed that the coefficient of variation in the yield of rice and wheat between farmers was decreased with PLL compared with TLL. From a long-term study, Rickman (2002) showed a 24% or 530 kg ha^{-1} increase in the yield of rice due to PLL over TLL at the same level of variety and fertilizer use. Sattar et al. (2003) reported a reduction in the yield of seed cotton up to 20% under TLL compared to PLL due to (1) low plant population in TLL, (2) greater variation in plant height from average plant height (tallness and shortness of the plants within the TLL field lowered the fruit-bearing capacity of the plants), and (3) late crop maturity or prolonged vegetative growth due to excessive water applied to the TLL fields. In an investigation in western Uttar Pradesh (UP), India, reported a significant improvement in the yield of direct-seeded rice due to PLL compared with TLL. Choudhary et al. (2002) demonstrated that as the time of sowing was delayed, the yield of wheat decreased, but the decrease in the yield due to delayed seeding was much higher in TLL fields compared with seeding under PLL. Tyagi (1984) reported that the rice and wheat yields were higher by 50% in PLL plots compared with TLL plots. Saharawat et al. (2009) reported a yield increase of 0.7 Mg ha^{-1} in puddle-transplanted rice and 0.4 Mg ha^{-1} in direct-seeded rice. From field studies conducted in Haryana, India, Singh (2007) reported that PLL enhanced the crop yields from 10% to 25%. Similarly, Sidhu et al. (2007) reported an increase in mean rice yield from 6% to 11% (0.42 to 0.78 Mg ha^{-1}) with PLL compared with TLL at different locations in Punjab, India.

13.3.2 Irrigation Water Use and WP

WP is the ratio of economic crop yield to depth of water applied having units of kg m^{-3} or kg mm^{-1} (Yadvinder-Singh et al. 2014). When the denominator is irrigation water and irrigation water + rainfall, WP will be denoted as WP$_I$ and WP$_{I+R}$, respectively. WP is a simple way to measure how effectively irrigation water has been used for crop production. Any effort that will augment crop yield or curtail the amount of water required without disturbing crop yield will enhance WP. PLL is highly useful in increasing application and distribution efficiencies of irrigation, which ultimately leads to higher crop yields and WP (Sidhu et al. 2007; Aggarwal et al. 2010). The application (Ea) and distribution efficiencies (Ed) of applied water were increased significantly from 60% and 80% in TLL to 85% and 92% under PLL, respectively (Sattar et al. 2003; Rajput et al. 2004). The distribution efficiency of applied water to wheat on sandy loam soil was significantly higher (10%–15%) in PLL compared with TLL (Jat et al. 2006).

Laser leveling technology offers great potential for water saving, increasing water productivity, better environmental quality, and higher grain yields (Playán 1996; Ahmed et al. 2001). Studies reported from India and Pakistan showed that irrigation water savings in rice and wheat ranged from 10 to 25 cm in rice and 3.5 to 14 cm in wheat (Table 13.2). The percent increase in irrigation water savings ranged from 14.7% to 25.1% in rice and 13.3% to 30.7% in wheat (Table 13.2). This data shows that while the absolute amount of irrigation water saved under PLL is higher in rice than in wheat, the percent savings in water is nearly similar under the two crops. Similarly, irrigation water-saving in sugarcane under PLL is 42 cm or 20.5% compared with TLL (Table 13.2). The irrigation water savings under PLL will depend on the size, shape, and slope of the conventional field, and the soil type and rainfall during the cropping season. In most of the studies the amount or depth of irrigation water applied under both PLL and TLL has been calculated from the data on

TABLE 13.2
Effect of Laser-Assisted PLL on Irrigation Water Savings in Rice, Wheat, and Sugarcane over TLL in South Asia

Crop	No. of Years/Sites/ Farmers	Irrigation Water Savings over TLL (ha-cm)		Yield Increase over TLL (%)		Reference
		Range	Mean	Range	Mean	
Rice	3			24.1–26.2	25.1	Jat et al. (2009b)
Rice	2 sites, 54 farmers	3–13	10			Jat et al. (2009b)
Rice	2		14.0		14.7	Naresh et al. (2014)
Wheat	2		5.2		13.3	
Sugarcane	2		42.0		20.5	
Rice	2	20–27	24	16–25	21	Jat et al. (2009a)
Wheat	2	3–4	3.5	12–15	13	
Rice + wheat	2	23–31	27	15.6–22.8	19.2	
Rice			20		17.8	Ambast (2006)
Wheat			10		30.7	
Wheat	59 farmers		14		31	Latif et al. (2013)
Rice			25		25	Kaur et al. (2012)

irrigation time and rate of tube well discharge (1 h^{-1}). In a few cases, a water flow meter has been attached to the delivery pipe of the tube wells or a Parshall flume installed at the field inlet to calculate volume of irrigation water. Jat et al. (2004) reported that in addition to higher yield, the savings of water from PLL is 35%–45% due to higher application under TLL. Results from 64 on-farm trials conducted in four districts of western UP, India, showed that farmers saved 5–10 cm (mean 6 cm) of irrigation water in wheat and 10–15 cm (mean 12 cm) of water in rice with PLL compared with TLL (Jat et al. 2009b). Data from 92 on-farm trials conducted in Haryana, India showed that RW productivity with PLL compared with TLL increased by 14% to 30% (mean 19%) with irrigation water savings ranging from 22%–40% (mean 30%) (Jat et al. 2009b). Time taken for irrigation ranged from 3.4–27.7 h (mean 7.2 h). On-farm studies conducted in Indian Punjab showed 24% (n = 10), 37% (n = 7), and 28% (n = 4) saving in irrigation time in rice for the PLL field compared with TLL at different locations in the Ludhiana and Moga districts (Sidhu et al. 2007). Studies conducted in Punjab, Pakistan showed irrigation water savings of around 25% and yield increases of 20%–35% for wheat (Kahlown et al. 2006). Rehman et al. (2009) reported that on average, 571 farmers participatory trials conducted in Pakistan Punjab saved irrigation water by 20%–30% in both rice and wheat using PLL compared with TLL saved irrigation water by 20%–30% in both rice and wheat.

On-farm investigations in Haryana, India, demonstrated a savings of 3.4 to 8.1 cm (15% to 20%) of total water use with PLL compared with TLL fields in wheat (Rajput et al. 2004). Similarly, on-farm studies from Punjab, India showed that PLL can save irrigation water by 24%–30% over TLL (Table 13.3) without having any adverse effect on the yields of different crops under similar conditions of crop management (Bhatt and Sharma 2009). While average values of percent saving in irrigation water are nearly similar among different crops, absolute values of water savings will be highest for rice, consuming about 100–120 cm of irrigation water. In on-farm trials conducted in northwestern IGP, it was found that PLL led to water savings of 23%, 25%, and 30% in western UP, Punjab, and Haryana, respectively (Jat et al. 2009b). Studies by Jat et al. (2009a) and Sidhu et al. (2010) suggested that the amount of irrigation water saved in rice is almost equivalent to irrigation water needed for wheat cultivation in the region. A total of 734 on-farm evaluation and demonstration trials conducted in IGP (163 in India and 571 in Pakistan) during 2005–2008 showed an average

TABLE 13.3

Irrigation Water Saving (%) Due to Laser-Assisted PLL for Different Crops

Crop	Average Water Saving	
	Range	Mean
Maize	22–33	27
Wheat	26–33	26
Cotton	26–43	27
Rice	26–30	26
Egyptian clover	27	27
Pea/potato	25	25
Overall mean	25–43	26

Source: Aggarwal R et al., *J Soil Water Conserv* 9: 182–185, 2010.

reduction in water use of 20%–30% and an increase in crop yields by 10%–20% (Ladha et al. 2009). Kaur et al. (2012) estimated that with adoption of PLL over 100% of the area under the four major crops (rice, wheat, cotton, and maize) in Punjab could save 1.466 million ha-cm of water annually and with the adoption of PLL in a RW system, groundwater draft could be reduced by 19 cm assuming specific yield for the state is 0.21.

The data from northwest IGP and Punjab show that the mean increase in WP_I under PLL over TLL ranged from 0.10 to 0.36 kg grain m^{-3} of water in rice and 0.20 to 0.89 kg grain m^{-3} of water in wheat (Table 13.4). The increase in WP_I with PLL ranged from 24.3% to 65.5% in rice and 16.1% to 127% in wheat (Table 13.4). The percent savings in WP_I under PLL is much higher in wheat than

TABLE 13.4

Effect of Laser-Assisted PLL on Increase in WP_I in Rice, Wheat, and Sugarcane over TLL in South Asia

Crop	No. of Years/Sites/ Farmers	Increase in WP_I over TLL (kg ha^{-1})		Increase in WP_I over TLL (%)		Reference
		Range	Mean	Range	Mean	
Rice	2		0.126		24.3	Naresh et al. (2014)
Wheat	2		0.25		22.7	
Sugarcane	2		0.21		25.6	
Rice	2	0.19–0.22	0.20	30–50	40	Jat et al. (2009a)
Wheat	2	0.14–0.46	0.30	8.4–23.8	16.1	
Rice + wheat	2	0.19–0.28	0.24	26.2–27.1	26.7	
Rice			0.36		65.5	Jat et al. (2006)
Wheat			0.49		59.8	
Rice	71 farmers	0.11–0.13	0.12	21.7–28.9	24.5	Jat et al. (2006)
Wheat		0.17–0.20	0.20	16.9–20.5	19.6	
Wheat	59 farmers		0.89		127	Latif et al. (2013)
Rice			0.10		27.0	Ambast (2006)
Wheat			0.94		62.7	
Rice			0.29		38.9	Kaur et al. (2012)

Note: WP_I = irrigation water productivity.

in rice. From other studies carried out in Pakistan, Sattar et al. (2003) reported that PLL significantly increased WP_I of seed cotton from 37% to 63%. The increase in WP_I was attributed to less requirement of irrigation under PLL (55 cm) as compared with TLL (75 cm). Similarly, on-farm investigations by Choudhary et al. (2002) showed higher WP_I of wheat under PLL (1.67 kg grain m^{-3}) compared with that under TLL (1.10 kg grain m^{-3}). The WP_I on laser-leveled fields has been found to be higher by 26% to 39% over the conventional field (Bhatt and Sharma 2009; Aggarwal et al. 2010). A field study in Egypt by El Yazal and Wissa (1990) showed that PLL decreased irrigation water requirements by 28% and increased cane yield by about 50%. It takes approximately 8 hours to level 1 hectare of land with a traditional leveler but about only 4 hours to level the same area with a laser leveler.

13.3.3 ECONOMIC PROFITABILITY

The cost of land leveling mainly depends on the length and slope of the field. Laser leveling of the fields with larger length and slope are more costly (U.S.$90–135 ha^{-1}) than gently sloping fields. For estimating net returns from PLL, Jat et al. (2009a) considered the cost of custom hiring a laser leveler at $5.60 h^{-1}. By and large, the cost of PLL on custom hiring in northwest IGP is about U.S.$40–45 ha^{-1} on gently sloping fields (Jat et al. 2006). Rickman (2002) reported that although the initial cost of land leveling is convincingly high, financial benefits ranged from U.S.$45 in the second year to U.S.$61 in year 3 through year 8. However, Rajput and Patel (2004) and Jat et al. (2009a) reported that on a custom hiring basis PLL provided net returns even in the first year itself. The estimations included extra costs of fertilizer in the first and second years. The benefits included increase in crop yields, reduction in weeding costs, and labor and water savings. Estimates presented in Table 13.5 show that net returns from rice and wheat range from U.S.$33 to 98 ha^{-1} and U.S.$69 to 533 ha^{-1}, respectively. The values of net returns (U.S.$533 ha^{-1}) reported by Latif et al. (2013) for PLL are exceptionally high. A part of the variation in net returns could be due to the values assigned to the costs and gross returns for different items in the calculations by different researchers. Jat et al. (2009a) reported that net returns were 68%–143% higher from wheat than from rice. In eastern India, the net benefit from irrigation water savings ranged from U.S.$11–32 ha^{-1} (mean 20.8) in rice and U.S.$14–44 ha^{-1} (mean 25.5) in wheat (Jat et al. 2009b). Net returns from PLL in sugarcane are U.S.$167 ha^{-1}, which is nearly similar to total net returns from an RW system (Table 13.5). Ambast (2006) reported financial benefits of Indian rupee (INR) 1000–1200 ha^{-1} in the first year and INR 4000–5000 ha^{-1} in the second year of PLL in wheat in Haryana. Rehman et al. (2009) demonstrated that PLL in an RW system in Punjab, Pakistan can lead to monetary gains of U.S.$107 ha^{-1} (U.S.$77 in rice and U.S.$30 in wheat). The gains included an increase in 3% area under PLL and increases in yield, labor, and water savings. Computations by Kaur et al. (2012) showed that compared with conventional fields, PLL in rice in Punjab, India saved INR 25 ha^{-1} in weeding costs, INR 720 ha^{-1} in irrigation costs, and Rs. 424 ha^{-1} in energy costs. The increase in rice yields under PLL was 0.32 Mg ha^{-1}, which is equivalent to INR 4192 ha^{-1} (based on the maximum support price of rice for 2013). Thus, net returns from PLL in rice are INR 5396 ha^{-1} excluding the cost of PLL.

13.4 INPUT-USE EFFICIENCY

13.4.1 FERTILIZER-USE EFFICIENCY

Soil cut during laser leveling may adversely affect soil fertility in the first year of leveling after exposure of infertile subsoil. Observations made from 71 field sites in northwest IGP, India have revealed that yield losses are minimal when soil cut is less than 10 cm but yield losses increase when the soil cut is >10 cm (Jat et al. 2006). The application of additional fertilizer may be necessary in areas from which soil is moved. Depending on the soil type and the volume of soil moved,

TABLE 13.5

Effect of Laser-Assisted PLL on Net Returns in Rice, Wheat, and Sugarcane over TLL in South Asia

Crop	Cost of Production	Net Returns	Net Returns over TLL	Reference
2005–2006				Jat et al. (2009a)
PLL-rice	624	452	33	
PLL-wheat	373	532	80	
TLL-rice	616	419		
TLL-wheat	387	452		
2006–2007				
PLL-rice	556	318	65	
PLL-wheat	376	1044	109	
TLL-rice	570	253		
TLL-wheat	383	935		
Rice				Naresh et al. (2014)[a]
PLL	403	695	98	
TLL	428	597		
Wheat				
PLL	312	510	69	
TLL	326	441		
Sugarcane				
PLL	827	1105	263	
	881	842		
Wheat				Choudhary et al. (2002)[b]
PLL	187	508	86	
TLL	207	422		
Wheat				
PLL	213	430	107	
TLL	202	323		
Wheat				Latif et al. (2013)[c]
PLL	771	1868	533	
TLL	873	1335		

[a] 1 U.S.$ = INR 60.
[b] 1 U.S.$ = Pakistan R 60.
[c] 1 U.S.$ = Pakistan Rs 65.

application of an extra 25 to 50 kg ha^{-1} of diammonium phosphate ha^{-1} may be needed to obtain normal crop yields (Rickman 2002). Many farmers in northwest India apply poultry manure or farmyard manure in the cut areas to reduce yield variability in the first season (Jat et al. 2006).

Flood irrigation on undulated fields may lead to leaching of certain nutrients due to excess water at lower elevations and inadequate availability of irrigated water at higher elevations. Improved use efficiency of the applied nutrients is obvious under PLL as uniform application of water leads to uniform distribution of nutrients and reduced N leaching leading to improved crop growth. A significant increase in the uptake efficiency as well as apparent recovery of the applied N, P, and K in rice and wheat was observed on laser-leveled fields compared to TLL (Choudhary et al. 2002). Jat et al. (2011) demonstrated higher nutrient (N, P, and K) uptake efficiency of wheat on raised beds with PLL compared with TLL (Table 13.6). Latif et al. (2013) also reported higher fertilizer use efficiency of wheat under PLL (18.2%) than under TLL (14%).

TABLE 13.6
Effect of Land Leveling on Nutrient Uptake Efficiency of Wheat
in Rice-Wheat System in Western Uttar Pradesh, India[a]

Treatment	Nutrient Uptake Efficiency (kg kg^{-1} Applied)		
	N	P	K
PLLRB	0.46	0.41	1.17
TLLRB	0.36	0.25	0.76

Source: Jat ML et al., *Am J Plant Sci* 2: 578–588, 2011.
Note: K = potassium, N = nitrogen, P = phosphorous, PLLRB = precision land leveling and raised beds, TLLRB = traditional land leveling and raised beds.
[a] Data averaged over two years.

13.4.2 Machine and Labor Use

Kaur et al. (2012) computed 152 h ha^{-1} of machine hours used on PLL fields compared with 194 h ha^{-1} on TLL fields. The machine hours used for irrigating rice were 131 h ha^{-1} on PLL fields and 180 h ha^{-1} on TLL fields, resulting in a savings of 27% time in machinery use. In spite of spending 5.4 h ha^{-1} on laser leveling, there was a savings of 28% on machine hours with the use of laser technology. Rickman (2002) reported a 10%–15% reduction in the operating time of agricultural machinery in the laser-leveled fields as compared with traditional leveling. Laser leveling thus may prove to be an important technology in reducing the consumption of fossil fuel for various farming operations, which will bring a direct benefit to farmers. Kaur et al. (2012) showed that the labor used for irrigation on laser-leveled fields was 35% lower (30 h compared with 46 h in conventional fields) and the irrigation cost was reduced by 44% over the conventional fields. For weeding, half of the labor time was saved by using PLL technology compared with the conventional practice. A total of 227 h ha^{-1} were spent on various crop operations on laser-leveled fields as compared with 217 h ha^{-1} on conventional fields. Naresh et al. (2014) computed a total energy use of 7266 MJ ha^{-1} under PLL and 8543 MJ ha^{-1} under TLL in an RW system.

13.4.3 Weed Control

Improved water coverage and uniform crop stand under PLL reduced weed density in wheat by 40% compared with TLL recorded after 30 days of sowing (Jat et al. 2003). Rickman (2002) reported that the time required for weed control in rice was reduced from 21 days in unleveled fields to 5 labor-days ha^{-1} in PLL fields—a 75% decrease in the labor required for weeding, which further reduced the cost of weedicide by about 13% over the farmers' practice. Latif et al. (2013) recorded lower weed intensity and about 11% lower herbicide cost in wheat on laser-leveled fields compared with TLL. Rehman et al. (2009) reported that PLL compared with TLL reduced weed pressure and labor use by 30% in rice and wheat.

13.4.4 Increase in Cultivable Area

The PLL system is also likely to enhance the cultivable area due to removal of extra bunds and channels in the field. The increase in area under PLL ranged from 1.5% to 6% with an average of 3.2% (Table 13.7, Jat et al. 2009b). Sidhu et al. (2007) recorded an 8% to 10% increase in cultivated area after PLL over TLL. In another investigation, a 2% to 3% addition in cultivable area was recorded due to PLL (Khan 1986). However, the area increase varies from field to field depending

TABLE 13.7

Estimated Additional Area Brought under Cultivation by PLL by Selected Farmers in Western Uttar Pradesh, India

	TLL		PLL	
Parameter	Canal Irrigated	Tube Well Irrigated	Canal Irrigated	Tube Well Irrigated
Total area of plot (m²)	1200	600	2500	1200
Area under bunds and channels (m² ha⁻¹)	600	1200	300	600
Additional area brought under cultivation (%)	–	–	3	6

Source: Jat ML et al., *Laser Land Leveling: A Precursor Technology for Resource Conservation.* Rice-Wheat Consortium Technical Bulletin Series 7, Rice-Wheat Consortium for the Indo-Gangetic Plains, New Delhi, India, 2006.

on plot size. Rickman (2002) reported that increasing field size from 0.1 ha to 0.5 ha resulted in an increase of the farming area between 5% and 7%. This increase in farming area gives the farmer the option to reshape the area, which can reduce operating time by 10% to 15%. Larger fields with PLL increase the farming area and improve operational efficiency. It has been reported that field sizes in different villages have increased from 33% to 80% when laser leveled as compared with the original plot sizes (Rajput and Patel 2004; Jat et al. 2006).

13.4.5 ENERGY SAVINGS

Kaur et al. (2012) estimated that PLL can save energy by 24% over TLL. In Punjab, India, the savings in electricity charges were U.S.$13.2 in rice and 8.3 in wheat ha⁻¹. Naresh et al. (2014) computed the total energy use of 7266 MJ ha⁻¹ under PLL and 8543 MJ ha⁻¹ under TLL in an RW system.

13.5 INTEGRATION OF PRECISION LAND LEVELING WITH CA-BASED MANAGEMENT PRACTICES

CA-based management practices such as zero-till planting, raised bed planting, and direct seeding of rice are being increasingly adopted by the farmers of SA for improving grain yields, WP, and income (Gupta and Seth 2007; Laxmi et al. 2007; Ladha et al. 2009; Ahmad et al. 2014). The integration of PLL with CA-based practices could provide a better option for sustainable crop production in cereal-based systems. Jat et al. (2011) studied the integrated effects of PLL and crop establishment methods (flat, raised beds) planting of wheat after rice on yield and water savings in the northwest IGP of India. Wheat yield was about 17% higher with nearly 50% less irrigation water and 132% higher WP_I with layering PLL and raised bed planting compared with TLL with flat planting (Table 13.8). The increase in wheat yield under raised beds with PLL over TLL with flat beds was ascribed to a higher spike density and number of grains per spike. The WP_I in wheat under PLL with raised beds was 33%–35% higher compared with PLL with flat sowing. Higher grain yield and less water use in raised bed planting with PLL compared with other treatments resulted in higher WP_I. Whereas the yield-enhancing effects of PLL alone under raised beds and flat beds were 8%–10%, PLL could save 31% water in flat planting and 23% in raised beds. Furthermore, PLL required less irrigation depth (about 5 cm) compared with TLL (7.5 cm), which facilitated good establishment of wheat in sodic soils leading to higher yields. The study showed that the raised bed planting technique is more advantageous on laser-leveled fields.

TABLE 13.8

Effect of Land Leveling and Tillage and Crop Establishment Techniques on Yield, Irrigation Water Used, and Water Productivity of Wheat in a Rice-Wheat System in Western Uttar Pradesh, India[a]

Treatment	Grain Yield (Mg ha^{-1})	Irrigation Water Use (cm)	Irrigation Water Productivity (kg Grain m^{-3})
PLLRB	5.10a	24d	2.13a
TLLRB	4.67b	31c	1.52b
PLLFB	4.69b	33b	1.44b
TLLFB	4.36c	48a	0.93c

Source: Jat ML et al., *Am J Plant Sci* 2: 578–588, 2011.

Note: PLLFB = precision land leveling and flat beds, PLLRB = laser-assisted precision land leveling and raised beds, TLLFB = traditional land leveling and flat beds, TLLRB = traditional land leveling and raised beds.

[a] Data averaged over 2 years.

Results from another study by Jat et al. (2009a) showed that RW system productivity under the permanent bed direct seeded rice (BDSR/BW) system was lower than the other methods of tillage-crop establishment (TCE) because DSR produced low yields on raised beds. There was, however, no interaction effect of leveling and TCE techniques on RW system productivity. The improvement in water application when shifting from TLL to PLL was highest in double zero till (ZT) plots and lowest in raised-bed and CT plots. PLL saved about 32% irrigation water in ZT-DSR/ZTW and BD-DSR/BW planting methods than in TLL, while the saving in irrigation water was 19% in reduced tillage-DSR/CTW method of planting in year 2. WP$_I$ was significantly higher (19%–29%) in PLL than in TLL plots in both the years. Among TCE methods, WP$_I$ was lower in conventional planting of rice and wheat under both PLL and TLL due to high inputs of irrigation water. Despite the low yields of permanent bed system, WP$_I$ was similar to that under conventional planting due to markedly lower amount of irrigation water applied in year 2. Results from field trials conducted on farmers' fields in Western Uttar Pradesh (India) indicated that PLL improved the yield and WP$_I$ of both puddled TPR and CT-DSR. The increase in yield of DSR was 3% and 8% of TPR was 17 and 10% with PLL over TLL in 2005 and 2006, respectively (Table 13.9). The increase in WP$_I$ under PLL compared to TLL was almost similar (by 18%–23%) in both CT-DSR and CT-TPR during the two years.

From a study conducted on a silty loam soil in NW IGP of India, Naresh et al. (2012) reported that yields of the RW system under PLL were similar in the puddled and non-puddled systems, but were lower in TLL plots. This indicated that puddling of soil, for which normally a large amount of water and labor are required, can be avoided without any penalty in rice yield on laser-leveled fields. The RW system productivity in TLL was lower by 16% compared with the PLL plots. They reported that savings in water use under beds with PLL and TLL were 33% and 25% as compared to conventional TPR. Similarly, the water application in wheat was remarkably lower with permanent beds compared to other practices. The total system water use was remarkably lower with permanent beds compared to other practices but the maximum water use was recorded with TPR-CTW. Net income in rice was higher with ZTDSR+PLL followed by CTDSR+PLL and TPR+PLL and the lowest being recorded with ZTDSR with TLL (Table 13.10). Profitability of wheat was remarkably higher with double ZT practices (ZTDSR/ZTW) with PLL and minimum with permanent beds with PLL due to higher productivity and less cost of production compared to conventional tillage practices. Net income of RW system was highest under ZTDSR/ZTW with PLL and lowest under ZTDSR/ZTW with TLL (Table 13.10). This study suggested that PLL can help in adoption of double ZT in RW system with improved irrigation savings and net income.

TABLE 13.9

Effect of Leveling and Tillage and Crop Establishment Techniques on Yield, Total Water Applied, and Water Productivity of an RW Cropping System in Western Uttar Pradesh, India

Treatment	Grain Yield		Irrigation Water Applied (cm)		WP_1 (kg Grain m^{-3} Water Applied)	
	2005 ($n = 17$)	2006 ($n = 15$)	2005	2006	2005	2006
PLL-CT-DSR	5.25ab	4.90a	112d	91d	0.50a	0.54a
PLL-CT-TPR	5.41a	4.94a	137c	102ab	0.39b	0.49b
TLL-CT-DSR	5.10b	4.18bc	125b	94c	0.41b	0.45b
TLL-CT-TPR	4.98bc	4.49b	151a	112a	0.33c	0.40c

Source: Jat ML et al., Laser-assisted precision land leveling: A potential technology for resource conservation in irrigated intensive production systems of the Indo-Gangetic plains. In: Ladha JK, Yadvinder-Singh, Erenstein O, Hardy B (eds). *Integrated Crop and Resource Management in the Rice-Wheat System in South Asia.* International Rice Research Institute, Los Banos, Philippines, pp. 223–238, 2009b.

Note: Within a column means followed by same letter are not significantly different at $P = 0.05$ by Duncan's multiple range test. CT-DSR = conventional till direct-seeded rice, CT-TPR = puddled-transplanted rice.

TABLE 13.10

Productivity, Water Application, and Profitability in Rice and Wheat with Different Tillage and Crop Establishment Techniques

Treatment	Grain Yield (Mg ha^{-1})			Irrigation Water Applied (cm)			Net Returns (U.S.$ ha^{-1})[a]		
	Rice	Wheat	R+W	Rice	Wheat	R+W	Rice	Wheat	R+W
PLL-TPR/ZTW+R	6.25	5.45	11.70	242	35	277	318	457	775
TLL-TPR/ZTW	5.69	5.15	10.84	295	43	338	308	432	740
PLL-ZTDSR/ZTW+R	4.85	5.60	10.45	235	32	267	364	469	833
TLL-ZTDSR/ZTW	4.25	5.25	9.50	277	42	319	259	440	699
PLL-PBTR/PBW+R	5.95	5.80	11.75	202	30	232	304	472	776
TLL-PBTR/PBW	5.71	4.15	9.86	219	36	255	274	440	714
LSD (0.05)	1.95	1.05	–	–	–	–	–	–	–

Source: Naresh RK et al., *Int J Life Sc Bt Pharm Res* 1: 1–13, 2012.

Note: PBTR = transplanted rice on permanent raised beds, TPR = puddled transplanted rice, ZTDSR = zero-till direct-seeded rice, ZTW = zero-till wheat.

[a] 1 U.S.$ = INR 60.

13.6 LANDSCAPE LEVEL IMPACTS OF PRECISION LAND LEVELING

Precision land leveling influences most of the farming operations, thereby affecting crop productivity, water productivity, and input use efficiency. On the one hand, laser assisted precision land leveling increases economic profitability of the farmers through increased crop productivity and reduced irrigation time and on the other, provides environmental benefits through conservation of water resource and reduced energy use for pumping water required for irrigation. To explain these components at a landscape scale, we draw specifically from the studies by Aryal et al. (2014), Gill (2014), and Ahmad et al. (2014). Using data collected from 192 farm households in Haryana and

Punjab states of India in 2011, Aryal et al. (2014) assessed the impact of laser assisted precision land leveling technology in rice-wheat systems. Their study compared the crop yields of rice and wheat and total duration of irrigation per season for each of these crops between laser-leveled and traditionally leveled field.

13.6.1 Crop Productivity

In the context of the increasing scarcity of water and land resources, closing the yield gap with modern technology can be a solution to achieve food security. The landscape-level study in northwest IGP by Aryal et al. (2014) showed that laser-assisted precision land leveling contributes toward closing the yield gap. Their study found a higher yield of both rice and wheat in the farm plots that are leveled using laser-assisted precision land leveling as compared with the farm plots that are leveled using traditional methods. For comparison, they used only those sample farmers who cultivated the two crops on both laser-leveled and traditionally leveled plots. The farmers in the study area reported that the level of undulation was low to medium, so the cut-and-fill operation might not impact the fertility of the soils in their farms. Table 13.11 presents the mean yield difference of wheat and rice between the laser-leveled field and the traditionally leveled field.

Table 13.11 reports that average yields of both wheat and rice are higher in the field leveled using laser-assisted precision land leveling as compared with the fields leveled using a traditional method. The average yields of wheat in Haryana with laser leveling and traditional leveling were 4576 kg ha^{-1} and 4291 kg ha^{-1}, respectively, and this difference is statistically significant at 1% level of significance. In Haryana, India, the average yields of rice under PLL and TLL were 5617 kg ha^{-1} and 5295 kg ha^{-1}, respectively; this difference is much higher (322 kg ha^{-1}) but not statistically significant as the variance in rice yield was much higher. This can be due to the knowledge gap among farmers who adopted PLL and can be overcome by designing appropriate policies to disseminate knowledge to farmers. Similarly, in Punjab, India, the average wheat yield was 4083 kg ha^{-1} in TLL and 4444 kg ha^{-1} in PLL, implying that use of PLL increased wheat yield by 360 kg ha^{-1} and the difference is statistically significant at the 1% level. Similar to Haryana, the yield difference in the

---

TABLE 13.11

Wheat and Rice Yield Difference between the Fields Leveled Using Laser-Assisted PLL and TLL

State and Crop	Average Yield (kg/ha)		Average Yield Difference (kg/ha)	t-Test
	PLL	TLL		
Haryana, wheat	4576	4291	285	2.46[a]
	(84.94)	(79.42)		
Haryana, rice	5617	5295	322	1.25
	(184.66)	(179.28)		
Punjab, wheat	4444	4083	361	4.36[a]
	(57.252)	(57.71)		
Punjab, rice	6168	5807	361	1.08
	(240.08)	(232.91)		

Source: Adapted from Aryal JP et al., Impacts of laser-assisted precision land leveling in rice-wheat rotation of northwestern Indo-Gangetic plains of India. In *World Congress of Environmental and Resource Economics*, June 28–July 2, 2014. Istanbul, Turkey. Retrieved from http://www.wcere2014.org/en/Programme-Overview.html, 2014.

Note: Observations are combined observations and standard errors are reported in parentheses.

[a] Significant at 99% confidence level.

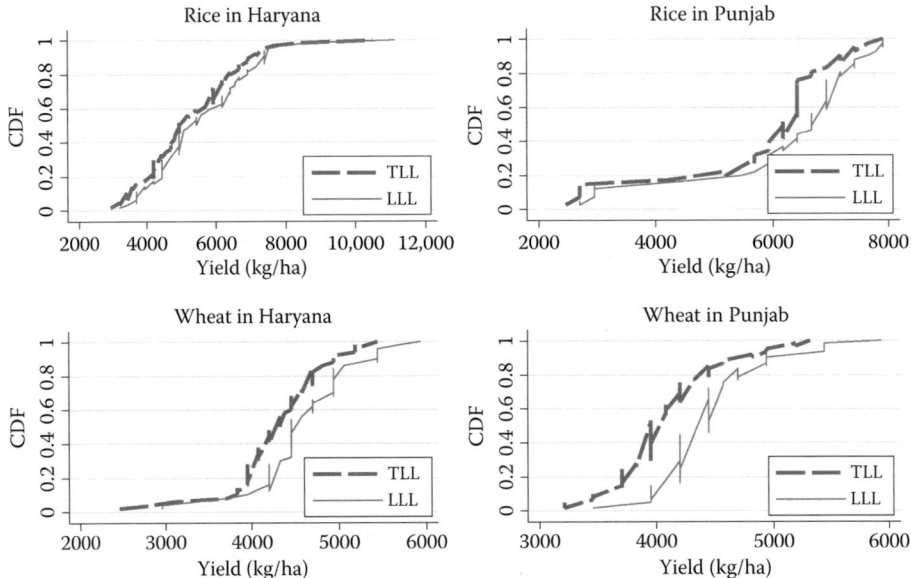

FIGURE 13.2 Yield difference between laser land leveling (LLL/PLL) and traditional land leveling (TLL) using stochastic dominance analysis. (Adapted from Aryal JP et al., Impacts of laser-assisted precision land leveling in rice-wheat rotation of northwestern Indo-Gangetic plains of India. In *World Congress of Environmental and Resource Economics*, June 28–July 2, 2014. Istanbul, Turkey. Retrieved from http://www.wcere2014.org/en/Programme-Overview.html, 2014.)

case of rice in Punjab was statistically not significant and the rice yield on average under PLL was higher (6168 kg ha^{-1}) compared to TLL (5807 kg ha^{-1}). Overall, on average, the difference in yields between laser leveling and traditional was slightly higher in Punjab than Haryana. These findings are also confirmed by the result of stochastic dominance analysis, which is presented in Figure 13.2. In the figure, we see that in all cases, the cumulative distribution function representing PLL lies below the cumulative distribution functions representing TLL, indicating that PLL dominates TLL in all cases. This means the impact of PLL is positive on crop yields in both Haryana and Punjab.

Investigations by Ahmad et al. (2014) from Punjab, Pakistan showed that at the current adoption rate of 9% the overall net increase in crop production was estimated at 24 and 4 million kg yr^{-1} of wheat and rice production, respectively. The anticipated increases in production with 30% adoption of PLL are estimated at 80 million kg yr^{-1} for wheat and 14 million kg yr^{-1} for rice.

13.6.2 ECONOMIC PROFITABILITY

Aryal et al. (2014) estimated economic profitability to farmers practicing a rice-wheat cropping system in a hectare of laser-leveled field. Although the time required differs by the level of undulation before the leveling, the average time to level 1 hectare of land using a laser-assisted precision land leveler in Haryana and Punjab was 5 hours. Almost 99% of the sampled farmers hired in laser-assisted precision land-leveling services from service providers. Hence, the average cost of hiring the service per hour is used for estimating the total cost. This included the entire laser-leveling service package comprising the laser leveler, driver, and fuel. The most common rate paid per hour of use of the laser-assisted precision land leveler was USD 10 per hour (USD 1 = INR 50 at the time of survey). Therefore, the total cost of leveling 1 hectare of land is USD 50. When a farmer levels 1 ha of land using a laser-assisted precision land leveler, the effect typically lasts for 4 years. Thus, the life of one leveling is assumed to be 4 years. Considering that the farmer practices the dominant cropping pattern of rice-wheat during the entire period, the net incremental benefit stream of the farmer is given by

$$\sum_{i=1}^{4}[p_{Ri}\Delta y_{Ri} + p_{Wi}\Delta y_{Wi}] - C_L H_L$$

Aryal et al. (2014) assumed that the increase in the price of rice and wheat and the discount rate over time would balance each other out. As a result, prices stay constant at the current minimum support prices (MSPs) as fixed by the government of India and the discount rate can be assumed as one. Therefore, they calculated the benefit stream using MSPs of rice and wheat for the year 2011. The MSPs of rice and wheat are USD 0.22 kg^{-1} and USD 0.234 kg^{-1}, respectively. Based on this, the net present value of the total revenue stream for the entire period of 4 years is given by

$$4*\{(0.22*341.5) + (0.234*323)\} - 50 = USD\ 522.85$$

Therefore, the net present value of the income stream resulting from 1 ha of laser-leveled land in a year then is USD 138.2. At the current conservative estimated level of adoption of PLL in South Asia (~3.5 mha), the economic gains are estimated at USD 484 million yr^{-1}. This economic profitability could be much higher if we consider the cost of water used for irrigation and under the scenario where there is no subsidy for electricity.

13.6.3 WATER PRODUCTIVITY

In their study, Aryal et al. (2014) included only those farmers who produced rice and wheat crops in their laser-leveled field and traditionally leveled field. This provides safe ground to assume that an individual sample household used the same tube well for irrigating the land. As a result, comparing the average duration of irrigation required per hectare per crop season was used to estimate water productivity at landscape level. The unit of output produced per hectare of land with an hour of irrigation per hectare of land under these two alternative methods of leveling land using the information provided in Aryal et al. (2014) was used for computation of WP and is summarized in Table 13.12.

TABLE 13.12

Water Productivity Difference between Laser-Assisted Precision Land Leveled and Traditionally Leveled Fields

State and Crop	Average Yield (kg ha^{-1})		Average Duration of Irrigation in a Season (h ha^{-1})			Water Productivity (kg h^{-1} of Irrigation)		
	PLL	TLL	PLL	TLL	t-Test[a]	PLL	TLL	t-Test[a]
Haryana, wheat	4576	4291	34.6	45.4	4.52[b]	132.3	94.5	4.27[b]
	(84.9)	(79.4)	(1.42)	(1.92)		(7.18)	(5.19)	
Haryana, rice	5617	5295	112.7	168.7	4.70[b]	49.8	31.4	5.41[b]
	(184.7)	(179.3)	(6.40)	(9.96)		(2.59)	(2.20)	
Punjab, wheat	4444	4083	33.12	43.0	3.42[b]	134.2	94.9	6.11[b]
	(52.3)	(57.7)	(1.59)	(2.52)		(4.31)	(4.77)	
Punjab, rice	6168	5807	106.8	158.9	3.47[b]	57.7	36.6	7.24[b]
	(240.1)	(232.9)	(4.88)	(11.07)		(2.42)	(1.62)	

Source: Redrawn from Aryal JP et al., Impacts of laser-assisted precision land leveling in rice-wheat rotation of northwestern Indo-Gangetic plains of India. In *World Congress of Environmental and Resource Economics*, June 28–July 2, 2014. Istanbul, Turkey. Retrieved from http://www.wcere2014.org/en/Programme-Overview.html, 2014, and the survey data collected in 2011.

Note: Values in parenthesis are standard errors of mean.

[a] Absolute values of *t*.

[b] Significant at $P < 0.1\%$.

From Table 13.12, it is clear that the average duration of irrigation per crop season for both rice and wheat are significantly higher in a traditionally leveled field as compared with the laser-leveled field. Similarly, water productivity is higher in the case of a laser-leveled field in both states and for both crops. For example, in Haryana, 1 hour of irrigation corresponds to 132 kg of wheat output in a laser-leveled field while this is only 95 kg in a traditionally leveled field. Gill (2014) in an impact study of PLL conducted under CIMMYT-CCAFS (CGIAR Research Program on Climate Change, Agriculture and Food Security) in Haryana revealed that at the current level of adoption of PLL in Haryana, India, the estimated water savings is ~1.0 billion m^3 yr^{-1}. In contrast to the Punjab, Pakistan RW zone (Ahmad et al. 2014), the adoption of PLL in the lower Indus basin with about 85% of irrigation supplies come from surface canal systems contributes to real water savings through reduced recharge to saline groundwater and increases overall agricultural production. There is a general scarcity of literature on the impact of CA-based management practices on water balance components at farm and higher spatial scales, which prohibits comparison of this study with other contexts outside of the Indus basin.

13.6.4 INPUT USE EFFICIENCY

In the survey data of 2011 used by the study of Aryal et al. (2014), only two of the 196 farmers interviewed reported changing the level of fertilizer and seed in a laser-leveled field. No difference is found in the average amount of fertilizer and seed used by all farmers in both types of field, either laser-leveled or traditionally leveled. However, a much higher average yield of rice and wheat was observed in a laser-leveled field. As farmers are able to produce additional units of rice and wheat in a laser-leveled field with the same amount of fertilizer and seed used per hectare of land, this indicates that input use efficiency is higher in a laser-leveled field than in a traditionally leveled field. Jat et al. (2009b) reported that nutrient use efficiency is higher in a laser-leveled field because of improved nutrient uptake. On-farm research in western Uttar Pradesh of India showed that nitrogen use efficiency is significantly improved in a rice-wheat cropping system after the field is laser leveled. Jat et al. (2006) observed that after PLL, nitrogen use efficiency in rice was increased from 45.11 to 48.37 kg grain per kg of applied nitrogen and in wheat, this increased from 34.71 to 36.9 kg grain per kg of applied nitrogen.

13.6.5 ENVIRONMENTAL IMPACTS

One of the major impacts of laser-assisted PLL is the mitigation of climate change by reducing GHG emissions in several farm operations, but the main contribution is undoubtedly from the reduction in demand for irrigation water and the resultant reduction in power requirements for pumping water.

13.6.5.1 Emission Reduction through Decreased Pumping Time

Irrigation in the study area is based on both groundwater and surface water. The former is pumped by grid-connected electric tube wells; the latter by diesel-powered low-lift pumps (LLPs). Electricity generation in India relies on a number of technologies, two of which are thermal. Coal-fired stations generate 59% of total supply and natural gas-fired stations 9% (Ramme et al. 2011), meaning that just over two-thirds of national generation capacity is thermal-powered and hence GHG-emitting. According to the latest available figures from the Indian government, the total installed capacity for electricity generation from thermal power plants in the country (in 2007) was 89,275.84 MW (INCCA 2010). The electricity generation sector's GHG emissions in the same year are shown in Table 13.13.

All of the respondents in the present study reported irrigating with electric tube wells, and a total of 54 have been installed by them. Almost 93% of these are deep tube wells (DTWs), and the rest are shallow tube wells (STWs). DTWs have motors ranging in capacity from 7 to 25 HP. The metric equivalents of these power ratings are 5.2 and 18.6 kW, respectively. The STWs have 3 HP (2.2 kW) to 7 HP (5.2 kW) electric motors. The most common rig in the area (42% of DTWs) has a 15-HP (11.2-kW) motor.

TABLE 13.13
GHG Emissions by India's Electricity Generation Sector (2007)

GHG	Chemical Name	Total Emissions (MT)
CO_2	Carbon dioxide	715,829,800
CH_4	Methane	8,140,000
N_2O	Nitrous oxide	10,666,000
CO_2 equivalent $(CO_2\text{-eq})$[a]		719,305,340

Source: INCCA 2010 (Table 5.3).
Note: GHG = greenhouse gas, MT = metric ton.
[a] CO_2-eq is calculated using coefficients of 21 for CH_4 and 310 for N_2O.

If generating 89,275.84 MW of electricity produces 719.31 million MT of carbon CO_2-eq per annum, generating 1 kilowatt-hour of electricity produces 10.19 kg of CO_2-eq. This suggests that the most commonly used tube well motor in the area requires sufficient electricity to produce 8.7 kg of CO_2-eq per hour. A weighted average of the power of all tube wells owned by farmer-service providers in the area is 10.7 kW (which is similar to that of a 15-HP motor), and equates to the generation of 7.0 kg of CO_2-eq per hour of thermally generated electricity. However, this first approximation needs to be refined by taking two further factors into account.

The first factor is the point noted earlier that only 68% of India's electricity supply derives from thermal sources, so that the above tube-well-level emissions estimate should be adjusted by a coefficient of 0.68 to convert them into GHG emissions at a national level. The other factor is transmission losses. India's network losses are exceptionally high. In 2010 they were 32% (including nontechnical losses), compared with a global average of less than 15% (Ramme et al. 2011). Thus on average 132 kW must be generated in order to supply 100 kW at point of use. Factoring both of these parameters into the equation and using the above weighted average of tube well power ratings produces an estimated mean of 6.3 kg of CO_2-eq emissions per tube well per hour.

Surface water irrigation is much less polluting than tube well irrigation because the total lift requirement is much less. Canals in the area are gravity-flow systems and therefore do not of themselves produce GHGs. However, diesel-powered LLPs must be used to raise the water to field channel level so that emissions occur at this point. The irrigation canal network in Karnal District is less well-developed than elsewhere in the state, and only two of the respondents in the present study had access to canal water for irrigation in addition to their tube wells. One had a 5-HP (3.7-kW) LLP that uses 2 liters of diesel h[-1], while the other had two 10-HP (7.4-kW) LLPs that use 1.5 liters diesel h[-1]. The three most common types of LLP used in India are shown in Table 13.14. On average their fuel consumption is fractionally below 1 liter h[-1] and the CO_2 emissions are therefore 2.5 kg h[-1].

It has been observed that the irrigated agriculture expanded rapidly in Haryana since the 1960s, but that the sources of irrigation have also changed markedly. Minor irrigation (which is virtually the same as groundwater irrigation) grew from less than a quarter of net irrigated area to 56% of the total by 2010–2011. Taking a weighted average of tube well and canal-based irrigation together, average emissions from the system in the state are therefore estimated to be 4.7 kg CO_2-eq h[-1] of irrigation time.

It was found that average savings in irrigation time following laser leveling was 12.2 hours/ha in wheat and 50.0 hours/ha in rice. The savings with rice are much higher, because farmers irrigate on average 13 times per season with rice, but only four times per season with wheat. However, Gill (2014) found the hourly savings to be much higher: 24 hours ha[-1] with wheat and 78 hours ha[-1] with rice. On average owners leveled 10.6 ha of their own land, but hired out leveling services on 201 ha in 2013. Using this difference as a weighting factor, the weighted average of time saved is therefore 64 hours ha[-1] yr[-1].

TABLE 13.14

Parameters of the Three Most Commonly Used LLPs in Indian Agriculture

Type of Pump[a]	HP (kW)[a]	Fuel Consumption (lt/h)[a]	Emissions (CO_2/h)[b]	Observations[a]
Older Indian-designed and manufactured pumps (Kirloskar, Bharat)	5 (3.7)	1.0–2.0	2.6–5.2	Green revolution model, very popular, and village mechanics can provide R&M
Newer Japanese-designed, Indian-manufactured (Honda)	2–3 (1.5–2.2)	0.5–1.0	1.3–2.6	Very popular and village mechanics can provide R&M
Newer Chinese-designed and manufactured (various)	2–3 (1.5–2.2)	0.4–0.9	1–2.3	Cheapest rig, but unreliable and have a short working life

Note: CO_2 = carbon dioxide, R&M = repair and maintenance.

[a] From Greenpeace India 2013. Available at http://www.greenpeace.org/india/en/publications/AnnualReport-2013/.

[b] Based on a coefficient of 2.6 kg CO_2 released into the atmosphere per liter of diesel consumed. From Grace PR et al., Long-term sustainability of the tropical and subtropical rice-wheat system: An environmental perspective. In JK Ladha, J Hill, RK Gupta, JM Duxbury, RJ Buresh (eds), *Improving the Productivity and Sustainability of Rice-Wheat Systems: Issues and Impact* (Vol. 146, pp. 1–18). Madison, WI: ASA Special Publications 65, 2003.

The estimated reduction in annual GHG emissions across the Haryana state as a result of precision land leveling is

$$Rghg = Thr/ha \times Ei \times All$$

where
   Rghg = reduction in GHG emissions in CO_2-eq
   Thr/ha = irrigation time saved per hectare/annum (64 hours)
   Ei = GHG emissions per hour of irrigation (4.7 kg CO_2-eq)
   All = area leveled across the Haryana state by PLLs (≈544,000 ha)

The estimated reduction in GHG emissions in Haryana as a result of reduced irrigation time stemming from the expansion of PLL is therefore 163,600 MT of CO_2-eq.

13.6.5.2 Emission Reduction through Decreased Cultivation Time

After diminished water requirements, reduction in time requirements for cultivation is the next most important advantage of PLL in terms of lowering GHG emissions. Before PLL farmers would typically plough or harrow the land three–four times and follow this by planking it once or twice. Each such operation took in the region of 50 min. ha^{-1}, so the total was in the area of 4¼ hours ha^{-1} for each of the two crops in a year. Postleveling, the need for harrowing is reduced to two–three and there is no need for planking, so the time required for cultivation falls to ≈2 hours ha^{-1} per crop. This translates into a time saving of 2¼ hours ha^{-1} crop^{-1}, or 4½ hours ha^{-1} yr^{-1}.

However, emissions during PLL must be factored into these figures. The norm, as reported by the owners, is that it takes 5 hours ha^{-1} the first time a field is leveled, but that subsequent leveling takes 50 min. ha^{-1}. For the PLL owners, who level all of their land at the outset and relevel it either every year or every second year, the additional time requirements are, on average, 75 min. ha^{-1} yr^{-1}, so that the net reduction in the time the tractor is on working land is 3¼ hours/ha.

In the case of hirers, the figure of 5 hours ha^{-1} must be spread over the norm of leveling only every fourth year, so that the figure again averages 75 minutes/ha/annum, and again the net reduction in tractor time is 3¼ hours ha^{-1}. Using the earlier-calculated estimate of 544,000 ha cultivated under PLL across the state, this translates into a total saving of 1.768 million hours. The hourly fuel consumption by the 50–55-HP tractors PLL owners use averages 4.25 liters, so that the fuel saved is 7.514 million liters. Based on the reported coefficient of 2.6 kg CO_2 released into the atmosphere per liter of diesel consumed, this equates to a reduction of 19,536 (say 19,500) MT of CO_2 emissions per annum.

13.6.5.3 Emission Reduction from Fertilizer Savings

One other source of savings noted by farmers was reduced fertilizer application levels. Although not all farmers reported reducing application levels, none reported increasing them. Although it has not been possible with available resources to quantify this information, it cannot be ignored when assessing the impact of this technology.

The link between chemical fertilizer and GHG emissions, particularly N_2O, is well established. Climate scientists have long understood that the cause of the increased nitrous oxide emissions was application of nitrogen-based fertilizer, because this stimulates microbes in the soil to convert N to N_2O at a faster-than-normal rate. However, it has only recently become possible to accurately identify the proportion of this GHG that is attributable to fertilizer use, distinguishing it from that arising naturally from forests and oceans.

One approach to mitigating such emissions is to time fertilizer application to avoid rain, because under wet conditions soil microbes produce large amounts of N_2O. Changes in the way fields are tilled, when they are fertilized, and how much is used can also affect N_2O production. It has been observed that by producing a uniformly flat field, PLL reduces the potential for both N_2O emissions and nutrient loss by improving runoff control, thus leading to improved fertilizer use efficiency and higher yields (Jat et al. 2006, 2009a, 2011).

It was noted earlier that empirical evidence presented in a number of papers on the subject indicated that PLL improves fertilizer use efficiency, yet only one of the 196 farmers interviewed in the Aryal et al. (2014) study reported changing the level of fertilizer use. The subject was therefore revisited and probed in some depth in the study by Gill (2014). As in the earlier study, the majority (in this case three-quarters) of the respondents reported no change in the level of fertilizer use. However, the remaining five all reported that they had reduced the amount of fertilizer applied as a direct result of PLL. Without exception, they reported that an important outcome of irrigating an undulating land surface is that the quality of the crop is not uniform, and that farmers therefore tend to apply additional doses of urea where the crop looked patchy and unhealthy, which tended to be in low spots where there was waterlogging. They did this on the assumption that the problem was lack of nitrogen. This, they reported, did not happen with laser-leveled fields. The reduction was far from negligible. One respondent reported reducing urea application from 3 bags per acre (7.4 bags/ha) to 2–2.5 bags/acre as a result, while another stated that he had reduced application of this fertilizer by 10%–15%. Interestingly, one of the farmers who reported reducing fertilizer application specifically noted that other farmers were wrong when they assumed that poor growth in low spots could be cured by applying an extra dose of fertilizer. In his view the correct solution was to eliminate the low spots. Discontinuing the practice of applying urea in low spots where there is standing water will reduce N_2O emissions, because, "wet and happy soil microbes can produce sudden bursts of nitrous oxide."

13.6.6 Social Impacts of PLL

The use of laser-assisted PLL is found to have several social impacts as it saves social goods such as groundwater, creates new areas of business investment for service providers, generates employment for local labor, enhances food security through increased water security, and enhances social as well as gender equity through increased crop diversification toward vegetable farming (Aryal et al. 2014; Gill 2014). Some of these major social impacts are discussed next.

13.6.6.1 Groundwater Savings

Given the current economic growth and increase in population, there would be a substantial increase in future water demand in India from all sectors including domestic, agricultural, and industrial. A study by Amarasinghe et al. (2007) projected that average domestic water demand would increase from 85 liters per capita per day (lpcd) in 2000 to 170 lpcd by 2050, livestock sector water demand would increase from 2.3 BCM in 2000 to 3.2 BCM by 2050, and industrial water demand would increase from 30 BCM in 2000 to 151 BCM by 2050. Therefore, water security can be one of the major obstacles for achieving food security (Hanjra and Qureshi 2010). Because of this situation, scaling up of PLL technology is crucial as it has the potential to save approximately 270,000 cubic meters of water per year in RW systems of northwestern IGP (Jat 2012). A recent study in the Haryana state of India by Gill (2014) reported that given that 544,000 ha of land in Haryana is laser leveled, water savings in Haryana through PLL is at least one BCM per annum.

13.6.6.2 Lower Subsidy Burden of the Indian Government through Reduced Energy Use for Agriculture

The irrigation system in India is becoming more dependent on energy due to the expansion of areas under groundwater irrigation (Kumar et al. 2011). The share of energy consumed for agriculture has increased in most of the Indian states, particularly in the states such as Haryana and Punjab where agricultural productivity is higher compared with the national average. For example, in 2009–10, the share of agricultural consumption to the total consumption of electricity was 21% at the national level in India while it was 40.3% in Haryana and 33.5% in Punjab (GOI 2013). This has considerably increased the amount spent on power subsidy for agriculture; in 2010–11, power subsidy for agriculture amounted to more than USD 45 million (Perveen et al. 2012). Recent estimates show that the reduction in the time required for irrigation in a laser-leveled field corresponds to electricity savings of about 558–762 kWh/ha/yr (if electric pump sets are used for pumping groundwater for irrigation) or diesel savings of about 300–410 liters ha^{-1} yr^{-1}. As a result, if 1.5 million ha of land under the RW system of the IGP is laser leveled, this would save electricity used for irrigation that is equivalent to USD 30 million/yr in India (Jat 2012). Hence, this reduces the fiscal burden of the government of India and makes the budget available for other purposes.

13.6.6.3 Expansion of Business Investment and Employment

The introduction of laser-assisted PLL technology creates an opportunity to invest in new sectors for local businesses, and thereby, it generates employment for local youths as machine operators and maintenance workers. Widespread adoption of laser-leveling technology has shown tremendous potential as an alternate source of employment for rural youth and income to the farmers through custom services on laser leveling (Jat et al. 2006). Due to a shorter window of 90–110 days in the irrigated ecoregion, custom service providers used to operate laser units for 24 hours a day through three shifts of operators. Thus, it is estimated that one laser unit can provide an employment opportunity of 270 to 330 person days in a year. The average savings through custom services is estimated at USD 16.6 per working day and if it works effectively for 100 days a year, a net profit of USD 1660 yr^{-1} can be earned by a custom service provider. Another estimation by Jat et al. (2006) showed that a service provider can earn a net profit of USD 4030 $year^{-1}$ (excluding depreciation cost) from one laser leveler. A recent study in Haryana by Gill (2014) reported that commercial profitability to the service provider is much higher, with an internal rate of return varying from 97% to 120%, depending on the assumptions used in the sensitivity analysis. Thus it is an extremely attractive technology from an investment viewpoint.

Using recent statistics on the sale of laser land levelers by different agencies, Jat (2012) stated that there are now 17,000 laser units in operation in northwestern IGP. Assuming that direct employment generation by a laser unit is 300 person days per year, the 17,000 laser units at present would generate the employment of 5.1 million person days annually. However, the employment generation for PLL operators is highly seasonal, since such operations can be carried out only when there are no crops in the field. Gill (2014) reported that service providers hired tractor drivers to operate the PLL rigs on a casual basis, and the season typically lasts 2–2½ months. The employment generation effect of PLL rigs was therefore around 80 person days per annum per machine. The reason for hiring at all is that the season is so short that the owners work their machines very intensively—on average 17½ hours/day—which is why they typically hire one or two tractor drivers. This extra labor is needed because of PLL service provision rather than one's own farm work, because the ratio of work done on a contract basis to work done on the owner's farm is 19:1. Contracting out is not the practice with traditional leveling techniques, so there is no direct labor displacement effect. It would be wrong, however, to assume that these machinery operatives are from marginalized groups. They are semiskilled workers with some degree of training, and they earn more than casual laborers. The typical wage for an 8-hour day is INR 500–550 compared to INR 300 for a male agricultural laborer.

There may, however, be indirect labor displacement effects. None of the informants mentioned savings in weeding time and costs as significant because they use herbicides for weed control. Herbicides are not entirely effective on high spots in the fields, and some degree of manual spot weeding is therefore required when the land is not level. However, none of the respondents regarded this as a significant savings resulting from PLL.

It should be noted that farmers do not tend to hire labor directly in Karnal. Instead, they engage labor contractors who will bring in laborers to perform the work. The PLL owners are not therefore the best sampling frame to use to investigate labor displacement issues. In the case of direct seeding in rice *vis-à-vis* transplanting, this is a technology that has been introduced only very recently, and there are as yet no reports concerning resulting labor displacement.

13.7 SOCIAL AND GENDER EQUITY

Studies by Aryal et al. (2014) did not find any evidence to support that PLL is biased toward large farmers. Small farmers (farmers operating up to 2 ha land) equally benefit from the use of PLL given the availability of the hiring services. However, this study did not focus on women and those who have traditionally suffered discrimination on grounds such as caste and religion. Gill (2014) was able to investigate the impact of PLL on women (discussed later in this section), but was unable to explore issues of caste and religion, partly because these are such sensitive issues and partly because the sampling frame (owners of laser leveling equipment) was not conducive to such an investigation. However, in general terms, Gill (2014) argues that the socially marginalized benefit from reduction in GHG emissions disproportionately to their numbers because they tend to live in marginal areas, which are especially prone to disasters, particularly drought and flood. Second, they also tend to benefit, again disproportionately, from an increase in food availability because in the competitive grain market of Haryana, increased availability tends to result in lower prices, and in accordance with Engels' law, the proportion of a household's income spent on food is inversely proportionate to its household's income level.

The same author found that the average size of holdings operated by PLL owners was 11.4 ha, which is in the official category of large farmers. However, not all of them could be described as such. Scale economies do exist for dealing with large units, and all service providers give a discount, typically INR 50/hour (7.7%) to large farmers, because it is easier to level larger fields and there is no need to constantly adjust the rig, as is the case with small plots. This is obviously evidence of scale economies rather than of discrimination against small farmers.

Two other pieces of evidence from the impact study (Gill 2014) in Haryana are relevant here. First, respondents were asked the smallest size of plot that could be leveled with this equipment. By far the most common response was 0.25 acres (0.1 ha), but with the caveat that it is more economical to level larger plots. It is of course necessary to distinguish between size of plot (which is more relevant from the viewpoint of feasibility of PLL) and size of farm (which is more relevant in terms of impact on small and marginal farmers), but the above finding does reinforce the view that any bias toward larger farmers is driven by economics rather than discrimination. Second, respondents were also asked the number of farmers to whom they provided PLL services and the area leveled, and the 2013 mean transpired to be 4 ha with a standard deviation of 3.2 ha. This suggests that while small farmers may dominate in terms of number of clients, larger farmers dominate in terms of area leveled. However the mean figure has been steadily falling since 2008, when it was 6.9 ha.

All of the above indicates that a competitive and economically rational market for PLLs has already developed and that the only factor that might reduce the scope for marginal farmers to access it is the technical problem that some of their fields may be too small. Even in this case, evidence is beginning to emerge that in some cases this particular scale diseconomy may be overcome through social organization. One farmer-service provider reported that he had begun to hire out PLL services to groups of marginal farmers who had taken to demolishing the boundaries between adjacent plots in order to create an area sufficiently large for economic leveling, before later reestablishing these boundaries. Some fragmentary evidence emerged from the same study of a very positive income generation effect following crop diversification from the RW rotation into vegetables.

All of the respondents in the study by Gill (2014)—as well as a great number of other resource persons in the district—reported that it was unusual for female-headed households (FHHs) with agricultural land to farm it themselves. The normal practice is to hire it out to male farmers. However, women farmers are far from unknown in the district. Almost half of the respondents reported having hired out their machines to FHHs, but the number was small—in the range of 1–5 per season, compared with an average of more than 70 male farmers, but conditions of hire were the same in each case. The others reported that they had never been asked to supply PLL services to such households but would have no objection to doing so if asked. The only difference—and it is instructive—is that all of the farmers who hired out their machines to FHHs reported that a woman would never approach a male PLL owner either in person or by mobile phone (the normal modes of communication), but would make contact either through one of her children or through a male relative.

It was noted that PLL enabled farmers to dispense with male laborers who were previously used for building and maintaining irrigation structures because PLL land eliminated the need for these. On the other hand, diversification into labor-intensive crops such as tomato and other vegetables makes it necessary to hire more laborers for tasks such as constructing trellises, harvesting, grading, and packing the crop. Women are hired for these tasks because their wage rate is much lower than that of men. Women are paid INR 120 for a 7-hour day, while men receive INR 300 for an 8-hour day. This wage differential is obviously a powerful incentive to hire female labor. It would be wrong, however, to assume that this pay differential is attributable to PLL technology. The Agricultural Census of India shows that significantly lower hourly wage rates for women are the norm in Indian agriculture across all of the operations for which data are available (GOI 2013).

13.8   FOOD SECURITY

PLL increases yields in irrigated cropping systems; for example, intensive RW rotation of IGP, and this increases food security by augmenting its food availability component (Aryal et al. 2014). This is primarily due to the fact that PLL eradicates the problem of low and high spots in the field, eliminating the problem of waterlogging in the former and moisture stress in the latter (Gill 2014). As a result, the crop stand is more uniform, there is more tillering, and the grain is better filled out, all of which are yield-increasing factors. Additionally, field bunds, which are traditionally used to terrace the field and hence keep it more level, are no longer required, so that the area under crops in each cultivated field is increased. In some areas, it promotes crop diversification into nutrient-rich foodstuffs such as vegetables, a process that not only increases the quantity of available food, but also make qualitative improvements in diet possible by supplying micronutrients that are either absent from, or in short supply, in cereals.

Aryal et al. (2014) showed that a shift from traditional to laser land leveling would on average increase the yields of rice and wheat by 342 kg ha^{-1} and 323 kg ha^{-1}, respectively. In Haryana, 2.5 million ha were under wheat and 1.2 million ha were under rice cultivation in 2011–12. Even if 50% of that area can be laser leveled, this would lead to an additional production of 0.357 MT of wheat per annum and 0.193 MT of rice per annum in Haryana. At the current level of adoption of PLL in Haryana (~0.55 mha), the annual increase in production of rice and wheat are, respectively, 0.32 and 0.29 MT yr^{-1} (Gill 2014). Similarly, in Punjab, 3.5 million ha of wheat and 2.8 million ha of rice were cultivated in 2011–12. If 50% of that area can be laser leveled (current level of adoption is ~40%), this would lead to an additional production of 0.63 MT of wheat and 0.506 MT of rice. The total value of additional outputs from adopting PLL in 50% of the land under an RW system in Haryana and Punjab is equivalent to USD 385 million yr^{-1}. Assuming that 650 g cereal per capita per day as a minimum consumption requirement, this additional output alone can feed approximately 7 million people for a year. As the 2008 food crisis showed most starkly that reductions in food availability quickly translate into rapidly increasing food prices, which have particularly adverse effects on the poor, increased production can substantially contribute to local food security.

13.9 CONCLUSIONS AND RECOMMENDATIONS

Over the past decade, researchers in association with farmers have been trying to overcome the problems of depleting water resources, diminishing input use efficiency, declining farm profitability, and deteriorating soil health by developing, evaluating, and adapting CA-based management practices in irrigated intensive production systems (for example RW systems) in the IGP of SA. Laser-assisted PLL of irrigated agricultural land is a relatively recent technology as a precursor to CA but is now being used on a large scale in SA. PLL has the potential for enhancing crop yields and resource-use efficiency of critical inputs and ensuring long-term sustainability of the resource base in intensively cultivated areas. Several studies conducted across SA demonstrated that PLL increased the area available for cultivation (1.5% to 6%) and increased crop yields and efficiency of water application resulting in large savings in irrigation water and in improved WP$_I$ and use efficiency of applied nutrients. Laser land leveling is a highly effective innovation in the field of agriculture especially for maximizing production with less water use. The main incentive for farmers to adopt PLL technology is its contribution to increased profitability mainly due to the reduced cost of production and increased crop productivity. Other major factors governing PLL adoption are accessibility and affordability along with their apparent ability to contribute to overcoming increasing water scarcity, and labor and energy savings. Where groundwater is saline, such as in the Sindh (Pakistan), irrigation efficiency improvements by reducing recharge to groundwater would contribute to real water savings, sustainable crop intensification and increase in food production. As a result, PLL has covered more than 3.5 mha in SA, mainly in northwest India and Punjab in Pakistan, where farmers are not willing to shift from the prevailing RW system, PLL can effectively help in sustaining the falling groundwater table by getting more economic yield with less irrigation water. Although adoption of precision farming has been modest in SA, the potential for its use to address environmental, food safety, and sustainability problems seems to be attracting political attention in the region.

The use of laser-assisted PLL equipment has become economically feasible and accessible through custom services, even to lower-income farmers. The holding size and shape of the field are important factors in the adoption of PLL in eastern parts of the IGP including Nepal and Bangladesh. Unlike other technologies, the cost of owning the equipment can be overcome by using custom hiring services by smallholder farmers. However, farm size matters for achieving operational efficiency of the laser leveler to manifest its full potential. Therefore, consolidation of land is

FIGURE 13.3   Recently developed two-wheel, tractor-drawn, laser-assisted precision land leveler for small-holder farmers.

an urgent need for this region to increase the plot size for better water and nutrient-use efficiencies and higher crop yields. Alternatively, efforts should be made to design and develop laser leveling technology more appropriate for small plot sizes (e.g., units that can be mounted onto smaller or two-wheel tractors) in the eastern IGP of India and Bangladesh and Nepal. Efforts are already in place to develop a two-wheel tractor-driven smaller version of a laser leveler and a prototype has been developed by the Borlaug Institute for South Asia (BISA)-CIMMYT, Ludhiana, India. The 2 wheel tractor operated laser land leveling in operation is depicted in Figure 13.3.

A judicious blending of appropriate research and policy strategies in conjunction with other CA-based management practices (e.g., zero tillage [ZT], direct seeded rice [DSR], bed planting) needs to the strengthened further for utilizing the full potential of PLL. The long-term effect of PLL on groundwater recharge and its quality, and its environmental impact on an ecoregional/regional basis needs greater attention. The long-term effects of PLL also need to be studied under varying agroecologies. Multidisciplinary analysis is needed to determine the adoption process and impacts of irrigation water savings technologies at field, farm, and higher spatial scales reagrding food production and real water savings. Studies demonstrated that the field level water savings by PLL and other CA-based technologies could not be linearly extrapolated to farm, cropping system, and landscape scales. PLL has a huge potential to improve and sustain agriculture production, ensure food security, save water, and reduce the environmental footprints of irrigated production systems in SA.

ACKNOWLEDGMENTS

The authors are grateful to the Department of Agriculture, Governments of Haryana and Punjab (India), Indian Council of Agriculture Research (ICAR) and Punjab Agricultural University, Ludhiana (India) for providing the necessary information for this synthesis. We acknowledge financial support from CGIAR Research Programs (CRPs) on Climate Change, Agriculture and Food Security (CCAFS), and the International Maize and Wheat Improvement Centre (CIMMYT) for synthesis of this work. The authors also acknowledge support from the Cereal Systems Initiative for South Asia (CSISA) project supported by USAID and BMGF. We are extremely thankful to the innovative farmers and service providers of laser levelers in South Asia.

REFERENCES

Abdullaev I, Husan MU, Jumaboev K. 2007. Water saving and economic impacts of land leveling: The case study of cotton production in Tajikistan. *Irrig Drainage System* 21: 251–263.

Aggarwal R, Kaur S, Singh A. 2010. Assessment of saving in water resources through precision land leveling in Punjab. *J Soil Water Conserv* 9: 182–185.

Ahmad MD, Masih I, Giordano M. 2014. Constraints and opportunities for water savings and increasing productivity through resource conservation technologies in Pakistan. *Agric Ecosystems Environ* 187: 106–115.

Ahmed B, Khokhar SB, Badar H. 2001. Economics of laser land leveling in district Faisalabad, Pakistan. *J Applied Sci* 1: 409–412.

Amarasinghe UA, Shah T, Anand BK. 2007. *India's Water Future to 2025–2050: Business-as-Usual Scenario and Deviations.* Colombo, Sri Lanka: International Water Management Institute, IWMI Research Report 123. Retrieved from http://books.google.co.in/books?hl=en&lr=&id=MXoME5HViUEC&oi =fnd&pg=PR4&dq=India%27s+water+demand&ots=VhBRDDRZ7B&sig=aQjJXB77mwbJY8p _LjStYsgHsqo#v=onepage&q=India's water demand&f = false.

Ambast SK. 2006. Land leveling: An on-farm water management strategy for improving crop productivity in saline environment. In Ambast SK, Gupta SK, Singh G (eds), *Agricultural Land Drainage: Reclamation of Waterlogged Saline Lands.* CSSRI, Karnal, India, pp. 70–81.

Ambast SK, Tyagi NK, Raul SK. 2006. Management of declining groundwater in the Trans Indo-Gangetic plain (India): Some options. *Agric Water Manage* 82: 279–296.

Anonymous. 2007. Statistical abstract of Punjab. The economic advisor to Govt of Punjab, Chandigarh, India.

Anuraja B, Kanannavar PS, Balakrishnan P, Pujari BT, Hadimani MB. 2013. Laser guided land leveler for precision land development. *Karnataka J Agric Sci* 26: 271–275.

Aryal JP, Bhatia M, Jat ML, Sidhu HS. 2014. Impacts of laser land leveling in rice-wheat rotation of northwestern Indo-Gangetic plains of India. In *World Congress of Environmental and Resource Economics*, June 28–July 2, 2014. Istanbul, Turkey. Retrieved from http://www.wcere2014.org/en/Programme-Overview .html.

Asif M, Ahmed M, Gafool A, Aslam Z. 2003. Wheat productivity, land and water use efficiency by traditional and laser land-leveling techniques. *J Biol Sci* 3: 141–146.

Babel MS, Wahid SW. 2008. *Freshwater under Threat in South Asia.* UNEP Report. Nairobi, Kenya, United Nations Environment Programme (UNEP).

Bhatt R, Sharma M. 2009. Laser leveller for precision land leveling for judicious use of water in Punjab, *Extension Bulletin*, Krishi Vigyan Kendra, Kapurthala, Punjab Agricultural University, Ludhiana, India.

Choudhary MA, Gill MA, Kahlown A, Hobbs PR. 2002. Evaluation of resource conservation technologies in rice-wheat system of Pakistan. In *Proceedings of the International Workshop on Developing an Action Program for Farm Level Impact in Rice-Wheat System of Indo-Gangetic Plains*, September 25–27, 2000. Paper Series 14, Rice-Wheat Consortium for the Indo-Gangetic Plains, New Delhi, India, p. 148.

Cook RL, Peikert FW. 1960. A comparison of tillage implement. *J Am Soc Agric Engg* 31: 221–214.

Corey G, Clyma W. 1973. Irrigation practices for traditional and precision leveled field in Pakistan. *Proceedings of Optimum Use of Water in Agriculture*, Scientific Paper No. 16.

El-Guindy AAM, Hasan E1, Sayd G, El-Banna O. 1994. Effect of precision land leveling system on wheat and maize production. Paper presented at 2nd International Conference on Laser and Applications, September 16–19, Cairo, Egypt.

El Yazal NS, Wissa ZHZ. 1990. Effect of laser land leveling on sugar cane yield and water requirements. *Sugar Cane* 4: 6–9, 12.

Erie LJ, Dedrick AR. 1979. Level basin irrigation: A method for conserving water and labor. *USDA Farmers' Bulletin* 2261.

Gill GJ. 2014. *An Assessment of the Impact of Laser-Assisted Precision Land Leveling Technology as a Component of Climate-Smart Agriculture in the State of Haryana, India.* CIMMYT-CCAFS, International Maize and Wheat Improvement Center (CIMMYT), New Delhi, India.

Gill MA. 1994. On-farm water management: A historical overview. In *Water and Community: An Assessment of On-Farm Water Management Programme*, Inayatullah C (ed). SDPI, Islamabad, Pakistan, pp. 24–39.

Gill MA. 2001. Agriculture in Pakistan, trends in crop production, issues and resource conservation strategies for improving efficiency. Paper presented at the International Workshop on Conservation Agriculture for Food Security and Environment Protection in Rice-Wheat Cropping Systems, February 6–9, 2001, Lahore, India.

GOI. 2013. *State of Indian Agriculture 2012–13*. Ministry of Agriculture, Government of India (GOI), New Delhi, India.

Gupta RK, Seth A. 2007. A review of resource conserving technologies for sustainable management of the rice-wheat cropping systems of the Indo-Gangetic plains. *Crop Protect* 26: 436–447.

Grace PR, Jain MC, Harrington L, Robertson GP, Antle J, Aggarwal PK, Basso B. 2003. Long-term sustainability of the tropical and subtropical rice-wheat system: An environmental perspective. In Ladha JK, Hill J, Gupta RK, Duxbury JM, Buresh RJ (eds), *Improving the Productivity and Sustainability of Rice-Wheat Systems: Issues and Impact*, Vol. 146. ASA Special Publications 65, Madison, WI, pp. 1–18.

Greenpeace India. 2013. Available at http://www.greenpeace.org/india/en/publications/AnnualReports-2013/.

Hanjra MA, Qureshi ME. 2010. Global water crisis and future food security in an era of climate change. *Food Policy* 35: 365–377.

Hill JE, Bayer DE, Bocchi S, Clampett WS. 1991. Direct seeded rice in the temperate climates of Australia. In *Direct Seeded Flooded Rice in the Tropics*. IRRI, Manila, Philippines, pp. 91–102.

Hira GS. 2009. Water management in northern states and the food security in India. *J Crop Improv* 23: 136–157.

Hobbs PR, Gupta RK. 2003. Rice–wheat cropping systems in the Indo-Gangetic plains: Issues of water productivity in relation to new resource conservation technologies. In Kijne JW, Barker R, Molden D (eds), *Water Productivity in Agriculture: Limits and Opportunity for Improvement*. CABI Publishing in association with International Water Management Institute, pp. 239–253.

Hobbs PR, Morris ML. 1996. Meeting South Asia's future food requirements from rice-wheat cropping systems: Priority issues facing researchers in the post green revolution era. NRG Paper 96-01. CIMMYT, Mexico, D.F.

Humphreys E, Kukal SS, Christen EW, Hira GS, Balwinder-Singh, Sudhir-Yadav, Sharma RK. 2010. Halting the groundwater decline in north-west India—Which crop technologies will be winners? *Adv Agron* 109: 155–217.

Jat ML. 2012. Laser land leveling in India: A success. Presentation given at the conference "Lessons Learned from Postharvest and Mechanization Projects, and Ways Forward." Asian Development Bank's Postharvest Projects' Post-production Workgroup of the Irrigated Rice Research Consortium (IRRC), held at the International Rice Research Institute, Los Banos, Manila, Philippines, May 22–24.

Jat ML, Chandna P, Gupta RK, Sharma SK, Gill MA. 2006. *Laser Land Leveling: A Precursor Technology for Resource Conservation*. Rice-Wheat Consortium Technical Bulletin Series 7, Rice-Wheat Consortium for the Indo-Gangetic Plains, New Delhi, India.

Jat ML, Gathala MK, Ladha JK, Saharawat YS, Jat AS, Kumar V, Sharma SK, Kumar V, Gupta R. 2009a. Evaluation of precision land leveling and double zero-till systems in the rice–wheat rotation: Water use, productivity, profitability and soil physical properties. *Soil Till Res* 105: 112–121.

Jat ML, Gupta R, Ramasundaram P, Gathala M, Sidhu HS, Singh S, Singh RG, Saharawat Y, Kumar V, Chandna P. 2009b. Laser-assisted precision land leveling: A potential technology for resource conservation in irrigated intensive production systems of the Indo-Gangetic plains. In: Ladha JK, Yadvinder-Singh, Erenstein O, Hardy B (eds). *Integrated Crop and Resource Management in the Rice-Wheat System in South Asia*. International Rice Research Institute, Los Banos, Philippines, pp. 223–238.

Jat ML, Gupta R, Saharawat YS, Khosla R. 2011. Layering precision land leveling and furrow irrigated raised bed planting: Productivity and input use efficiency of irrigated bread wheat in Indo-Gangetic plains. *Am J Plant Sci* 2: 578–588.

Jat ML, Pal SS, Subba Rao AVM, Sirohi K, Sharma SK, Gupta RK. 2004. Laser land leveling: The precursor technology for resource conservation in irrigated eco-system of India. *Proceedings of the National Conference on Conservation Agriculture*, New Delhi, India, 2004, pp. 9–10.

Kahlown MA, Azam M, Kemper WD. 2006. Soil management strategies for rice–wheat rotations in Pakistan's Punjab. *J Soil Water Conserv* 61: 40–44.

Kaur B, Singh S, Garg BR, Singh JM, Singh J. 2012. Enhancing water productivity through on-farm resource conservation technology in Punjab agriculture. *Agric Econ Res Rev* 25: 79–85.

Khan BM. 1986. Overview of water management in Pakistan. *Proceedings of Regional Seminar for SAARC Member Countries on Farm Water Management*. Govt. of Pakistan.

Khattak JK, Larsen KE, Rashid A, Khattak RA, Khan SU. 1981. Effect of land leveling and irrigation on wheat yield. *JAMA* 12: 11–14.

Khepar SD, Chaturvedi MC, Sinha BK. 1982. Effect of precise leveling on the increase of crop yield and related economic decision. *J Agric Engg* 19: 23–30.

Kukal SS, Hira GS, Sidhu AS. 2005. Soil matric potential-based irrigation scheduling to rice (*Oryza sativa*). *Irrig Sci* 23: 153–159.

Kumar MD, Scott CA, Singh OP. 2011. Inducing the shift from flat-rate or free agricultural power to metered supply: Implications for groundwater depletion and power sector viability in India. *J Hydrol* 409: 382–394.

Ladha JK, Fisher KS, Hossain M, Hobbs PR, Hardy B (eds). 2000. *Improving the Productivity and Sustainability of Rice-Wheat Systems of the Indo-Gangetic Plains: A Synthesis of NARS-IRRI Partnership Research.* IRRI Discussion Paper Series 40. IRRI, Los Banos, Philippines.

Ladha JK, Pathak H, Padre AT, Dawe D, Gupta RK. 2003. Productivity trends in intensive rice–wheat cropping systems in Asia. In Ladha JK et al. (eds), *Improving the Productivity and Sustainability of Rice–Wheat Systems: Issues and Impacts.* ASA Special Publication 65, ASA, CSSA, and SSSA, Madison, WI, pp. 45–76.

Ladha JK, Kumar V, Alam MM, Sharma S, Gathala MK, Chandna P, Saharawat YS, Balasubramanian V. 2009. Integrating crop and resource management technologies for enhanced productivity, profitability and sustainability of the rice-wheat system in South Asia. In Ladha JK, Yadvinder-Singh, Erenstein O, Hardy B. (eds). *Integrated Crop and Resource Management in the Rice-Wheat System in South Asia.* International Rice Research Institute, Los Banos, Philippines, pp. 69–108.

Latif A, Shakir AS, Rashid MU. 2013. Appraisal of economic impact of zero tillage, laser land leveling and bed-furrow interventions in Punjab, Pakistan. *Pak J Engg Appl Sci* 1: 65–81.

Mathankar SK, Chaudhuri D, Singh VV, Shirsat NA. 2005. Laser guided land leveling for rice crop production. Paper presented at the 39th Annual Convention of ISAE held at Acharya N. G. Ranga Agricultural University, Hyderabad, India, March 9–11, 2005.

Naresh RK, Gupta RK, Kumar A, Prakesh S, Tomar SS, Singh A, Rathi RC, Misra AK, Singh M. 2011. Impact of laser leveler for enhancing water productivity in Western Uttar Pradesh. *Intern J Agric Engg* 4: 133–147.

Naresh RK, Kumar Y, Chauhan P, Kumar D. 2012. Role of precision farming for sustainability of rice-wheat cropping system in western Indo-Gangetic plains. *Int J Life Sc Bt Pharm Res* 1: 1–13.

Naresh RK, Singh SP, Misra AK, Tomar SS, Kumar P, Kumar V, Kumar S. 2014. Evaluation of the laser leveled land leveling technology on crop yield and water use productivity in Western Uttar Pradesh. *Afr J Agric Res* 9: 473–478.

Perveen S, Krishnamurthy CK, Sidhu RS, Vatta K, Kaur B, Modi V, Lall U. 2012. *Restoring Groundwater in Punjab, India's Breadbasket: Finding Agricultural Solutions for Water Sustainability.* Columbia Water Center, Earth Institute, Columbia University, New York.

Playán E, Faci JM, Serreta A. 1996. Modeling micro topography in basin irrigation. *J Irrig Drainage Eng, ASCE* 122: 339–347.

Qureshi AS, Shah T, Akhtar M. 2003. *The Groundwater Economy of Pakistan.* IWMI Working Paper No. 64, International Water Management Institute, Colombo, Sri Lanka.

Rajput TBS, Patel N. 2004. Effect of land leveling on irrigation efficiencies and wheat yield. *J Soil Water Conserv* 3: 86–96.

Rajput TBS, Patel N, Agarwal G. 2004. Laser leveling—A tool to increase irrigation efficiency at field level. *J Agric Engg* 41: 20–25.

Ramme U, Trudean N, Graczyk D, Taylor P. 2011. *Technology Developments and Prospects for the Indian Power Sector;* International Energy Agency, Organization for Economic Co-operation and Development, Paris.

Rehman HM, Gill MA, Awan NA, Ladha JK. 2009. Evaluation and promotion of integrated crop and resource management technologies in rice-wheat system in Pakistan. In Ladha JK, Yadvinder-Singh, Erenstein O, Hardy B (eds), *Integrated Crop and Resource Management in the Rice-Wheat System in South Asia.* International Rice Research Institute, Los Banos, Philippines, pp. 111–132.

Rickman JF. 2002. *Manual for Laser Land Leveling, Rice-Wheat Consortium Technical Bulletin Series 5.* Rice-Wheat Consortium for the Indo-Gangetic Plains, New Delhi, India.

Rodell M, Velicogna I, Famiglietti JS. 2009. Satellite-based estimates of groundwater depletion in India. *Nature* 460: 999–1002.

Rudragouda C, Ravindra Y, Kanannavar PS, Vasantagouda BR, Kumar M. 2012. Precision leveling: Its impact on slope variation in vertisols of Karnataka. *Proc Agro-Informatics Precision Agric, India.* pp. 332–334.

Saharawat YS, Gathala MK, Ladha JK, Malik RK, Singh S, Jat ML, Gupta RK, Pathak H, Singh K. 2009. Evaluation and promotion of integrating crop and resource management technologies in the rice-wheat system in northwest India. In: Ladha JK, Yadvinder-Singh, Erenstein O, Hardy B (eds), *Integrated Crop and Resource Management in the Rice-Wheat System in South Asia.* International Rice Research Institute, Los Banos, Philippines, pp. 133–150.

Sattar A, Khan FH, Tahir AR. 2003. Impact of precision land leveling on water saving and drainage requirement. *JAMA* 34: 39–41.

Sidhu HS, Mahal JS, Dhaliwal IS, Bector V, Manpreet-Singh, Sharda A, Singh T. 2007. *Laser Land Leveling–A Boon for Sustaining Punjab Agriculture*. Dept. of FPM, Punjab Agricultural University, Ludhiana, India. Farm Machinery Bulletin-2007/01:13.

Sidhu RS, Vatta K, Dhaliwal HS. 2010. Conservation agriculture in Punjab—Economic implications of technologies and practices. *Indian J Agric Econ* 65: 413–427.

Singh G. 2007. Conservation Agriculture for Managing Soil and Water Resources. In Kaledhonkar MJ, Gupta SK, Bundela DS, Singh G (eds), *On-Farm Land and Water Management*. Central Soil Salinity Research Institute, Karnal, India, pp. 1–8.

Timsina J, Connor DJ. 2001. The productivity and sustainability of rice wheat cropping systems: Issues & challenges. *Field Crops Res* 69: 93–132.

Tyagi NK. 1984. Effect of land surface uniformity on irrigation quality and economic parameters on sodic soils under reclamation. *Irrig Sci* 5: 151–166.

Yadvinder-Singh, Kukal SS, Jat ML, Sidhu HS. 2014. Improving water productivity of wheat-based cropping systems in South Asia for sustained productivity. *Adv Agron* 127: 157–258.

14 Data-Driven Precision Agriculture
Opportunities and Challenges

Wenxuan Guo, Song Cui, Jessica Torrion, and Nithya Rajan

CONTENTS

14.1 INTRODUCTION

Agriculture is facing the greatest challenge of feeding more than 9 billion people by 2050 in a manner that advances economic development and a healthy environment. A 70% increase from 2006 in food production is required to meet this demand (World Resources Institute 2013). With limited resources of land and water, increased agricultural production is projected to come primarily from intensification on existing arable land (FAO 2011). Conventional farming practices treat an agricultural field uniformly despite the inherent variability in soil properties and crop growth conditions. Uniform management may result in over- or under-application of resources in specific locations within a field, which may have a negative impact on the environment and profitability (McKinion et al. 2001; Plant 2001). Sustainable agriculture is a viable means of meeting the food demand while balancing crop production and minimizing environmental impacts. Precision agriculture (PA) is a promising approach to attain sustainable agriculture.

PA is the management of soil and crops at subfield scale using information and technology for optimum profitability, sustainability, and protection of the environment (Robert et al. 1995, 1996; National Academy of Sciences 1997). Also called site-specific management, PA involves farming practices to apply the right amount of right resource at the right place, at the right time, and in the right manner (Khosla 2010; Robert et al. 1995). For PA to be applicable, as indicated in the definition, significant within-field spatial variability must exist in soil properties and crop growth; such variability can be identified and measured and information from these measurements can be used to improve crop production and environment (Miller et al. 1988). In order to determine the variability, intensive soil and plant samples may be collected and analyzed in the laboratory, but the costs of sampling and analyses will potentially exceed the benefit from the site-specific management (Swinton and Lowenberg-DeBoer 1998). Thus, efficient methods for accurately measuring within-field variability in soil properties and plant growth are important for PA (Bullock and Bullock 2000). The use of advanced technologies, such as the Global Positioning System (GPS), Geographic Information System (GIS), yield monitors, and remote sensing, enables efficient quantification of spatial variability in soil properties, crop growth, and crop yield within fields. With variable rate technologies (VRT) together with precision guidance systems, inputs (i.e., fertilizers, water, pesticides, seeds) can be applied according to the spatial variability in soil properties and crop yield potential.

The main components of PA include data collection, data interpretation and analysis, and implementation of management at an appropriate scale and time (National Academy of Sciences 1997). As technology advances, numerous tools and sensors have been developed to collect spatial data and information, such as yield monitors, satellites, combine-mounted crop scanners, handheld sensors, and on-the-go soil property sensors. Massive data and information can be collected within a relatively short period of time. It remains a challenge to appropriately collect, analyze, and interpret the data, as well as applying the derived information and knowledge to site-specific management. The objective of this chapter is to review the opportunities and challenges of PA technologies used to collect, analyze, and apply spatial and temporal data and information for optimized agricultural production.

14.2 TECHNOLOGIES AND DATA COLLECTION

14.2.1 YIELD MONITORING AND YIELD DATA

Yield monitoring and mapping using combine-mounted yield monitors, often recommended as a first step, has become one of the most widely adopted PA technologies (Figure 14.1). Yield monitors include sensors to measure the mass or volume of product flow, ground speed, moisture, and header position. Yield is derived as a product of these parameters being sensed. The combine ground speed and cut width are used to determine the harvest area per unit time. A submeter accurate differential GPS (DGPS) receiver is usually installed together with a yield monitor to record the geographic position (latitude and longitude), time, and elevation of each yield data point.

The yield monitor plays two roles in PA. On the one hand, yield monitors generate spatially dense data with relatively low cost, allowing characterization of the spatial and temporal yield variability (Dobermann et al. 2003; Pierce and Nowak 1999). Based on the yield variability, site-specific

FIGURE 14.1 Cotton yield monitor mounted on a harvester and the yield map generated from the yield data.

management can be implemented to match the yield potential at different locations of the field. On the other hand, yield monitor data can also be utilized to evaluate the effectiveness or the response of PA practices. Producers have used yield data and maps for crop moisture monitoring, documenting yields, field experiments, drainage tiles installation, new crop lease negotiation, dividing crop production, and bottom-line considerations (Griffin 2009). In addition, yield maps can be used to guide field scouting, design soil sampling schemes, and calculate nutrient requirements for variable rate fertilizer applications.

Users need to be cautious when applying yield data and yield maps for site-specific management. Yield data contains various systematic and random errors. The sources of errors can be classified into four categories: sensor errors, errors due to operating conditions, operator errors, and yield mapping errors (Thylen et al. 1997). The main errors involved in yield mapping are unknown crop cutting width, grain lag time, GPS error, grain mixing through combine components, combine grain losses, and calibration (Blackmore and Marshall 1996). Postharvest processing is needed to remove these errors, especially when yield data is compared to other data layers within the decision support system. Various techniques have been applied to remove errors related to unrealistic cycle distance, moisture, combine speed surge, wrong cutting widths, overlapping yield points, and so forth. A software tool called Yield Editor (Sudduth and Drummond 2007) was developed and widely adopted to simplify the process of applying filtering techniques for yield data outlier detection and removal. However, when using the tool, users must make sure that the parameter settings are appropriate, because these parameters may be specific to each field. The user's experience and knowledge should be incorporated in such data cleaning.

Another challenge is interpreting yield variability and the underlying causes of such variability in yield maps. Yield variability may be caused by many factors, including spatial variability in soil type, landscape position, crop history, soil physical and chemical properties, and nutrient variability (Wibawa et al. 1993). Interactions among biotic (plant genotype, soil fauna, pests, and diseases) and abiotic factors (soil physical, chemical, moisture characteristics, and climatic conditions) influence spatial yield variability. Effects of crop stress, pests, and diseases on crop yield are temporal factors that could explain up to 50% of crop yield variability across years and sites (Machado et al. 2002). As a result, yield maps tend to vary from year to year, which makes it more difficult for producers to make decisions on site-specific management. A single-year yield map is useful for interpretation of possible causes of yield variation but may be of limited value for more strategic and long-term site-specific management. With multiple years of yield data, repeating patterns and their more stable natural causes may be separated from random variation in each year, providing a basis for spatially varying yield goals or other site-specific management practices (Dobermann et al. 2003). For example, Blackmore (2000) proposed a method to use multiple years of yield maps to classify a field into three categories: high-yielding and stable, low-yielding and stable, and unstable. Based on the spatial and temporal trends of the yield maps, the economic significance of these areas can be assessed and site-specific management can be implemented based on the economic return of each area. In summary, yield maps alone cannot give clear guidelines for site-specific management unless the sources of variation are identified, especially those related to soil physical and chemical properties and seasonal weather conditions.

14.2.2 Remote Sensing

Remote sensing data from various platforms, including satellites, aircrafts, unmanned aerial systems, and field vehicles, has been used as an important source for obtaining spatial information of soil and crops for PA. Remote sensing technology is a non-destructive method that can systematically collect information about agricultural fields over a large geographical area. Remote sensing data can reveal unbiased information about areas that are sometimes inaccessible to humans (Liaghat and Balasundram 2010). Remote sensing has been applied to evaluate crop growth (Clevers and Van Leeuwen 1996; Moran et al. 1997; Sakamoto et al. 2005), leaf area index (Delegido et al.

2013; Papadavid et al. 2013; Zhao et al. 2012), chlorophyll content (Daughtry et al. 2000; Gitelson and Merzlyak 1996; Zheng and Moskal 2009) ground cover (Maas 1998; Rajan and Maas 2009; Rajan et al. 2014), pest and disease infestation (Curran et al. 2000; Kelly and Guo 2007; Prabhakar et al. 2012; Qin and Zhang 2005), and soil physical and chemical properties (Bausch et al. 2004).

For many years, remote sensing relied on spectral reflectance data in the visible and infrared wavelengths of the electromagnetic spectrum for the quantitative and qualitative analysis of soil and crop characteristics. In recent years, remote sensing technology has advanced beyond the commonly used visible and infrared sensors and includes hyperspectral sensing, thermal sensing, and light detection and ranging (lidar). Hyperspectral remote sensing in narrow bandwidths has allowed the application of remote sensing in identifying specific biophysical and biochemical characteristics of crops (Pacheco et al. 2001; Thenkabail et al. 2013; Zhang et al. 2003). Lidar holds promise for many PA applications (Eitel et al. 2014; Tang et al. 2014). For example, Andujar et al. (2013) successfully used lidar for weed detection and discrimination in a maize (*Zea mays* L.) cropping system. Remote sensing in thermal wavelengths is being used to detect soil moisture status and water stress in plants. This offers the opportunity of assessing crop water requirement at a subfield scale for variable rate irrigation management (Torrion et al. 2014).

The application of unmanned aerial vehicles (UAVs) has tremendous potential for acquiring aerial images because of the low operational cost, high temporal and spatial resolutions, easy-to-use controlling system, and high flexibility in image acquisition planning. The majority of UAVs currently in use are designed to operate at low altitudes (less than 150 m) and adverse weather conditions. For UAV-based image acquisition systems, clouds and other kinds of atmospheric interference are not as influential as they are to satellite and piloted airplanes. Lower flying height also means much higher spatial resolution, enabling finer-scale interpretation of soil and crop characteristics from images. However, UAV-based image processing and interpretation at the producer level still require development of analyzing and processing tools before they can be easily adopted by farmers. Additionally, short flying duration, lack of stability, limited communication distance, scarcity of affordable lightweight camera systems, and public skepticism and misinformation greatly challenge the application of UAVs in PA.

The latest advancements in remote sensing data collection for precision management are geared at measuring individual plant characteristics in real time (Mulla 2013). Sensors mounted on field vehicles and stationary sensors are becoming valuable tools for measuring soil and crop characteristics. The considerable interest in collecting this data at high spatial, spectral, and temporal frequencies is posing new challenges in processing remote sensing data for PA applications because it requires sophisticated computing and analyzing capabilities for processing massive amount of data.

14.2.3 Soil Sampling and Soil Data

The purpose of soil sampling in PA is to assess soil fertility and determine the amount of fertilizer required to produce certain yield goals at different locations in the field. In conventional agriculture, soil samples are collected, the average of a certain nutrient is calculated, and the amount of fertilizer for the whole field is based on this average. In PA, soil samples are collected with geographic information attached to each soil sample using a GPS receiver. The amounts of fertilizer applied vary according to the needs at specific locations. There are two main methods of soil sampling in PA: grid sampling and directed sampling. Grid sampling is an ideal approach if there is no prior knowledge of the fertility variability within the field. Soil test results from a well-designed grid sampling scheme provide an accurate base nutrient map for long-term management. However, a great number of samples are required, which can be very expensive and time-consuming. Directed soil sampling, on the other hand, requires prior knowledge of the field characteristics that may be limiting crop yield. A background layer containing subfield regions with different characteristics can be constructed for directed soil sampling. The background layer can be a yield map, soil types, cropping systems, remote sensing imagery, and so forth. A standard test result usually includes available

FIGURE 14.2   Veris soil EC mapping system (left: Veris Technologies) and an apparent soil electrical conductivity map.

phosphorous (P), exchangeable potassium (K), calcium (Ca), magnesium (Mg), cation exchange capacity (CEC), pH, and so forth. Some laboratories may also test for organic matter, nitrate, salinity, sulfate, certain micronutrients, and heavy metals (Foth and Ellis 1988).

Soil sampling and soil laboratory analysis can be labor-intensive, costly, and time-consuming. As a result, usually only a limited number of soil samples are collected and analyzed. The sparse spatial distribution of soil test data, such as the levels of macronutrients and micronutrients, limits the scale of site-specific management. Efficient methods for accurately measuring within-field variations in soil physical and chemical properties are critical for PA (Bullock and Bullock 2000). Various on-the-go soil sensors have been or are being developed to measure soil texture, organic matter, moisture content, salinity, bulk density, topsoil depth, pH, nitrate, CEC, and so forth (Adamchuk et al. 2004). The measurement of soil apparent electrical conductivity (EC_a) has been widely adopted and increasingly used in PA (Clay et al. 2001; Corwin et al. 2003; Johnson et al. 2001). There are two types of sensors commercially available: contact and noncontact. One commonly used contact type is the Veris mapping system (Figure 14.2) (Veris Technologies, Salina, KS) that consists of a Wenner array (coulters) and records EC_a by electrical resistivity at a shallow depth (0–30 cm) and a deep depth (0–90 cm) simultaneously. Noncontact EC_a sensors are typically composed of a transmitter and a receiver coil. They measure EC_a without contacting the soil surface via electromagnetic induction. Examples of these types of sensors are EM38 (Geonics Limited, Mississauga, Ontario, Canada) and GEM-2 (Geophex, Raleigh, NC).

The purpose of the field-scale EC_a survey used in site-specific management is to establish the within-field variability in soil properties influencing the variability in crop yield (Figure 14.2). However, directly relating crop yield and EC_a measurement has resulted in inconsistent results due to the fact that EC_a measurements are affected by multiple factors. Spatial variability in soil electrical conductivity is related to such factors as texture, organic matter, CEC, landscape positions, salinity, subsoil characteristics, soil water content, and depth to claypan (Clay et al. 2001; Rhoades 1993). To use EC_a in PA, it is necessary to understand the factors that most significantly influence the EC_a measurement. Simple statistics or wavelet analysis can be used to determine the dominant factors influencing EC_a measurement (Corwin and Lesch 2003). Although temporal variability exists, the relative spatial pattern of EC_a distribution within a field is considerably stable (Clay et al. 2001). As a result, an EC_a map provides useful spatial information to identify potential areas in need of improved irrigation, drainage, fertilizer, and pest management (Corwin and Lesch 2003).

14.2.4   TOPOGRAPHIC DATA AND DIGITAL TERRAIN ANALYSIS

Topographic and hydrological attributes, including primary and secondary attributes, have been widely used in PA, especially in site-specific management of seed, irrigation water, and fertilizer (Iqbal et al. 2005). A digital elevation model (DEM), a three-dimensional (3-D) representation of a terrain's surface, is the most commonly used format to represent the elevation. A DEM is

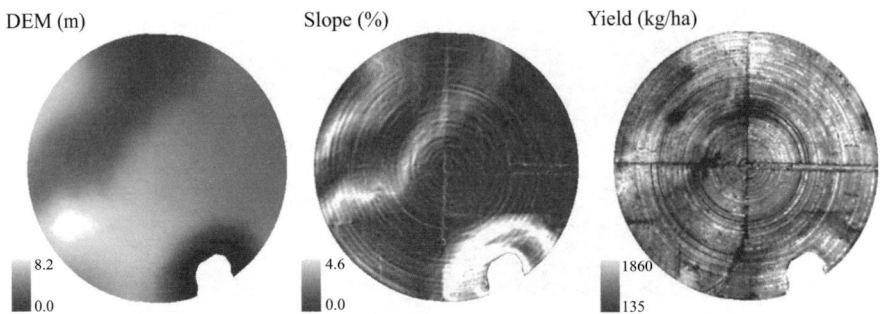

FIGURE 14.3 Digital elevation model (DEM), slope, and yield map of a field in Texas.

usually derived from point elevation data through various interpolation methods. After the DEM is obtained, primary and secondary terrain attributes can be derived (Figure 14.3). Primary attributes include slope, aspect, plan and profile curvature, flow path length, and upslope contributing area (Wilson and Gallant 2000). Elevation, slope, and curvature have a direct effect on infiltration and runoff through their influence on surface and subsurface water flow. Water tends to move downslope causing lower positions to receive water from adjacent higher positions (Kaspar et al. 2003). Secondary attributes are computed from two or more primary attributes. Secondary attributes mainly include topographic wetness index, sediment transport index, stream power index, flow direction, and flow length. These attributes may affect soil characteristics, distribution and abundance of soil water, susceptibility of landscapes to erosion by water, and the distribution and condition of plant growth (Wilson and Gallant 2000). In addition, topography influences the redistribution of soil particles, organic matter, and soil nutrients. Topography has been reported to be related with the yield of various crops, such as corn (Kaspar et al. 2003; Kravchenko and Bullock 2000), wheat (*Triticum aestivum* L.) (Sinai et al. 1981), cotton (*Gossypium hirsutum* L.) (Bronson et al. 2003; Guo et al. 2012; Iqbal et al. 2005; Li et al. 2001), sorghum (*Sorghum bicolor* L.) (Machado et al. 2002), soybean (*Glycine max* L.) (Kravchenko and Bullock 2000), and dry pea (*Pisum sativum* L.) (Mahler et al. 1979). In PA applications, seed, fertilizer, and irrigation water rates can be varied according to topographic properties. For example, consistently low-yielding areas (on summit or steep slope) would receive lower seed, fertilizer, and water rates to reduce the waste of recourse and improve overall profit.

In principle, any data that contains elevation information can be used as source for digital terrain analysis. However, for PA applications, highly accurate elevation data is required. Lidar can be used to effectively survey elevation of a field (Galzki et al. 2011). However, this is usually difficult to set up and may be cost-prohibitive. Automated guidance equipments with real-time kinematic (RTK) GPS receivers have gained rapid and widespread adoption. Real-time kinematic is a carrier-phase-based survey method of determining relative positions between receivers simultaneously tracking the same satellites (Sickle 2001). An RTK GPS receiver has a nominal positioning accuracy of 1 cm in horizontal directions and 2 cm in vertical directions (Tamura et al. 2002). With RTK autosteering systems, dense and accurate elevation data are collected during any field operations at no additional cost. However, the user needs to pay close attention to the quality of raw elevation data. RTK GPS receivers may sometimes lose correction, resulting in low accuracy in elevation. These errors must be filtered out before conducting digital terrain analysis.

14.2.5 TEMPORAL VARIABILITY

Many studies have documented spatial variability of soil and crop characteristics and PA management recommendations are provided. The time factor or temporal variability is less documented and mostly ignored. In reality, spatial variability of soil and crop characteristics is dynamic within each growing season and between growing seasons. Temporal variability occurs both intraseasonally and interseasonally. Soil water content, nitrogen (N) status, and climatic parameters change day to day within a season, whereas crop yield and weed infestation patterns vary from season to season (Zhang et al. 2003). Temporal variability in crop yield or some soil characteristics at the within-field scale is often larger in magnitude than spatial variability. This will increase the risk of economically and environmentally inappropriate actions if PA practices are solely based on spatial information (Whelan and McBratney 2000).

Incorporating the temporal variability and applying crop inputs and management at the right time is a key component of PA. Management considerations such as scouting for pests and diseases, application of chemical control, fertilizer application, and irrigation scheduling is known to be more effective and responsive when applied at a specific growth stage (Darby and Lauer 2000; Kranz et al. 2008; Specht et al. 1986; Torrion et al. 2011; Wise et al. 2011). When a crop input is suboptimal, particularly supplemental irrigation water, prioritization of such application can increase water use efficiency (WUE). Under extreme water supply deficit, applying the limited irrigation water to only the productive portion of the field was reported to be economical instead of delivering irrigation to the entire field by variable irrigation (Nair et al. 2012). The economic return of this approach can be attributed to a reduction of fuel and man-hours while increasing yield on the more productive part of a field.

Implementing timely irrigation in a variable rate manner requires intensive temporal data about soil moisture and plant conditions at different parts of the field. Recent advances in sensor and wireless radio frequency (RF) technologies along with the Internet offer great opportunities for development and application of sensor systems for agriculture (Pierce and Elliott 2008). Vellidis et al. (2008) discussed linking soil moisture sensors to RF identification (RFID) tags. Data transmitted to a local receiver monitors variable water needs of crops within fields. Camilli et al. (2007) proposed to apply wireless sensor networks (WSNs) consisting of sensor nodes to continuously measure soil moisture, temperature, solar radiation, and other environmental factors. For example, the granular matrix seems to be preferred due to affordability, life of operation, and ease of installation and maintenance (Irmak et al. 2014). The relatively low cost of the devices allows the installation of a dense population of nodes to adequately represent the variability present in the environment of the field. The real-time information obtained by the sensors from the fields can provide a solid base for farmers to adjust strategies at any time (Hwang et al. 2010).

Another way of handling temporal variability is through the use of crop models to predict critical stages for optimal timing of crop inputs and management. Commonly, these growth stages are simulated with accumulation of growing degree units and often integrated with environmental factors, such as stress, soil types, and microenvironment, as well as crop variety and management. Predicting crop stages requires historical and current weather data. To simplify the access of crop models to farmers, many models or tools are posted on the internet. One such example is the Nebraska *SoyWater* program (http://hprcc3.unl.edu/soywater) that integrates the SoySim model (Setiyono et al. 2010). In this model, daily crop water use is calculated and used for the soil-water balance to determine the daily irrigation requirement. When the occurrence of crop stages is known, it assists researchers, agronomists, and farmers to have a lead time planning for precise application of farm inputs (Torrion et al. 2011).

14.3 DATA INTEGRATION AND ANALYSIS

14.3.1 Data Aggregation and Integration

Geospatial data of different sources and formats at different scales, such as soil physical and chemical properties, crop development, remote sensing images, yield data, and as-applied planting data, are often stored in pieces without systematic integration. This limits data access and data use efficiency, potentially resulting in inadequate management decisions based on incomplete information.

The GIS and spatial databases play a key role in organizing and integrate different layers of data and information. A GIS is a computer-based system for capturing, storing, analyzing, and managing data and associated attributes that are spatially referenced to the earth. One of the advantages of GIS is its capability of overlaying different information and data and relating them in the same spatial context. For example, yield data may be compared with other data layers, such as the soil test data, landscape position, remote sensing crop canopy, soil electrical conductivity, and perhaps previous years' yield data. This may reveal some information as to why yields are high in one location and low in other locations. Based on the relationship of these different layers, better decisions can be made on site-specific application of fertilizers, water, and other crop inputs.

One of the challenges facing PA is data incompatibility. Agronomic data generated from one manufacturer's device often do not match devices from other manufacturers. A prescription map applied on one system cannot be directly transferred to another. Special software and often proprietary software programs have to be used to conduct such transitions. This is not only inconvenient, but also wastes time and increases the cost of operations. Various efforts have been pursued to solve this issue. In 2012, a nonprofit organization, AgGateway, formed a Standardized Precision Ag Data Exchange (SPADE) project to improve data sharing and interoperability (www.aggateway.org). This project is collaboration among agricultural suppliers of hardware, software, inputs, services, implements, and vehicles. Its goals are to establish a framework of standards to simplify data share and exchange among advisors, suppliers, and other partners who provide services with different system components. To further increase data transparency and integration, the Open Ag Data Alliance (OADA) is building open application programming interfaces (APIs) to allow different hardware and software systems to communicate automatically through secure cloud services. In addition, OADA is building guidelines regarding data privacy and use standards to ensure compliance with OADA principles. Many agribusiness companies have agreed or planned to adopt such standards. However, when it comes to implementing the standards, few companies are willing to abandon their proprietary data format. As a result, the process can take a long time.

Systematic approaches on data integration are greatly needed to store, analyze, and provide suggestions for site-specific management services. Just like an electronic health record (EHR), the digital data of a field can be shared so that any crop advisor can diagnose a field and provide service based on all its previous records. Databases with these shared data and information can help make recommendations on guiding site-specific management. Establishing such databases requires collaboration across different government agencies, research institutions, agribusiness companies, and farmers. The outcome database will not only be in technical formats for academic communities, but also as accessible summaries that crop advisors, policymakers, and producers can use to guide agronomic decisions and create managerial guidelines. An agribusiness, SST Software, is adopting such a systematic approach to build a database that is compatible across different hardware and software industries. They provide services including data storage and data analysis as well as management recommendations.

14.3.2 Spatial Analysis and Geostatistics

Tobler (1970) called this the first law of geography: "everything is related to everything else, but near things are more related than distant things." Agronomic data, such as soil clay content, soil nutrient status, electrical conductivity, and yield, is spatially distributed data that is more related close together than far apart. Classical statistical analysis assumes that data is independent. Geostatistics has been widely adopted in PA because it provides a collection of statistical methods to analyze spatially dependent data. Geostatistics treats a spatial variable as a random variable. For each point of a population, x, there is a series of values for a property, $Z(x)$, and the observed value, $z(x)$, is drawn at random according to a probability distribution function. The series of random variables is a random process and the actual value of Z observed is one number of realizations of that process (Oliver 2010).

Spatial patterns are usually described using the semivariogram, which measures the average dissimilarity between data separated by a distance h. It is calculated as

$$\gamma(h) = \frac{1}{2N} \sum_{i=1}^{N} [Z(x_i) - Z(x_i + h)]^2$$

where, x_i is a data location, h is a lag distance between samples, $Z(x_i)$ is the data value at the location x, and N is the number of data pairs at distance h. The semivariogram is usually modeled using several functions, which then fit the semivariogram data. Semivariogram models are described with three parameters, range, sill, and nugget (Figure 14.4). Range is the distance over which data are correlated. This means when separation distances are greater than the range, sampled points are no longer spatially correlated (i.e., random). Sill refers to the semivariance at which range is reached. Nugget represents the semivariance at zero separation distance (lag = 0), usually due to errors in sampling, measurement, or other unexplained sources of variance (Isaaks and Srivastava 1989).

There are two main applications of geostatistics in PA: modeling spatial dependence and predicting variable values at unsampled locations. A spatial description combined with good knowledge about the phenomenon can improve understanding the underlying physical mechanisms controlling the spatial patterns (Goovaerts 1998). Evaluating the spatial dependence of soil or plant variables helps to minimize the number of samples to make appropriate estimation of variable distribution without significant loss of information (Vachaud et al. 1985). The characterization of spatial structure has also been used to filter and remove errors in yield monitor data (Ping and Dobermann 2005) and serves as a tool to assess the response of error filtering (Simbahan et al. 2004). In these studies, yield data not spatially correlated within small regions are considered outliers and removed. In addition, the spatial structure of yield data improved after outlier removal. The analysis of spatial variation can also be used to improve analysis of treatment effects in experimental design and analysis. Most variety evaluation trials are analyzed using classical analysis of variance without adequately accounting for spatial variability. Stroup et al. (1994) found the analysis of breeding trials is improved by removing the spatial variance in the data.

The second application of geostatistics in PA is the spatial prediction of a variable at unsampled locations. Kriging is a generic term for a set of generalized least-squares regression algorithms for spatial prediction (Goovaerts 1998; Webster and Oliver 2001). The most commonly used kriging methods are the ordinary kriging and cokriging. Ordinary kriging unbiasedly estimates an unknown value as a linear combination of neighboring observations using weights applied to each observation. These weights are chosen to minimize the estimation error. Ordinary cokriging is an extension of the ordinary kriging method to incorporate additional information of another variable.

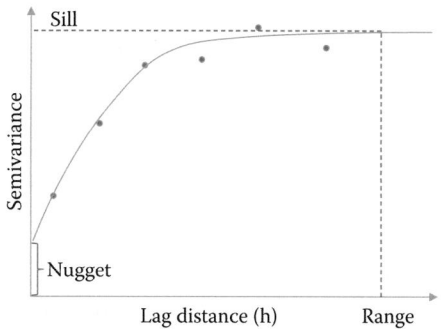

FIGURE 14.4 Example of a conceptual semivariogram (dots are observed semivariance).

For example, the sparsely distributed soil test P levels can be better estimated by taking into account secondary information originating from other correlated information, such as EC_a.

There are several challenges encountering geostatistics application in PA. Many professionals working in PA do not have a strong geostatistical background. More training opportunities or university courses need to be readily available for them to improve their knowledge and skills in geostatistics. Geostatistical software programs need to provide better documentation and explanation for each analysis method, instead of only providing a black-box-based final result. The time factor in PA is also not being handled adequately. Fundamental research is required to develop geostatistics to analyze the integrated dynamics of spatial and temporal variability (Schueller 2010).

14.3.3 DECISION SUPPORT AND MANAGEMENT ZONES

The purpose of collecting information and data is to have site-specific management. Dividing the field into subfields or management zones is a natural first step. Management zones are field areas possessing approximately homogenous attributes in landscape and soil condition and are used as the smallest units for site-specific management, such as variable rate fertilizer application (Doerge 1999). Creation of management zones provides a convenient means of capturing the spatial distribution of yield-influencing factors in a season. Two criteria may be used to evaluate the appropriateness of management zones: (1) yield differences between zones should be substantially greater than those within zones, and (2) the major factors that influence yield within a zone must be approximately homogeneous (Plant et al. 1999).

Yield maps have often been used to delineate management zone delineation. For example, Kitchen et al. (1995) used a classified corn yield map from the previous year to determine yield potential zones for variable rate application of N fertilizer. The application of yield maps to identify zones is challenged by spatial and temporal variation in yield because it is affected by many interacting factors (Huggins and Alderfer 1995). Soil physical and chemical properties have been extensively used to identify management zones, including soil texture, bare soil brightness, and apparent electrical conductivity. Mzuku et al. (2005) delineated management zones based on bare soil aerial imagery, a farmer's perception of field topography, and past crop and management practices. Georeferenced apparent electrical conductivity in a field provides spatial information about soil salinity, texture, and water content, enabling producers to identify zones for particular management practices (Bullock and Bullock 2000; Clay et al. 2001; Corwin and Lesch 2003).

Many of the previous studies about the delineation of management zones did not integrate crop yield and soil properties. According to Lund et al. (2000), producers are unwilling to adjust inputs on different yielding areas until they have some evidence about the underlying soil properties. Any management zones produced without considering crop yield would be of little value (Doerge 1999). Delineation of management zones using yield maps along with soil properties can be used to directly associate crop production with soil properties so that management zones are more meaningful and interpretable. However, because yield varies from year to year due to complex factors, potential management zones are likely to be different from year to year. Taking temporal stability into account allows better management of weather and climatic risk (McBratney et al. 2005). Potentially stable management zones in a field provide important decision support for crop producers to apply inputs such as water, nitrogen, and chemicals in a site-specific manner.

Producers or crop advisors must identify the purpose of site-specific management before proceeding to collecting data and preparing for management zones delineation. Factors such as terrain elevation, soil physical properties, and soil nutrient levels have the most direct impact on yield, and hence should be included for zone delineation. Data with a stable temporal pattern can be the most cost efficient in applying management zones for site-specific management. Topography, EC_a, soil physical properties, or multi-year yield data along with supplemental information (e.g., soil electrical conductivity or elevation) is recommended for zone delineation in order to identify consistent yield patterns (Ortiz et al. 2011). For many in-season management practices, such as application of growth regulator and defoliants, a high resolution remote sensing image is a perfect choice for management zone delineation. A general guideline is provided in Table 14.1 to associate the agricultural inputs suitable for site-specific management and the suggested data for management zone delineation.

Management zone delineation is usually a multivariate clustering procedure to divide the field into different zones based on the input variables. These clustering methods used in agriculture include hierarchical agglomerative clustering, fuzzy clustering, hierarchical divisive clustering, and Kohonen self-organizing feature maps (Tiwari and Misra 2011). Fridgen et al. (2004) developed a software program, Management Zone Analyst (MZA), to automatically analyze data and output clusters using the fuzzy c-means theory. Fuzzy c-means applies a weighting exponent to control the degree of membership sharing between clusters (Bezdek 1981). This software program not only classifies the data set into zones, but also suggests the best number of zones that should be created.

Management zones are delineated using historical data. Therefore, the user needs to assess the performance of management zones or management strategies for a field during the growing season and after the season, depending on the management purpose. A grower can make a historical comparison to yield or profitability attained with a previous variable rate or uniform rate input strategy. Another method will be to directly compare the value of two management zone strategies using multiple side-by-side comparisons using yield monitor data (Doerge 1999). Depending on the comparison results, it may be necessary to adjust the management zones by combining or splitting for the next crop or season (USDA-NRCS 2010).

With the increasing applications of real time sensors, management zones may eventually become obsolete for some variable rate applications and management. One such sensor, for example, is the GreenSeeker (Trimble Navigation, Sunnyvale, CA) mounted on the sprayer. It uses an active light source to measure spectral reflectance from the crop canopy to calculate normalized difference vegetation index (NDVI) for determining the amount of N required at different locations of a field. As the sprayer is traveling and recording this data, a prescription is created to apply N in the right amount needed in that particular portion of the field. Another product, the WeedSeeker spray system, applies a similar principle to sense the presence of weeds and triggers spray nozzles to deliver a precise amount of chemical to spray the weeds. Maleki et al. (2008) demonstrated the feasibility

---

TABLE 14.1

Agricultural Inputs Suitable for Variable Rate Application and Data Used to Delineate Management Zones for these Inputs

Input Suitable for Variable Rate Application	Data Used to Delineate Management Zones
Seeding rate	Soil EC_a, topography, soil survey, historic yield data, soil organic matter
Nematicides	Soil EC_a, topography
Lime	Soil EC_a, grid or zone-sampled soil pH, buffer pH, soil survey, topography, bare soil imagery, historic yield data
Nitrogen	Soil texture, soil organic matter, soil color imagery, crop spectral reflectance using sensors like GreenSeeker or Crop Circle, yield data
Other nutrients	Soil EC_a, crop spectral reflectance, CEC, soil survey
Irrigation	Soil EC_a, soil color imagery, yield data, canopy imagery, soil survey, farmer's knowledge
Plant growth regulators, defoliants	Plant spectral reflectance of visible and infrared bands from in-season aerial or high-resolution satellite imagery

Source: Adapted with permission from Ortiz, B.V. et al., Management zones II–basic steps for delineation. Precision Agriculture Series–Timely Information. Agriculture, Natural Resources & Forestry. Alabama Cooperative Extension System, 2011.

and effectiveness of on-the-go variable-rate P fertilizer application using a visible (VIS) and near-infrared (NIR) soil sensor for on-the-go measurement of soil P. Variable rate irrigation has been also proved feasible without using management zones. For example, Kim et al. (2008) proposed a real-time variable rate irrigation system consisting of a WSN, software for real-time information, and an irrigation control system. The system site-specifically operates individual sprinklers to apply a specified amount of water based on the real-time information collected by the sensors distributed across the field.

14.4 BIG DATA CHALLENGE

Big data refers to data sets that cannot be processed, analyzed, and managed using traditional analyzing algorithms or tools due to its large volume, structural variety, and high data accessing and retrieving velocity (Fan and Bifet 2012; Zikopoulos et al. 2011). Data preprocessing and advanced multivariate data analysis techniques become crucial in analyzing high-dimensional data and building accurate analysis pipeline for applications in PA. The accumulated massive amount of agricultural data will require robust information technology (IT) infrastructure and complicated data-analyzing algorithms to develop weather forecast models, yield prediction algorithms, and offer decision supporting tools to farmers. For example, John Deere's FarmSight and Pioneer's Field360 are among the tools that provide detailed management prescriptions based on the multiple sources of data collected including those by farmers from their farm equipment.

14.4.1 DATA ANALYTICS USING MACHINE LEARNING

Machine learning is a scientific and engineering discipline based on programming computers to optimize a performance criterion using example data or previous experience (Alpaydin 2004). Machine learning serves as a proxy for handling complex data gathered across large spatial and temporal scales. The kernel machine method developed over the last decade is a powerful machine-learning technique that has found wide applications in biological and agricultural studies (Cui et al. 2014a,b; Mirik et al. 2014a,b). To date, the most popular machine-learning techniques widely used by researchers and industry teams include support vector machine (SVM), relevance vector machine (RVM), and artificial neural networks (ANN).

We focus our discussion on SVM and RVM in this section because of their superior performances and great adaptability. SVM is a popular kernel machine method used in bioinformatic and multispectral and hyperspectral image analysis (Bazi and Melgani 2006; Huang et al. 2002; Melgani and Bruzzone 2004). In a typical machine-learning paradigm, a variable of interest is treated as one dimension in a multidimensional input space, termed a feature. Take remote sensing data analysis for example; each spectral band could be treated as a feature and the corresponding reflectance value is the feature value. Each pixel (sample) associated with a specific geospatial location will be represented by its reflectance values (feature values) of different spectral bands. Figure 14.5 illustrates a simple scenario of a two-dimensional–two-class classification problem using SVM. The greatest advantage of SVM classification is that once the classification model has been constructed, only the samples that lie on the margin (called support vectors [SVs]) determine the position of the decision boundary, which provides the maximum margin that separates those two classes with minimized measures of errors.

The major trend of machine learning has led its way into the application of Gaussian processes. RVM based on probabilistic predictions introduced by Tipping (2001) is regarded as a successful alternative to SVM. RVM utilizes fewer kernel functions and generates fewer relevance vectors (RVs) compared with the kernel functions and SV generated by SVM (Cui et al. 2014b; Tipping 2001). Therefore, the classification processes could be finished much faster than SVM. In terms of classification accuracy, performances of RVM are similar to SVM (Demir and Ertürk 2007; Psorakis et al. 2010).

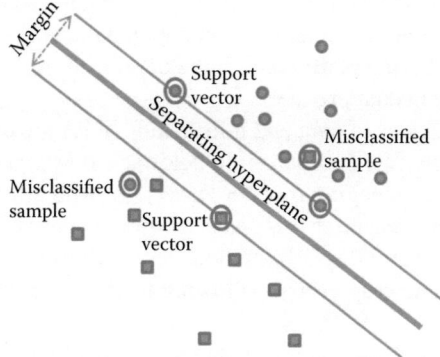

FIGURE 14.5 Linear support vector machine with two-dimensional input space and two classes.

14.4.2 APPLICATION OF BIG DATA AND ALGORITHMS IN AGRICULTURE

A large amount of agronomic studies have applied data mining approaches on big data sets (Mucherino et al. 2009). Different machine-learning methods have been widely used in detecting biotic stresses in agricultural production, including weed encroachment (Ahmed et al. 2012; Longchamps et al. 2010; Lopez-Granados et al. 2008; Mirik et al. 2014a,b; Nieuwenhuizen et al. 2007), drought stress (Behmann et al. 2014), herbicide injury (Zhao et al. 2014), and plant diseases (Berdugo et al. 2014; Bock et al. 2010; Mahlein et al. 2012a,b; Rumpf et al. 2010). Gianpiero et al. (2014) conducted a study using discrete optimization procedures for automatic error correction on a large-scale data set in Italy and showed satisfactory results. Ananthara et al. (2013) proposed an improved crop yield prediction model called CRY, which used the bee hive clustering approach based on a big crop yield database (Crop Knowledge Base) in India, and compared CRY with other popular big-data analyzing algorithms. Likewise, another study conducted by Ramesh and Vardhan (2013) compared various data mining techniques such as K-Means, K-Nearest Neighbor, ANN, and SVM for crop yield prediction on a large data set obtained from 1965 to 2009 in the East Godavari District of Andhra Pradesh in India. Most recently, Holman et al. (2014) conducted a study using both Gaussian process models and ANN to estimate reference evapotranspiration (ET) for irrigation management in the Texas High Plains. The entire data set consists of daily reference ET and other critical climatic variable data over a period of 10 years (2001–2010) across 15 different ET network stations and five national weather service weather stations in the Texas High Plains. The results indicated significant predicting accuracy of reference ET using sophisticated machine-learning algorithms that could be used for irrigation management. In another study, SVM-based classification and regression approaches were both applied to surface water quality monitoring and analysis based on a data set collected from 1500 water samples representing 10 different sites over 10 years (Singh et al. 2011). The results indicated adequacy of constructed models and great predictive capacities.

14.5 SUMMARY AND OUTLOOK

PA has been increasingly dependent on collecting, analyzing, and utilizing data and information for optimal management practices. Various sensors along with platforms have been developed to efficiently and effectively collect spatial data and information of soil properties, crop growth status, crop yield, and environmental factors. With the increasing capabilities of hardware and software of these sensors, extremely large volumes of data and information are being collected, archived, and processed. Effectively and efficiently storing, processing, and analyzing big data for site-specific application and management is becoming one of the greatest challenges facing PA. Additionally, big data acquisition can also cause data ownership controversial issues. Who owns the data? How will the data be used? Proprietary data sharing rules and the law and unintended infringement are premature. However, with the advancement of computational science and innovation of PA, new management tools and data sharing platforms will be developed for the agricultural community to share big data, while protecting data privacy.

The spatial and temporal scales of data and information in PA are continuing to increase thanks to technological advancements. We are managing fields at a finer scale, eventually plant-by-plant, or even at subplant levels (i.e., managing roots or leaves), incorporating data and information from advanced sensors. Field management at this scale requires intensive data and information input. Decision support systems incorporating information from various sensors enable continuous monitoring the conditions of soil and crop growth, allowing farmers to adjust site-specific management at any time.

An integrated and systematic approach of data management and field operations is becoming a trend in PA. We expect more cloud-based data processing to increase cost effectiveness, agility, productivity, and scalability. At the same time, systematic approaches on data integration is becoming urgent and requires collaboration across different government agencies, research institutions, agribusiness companies, and farmers. New methods of data analytics such as machine learning are becoming popular. Computer technologies continue to evolve. Moore's law continues to predict the future storage and computational capabilities of computers. This eventually allows instantaneous decision support based on big data and information and on-the-fly site-specific management and application. High-speed wireless Internet allows virtually all devices and equipment connected for integrated information transfer, on-the-go data processing, and real-time decision support, leading to much more efficient farming operations.

It is also noteworthy that despite technological advancements in computer science, engineering, sensing technologies, and advanced analytics tools, the science involved in PA is still lagging behind the requirements of the technologies. The advancement of technology has to be integrated with agronomic background science to realize the full potential of PA. More interdisciplinary and innovative studies are needed to enhance the understanding of the interactions among crop growth, soil, and environmental factors. Based on this better understanding, more effective management of the soil and crops can further improve agricultural production while protecting the environment.

REFERENCES

Adamchuk, V.I., J.W. Hummel, M.T. Morgan, and S.K. Upadhyaya. 2004. On-the-go soil sensors for precision agriculture. *Computers and Electronics in Agriculture* 44:71–91.

Ahmed, F., H.A. Al-Mamun, H.A.S.M. Bari, E. Hossain, and P. Kwan. 2012. Classification of crops and weeds from digital images: A support vector machine approach. *Crop Protection* 40:98–104.

Alpaydin, E. 2004. *Introduction to Machine Learning (Adaptive Computation and Machine Learning)*, Cambridge, MA: MIT Press.

Ananthara, M.G., T. Arunkumar, and R. Hemavathy. 2013. CRY–An improved crop yield prediction model using bee hive clustering approach for agricultural data sets. *IEEE International Conference on Pattern Recognition, Informatics and Medical Engineering*, 473–478.

Andujar, D., A. Escola, J.R. Rosell-Polo, C. Fernandez-Quintanilla, and J. Dorado. 2013. Potential of a terrestrial LiDAR-based system to characterize weed vegetation in maize crops. *Computers and Electronics in Agriculture* 92:11–15.

Bausch, W.C., K. Diker, R. Khosla, and J.F. Paris. 2004. Estimating corn nitrogen status using ground-based and satellite multispectral data. In *Optical Science and Technology, the SPIE 49th Annual Meeting*, 489–498. International Society for Optics and Photonics.

Bazi, Y., and F. Melgani. 2006. Toward an optimal SVM classification system for hyperspectral remote sensing images. *IEEE Transactions of Geosciences and Remote Sensing* 44:3374–3385.

Behmann, J., J. Steinrücken, and L. Plümer. 2014. Detection of early plant stress responses in hyperspectral images. *ISPRS Journal of Photogrammetry and Remote Sensing* 93:98–111.

Berdugo, C., R. Zito, S. Paulus, and A.K. Mahlein. 2014. Fusion of sensor data for the detection and differentiation of plant diseases in cucumber. *Plant Pathology* doi: 10.1111/ppa.12219.

Bezdek, J.C. 1981. *Pattern Recognition with Fuzzy Objective Function Algorithms*. New York: Plenum Press.

Blackmore, B.S., and C.J. Marshall. 1996. Yield mapping: Errors and algorithms. In *Precision Agriculture*, ed. P.C. Robert, R.H. Rust, and W.E. Larson, 403–415. Madison, WI: ASA, CSSA, and SSSA.

Blackmore, B.S. 2000. The interpretation of trends from multiple yield maps. *Computer and Electronics in Agriculture* 26:37–51.

Bock, C.H., G.H. Poole, P.E. Parker, and T.R. Gottwald. 2010. Plant disease severity estimated visually, by digital photography and image analysis, and by hyperspectral imaging. *Critical Reviews in Plant Sciences* 29:59–107.

Bronson, K.F., J.W. Keeling, J.D. Booker, T.T. Chua, T.A. Wheeler, R.K. Boman, and R.J. Lascano. 2003. Influence of landscape position, soil series, and phosphorus fertilizer on cotton lint yield. *Agronomy Journal* 95:949–957.

Bullock, D.S., and D.G. Bullock. 2000. Economic optimality of input application rates in precision farming. *Precision Agriculture* 2:71–101.

Camilli, A., C.E. Cugnasca, A.M. Saraiva, A.R. Hirakawa, and P.L.P. Correa. 2007. From wireless sensors to field mapping: Anatomy of an application for precision agriculture. *Computers and Electronics in Agriculture* 58:25–36.

Clay, D.E., J. Change, D.D. Malo, C.G. Carlson, C. Reese, S.A. Clay, M. Ellsbury, and B. Berg. 2001. Factors influencing spatial variability of soil apparent electrical conductivity. *Communications in Soil Science and Plant Analysis* 32:2993–3008.

Clevers, J.G.P.W., and H.J.C. Van Leeuwen. 1996. Combined use of optical and microwave remote sensing data for crop growth monitoring. *Remote Sensing of Environment* 561:42–51.

Corwin, D.L., and S.M. Lesch. 2003. Application of soil electrical conductivity to precision agriculture: Theory, principles, and guidelines. *Agronomy Journal* 95:455–471.

Corwin, D.L., S.M. Lesch, P.J. Shouse, R. Soppe, and J.E. Ayars. 2003. Identifying soil properties that influence cotton yield using soil sampling directed by apparent soil electrical conductivity. *Agronomy Journal* 95:352–364.

Cui, S., E. Youn, J. Lee, and S.J. Maas. 2014a. An improved systematic approach to predicting transcription factor target genes using support vector machine. *PLoS ONE* 9(4):e94519. doi:10.1371/journal.pone.0094519.

Cui, S., N. Rajan, S.J. Maas, and E. Youn. 2014b. An automated soil line identification method using relevance vector machine. *Remote Sensing Letters* 5:175–184.

Curran, P.J., P.M. Atkinson, G.M. Foody, and E.J. Milton. 2000. Linking remote sensing, land cover and disease. *Advances in Parasitology* 47:37–80.

Darby, H., and J. Lauer. 2000. Plant physiology: Critical stages in the life of a corn plant. http://corn.agronomy.wisc.edu/Management/pdfs/CriticalStages.pdf (accessed October 1, 2014).

Daughtry, C.S.T., C.L. Walthall, M.S. Kim, E.B. De Colstoun, and J.E. McMurtrey III. 2000. Estimating corn leaf chlorophyll concentration from leaf and canopy reflectance. *Remote Sensing of Environment* 74:229–239.

Delegido, J., J. Verrelst, C.M. Meza, J.P. Rivera, L. Alonso, and J. Moreno. 2013. A red-edge spectral index for remote sensing estimation of green LAI over agroecosystems. *European Journal of Agronomy* 46:42–52.

Demir, B., and S. Ertürk. 2007. Hyperspectral image classification using relevance vector machines. *IEEE Geoscience and Remote Sensing Letters* 4:586–590.

Dobermann, A., J.L. Ping, V.I. Adamchuk, G.C. Simbahan, and R.B. Ferguson. 2003. Classification of crop yield variability in irrigated production fields. *Agronomy Journal* 95:1105–1120.

Doerge, T.A. 1999. Site-specific management guidelines: Management zone concepts. http://www.ipni.net/publication/ssmg.nsf/0/C0D052F04A53E0BF852579E500761AE3/$FILE/SSMG-02.pdf (accessed October 1, 2014).

Eitel, J.U., T.S. Magney, L.A. Vierling, T.T. Brown, and D.R. Huggins. 2014. LiDAR based biomass and crop nitrogen estimates for rapid, non-destructive assessment of wheat nitrogen status. *Field Crops Research* 159:21–32.

Fan, W., and A. Bifet. 2012. Mining big data: Current status, and forecast to the future. *ACM SIGKDD Explorations Newsletter* 14:1–5.

FAO. 2011. The state of the world's land and water resources for food and agriculture (SOLAW)–Managing systems at risk. Rome: Food and Agriculture Organization of the United Nations.

Foth, H.D., and B.G. Ellis. 1988. *Soil Fertility*. New York: Wiley.

Fridgen, J.J., N.R. Kitchen, K.A. Sudduth, S.T. Drummond, W.J. Wiebold, and C.W. Fraisse. 2004. Management Zone Analyst (MZA): Software for subfield management zone delineation. *Agronomy Journal* 96:100–108.

Galzki, J.C., A.S. Birr, and D.J. Mulla. 2011. Identifying critical agricultural areas with three-meter LiDAR elevation data for precision conservation. *Journal of Soil and Water Conservation Society* 66:423–430.

Gianpiero, B., B. Renato, and R. Alessandra. 2014. Balancing of agricultural census data using discrete optimization. *Optimization Letters* 8:1553–1565.

Gitelson, A.A., and M.N. Merzlyak. 1996. Signature analysis of leaf reflectance spectra: Algorithm development for remote sensing of chlorophyll. *Journal of Plant Physiology* 148:494–500.

Goovaerts, P. 1998. Geostatistical tools for characterizing the spatial variability of microbiological and physicochemical soil properties. *Biology and Fertility of Soils* 27:315–334.

Griffin, T.W. 2009. Farmers' use of yield monitors. http://www.uaex.edu/publications/pdf/FSA-36.pdf (accessed October 1, 2014).

Guo, W., S.J. Maas, and K.F. Bronson. 2012. Relationship between cotton yield and soil electrical conductivity, topography, and Landsat imagery. *Precision Agriculture* 13:678–692.

Holman, D., M. Sridharan, P. Gowda et al. 2014. Gaussian process models for reference ET estimation from alternative meteorological data sources. *Journal of Hydrology* 517:28–35.

Huang, C., L.S. Davis, and J.R.G. Townshend. 2002. An assessment of support vector machines for land cover classification. *International Journal of Remote Sensing* 23:725–749.

Huggins, D.R., and R.D. Alderfer. 1995. Yield variability within a long-term corn management study: Implications for precision farming. In *Site-Specific Management for Agricultural Systems*, ed. P.C. Robert, R.H. Rust, and W.E. Larson, 417–426. Madison, WI: ASA, CSSA, SSSA.

Hwang, J., C. Shin, and H. Yoe. 2010. Study on an agricultural environment monitoring server system using wireless sensor networks. *Sensors* 10:11189–11211.

Iqbal, J., J.J. Read, A.J. Thomasson, and J.N. Jenkins. 2005. Relationships between soil–landscape and dryland cotton lint yield. *Soil Science Society America Journal* 69:872–882.

Irmak, S., J.O. Payero, B. VandeWalle, J. Rees, and G. Zoubek. 2014. Principles and operational characteristics of watermark granular matrix sensors to measure soil water status and its practical applications for irrigation management in various soil textures. University of Nebraska Extension EC783.

Isaaks, E., and R. Srivastava. 1989. *An Introduction to Applied Geostatistics*. New York: Oxford University Press.

Johnson, C.K., J.W. Doran, H.R. Duke, B.J. Wienhold, K.M. Eskridge, and J.F. Shanahan. 2001. Field scale electrical conductivity mapping for delineating soil conditions. *Soil Science Society America Journal* 65:1829–1837.

Kaspar, T.C., T.S. Colvin, D.B. Jaynes et al. 2003. Relationship between six years of corn yields and terrain attributes. *Precision Agriculture* 4:87–101.

Kelly, M., and R. Guo. 2007. Integrated agricultural pest management through remote sensing and spatial analyses. In *General Concepts in Integrated Pest and Disease Management*. ed. A. Ciancio and K.G. Mukerji, 191–207. Netherlands: Springer.

Khosla, R. 2010. Precision agriculture: Challenges and opportunities in a flat world. *Proceedings of the 19th World Congress of Soil Science, Soil Solutions for a Changing World*, 1–6 August 2010, Brisbane, Australia.

Kim, Y., R.G. Evans, and W.M. Iversen. 2008. Remote sensing and control of an irrigation system using a distributed wireless sensor network. *IEEE Transactions on Instrumentation and Measurement* 57:1379–1387.

Kitchen, N.R., D.F. Hughes, K.A. Sudduth, and S.J. Birrel. 1995. Comparison of variable rate to single rate nitrogen fertilizer applications: Corn production and residual soil NO_3-N. In *Proceedings of Site-Specific Management for Agricultural Systems*, ed. P.C. Robert, 427–439. Madison, WI: ASA, CSSA, SSSA.

Kranz, W.L., S. Irmak, S.J. van Donk, C.D. Yonts, and D.L. Martin. 2008. Irrigation management for corn. NebGuide G1850. University of Nebraska Extension.

Kravchenko, A.N., and D.G. Bullock. 2000. Correlation of corn and soybean grain yield with topography and soil properties. *Agronomy Journal* 92:75–83.

Li, H., R.J. Lascano, J. Booker, T. Wilson, and K.F. Bronson. 2001. Cotton lint yield variability in a heterogeneous soil on a landscape-scale. *Soil & Tillage Research* 58:245–258.

Liaghat, S., and S.K. Balasundram. 2010. A review: The role of remote sensing in precision agriculture. *American Journal of Agricultural and Biological Sciences* 5:50–55.

Longchamps, L., B. Panneton, G. Samson, G. Leroux, and R. Thériault. 2010. Discrimination of corn, grasses and dicot weeds by their UV-induced fluorescence spectral signature. *Precision Agriculture* 11:181–197.

Lopez-Granados, F., J.M. Pena-Barragan, M. Jurado-Exposita, M. Francisco-Fernandez, R. Cao, and A. Alosno-Betanzos. 2008. Multispectral classification of grass weeds and wheat (*Triticum durum*) using linear and nonparametric functional discriminant analysis and neural networks. *Weed Research* 48:28–37.

Lund, E.D., C.D. Christy, and P.E. Drummond. 2000. Using yield and soil electrical conductivity (EC) maps to derive crop production performance information. *Proceedings of the 5th International Conference on Precision Agriculture*, Bloomington, MN, July 16–19, 2000.

Maas, S.J. 1998. Estimating cotton ground cover from remotely sensed scene reflectance. *Agronomy Journal* 90:384–388.

Machado, S., E.D. Bynum, Jr., T.L. Archer et al. 2002. Spatial and temporal variability of sorghum grain yield: Influence of soil, water, pests, and disease relationships. *Precision Agriculture* 3:389–406.

Mahlein, A.K., E.C. Oerke, U. Steiner, and H.W. Dehne. 2012a. Recent advances in sensing plant diseases for precision crop protection. *European Journal of Plant Pathology* 133:197–209.

Mahlein, A.K., U. Steiner, C. Hillnhütter, H.W. Dehne, and E.C. Oerke. 2012b. Hyperspectral imaging for small-scale analysis of symptoms caused by different sugar beet diseases. *Plant Methods* 8:3, doi:10.1186/1746-4811-8-3.

Mahler, R.L., D.F. Benzdicek, and R.E. Witters. 1979. Influence of slope position on nitrogen fixation and yield of dry peas. *Agronomy Journal* 71:348–354.

Maleki, M.R., A.M. Mouazen, B.D. Ketelaere, H. Ramon, and J.D. Baerdemaeker. 2008. On-the-go variable-rate phosphorus fertilisation based on a visible and near-infrared soil sensor. *Biosystems Engineering* 99:35–46.

McBratney, A., B. Whelan, T. Ancev, and J. Bouma. 2005. Future directions of precision agriculture. *Precision Agriculture* 6:7–23.

McKinion, J.M., J.N. Jenkins, D. Akins et al. 2001. Analysis of a precision agriculture approach to cotton production. *Computers and Electronics in Agriculture* 32:213–228.

Melgani, F., and L. Bruzzone. 2004. Classification of hyperspectral remote sensing images with support vector machines. *IEEE Transactions of Geosciences and Remote Sensing* 42:1778–1790.

Miller, M.P., M.J. Singer, and D.R. Nielsen. 1988. Spatial variability of wheat yield and soil properties on complex hill. *Soil Science Society of America Journal* 52:1133–1141.

Mirik, M., R.J. Ansley, K. Steddom et al. 2014a. High spectral and spatial resolution hyperspectral imagery for quantifying Russian wheat aphid infestation in wheat using the constrained energy minimization classifier. *Journal of Applied Remote Sensing* 8(1):083661.

Mirik, M., Y. Emendack, A. Attia et al. 2014b. Detecting musk thistle (*Carduus nutans*) infection using a target recognition algorithm. *Advances in Remote Sensing* 3:95–105.

Moran, M.S., Y. Inoue, and E.M. Barnes. 1997. Opportunities and limitations for image-based remote sensing in precision crop management. *Remote Sensing of Environment* 61:319–346.

Mucherino, A., P. Papajorgji, and P.M. Pardalos. 2009. A survey of data mining techniques applied to agriculture. *Operational Research* 9:121–140.

Mulla, D.J. 2013. Twenty five years of remote sensing in precision agriculture: Key advances and remaining knowledge gaps. *Biosystems Engineering* 114:358–371.

Mzuku, M., R. Khosla, R. Reich, D. Inman, F. Smith, and L. MacDonald. 2005. Spatial variability of measured soil properties across site-specific management zones. *Soil Science Society of America Journal* 69:1572–1579.

Nair, S., S.J. Maas, C. Wang, and S. Mauget. 2012. Optimal field partitioning for center-pivot-irrigated cotton in the Texas High Plains. *Agronomy Journal* 105:124–133.

National Academy of Sciences. 1997. *Precision Agriculture in the 21st Century: Geospatial and Technologies in Crop Management*. Washington, DC: National Academy Press.

Nieuwenhuizen, A.T., L. Tang, J.W. Hofstee, J. Müller, and E.J. Van Henten. 2007. Colour based detection of volunteer potatoes as weeds in sugar beet fields using machine vision. *Precision Agriculture* 8:267–278.

Oliver, M.A. 2010. An overview of geostatistics and precision agriculture. In *Geostatistical Applications for Precision Agriculture*, ed. M.A. Oliver, 1–34. New York: Springer.

Ortiz, B.V., J. Shaw, J.P. Fulton, and A. Winstead. 2011. Management zones II–Basic steps for delineation. Precision Agriculture Series–Timely Information. Agriculture, Natural Resources & Forestry. Alabama Cooperative Extension System.

Pacheco, A., A. Bannari, J.C. Deguise, H. McNairn, and K. Staenz. 2001. Application of hyperspectral remote sensing for LAI estimation in precision farming. *Proceedings of 23rd Canadian Symposium on Remote Sensing*, 281–287.

Papadavid, G., D. Fasoula, M. Hadjimitsis, P.S. Perdikou, and D.G. Hadjimitsis. 2013. Image based remote sensing method for modeling black-eyed beans (Vigna unguiculata) Leaf Area Index (LAI) and Crop Height (CH) over Cyprus. *Central European Journal of Geosciences* 5:1–11.

Pierce, F.J., and P. Nowak. 1999. Aspects of precision agriculture. *Advances in Agronomy* 67:1–85.

Pierce, F.J., and T.V. Elliott. 2008. Regional and on-farm wireless sensor networks for agricultural systems in Eastern Washington. *Computers and Electronics in Agriculture* 61:32–43.

Ping, J.L., and A. Dobermann. 2005. Processing of yield map data. *Precision Agriculture* 6:193–212.

Plant, R.E., A. Mermer, G.S. Pettygrove et al. 1999. Factors underlying grain yield spatial variability in three irrigated wheat fields. *Transactions of the ASABE* 42:1187–1202.

Plant, R.E. 2001. Site-specific management: The application of information technology to crop production. *Computers and Electronics in Agriculture* 30:9–29.

Prabhakar, M., Y.G. Prasad, and M.N. Rao. 2012. Remote sensing of biotic stress in crop plants and its applications for pest management. In *Crop Stress and its Management: Perspectives and Strategies*, ed. B. Venkateswarlu, A.K. Shanker, C. Shanker, and M. Maheswari, 517–545. Netherlands: Springer.

Psorakis, I., T. Damoulas, and M.A. Girolami. 2010. Multiple relevance vector machines: Sparsity and accuracy. *IEEE Transaction on Neural Networks* 21:1588–1598.

Qin, Z., and M. Zhang. 2005. Detection of rice sheath blight for in-season disease management using multispectral remote sensing. *International Journal of Applied Earth Observation and Geoinformation* 7:115–128.

Rajan, N., and S.J. Maas. 2009. Mapping crop ground cover using airborne multispectral digital imagery. *Precision Agriculture* 10:304–318.

Rajan, N., N. Puppala, S. Maas, P. Payton, and R. Nuti. 2014. Aerial remote sensing of peanut ground cover. *Agronomy Journal* 106:1358–1364.

Ramesh, D., and B.V. Vardhan. 2013. Data mining techniques and applications to agricultural yield data. *International Journal of Advanced Research in Computer and Communication Engineering* 2:3477–3480.

Rhoades, J.D. 1993. Electrical conductivity methods for measuring and mapping soil salinity. *Advances in Agronomy* 49:201–251.

Robert, P.C., R.H. Rust, and W.E. Larson. 1995. *Site-Specific Management for Agricultural Systems*. Madison, WI: ASA, CSSA, and SSSA.

Robert, P.C., R.H. Rust, and W.E. Larson. 1996. *Precision Agriculture*. Madison, WI: ASA, CSSA, and SSSA.

Rumpf, T., A.K. Mahlein, U. Steiner, E.C. Oerke, H.W. Dehne, and L. Plümer. 2010. Early detection and classification of plant diseases with support vector machines based on hyperspectral reflectance. *Computers and Electronics in Agriculture* 74:91–99.

Sakamoto, T., M. Yokozawa, H. Toritani, M. Shibayama, N. Ishitsuka, and H. Ohno. 2005. A crop phenology detection method using time-series MODIS data. *Remote Sensing of Environment* 96:366–374.

Schueller, J.K. 2010. Geostatistics and precision agriculture: A way forward. In *Geostatistical Application for Precision Agriculture*, ed. M.A. Oliver, 305–312. New York: Springer.

Setiyono, T.D., K.G. Cassman, J.E. Specht et al. 2010. Simulation of soybean growth and yield in near-optimal growth conditions. *Field Crops Research* 119:161–174.

Sickle, J.V. 2001. *GPS for Land Surveyors*, 2nd Edition. New York: Taylor & Francis.

Simbahan, G.C., A. Dobermann, and J.L. Ping. 2004. Screening yield monitor data improves grain yield maps. *Agronomy Journal* 96:1091–1102.

Sinai, G., D.Z. Zaslavsky, and P. Golany. 1981. The effect of soil surface curvature on moisture and yield–Beer Sheba observation. *Soil Science* 132:367–375.

Singh, K.P., N. Basant, and S. Gupta. 2011. Support vector machines in water quality management. *Analytica Chimica Acta* 703:152–162.

Specht, J.E., J.H. Williams, and C.J. Weidenbenner. 1986. Differential responses of soybean genotypes subjected to a seasonal soil water gradient. *Crop Science* 26:922–934.

Stroup, W.W., P.S. Baenzier, and D.K. Mulitze. 1994. Removing spatial variation from wheat yield trials: A comparison of methods. *Crop Science* 34:62–66.

Sudduth, K.A., and S.T. Drummond. 2007. Yield Editor: Software for removing errors from crop yield maps. *Agronomy Journal* 99:1471–1482.

Swinton, S.M., and J. Lowenberg-DeBoer. 1998. Evaluating the profitability of site specific farming. *Journal of Production Agriculture* 11:439–446.

Tamura, Y., M. Matsui, L.G. Pagnini, R. Ishibashi, and A. Yoshida. 2002. Measurement of wind-induced response of buildings using RTK-GPS. *Journal of Wind Engineering and Industrial Aerodynamics* 90:1783–1793.

Tang, H., M. Brolly, F. Zhao et al. 2014. Deriving and validating leaf area index (LAI) at multiple spatial scales through lidar remote sensing: A case study in Sierra National Forest, CA. *Remote Sensing of Environment* 143:131–141.

Thenkabail, P.S., I. Mariotto, M.K. Gumma, E.M. Middleton, D.R. Landis, and K.F. Huemmrich. 2013. Selection of hyperspectral narrowbands (HNBs) and composition of hyperspectral twoband vegetation indices (HVIs) for biophysical characterization and discrimination of crop types using field reflectance and Hyperion/EO-1 data. *IEEE Journal of Selected Topics in Applied Earth Observations and Remote Sensing* 6:427–439.

Thylen, L., P. Jurschik, and D.P.L. Murphy. 1997. Improving the quality of yield data. In *Precision Agriculture: Proceedings of the 1st European Conference on Precision Agriculture*, ed. J.V. Stafford, 743–750. Oxford, UK: BIOS Scientific Publishers Ltd.

Tipping, M.E. 2001. Sparse Bayesian learning and the relevance vector machine. *Journal of Machine Learning Research* 1:211–244.

Tiwari, M., and B. Misra. 2011. Application of cluster analysis in agriculture–A review article. *International Journal of Computer Applications* 36:43–47.

Tobler, W.R. 1970. A computer model simulation of urban growth in the Detroit region. *Economic Geography* 46:234–240.

Torrion, J.A., T.D. Setiyono, K.G. Cassman, and J.E. Specht. 2011. Soybean phenology simulation in the North-Central USA. *Agronomy Journal* 103:1661–1667.

Torrion, J.A., S.J. Maas, W. Guo, J.P. Bordovsky, and A.M. Cranmer. 2014. A three-dimensional index for characterizing crop water stress. *Remote Sensing* 6:4025–4042.

USDA-NRCS. 2010. Precision nutrient management planning. Agronomy Technical Note No. 3, February 2010.

Vachaud, G., A. Passerat de Silans, P. Balabanis, and M.Vauclin. 1985. Temporal stability of spatially measured soil water probability density function. *Soil Science Society of America Journal* 49:822–828.

Vellidis, G., M. Tucker, C. Perry, C. Kvien, and C. Bednarz. 2008. A real-time wireless smart sensor array for scheduling irrigation. *Computers and Electronics in Agriculture* 61:44–50.

Webster, R., and M.A. Oliver. 2001. *Geostatistics for Environmental Scientists*. New York: John Wiley & Sons, Ltd.

Whelan, B.M., and A.B. McBratney. 2000. The "null hypothesis" of precision agriculture management. *Precision Agriculture* 2:265–279.

Wibawa, W.D., D.L. Dludlu, L.J. Swenson, D.G. Hopkins, and W.C. Dahnke. 1993. Variable fertilizer application based on yield goal, soil fertility, and map unit. *Journal of Production Agriculture* 6:255–261.

Wilson, J.P., and J.C. Gallant. 2000. Digital terrain analysis. In *Terrain Analysis: Principles and Applications*, ed. J.P. Wilson, and J.C. Gallant, 1–28. New York: John Wiley & Sons, Ltd.

Wise, K., B. Johnson, C. Mansfield, and C. Krupke. 2011. Managing wheat by growth stage. Purdue Ext. 422. Purdue University, IN. https://www.extension.purdue.edu/extmedia/ID/ID-422.pdf (accessed October 1, 2014).

World Resources Institute. 2013. *Creating a Sustainable Food Future: Interim Findings*. Washington, DC: World Resources Institute.

Zhang, M., Z. Qin, X. Liu, and S.L. Ustin. 2003. Detection of stress in tomatoes induced by late blight disease in California, USA, using hyperspectral remote sensing. *International Journal of Applied Earth Observation and Geoinformation* 4:295–310.

Zhao, D., T. Yang, and S. An. 2012. Effects of crop residue cover resulting from tillage practices on LAI estimation of wheat canopies using remote sensing. *International Journal of Applied Earth Observation and Geoinformation* 14:169–177.

Zhao, F., Y. Huang, Y. Guo, K.N. Reddy et al. 2014. Early detection of crop injury from glyphosate on soybean and cotton using plant leaf hyperspectral data. *Remote Sensing* 6:1538–1563.

Zheng, G., and L.M. Moskal. 2009. Retrieving leaf area index (LAI) using remote sensing: Theories, methods and sensors. *Sensors* 94:2719–2745.

Zikopoulos, P., C. Eaton, D. deRoos, T. Deutsch, and G. Lapis. 2011. *Understanding Big Data: Analytics for Enterprise Class Hadoop and Streaming Data*. New York: McGraw-Hill Companies, Inc.

15 Role of Soil-Specific Farming in Converting Blue Water into Green Water

Jianbin Lai, Hailong Yu, and Henry Lin

CONTENTS

15.1 INTRODUCTION

Food and water security and sustainable development have been a major concern around the world. However, the results of worldwide studies of land and water use, along with their ecological impacts, lead to three constraints that will narrow the range of possible alternatives to increase food production sufficiently to meet global demand (Sposito 2013): (1) land conversion for crop cultivation is nearing its planetary limit, (2) use of blue water (i.e., liquid water in rivers, lakes, wetlands, and aquifers) by farmlands is also nearing its planetary limit, and (3) most of the water consumed by farmlands is green water (i.e., soil water held in the unsaturated zone and available to plants). Among these three constraints, the first one is related to limited land availability, the second one is linked to limited agricultural water availability, and the third one reveals an important facet of cropland water use. Feeding an additional 3.3 billion people by 2050 and at the same time eradicating malnutrition would require an additional 5600 km^3 per year in consumptive water use (Falkenmark and Rockström 2004). However, where this water will come from is hardly clear. It is becoming increasingly clear that only if we improve water use efficiency in farming can we hope to meet the intense challenge ahead (Molden 2007).

Water scarcity has become a major threat to sustainable development including global food security. Sustainable and economical farming requires precise adaptation to local conditions including soil and hydrology. In this chapter, various field measures, including rain water harvesting, mulching, and tillage, are discussed to enhance the conversion of blue water (rainwater and irrigation water) into green water (water stored in soils and utilized by plants). The application of soil-specific irrigation (termed precision irrigation) is also examined to develop effective means of increasing irrigation efficiency. The soil-specific farming measures, through which water scarcity should be intimately considered, show significant potential in meeting the grand challenge of global food and water security.

Water is the bloodstream of the biosphere on Earth. Among the total continental precipitation of 110,305 km^3 per year, two-thirds is involved in biomass production in terrestrial ecosystems (i.e., green water) and another one-third goes to rivers, lakes, wetlands, and eventually runs into the sea (i.e., blue water) (Rockström 1999). The blue water flow can be divided in two flow components: the potentially accessible but presently uncommitted flow and the nonutilizable flow. The uncommitted but potentially utilizable blue water flow amounts to 9% of the total freshwater cycle, while the nonutilizable blue water flow represents 26% of global freshwater resources. Accessible blue water constitutes 11% of the terrestrial freshwater flow. Humans withdraw 3.5% of rainfall on land, of which 45% is nonconsumptive use and returns to the river afterward. During the peak of the Green Revolution (1967–1982), average growth rates of cereal yields attained an impressive 2.9% per year. This was achieved through a blue water supply for irrigation at all scales, from the adoption of small petrol pumps to lift water from rivers and wells to large-scale irrigation schemes. In a sense, the Green Revolution was thus a blue revolution, because each increase in crop yields required more consumptive crop water use (Falkenmark and Rockström 2004). The degrees of freedom in increasing human blue water withdrawals are limited by the ceiling of realistic withdrawals of 12,500 km^3 per year. Among them, 40% (5000 km^3/year) of the stable runoff has been suggested to be secured for terrestrial and aquatic ecosystems, plus global withdrawal amount of 4000 km^3/year at the present time. This leaves us with a remaining blue water flow of 3500 km^3/year (Falkenmark and Rockström 2004). This is the additional blue water flow that can be theoretically appropriated in the future.

A large portion of the water consumed by crops (an estimated 78%) comes directly from rainfall that infiltrates into the soil to generate soil moisture (green water). The other 22% is from surface and groundwater sources. Assuming an estimated delivery efficiency of 60%, 2630 km^3 is withdrawn each year from surface and groundwater sources and delivered to farm fields to provide the 1570 km^3 of crop water consumption at the global scale (De Fraiture and Wichelns 2010). The green water flow has two components: (1) the productive part, or transpiration (T), involved in biomass production in terrestrial ecosystems, and (2) the nonproductive part, or evaporation (E). Thus, it is important to ensure the maximum possibility of productive use of green water.

It is possible to meet the food security and sustainability challenges of the coming decades, but this will require considerable changes in water management (Rost et al. 2008; Mueller et al. 2012). Studies have shown that global yield variability is heavily controlled by irrigation, climate, and fertilizer use (Rost et al. 2008; Mueller et al. 2012). Large production increases are possible from closing yield gaps to 100% of attainable yields, but the changes to management practices needed to close yield gaps vary considerably by region.

From an agrohydrological perspective, there is enough rainfall even in semiarid and dry subhumid savanna agroecosystems to allow significantly increased yield levels (Rockström et al. 2007). The current low yields in semiarid agroecosystems are explained and manifested by on-farm blue water losses in terms of surface runoff, limited infiltration to the root zone, and percolation loss to groundwater, and by nonproductive green water loss (i.e., soil evaporation).

Based on a farm-level blue and green water balance analysis, Rockström et al. (2007) illustrated the gap between actual and potential crop yields (Figure 15.1). The upper limit of the productive flow of green water is 85%, as constrained by unavailable soil evaporation occurring at the beginning of the growing season. Under typical on-farm conditions, because combined rainfall losses to blue water flow as runoff into streams and deep percolation into ground water, green water availability is less than 40% and only 30% of the green water flow is productive, leading to a large green water loss and consequently a low yield (Sposito 2014). However, if all local rain could be put to use without any farm-level water losses, the potential maize yield could rise to 7.5 t/ha and it could be several times higher than the present yield.

Pala et al. (2011) showed the importance of improved soil and crop management practices in reducing the yield gap for wheat crops. They showed that wheat yields can be doubled in experimental or demonstration fields with runoff water-harvesting, supplemental irrigation, and increased efficiency of water use, along with improved agronomic management. In reality, among smallholder farmers, only a fraction of rainfall typically infiltrates, and only a small fraction of this water is taken up by crop, resulting in low on-farm crop yields. If green water availability drops to 20% because of scarcity, the maize yield would not improve even if 85% of the green water flow could be made productive.

On the other hand, green water scarcity may also have man-made origin, for instance through soil mismanagement on the farmer's field, causing disturbances in local rainfall partitioning (Falkenmark 2013). As illustrated in Figure 15.2, only a limited part of incoming rain is taken up and transformed into biomass, typically resulting in crop yields in 1 t/ha level only, far below the potential yield level for that particular hydroclimate. There is, however, enough water available for considerably higher yields, but it is not accessible to plants because of low infiltration, disturbed water holding capacity of the soil, and many other factors.

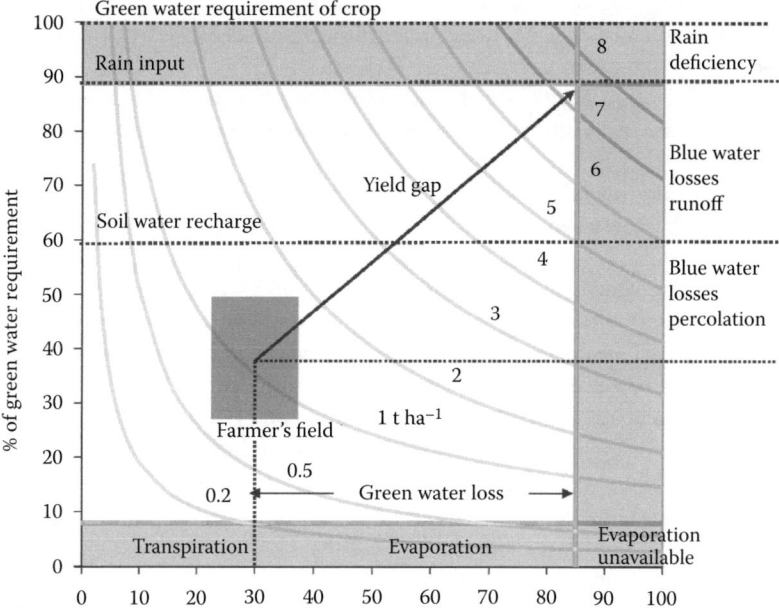

FIGURE 15.1 The yield gap and green water use efficiency. The Y-axis indicates the percentage of rainfall that infiltrates into soil and returns as green vapor flow and the X-axis is the percentage of transpiration of total vapor flow. The shaded square shows the observed hydrological range for on-farm rain-fed maize farming in sub-Saharan Africa, yielding an average of 1 t/ha. Isolines of equal yield indicate the potential yield with different combinations of flows in the on-farm water balance. (Adapted from Rockström, J. et al., *Proceedings of the National Academy of Sciences* 104: 6253–6260, 2007.)

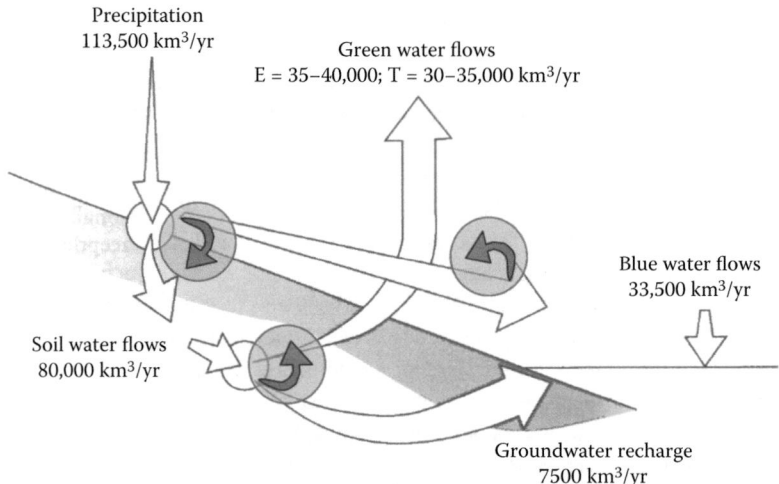

Precipitation
113,500 km³/yr

Green water flows
E = 35–40,000; T = 30–35,000 km³/yr

Blue water flows
33,500 km³/yr

Soil water flows
80,000 km³/yr

Groundwater recharge
7500 km³/yr

FIGURE 15.2 Overview of annual flow partitioning in the global terrestrial hydrologic cycle. E = evaporation volume from soil, T = vegetation transportation. (Adapted from Falkenmark, M. and J. Rockström, *Balancing Water for Humans and Nature: The New Approach in Ecohydrology.* Earthscan: London, 2004.)

Humans now alter global runoff flows through withdrawals of blue water and changes in green water flows, which affect water partitioning and moisture feedback. The partitioning points show the soil link in the hydrological cycle and the distinction between productive and nonproductive green water flows (Figure 15.2). Total green water flow amounts to 72,500 km³ per year, of which at least 50% is productive green water flow as T (L'vovich 1979). To address the conflict between tomorrow's food production needs and the need to save more water for ecosystems, two broad investment strategies are to (1) improve productivity in rainfed settings, and (2) increase production in irrigated areas.

The rainwater partitioning takes place at two different partitioning points: the upper partitioning point at the land surface between surface runoff and infiltration, and the lower partitioning point between root water uptake, evaporation from wet soil, and percolation to groundwater.

Irrigated agriculture receives blue water (from irrigation) as well as green water (from precipitation), while rainfed agriculture only receives green water. The availability of blue water flow is influenced by (1) soil surface conditions, which determine the partitioning between surface runoff and infiltration, (2) soil properties that influence water-holding capacity, (3) vegetation characteristics that determine the capacity to take up soil moisture, and (4) climatic conditions that influence the atmospheric thirst for green water. Therefore, human land use directly affects partitioning of rainfall at the land surface.

It is widely recognized that soils, landscapes, and water in the field vary within a farm. Farmers have often adapted their practices to such field variability. However, industrialized machinery operation has fundamentally changed the farm operation that generally treated large fields in a uniform way. In reality, neither the soil nor the crop is uniform within a field. Precision adjustments to site-specific soil and crop conditions in the field can address environmental concerns as well as ensure economic and high yields. This is the essential rationale for site-specific farming also called precision agriculture (Heege 2012).

Precision agriculture has emerged and developed as a way to apply the right treatment in the right place at the right time and with the right amount (Srinivasan 2006). In general, there are two major features for each specific site: the landscape setting and the soil conditions. Soil-specific farming is based on soil texture, organic matter content, nutrient level, soil structure, bulk density, and other soil properties to adopt reasonable irrigation scheme and other management practices according to local soil conditions (Larson and Robert 1991). Recent development in technology has allowed on-the-go sensing of various soil properties, which has stimulated a whole new approach in agricultural production, often referred to as farming by soils (FBS). Most of the FBS applications to date have been in the application of chemicals such as fertilizer and herbicides and have focused on crop yield and profitability (Tebrügge and Düring 1999; De la Rosa et al. 2009). The most obvious benefits from the FBS are higher net income or less environmental pollution due to better matching of inputs to the productivity potential of local soils. However, another area of agricultural production that would benefit from the FBS but that has been largely ignored is water use efficiency in agriculture. Munson and Runge (1990) suggested that FBS requires integrated soils information to help identify soil sensitivity to nutrient leaching or erosion and runoff.

Agricultural systems have never been strictly rainfed or irrigated around the world. Even if farmers master some level of irrigation technology, they are not operating under full irrigation, nor are they cultivating using just rainwater. Between irrigated and rainfed agriculture, farmers' reality has been that they have stored, mobilized, and applied water to plants through a variety of methods depending on the nature of resources available. Irrigated systems typically also use green as well as blue water, and rainfed systems sometimes also use blue in addition to green water even in the absence of formal irrigation systems. In a nutshell, farmers' coping strategies worldwide have always been to deal with a blue to green water continuum.

Figure 15.3 shows the scheme of converting blue water to green water as discussed in this chapter. On the right side, soil water storage is depleted by transpiration and evaporation. In rainfed agriculture, the partitioning of evaporation and transpiration takes on special significance. Rockström (2003) argued that much of evaporation can be transferred into crop transpiration, thus contributing to increased crop yield and increased water productivity. In an irrigation system, inputs include rainfall and surface and subsurface water supplies. Irrigation infrastructure is primarily built to provide water for crop transpiration, but in many irrigated areas, infrastructure also provides water for domestic and industrial uses and for fishing and livestock. In addition to the intended depletion by crop transpiration, water is also depleted by evaporation from weeds, trees, fallow land, and water bodies. Drainage water through percolation is sometimes directed to sinks such as aquifers. Other outflows can be recaptured for use. However, disentangling these various components or processes of the field scale hydrologic cycle demands a clear analytical framework of water accounting.

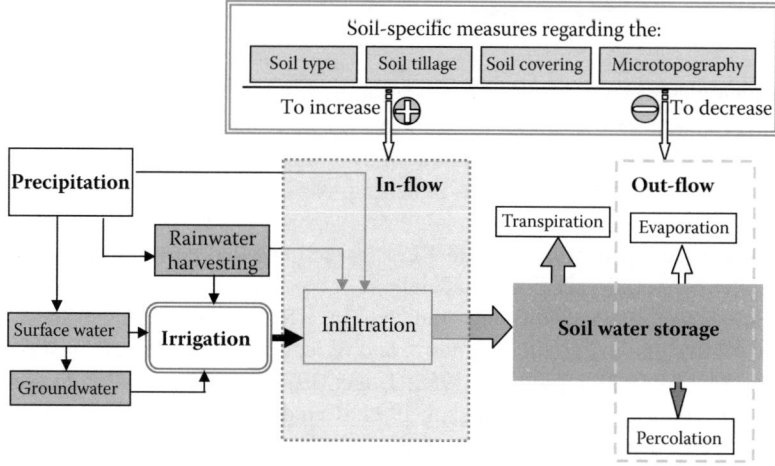

FIGURE 15.3 General scheme of converting blue water into green water discussed in this chapter.

15.2   SOIL-SPECIFIC FARMING: NONIRRIGATION FIELD MEASURES

15.2.1   RAINWATER HARVESTING

The basic principle of agriculture water harvesting is to capture precipitation falling on land. Water harvesting may occur naturally or through human intervention. Water harvesting system in dry area can provide water for domestic consumption, including drinking water and production of agriculture crops.

Rainwater harvesting encompasses methods to induce, collect, and store rainfall and runoff from various sources and for various purposes. The success or failure of rainwater harvesting depends on the quantity of water that can be harvested from an area under given climatic conditions. Moreover, the methods applied depend strongly on local conditions (Boers and Ben-Asher 1982).

The conventional water resource management has focused on blue water only. However, most rain goes back to the atmosphere through vapor flow, dominated by consumptive water use by vegetation. As suggested by Falkenmark and Rockström (2004), rainfed agriculture entails the increase in the use of rainfall that infiltrates into the soil and becomes accessible by plants to generate vapor flow in support of biomass growth. When analyzing food production, we need to incorporate this green water resource. Rainfed agriculture covers >95% of croplands in water-scarce tropical regions and will be the dominating source of food for growing populations in the tropics in the foreseeable future (Rockström 1999). Water harvesting is also an option for increasing the availability of water to crop in dry area. This increases the amount of water per unit cropping area, reduces the impact of drought, and uses runoff beneficially.

The most important soil properties relevant to water harvesting are texture, structure, depth, and infiltration. Selection of appropriate sites and suitable methods under the prevailing conditions are among the most important prerequisites for successful water harvesting systems. Although water harvesting can be implemented in all dry land areas, there are several parameters that should be considered in identifying areas suitable for water harvesting, such as the prevailing climatic characteristics, the topography of the land, the type of soils and vegetation, and agriculture production activities.

Table 15.1 illustrates some rainwater harvesting systems used in various regions of China. For example, Wu and Feng (2009) documented several rainwater harvesting techniques applied in the Loess Plateau, such as ridge and furrow planting, contour terrace, and slope terrace farming. These measures can significantly reduce runoff, increase the effective rainwater stored in soils, and thus increase crop yield per unit area.

15.2.2   SOIL COVER AND SURFACE TREATMENT

Reducing evaporation from soil surface while increasing productive transpiration can enhance the transfer from blue water to green water. Evaporation varies with agricultural practices (Burt et al. 2005) and could range from 4% to 15%–25% in sprinkle irrigation systems and up to 40% or more in rainfed systems (Molden 2007). The amount of evaporation strongly depends on soil properties and soil cover conditions. Practices such as mulching, plowing, or breeding for fast leaf expansion to shade the ground as rapidly as possible can reduce evaporation and increase productive consumption of green water (see examples in Table 15.2). Targeting rapid early growth to shade the soil surface and reduce evaporation is also beneficial.

Many studies showed a gravel and straw mulch to be effective in reducing evaporation and runoff, improving infiltration and soil temperature, and reducing soil erosion and salinization (Lamb and Chapman 1943; Hide 1954; Adams 1966; Unger 1971; Fairbourn 1973; Poesen and Lavee 1991; Kemper et al. 1994; Van Wesemael et al. 1995; Roundy et al. 1997; Li 2003). Gravel mulches have been used to conserve soil moisture in low rainfall regions around the world, such as France (Lamb and Chapman 1943), South Africa (Adams 1966), the United States (Fairbourn 1973), and

TABLE 15.1

Examples of Rainwater Harvesting Systems in Different Regions of China and Their Effects on Crop Yields

Region	Climate Type	Soil Type	Annual Rainfall (mm)	Pan Evaporation (mm)	Rain Harvesting Measures	Crop Type	Yield Increase (%)
Henan Huixian	Warm temperate continental monsoon	Cinnamon soil	589	1500	Straw mulch, deep tillage	Wheat, maize	17–41
Hebei Taihang Mt.	Semiarid continental monsoon	Cinnamon soil	400–600	1934	Biological crust, blasting ripper	Fruit tree	20–30
Sichuan Jianyang	Subtropical humid monsoon	Purple alluvial soil	684–859	(Unknown)	Covering, slope hedgerow	Rice, maize, fruit tree	26–34
Shaanxi Yan'an	Humid continental	Loess	539	(Unknown)	Level terrace, cistern	Fruit tree, cereal	28–33
Inner Mongolia Zunger	Semiarid continental	Chestnut soil, loess	143–637	(Unknown)	Ridge and furrow mulching, curing agent	Cereal, maize	5–119
Gansu Dingxi	Temperate semiarid	Sierozem	427	1526	Cistern	Wheat, maize	34–61
Ningxia Pengyang	Temperate semiarid	Loess, black clay soil	327	1360	Eyebrow terrace, semicircular bounds, contour infiltration ditch	Alfalfa	20–38

Source: Adopted from Wu, P., and H. Feng, *Rainwater Harvesting in China.* Huanghe Hydraulic Publisher (in Chinese).

Switzerland (Nachtergaele et al. 1998). Straw mulch also protects soil structure, prevents ground compaction, and cuts the capillary contact between surface soil and subsoil to reduce soil evaporation. In addition, straw can effectively improve soil organic matter content, soil fertility, soil aggregate structure, and soil infiltration rate.

15.2.3 Soil Tillage

Common high-intensity tillage systems in dryland farming include moldboard plowing to break apart hardened soil surface and surface disking and harrowing to reduce soil clod size and to control weeds. This can improve infiltration and the storage of water in soils. However, repeated tillage can also cause compaction in subsoils and thus restrict root growth and reduce actual transpiration. In comparison, reduced or no-tillage system can have a continuous increase in organic matter content and improvement in soil structure, leading to improved infiltration and enhancement of plant water update and transpiration. Table 15.3 illustrates varying outcomes associated with tillage effects on runoff and water use efficiency.

TABLE 15.2

Examples of Mulching Effects on Runoff Reduction or Rainwater Use Efficiency in Various Parts of the World

Country	Soil Type	Runoff Reduction	Reference
Burkina Faso	Chromic Luvisol	65% (plus evaporation reduction 25%)	Stroosnijder and Hoogmoed 2002; Stroosnijder and Rheenen 2001
Burkina Faso	(Unknown)	60%	Zougmoré et al. 2000
Burkina Faso	Lixisols and Cambisols	15%–50%	Mando 1997
China	Sandy loam (loess origin)	100%	Li et al. 2000
Ghana	(Unknown)	90%	FAO 1993
Nigeria	(Unknown)	90%	FAO 1993
Mauritius	(Unknown)	85%	Facknath and Lalljee 1999
Morocco	Calcic Chromoxerert	RUE: 5.7–6.5	Mrabet 2002
Mozambique	(Unknown)	RUE: 6.8–8.4	Rothert and Macy 2002
Niger	Ustifluvent sandy loam	RUE: 2.48–3.14	Zaongo et al. 1997
Canary Islands	Various	90%	Tejedor et al. 2002
United States	Clay loam and sandy loam	RUE (biomass): 2.6–2.93	Tolk et al. 1999

Source: Adapted from Ringersma, J. et al., *Green Water: Definitions and Data for Assessment*. No. 2003/02. ISRIC-World Soil Information, 2003.

Note: RUE = rainwater use efficiency, in kg ha^{-1} mm^{-1}.

TABLE 15.3

Examples of Tillage Effects on Runoff or Rainwater Use Efficiency

Country	Soil Type	Result	Reference
Ghana	(Unknown)	Runoff reduction up to 90%	FAO 1993
India	Fluventic Ustochrept	Minimum tillage plus residues most effective	Ghuman and Sur 2001
Mauritius	(Unknown)	Runoff reduction 30%–40%	Facknath and Laljee 1999
Morocco	Calcic Chromoxerert	RUE: 5.7–6.6	Mrabet 2000, 2002
Ontario	Clay loam	Runoff increased	Tan et al. 2002
Sudan	Vertisol	Subsoil tillage most effective	Salih et al. 1998
Zimbabwe	Fersiallitic	RUE: 12–24	Riches et al. 1997

Source: Adapted from Ringersma, J. et al., *Green Water: Definitions and Data for Assessment*. No. 2003/02. ISRIC-World Soil Information, 2003.

Note: RUE = rainwater use efficiency, in kg ha^{-1} mm^{-1}.

15.3   SOIL-SPECIFIC IRRIGATION MEASURES

Irrigation is an essential practice in many cropping systems in semiarid and arid areas, and efficient water applications and management are major concerns. In semiarid and arid areas, rainfall is characterized by high spatial and temporal variability, requiring farmers to use irrigation to supplement water during dry periods. However, in these areas, irrigation water is limited. Methods are needed to optimize the timing and amount of irrigation water applied to supplement rainwater.

Ideally, all irrigation water would be delivered to crops without loss and at the precise time to provide the greatest benefit. Goals for soil-specific irrigation are to achieve the greatest economic benefits from the water applied and to minimize water loss and negative environmental impacts. A precision irrigation system is expected to have the ability to apply the right amount of water directly where it is needed, thereby saving water through preventing excessive water runoff and leaching.

Soil-specific irrigation is defined as timely and accurate water application in accordance with spatial and temporal soil properties and in response to plant demand during different growth stages. Two important aspects for soil-specific irrigation are (1) information about how much to put where and when, and (2) method of controlling the application rate. Wall and King (2004) explored the design for smart soil moisture sensors and sprinkler valve controllers to implement plug-and-play technology and proposed architectures of distributed sensor networks for site-specific irrigation automation.

15.3.1 ADVANTAGES AND FUNCTION OF PRECISION IRRIGATION

Soil-specific irrigation makes it possible for farmers to monitor water requirement at all growth stages throughout their fields and to adjust water application rates to achieve production goals while minimizing the amount of water applied. This would save water, use less energy, reduce pumping cost, increase yields, and help protect the environment from excess irrigation (which tends to cause runoff, waterlogging, and leaching of soil nutrients and other chemicals that may contaminate water sources).

Traditional field irrigation usually needs massive manpower and material resources but has a deficiency of real-time monitoring and accuracy, whereas soil-specific irrigation has the potential to help farmers improve input allocation decisions, thereby lowering production costs or increasing outputs and profits. This has the potential to increase economic efficiencies of operations by optimally matching inputs to yields in each area of a field and reducing costs.

Spatial-based information will enable farmers to apply inputs only to specific locations and in the amounts needed. Farmers can also benefit from the increasing availability of better and more timely information about their crops for decision-making. They will be able to denote problem areas in a field before problems become visible and before it is too late to take corrective action. Soil-specific irrigation delivers water more accurately, which can better overcome the negative effect of traditional irrigation such as water wasting and water leaching.

In a study conducted by Sadler et al. (2005), the potential for water conservation using precision irrigation ranges from marginal to nearly 50% in single years and averages from 8% to 20% (Figure 15.4). In specific examples where quantitative estimates of water savings could be made, precision irrigation could save 10% to 15% of the water used in conventional irrigation practice. In some cases, there are also benefits beyond the saving of water, such as increased harvestable area, decreased incidence of disease, and reduced leaching. Improved decision support systems and technology for real-time monitoring and control can further increase the utility and benefits of precision irrigation.

15.3.2 COMPONENTS OF SOIL-SPECIFIC IRRIGATION

Soil-specific irrigation involves a set of geospatial technologies (remote sensing, geographic information systems [GISs], and Global Positioning Systems [GPSs]) that link locations of a field with the most appropriate decision regarding irrigation (Figure 15.5). These georeferenced technologies help farmers settle the problem of how much water to put in where and at what time, as well as control application rates using variable rate technologies.

There are two sets of methodologies for precision farming: map-based and sensor-based (Moran et al. 2004). Map-based methods involve grid sampling, laboratory analysis of soil samples, generating soil fertility, and nutrient distribution maps or maps that depict soil properties such as pH, cation exchange capacity, electrical conductivity, and others. The maps generated can then be used to guide a variable applicator. During both soil sampling and variable rate inputs, a positioning system is needed to identify locations. In comparison, sensor-based methods use

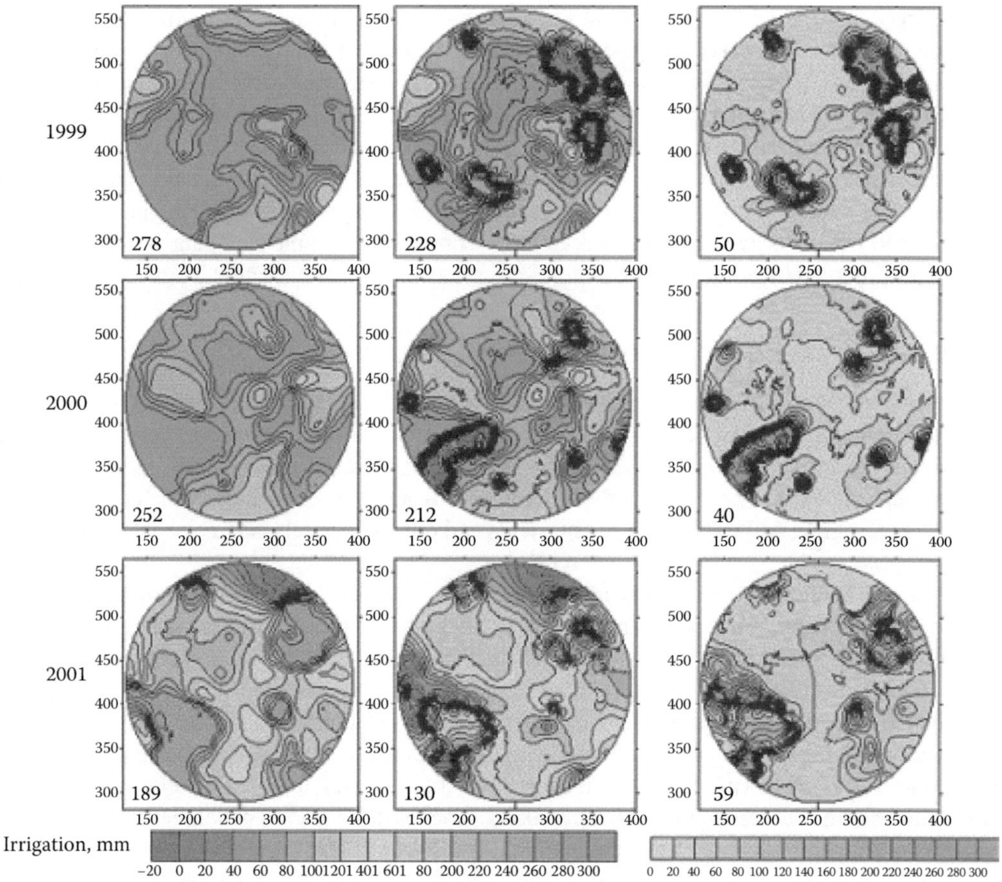

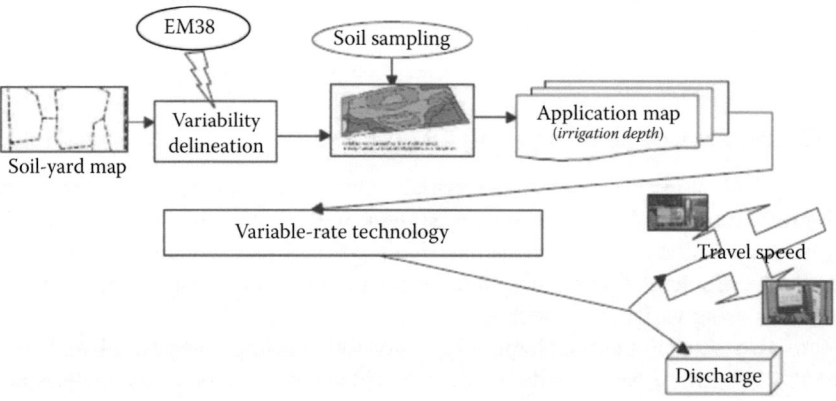

FIGURE 15.4 Comparison of spatial maps between irrigation that would have achieved maximum yield (left) and irrigation that would have achieved maximum profit (middle) and their difference (right) for 3 years (1999–2001) in Florence, SC. (Adapted from Sadler, E. J. et al., *Journal of Soil and Water Conservation* 60: 371–378, 2005.)

FIGURE 15.5 A general strategy for soil-specific irrigation. (Adapted from Al-Karadsheh, E. et al., Precision Irrigation: New strategy irrigation water management. In *Conference on International Agricultural Research for Development,* 9–11. Deutscher Tropentag, Witzenhausen, 2002.)

real-time sensors and a feedback control system to measure soil properties rapidly on the go, then immediately use these signals to direct a variable rate applicator.

Based on the requirements of precision agriculture, soil-specific irrigation practices include field scouting, grid soil sampling, and variable rate technology and its applications. Many crop fields have a high degree of variability in soil type, topography, soil moisture, and other major factors that affect crop production. Soil-specific irrigation can enable the development of an irrigation system to effectively manage fields to account for this variability.

15.3.2.1 Field Scouting and Management Zones

Precision irrigation is an approach for subdividing fields into smaller, relatively homogeneous management zones where irrigation is custom-managed according to the unique mean characteristics of each management zone. The first information needed to delineate in-field spatial variability is a soil map of the field. The number of different management zones within a field is a function of the natural variability within the field, the size of the field, and certain management factors. Water transport characteristics of a soil strongly influence water infiltration, evaporation, and deep percolation rates. Because of spatial variability (topography, soil texture, water-holding capacity, infiltration, and drainage rate), the need for irrigation may differ between different zones of a particular field. An example of irrigation management zones is presented in Figure 15.6.

15.3.2.2 Soil Sampling or Monitoring

In general, there are two sampling strategies (grid or zone) that can be used to direct site-specific field management. Grid sampling uses a systematic approach that divides the field into squares or rectangles of equal size (called grid cells). Soil samples are collected from within each of these cells. Grid soil sampling provides an initial base of information to develop variable rate applications plans. Sampling by zone, on the other hand, uses a more subjective but intuitive approach to divide a field into smaller units based on accumulated knowledge and experience. Soil samples collected at random from within each zone are bulked together and analyzed to provide an average sample value for each zone. In either sampling strategy, soil sampling aims at identifying the current status of soil conditions and producing maps of potential water requirements. Common soil properties analyzed included soil moisture, clay content, organic matter content, nutrient availability, pH, and bulk density. An *in situ* soil-specific irrigation system is illustrated in Figure 15.7.

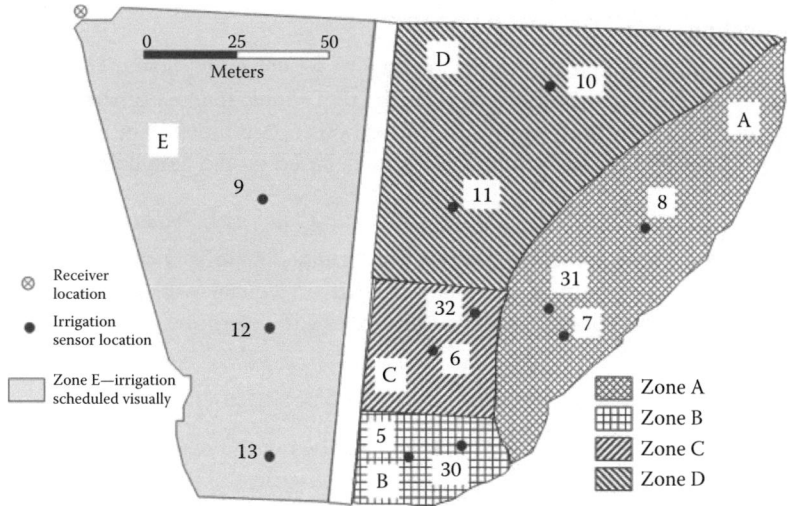

FIGURE 15.6 Example of irrigation management zones based on scheduling strategies and inherent field variability. (Adapted from Vellidis, G. et al., *Computers and Electronics in Agriculture* 61(1): 44–50, 2008.)

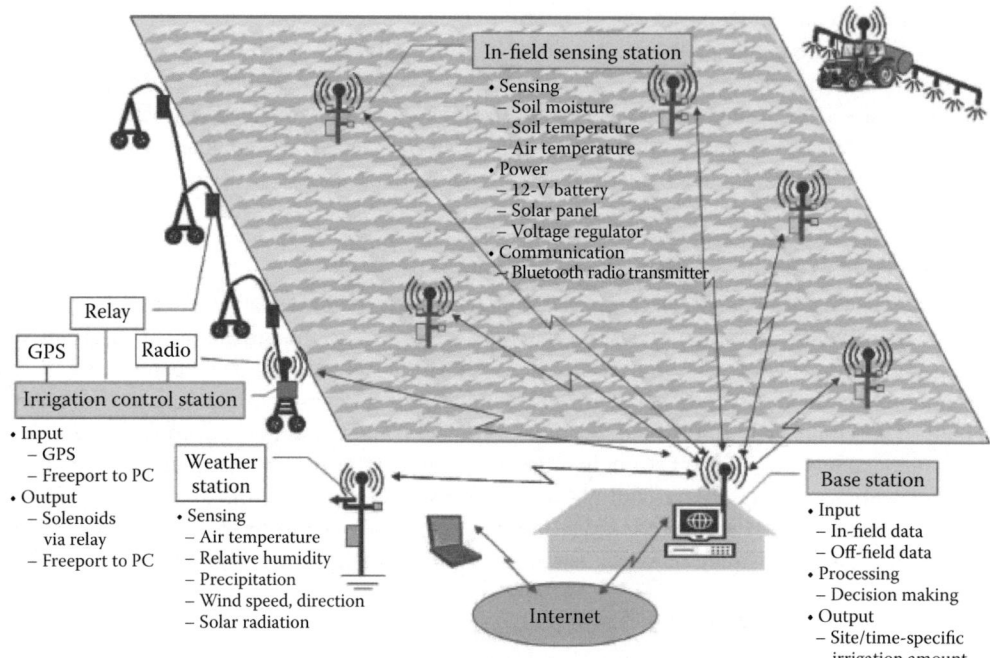

In-field sensing station
• Sensing
  – Soil moisture
  – Soil temperature
  – Air temperature
• Power
  – 12-V battery
  – Solar panel
  – Voltage regulator
• Communication
  – Bluetooth radio transmitter

Relay

GPS   Radio

Irrigation control station
• Input
  – GPS
  – Freeport to PC
• Output
  – Solenoids
    via relay
  – Freeport to PC

Weather
station
• Sensing
  – Air temperature
  – Relative humidity
  – Precipitation
  – Wind speed, direction
  – Solar radiation

Internet

Base station
• Input
  – In-field data
  – Off-field data
• Processing
  – Decision making
• Output
  – Site/time-specific
    irrigation amount

FIGURE 15.7   General workflow layout for soil-specific irrigation. (Adapted from Beeri, O., and A. Peled, *ISPRS Journal of Photogrammetry and Remote Sensing* 64: 47–54, 2009.)

15.3.2.3 Variable Rate Technology

A variable rate irrigation (VRI) control system enables a center pivot irrigation system to supply water at rates relative to the needs of individual areas within the field. The VRI system varies application rate by cycling sprinklers on and off and by varying the center pivot travel speed. However, to optimize irrigation applications, a smart sensor array is best used in conjunction with a VRI system. Irrigation rates are based on perceived or measured water requirement by crops, which vary with field soil and topographic features (Vellidis et al. 2008).

15.3.2.4 Irrigation Controller

An irrigation controller is an electronic device used to set when and how much to irrigate. There are many options for irrigation controllers, ranging from simple manual timers to more intelligent systems that can decide irrigation applications based on certain inputs. The inputs from georeferenced tools give irrigation controller application rates based on the data collected and prescribed recommendation.

VRI can be achieved by fitting each sprinkler or nozzle with a remotely controlled on-off automatic valve. The choice of sprinklers or nozzles is determined mainly by considerations of drop size to avoid crop damage while minimizing wind drift, to give adequate overlap, and to spread the water to avoid runoff. The relative duration of the on-off cycles will determine the depth of water applied.

15.3.3 Optimization of Soil-Specific Irrigation Management

Soil moisture conditions and crop water requirements, two of the most important factors for soil-specific irrigation decision-making, can be monitored by a variety of methods. Measuring soil moisture is one way of determining when crops need irrigation and how much irrigation water should be applied. There are at least four steps to optimize irrigation management: gathering information of inputs such as soil moisture map, processing input information, prescribing recommendations for input applications, and controlling VRI according to the set recommendations. In the following sections, these individual steps are further described.

15.3.3.1 Gathering Information of Inputs

To collect data, farmers could use a GPS to collect information related to crop production, including soil sampling, plant growth parameters, and crop scouting, all of which provide information for management decisions (Hrubovcak et al. 1999).

A conventional soil survey is gradually being replaced by newer and more cost-effective approaches (Brown et al. 2006), such as the use of mid-infrared reflectance (MIR) spectroscopy, visible and near-infrared reflectance (VNIR) spectroscopy, and other sensing methods (Tables 15.4 and 15.5). A smart soil sensor array can also be used to determine the optimum amount of water to apply across the field. The development of soil sensors is expected to increase the effectiveness of precision agriculture (Pierce and Nowak 1999). Traditionally, soil properties have been measured by soil sampling and offsite laboratory analysis or by on-the-spot measurement. But the relatively coarse sampling/measurement density of these conventional strategies may not be sufficient to reveal soil spatial variation. Some spectroscopic or other sensing techniques (Table 15.4) are being used as possible alternatives (or surrogates) to enhance or replace conventional laboratory methods of soil analysis (Janik and Skjemstad 1995). These techniques are nondestructive and can allow repeated measurements without disturbing soil and crop systems.

15.3.3.2 Processing Soil-Specific Irrigation

Soil-specific irrigation is information-intense, with a lot of position-tagged, sensed data, and is required to generate treatment maps. The objective is to establish proper timing and amount of irrigation for the greatest effectiveness. A desirable solution is a real-time, robust, and low-cost monitoring and mapping system for soil, crop, and environmental variables.

A mapping approach is the main way to process precision information. For instance, a map of soil water availability may highlight areas of deficiency but cannot be used directly for variable application of water. Sensing and mapping soil variables provide decision support information by identifying factors limiting to growth and yield in various parts of the field. The aim is to divide the irrigated field into different zones for varying rates of application water.

TABLE 15.4

Examples of Soil Properties That May Be Estimated by Mid-Infrared Reflectance Spectroscopy Based on Various Studies

Soil Physical Properties	Soil Chemical Properties
Particle size (clay, silt, sand)	Exchangeable cations (Ca, Mg, K, Na) and CEC
Bulk density	Carbon pools (total organic, particulate organic, charcoal, inorganic C) and total nitrogen
Volumetric water content (at various tensions from 0 to 15,000 kPa)	Phosphorus buffering index
Quantitative x-ray diffraction (quartz, kaolinite, smectite)	Soil reaction (pH-water, pH-CaCl$_2$)
Electrical conductivity	Quantitative x-ray fluorescence (Ca, Mg, Fe, Al, Si)
Water stable aggregates	Exchangeable sodium percentage

Source: Merry, R.H., and L.J. Janik, Mid infrared spectroscopy for rapid and cheap analysis of soils. *Proceedings of 10th Australian Agronomy Conference,* CD-ROM. Hobart, Australia: Australian Society of Agronomy, 2001; Janik, L.J. et al., *Soil Research* 45: 73–81, 2007; Viscarra et al. 2010.

Note: C = carbon, Ca = calcium, CaCl$_2$ = calcium chloride, CEC = cation exchange capacity, Fe = iron, K = potassium, Mg = magnesium, Na = sodium, Si = silicon.

TABLE 15.5

Summary of Common Sensing Methods Used for Estimating Volumetric Soil Moisture Content (θ)

Sensing Method	Advantages	Limitations
Visible, NIR, SWIR reflectance Spectral information in visible, NIR, and SWIR wavelengths is related to θ as a function of spectral absorption features; for bare soils, increase in θ generally leads to a decrease in soil reflectance	• Fine spatial resolution • Broad coverage • Multiple satellite sensors available • Hyperspectral sensors show promise	• Weak relation to θ • Minimal surface penetration (~1 mm) • Limited ability to penetrate clouds and vegetation; attenuated by earth's atmosphere • Infrequent repeat coverage • Strongly perturbed by vegetation biomass
TIR emittance Soil moisture directly influences soil temperatures by increasing both specific heat and thermal conductivity, thus thermal inertia of soils	• Fine spatial resolution • Broad coverage • Multiple satellite sensors available • Strong relation to θ • Surface TR primarily due to varying θ • Surface temperature-vegetation index approaches show promise	• Minimal surface penetration (~1 mm) • Limited ability to penetrate clouds and vegetation; attenuated by earth's atmosphere • Infrequent repeat coverage
Microwave TB Intensity of microwave emission (at $\sigma^0 = 1$–30 cm) from soil is related to θ because of large differences in dielectric constant of dry soil (~3.5) and water (~80); for bare soils, increase in θ generally leads to increase in radiative temperature	• Broad coverage • Satellite sensor recently available • Strong relation to θ • Surface penetration up to ~5 cm • Insensitive to clouds and earth's atmosphere	• Perturbed primarily by surface roughness and vegetation biomass • Coarse spatial resolution (~30 km)
Radar σ^0 As with passive microwave sensing, magnitude of σ^0 is related to θ through contrast of dielectric constants of bare soil and water; for bare soils, increase in θ generally leads to increase in σ^0	• Fine spatial resolution • Multiple satellite sensors available • Strong relation to θ • Surface penetration up to ~5 cm • Insensitive to clouds and earth's atmosphere	• Infrequent repeat coverage • Perturbed primarily by surface roughness and vegetation biomass

Source: Adapted from Moran, M.S. et al., *Canadian Journal of Remote Sensing* 30: 805–826, 2004.

High-resolution mapping of soil variability for precision management is enabled by various sensing techniques such as electromagnetic (EM) surveys together with geostatistical interpolation and ground-truthing. On-the-go EM mapping, for instance, provides an opportunity for simultaneous collection of georeferenced apparent soil electrical conductivity (EC) and elevation data, which, with interpolation and correlation with other geospatial data of the field, can allow quantitative assessment of soil and topographic factors to guide irrigation and other management decisions.

15.3.3.3 Prescribing Recommendations for Input Applications

Many models have been developed for providing agricultural management recommendations. To meet the demands of precision agriculture, models must be based on data collected and processed by the system and then give recommendations quickly.

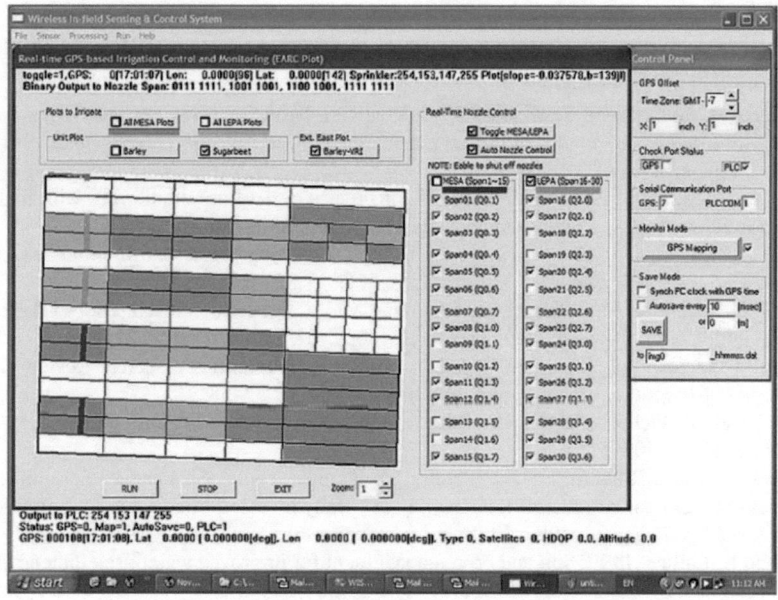

FIGURE 15.8 Screen shot of a real-time GPS-based irrigation control and monitoring system. (Adapted from Kim, Y. et al., *Instrumentation and Measurement, IEEE Transactions* 57: 1379–1387, 2008.)

15.3.3.4 Georeferenced VRI

The irrigation control station updates and sends georeferenced locations of the machine from a differential GPS mounted on a cart to the base station for real-time monitoring and control of the irrigation system (Figure 15.8). Based on sprinkler head GPS locations, the base station feeds control signals back to the irrigation control station to site-specifically operate individual sprinklers to apply a specified amount of water. Sprinklers cycle on and off to meet target applications. For example, for a 100% rate, sprinklers are turned on all the time; for 50% rate, sprinklers are on for 30 seconds of every minute in 30-second intervals. Variable pivot speed can be adjusted to meet different requirements.

15.4 SUMMARY

Future farming has to provide high yields in order to feed the growing world population, but should do so with minimum environmental impact so that it can be sustained. With increasing freshwater scarcity in many parts of the world, this challenge can only be met by precise adaptation to locally varying soil conditions. Green water holds a key to current and future sustainable agriculture.

The prospects in soil-specific farming depend largely on the possibilities to sense the properties of soils, ideally in real time. Automatic sensing of the spatial variability in soil properties is an important component for successful soil-specific farming including precision irrigation. Many of these properties can change significantly over a small area and over time, making manual sample collection or processing prohibitive in terms of cost and timeliness. Therefore, for site-specific farming, it is necessary to develop a system to sense the variation in important soil properties and ideally in real time. It can be expected that many farming operations in the future will include a soil-specific control system that can enhance the capture of rainwater as well as the conversion of blue water to green water through precision irrigation and other non-irrigation measures as discussed in this chapter.

REFERENCES

Adams, J.E. 1966. Influence of mulches on runoff, erosion, and soil moisture depletion. *Soil Science Society of America Journal* 30(1): 110–114.

Al-Karadsheh, E., H. Sourell, and R. Krause. 2002. Precision Irrigation: New strategy irrigation water management. In *Conference on International Agricultural Research for Development,* 9–11. Deutscher Tropentag, Witzenhausen.

Beeri, O., and A. Peled. 2009. Geographical model for precise agriculture monitoring with real-time remote sensing. *ISPRS Journal of Photogrammetry and Remote Sensing* 64: 47–54.

Boers, M., and J. Ben-Asher. 1982. A review of rainwater harvesting. *Agricultural Water Management* 5(2): 145–158.

Brown, D.J., K.D. Shepherd, M.G. Walsh, M. Dewayne Mays, and T.G. Reinsch. 2006. Global soil characterization with VNIR diffuse reflectance spectroscopy. *Geoderma* 132(3): 273–290.

Burt, C.M., A.J. Mutziger, R.G. Allen, and T.A. Howell. 2005. Evaporation research: Review and interpretation. *Journal of Irrigation and Drainage Engineering* 131(1): 37–58.

De Fraiture, C., and D. Wichelns. 2010. Satisfying future water demands for agriculture. *Agricultural Water Management* 97(4): 502–511.

De la Rosa, D., M. Anaya-Romero, E. Diaz-Pereira, N. Heredia, and F. Shahbazi. 2009. Soil-specific agro-ecological strategies for sustainable land use–A case study by using MicroLEIS DSS in Sevilla Province (Spain). *Land Use Policy* 26: 1055–1065.

Facknath, S., and B. Lalljee. 1999. Soil and crop management for improved water use efficiency in agriculture under conditions of water stress. *PROSI Magazine* 31(367): 44–49.

Fairbourn, M.L. 1973. Effect of gravel mulch on crop yield. *Agronomy Journal* 65(6): 925–928.

Falkenmark, M., and J. Rockström. 2004. *Balancing Water for Humans and Nature: The New Approach in Ecohydrology.* Earthscan: London.

Falkenmark, M. 2013. Growing water scarcity in agriculture: Future challenge to global water security. *Philosophical Transactions of the Royal Society A* 371: 20120410.

FAO 1993. Soil tillage in Africa, needs and challenges. *Soil Bull* 69, Rome.

Ghuman, B.S., and H.S. Sur. 2001. Tillage and residue management effects on soil properties and yields of rainfed maize and wheat in a subhumid subtropical climate. *Soil and Tillage Research* 58(1): 1–10.

Heege, H.J. 2012. *Precision in Crop Farming.* Springer, doi 10.1007/978-94-007-6760-7.

Hide, J.C. 1954. Observations on factors influencing the evaporation of soil moisture. *Soil Science Society of America Journal* 18(3): 234–239.

Hrubovcak, J., U. Vasavada, and J.E. Aldy. 1999. Green technologies for a more sustainable agriculture. No. 33721. United States Department of Agriculture, Economic Research Service.

Janik, L.J., and J.O. Skjemstad. 1995. Characterization and analysis of soils using mid-infrared partial least-squares. 2. Correlations with some laboratory data. *Soil Research* 33: 637–650.

Janik, L.J., J.O. Skjemstad, K.D. Shepherd, and L.R. Spouncer. 2007. The prediction of soil carbon fractions using mid-infrared-partial least square analysis. *Soil Research* 45: 73–81.

Kemper, W.D., A.D. Nicks, and A.T. Corey. 1994. Accumulation of water in soils under gravel and sand mulches. *Soil Science Society of America Journal* 58(1): 56–63.

Kim, Y., R.G. Evans, and W.M. Iversen. 2008. Remote sensing and control of an irrigation system using a distributed wireless sensor network. *Instrumentation and Measurement, IEEE Transactions* 57: 1379–1387.

L'vovich, M.I. 1979. *World Water Resouces and Their Future.* American Geophysical Union. Washington, DC.

Lamb Jr, J., and J.E. Chapman. 1943. Effect of surface stones on erosion, evaporation, soil temperature, and soil moisture. *Journal of the American Society of Agronomy.*

Larson, W.E., and P.C. Robert. 1991. Farming by soil. In R. Lal and F.J. Pierce (eds.), *Soil Management for Sustainability,* pp. 103–112. Soil and Water Conservation Society: Ankeny, IA.

Li, X.-Y., J.-D. Gong, and X.-H. Wei. 2000. In-situ rainwater harvesting and gravel mulch combination for corn production in the dry semi-arid region of China. *Journal of Arid Environments* 46(4): 371–382.

Li, X.-Y. 2003. Gravel–sand mulch for soil and water conservation in the semiarid loess region of northwest China. *Catena* 52(2): 105–127.

Mando, A. 1997. The role of termites and mulch in the rehabilitation of crusted Sahelian soils. No. 16. Wageningen, The Netherlands: Wageningen Agricultural University.

Merry, R.H., and L.J. Janik. 2001. Mid infrared spectroscopy for rapid and cheap analysis of soils. *Proceedings of 10th Australian Agronomy Conference,* CD-ROM. Australian Society of Agronomy: Hobart, Australia.

Molden, D. 2007. *Water for Food, Water for Life: A Comprehensive Assessment of Water Management in Agriculture*. Rome: International Water Management Institute (IWMI) and FAO.

Moran, M.S., C.D. Peters-Lidard, J.M. Watts, and S. McElroy. 2004. Estimating soil moisture at the watershed scale with satellite-based radar and land surface models. *Canadian Journal of Remote Sensing* 30: 805–826.

Mrabet, R. 2000. Differential response of wheat to tillage management systems in a semiarid area of Morocco. *Field Crops Research* 66(2): 165–174.

Mrabet, R. 2002. Wheat yield and water use efficiency under contrasting residue and tillage management systems in a semiarid area of Morocco. *Experimental Agriculture* 38(02): 237–248.

Mueller, N.D., J.S. Gerber, M. Johnston, D.K. Ray, N. Ramankutty, and J.A. Foley. 2012. Closing yield gaps through nutrient and water management. *Nature* 490: 254–257.

Munson, R.D., and C.F. Runge. 1990. Improving fertilizer and chemical efficiency through "high precision farming." Center for International Food and Agricultural Policy, University of Minnesota.

Nachtergaele, J., J. Poesen, and B. Van Wesemael. 1998. Gravel mulching in vineyards of southern Switzerland. *Soil and Tillage Research* 46(1): 51–59.

Pala, M., T. Oweis, B. Benli, E. De Pauw, M. El Mourid, M. Karrou, M. Jamal, and N. Zencirci. 2011. Assessment of wheat yield gap in the Mediterranean: Case studies from Morocco, Syria and Turkey. Aleppo, Syria: International Center for Agricultural Research in the Dry Areas (ICARDA).

Pierce, F.J., and P. Nowak. 1999. Aspects of precision agriculture. *Advances in Agronomy* 67: 1–85.

Poesen, J.W.A., and H. Lavee. 1991. Effects of size and incorporation of synthetic mulch on runoff and sediment yield from interrils in a laboratory study with simulated rainfall. *Soil and Tillage Research* 21(3): 209–223.

Riches, C.R., S.J. Twomlow, and H. Dhliwayo. 1997. Low-input weed management and conservation tillage in semiarid Zimbabwe. *Experimental Agriculture* 33(2): 173–187.

Ringersma, J., N. H. Batjes, and D. Dent. 2003. *Green Water: Definitions and Data for Assessment*. No. 2003/02. ISRIC-World Soil Information.

Rockström, J. 1999. On-farm green water estimates as a tool for increased food production in water scarce regions. *Physics and Chemistry of the Earth, Part B: Hydrology, Oceans and Atmosphere* 244: 375–383.

Rockström, J. 2003. Water for food and nature in drought–prone tropics: Vapour shift in rain-fed agriculture. *Philosophical Transactions of the Royal Society B: Biological Sciences* 358: 1997–2009.

Rockström, J., M. Lannerstad, and M. Falkenmark. 2007. Assessing the water challenge of a new green revolution in developing countries. *Proceedings of the National Academy of Sciences* 104: 6253–6260.

Rost, S., D. Gerten, A. Bondeau, W. Lucht, J. Rohwer, and S. Schaphoff. 2008. Agricultural green and blue water consumption and its influence on the global water system. *Water Resources Research* 44. doi:10.1029/2007WR006331.

Rothert, S., and P. Macy. 2000. "The potential of water conservation and demand management in Southern Africa: An untapped river, submission to the World Commission on Dams."

Roundy, B.A., L.B. Abbott, and M. Livingston. 1997. Surface soil water loss after summer rainfall in a semidesert grassland. *Arid Land Research and Management* 11(1): 49–62.

Sadler, E.J., R. G. Evans, K. C. Stone, and C. R. Camp. 2005. Opportunities for conservation with precision irrigation. *Journal of Soil and Water Conservation* 60: 371–378.

Salih, A.A., H.M. Babikir, and S.A.M. Ali. 1998. Preliminary observations on effects of tillage systems on soil physical properties, cotton root growth and yield in Gezira Scheme, Sudan. *Soil and Tillage Research* 46(3): 187–191.

Sposito, G. 2013. Green water and global food security. *Vadose Zone Journal* 12. doi:10.2136/vzj2013.02.0041.

Sposito, G. 2014. Sustaining the genius of soils. In Churchman, G. Jock, and Edward R. Landa (eds.), *The Soil Underfoot: Infinite Possibilities for a Finite Resource*. CRC Press: Boca Raton, FL.

Srinivasan, A. 2006. *Handbook of Precision Agriculture: Principles and Applications*. Food Products Press: Binghamton, NY.

Stroosnijder, L., and W.B. Hoogmoed. 2002. The contribution of soil and water conservation to carbon sequestration in semiarid regions. *International Colloquium on Land Use Management, Erosion and Carbon Sequestration,* Montpeller.

Stroosnijder, L., and T. van Rheene. 2001. *Agno-Silvo-Pastoral Land Use in Sahelian Villages*. Catena Verlag.

Tan, C.S., C.F. Drury, J.D. Gaynor, T.W. Welacky, and W.D. Reynolds. 2002. Effect of tillage and water table control an evapotranspiration, surface runoff, tile drainage and soil water content under maize on a clay loam soil. *Agricultural Water Management* 54(3).

Tebrügge, F., and R-A. Düring. 1999. Reducing tillage intensity—A review of results from a long-term study in Germany. *Soil and Tillage Research* 53(1): 15–28.

Tejedor, M., C.C. Jiménez, and F. Díaz. 2002. Traditional agricultural practices in the Canaries as soil and water conservation techniques. In J. Graaff and M. Ouessar (eds.), Water harvesting in Mediterranean zones: An impact assessment and economic evaluation. *Tropical Resource Management Papers 40*, pp. 3–11. WUR: Wageningen.

Tolk, J.A., T.A. Howell, and S.R. Evett. 1999. Effect of mulch, irrigation, and soil type on water use and yield of maize. *Soil and Tillage Research* 50(2): 137–147.

Unger, P.W. 1971. Soil profile gravel layers: I. Effect on water storage, distribution, and evaporation. *Soil Science Society of America Journal* 35(4): 631–634.

Van Wesemael, B., J. Poesen, and T. de Figueiredo. 1995. Effects of rock fragments on physical degradation of cultivated soils by rainfall. *Soil and Tillage Research* 33(3): 229–250.

Vellidis, G., M. Tucker, C. Perry, C. Kvien, and C. Bednarz. 2008. A real-time wireless smart sensor array for scheduling irrigation. *Computers and Electronics in Agriculture* 61(1): 44–50.

Viscarra Rossel, R.A., and T. Behrens. 2010. Using data mining to model and interpret soil diffuse reflectance spectra. *Geoderma* 158: 46–54.

Wall, R.W., and B.A. King. 2004. Incorporating plug and play technology into measurement and control systems for irrigation management. ASABE Paper 042189.

Wu, P., and H. Feng. 2009. *Rainwater Harvesting in China*. Huanghe Hydraulic Publisher (in Chinese).

Zaongo, C.G.L., C.W. Wendt, R.J. Lascano, and A.S.R. Juo. 1997. Interactions of water, mulch and nitrogen on sorghum in Niger. *Plant and Soil* 197(1): 119–126.

Zougmore, R., F.N. Kambou, K. Ouattara, and S. Guillobez. 2000. Sorghum-cowpea intercropping: An effective technique against runoff and soil erosion in the Sahel (Saria, Burkina Faso). *Arid Soil Research and Rehabilitation* 14(4): 329–342.

16 Challenges and Opportunities in Precision Agriculture

Rattan Lal

CONTENTS

16.1 INTRODUCTION

Precision agriculture (PA), also called satellite farming, is based on space-age technology including remote sensing (RS), Geographic Information System (GIS), and Global Positioning System (GPS) to use soil- and site-specific inputs and other farming operations to optimize resource use. Information technology, based on computers and electronics, is used to target management of plant nutrients, water, pests, and tillage and to address spatial and temporal variability. The French term *terroira* refers to "sense of place" or special geographic quality relevant to the PA concept. Simply put, PA involves site-specific management based on information technology with regard to within-field spatial and temporal variability in edaphic properties and processes. It is a variable-rate technology (VRT) for optimum use of nutrients, seeding, pesticides, herbicides, irrigation, and tillage-related input. An objective of PA is to use information about soil, terrain, and weather to optimize the use efficiency of inputs and enhance economic and environmental benefits. The strategy is to increase soil/land productivity from ever-decreasing resources to meet the continuously increasing demands of the growing and affluent population for feed, fiber, fuel, and other necessities. The strategy is to address global issues of the twenty-first century (e.g., increasing population and its affluence, growing energy needs, increasing atmospheric concentrations of CO_2 and other greenhouse gases, tropical deforestation, exacerbating risks of soil degradation and desertification, decreasing per capita arable land area and renewable freshwater supply, increasing risks of hidden hunger and malnutrition, and civil strife and political instability) through a judicious use of finite resources and innovative management systems. Conceptualizing sustainability of agroecosystems in view of a holistic approach with a focus on interconnectivity (Figure 16.1) is more important now than ever before. Some of the global issues are directly or indirectly related to soil resources and

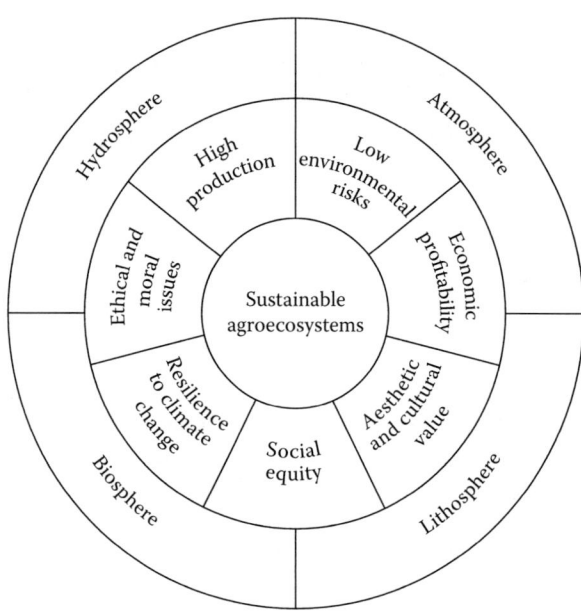

FIGURE 16.1 A holistic approach to advancing sustainability of agroecosystems in view of the interconnectivity of global issues related to natural resources.

the management of their quality and productivity. Because of competing uses (i.e., urbanization, industrial use, recreation, and nature conservancy), per capita availability of basic resources needed for agricultural production (e.g., land, water, energy) is progressively decreasing. While agricultural production must be increased, its ecological/environmental footprint must be decreased.

Thus, the objective of this synthesis chapter is deliberate potential and challenges of PA being integral to any strategy as a viable alternative to conventional systems of agricultural intensification within the framework of sustainable and ecological intensification.

16.2 SCOPE OF PA IN THE CONTEXT OF ECOLOGICAL INTENSIFICATION AND SUSTAINABLE INTENSIFICATION

The strategy discussed above is of sustainable and ecological intensification of agroecosystems. The term sustainable intensification (SI) implies producing more agricultural goods and services from less land area, fertilizers and pesticides, energy, and water. Indeed, the ecological and environmental footprint of agriculture must be minimized through ecological intensification. The Green Revolution of the 1960s and 1970s was based on agricultural intensification (AI) with intensive use of nonrenewable resources and agrochemicals. Thus, there are concerns about alternatives that minimize the environmental hazards and yet enhance and sustain production of food, feed, fuel, fiber, and other ecosystem services.

Some desirable attributes of SI are (1) high productivity, (2) low input of new land area, (3) low use of agricultural chemicals, (4) low input of fossil fuel energy, (5) low risks of soil erosion, (6) low emissions of greenhouse gases (GHGs), (7) low impact on water quality and nonpoint source pollution, (8) high resilience to climatic change and other extreme events, (9) high biodiversity, and (10) high soil equity.

Another option is that of the so-called ecological intensification (EI), which is defined as producing more from existing and already appropriated agroecosystems while reducing the environmental footprint. Desirable attributes of EI (Hochman et al. 2013) are (1) high productivity, (2) high use

TABLE 16.1

Comparison of Agricultural, Ecological, and Sustainable Intensification of Agroecosystems

Merits	Challenges/Concerns
Agricultural Intensification	
• Reduced input	• Enhancing and sustaining high productivity
• High profits	• Decreasing positive feedbacks to climate change
• Resource/soil conservation	• Reducing off-site and on-site impacts
• High biodiversity	• Minimizing trade-offs in ecosystem services
• High resilience to extreme events	• Improving societal/gender equity
• Low risks	
Ecological Intensification	
• Recycling materials	• Improving and sustaining productivity
• Minimizing material fluxes into the environment	• Decreasing gaseous emissions
• Reducing waste	• Reducing human drudgery
• Decreasing risks	• Enhancing nutritional security
Sustainable Intensification	
• High production	• High inputs
• Increase income	• Environmental pollution
• Development of agroindustries	• Soil degradation
• Rural employment	• Gaseous emissions

efficiency, (3) decreased emission (GHGs), (4) minimal adverse on-site and off-site impacts, (5) low risks, (6) high resilience, (7) sustain/enhance biodiversity, and (8) social equity.

Comparisons among three scenarios are outlined in Table 16.1. Adoption of PA is relevant to both SI and EI strategies by increasing precision in the use of inputs on the basis of soil-specific requirements thereby reducing losses and improving the use efficiency of finite but essential inputs (Hochman et al. 2013).

Thus, use of PA techniques can reduce the ecological footprint. For example, two nutrients with high impact on crop productivity and also high risks of environmental pollution are nitrogen (N) and phosphorus (P). Application of PA can enhance the use efficiency and reduce environmental hazard of N (Khosla et al. 2002, 2008) and P (Dao et al. 2011). In some cases geostatistics and kriging methods can be used to guide soil sampling for improving application of nutrient-related PA systems (Liu et al. 2010; Motomiya et al. 2006), and an x-ray fluorescence spectrometry-based approach can be useful in optimizing the use of P in agricultural soils (Dao et al. 2011). In arid and semiarid conditions, measurement of electrical conductivity of soil can provide information on within-field variability of edaphic properties and processes (Mertens et al. 2008). Crop models are also a useful approach in understanding the basic causes of spatial and temporal variability in productivity (Irmak et al. 2001). Availability and adoption of these innovative methodologies have expanded the scope and potential of PA technologies.

16.3 POTENTIAL AND OPPORTUNITIES

Site- or soil-specific information, using computers and electronics in agroecosystems, and space-age technology, has created numerous opportunities to improve inputs into agroecosytems, facilitate farm operations, and improve productivity while at the same time reduce environmental pollution and ecological footprint. Examples of specific opportunities created by PA are discussed next.

16.3.1 Nutrient Management

Nitrogen management is vastly improved with PA technology (Khosla et al. 2002, 2008). While improving efficiency and productivity, PA technology also reduces the magnitude of pollutants released by agriculture into the environment. The impact is especially notable in reducing N-related pollutants (including ammonia [NH_3^+], nitrate [NO_3^-], nitrous oxide [N_2O], nitrogen mono oxide [NO], nitrogen dioxide [NO_2] and nitrogen oxides [NO_x]; Bailey et al. 2001). Precision conservation can drastically reduce the risks of N leaching in irrigated crops (Delgado and Bausch 2005). Infrared spectroscopy is a pertinent technology to evaluate soil fertility (Du and Zhou 2009), and useful to soil-specific input of nutrients to meet the plant requirement. GIS-based fertilizer decision support systems have been developed for use by farmers to precisely determine the fertilizer requirements (Xie et al. 2012). In addition to reducing inputs, PA technology can also increase the profit margin (Khosla et al. 2008).

16.3.2 Pest Management

Management of weeds, pests, and pathogens is important to obtaining high yields. Economic and environmental costs of pest management can be reduced by RS techniques involving GPS and GIS (Lan et al. 2009). VRT optimizes the use of agrochemicals. Modern technologies allow us to record disease-related weather data, spatial variability of microclimates, and disease-specific systems (Steiner et al. 2008), and airborne RS systems of pest management (Lan et al. 2009) can be used to design targeted and timely interventions.

Microbial control of insects is also gaining momentum in agroecosystems (Lord 2005). Friendly fungi are useful to controlling insects. Utilization of pathogen genes for their toxins is another strategy to control insects. As with the process of making disease-control of crops more precise, autonomous robotic weed control measures also involve PA-based innovations (Slaughter et al. 2008). For resource-poor small landholders of the tropics and subtropics, manual weed control is among the most drudging tasks. For large-scale commercial farms, dependency on herbicides is expensive and has high environmental impact. Thus, effective weed control using PA technology has strong economic, environmental, and agronomic benefits for small landholders and large-scale commercial farms. Weed mapping by color images is useful to site-specific application of herbicides. Color images can be used for mapping the percentage of weed cover in no-till (NT) and infrared images for weed mapping in conventional tillage (Silva et al. 2012). Row orientation can also reduce weed growth (Borger et al. 2010).

16.3.3 Water Management and Precision Irrigation

Drought stress, a serious constraint to achieving high yields in semiarid and arid regions, is likely to be exaggerated by changing and uncertain climate characterized by an increase in the frequency of extreme events. Thus, there are opportunities for using PA technologies in soil-water conservation and saving water. For example, there are over 150,000 center pivot irrigation systems in the United States irrigating >8.5 million hectare of cropland. The groundwater depletion is a serious issue in the Ogallala aquifer, beneath the Great Plains (USA) and PA technology can save the water. For example, field variability can be matched with a variable-rate irrigation (VRI) system to save water. The VRI system applies water according to the specific water needs of a management zone, which can save millions of gallons of water. Precision irrigation can lead to savings in water from 8% to 20% and often as much as 50% (Sadler et al. 2005). In addition, precision irrigation can reduce leaching of N in irrigated crops (Delgado and Bausch 2005). Precision land leveling of the irrigated rice-wheat system in the Indo-Gangetic Plains can enhance water productivity and profitability (Jat et al. 2009). Precision land leveling, in conjunction with permanent raised-bed planting and recommended doses of fertilizer, can also optimize use efficiency of inputs and increase yields (Naresh et al. 2014).

16.3.4 PRECISION CONSERVATION

Precision conservation implies increase in use of conservation-effective practices to intensively managed systems for advancing and promoting conservation of natural areas. In this context, precision conservation is also technologically based (e.g., GPS, RS, GIS, and the variable rate of input) involving surface modeling, spatial data mining, and map analysis for better understanding and management of agricultural and natural ecosystems (Delgado and Berry 2008). Precision conservation is specifically useful to identify locations to install/establish riparian buffers, grass waterways, diversion ditches, and water ponds for effective conservation of soil and water. For example, estimating the spatial distribution of water runoff for the most likely rainfall events and siting the structures/land-forming devices is useful for minimizing runoff and conserving water (Murthy et al. 2004). Using yield maps and production records, eroded areas on shoulder and side slope positions can be identified for restorative purposes. Crop production assessment and plan implementation can increase farm income while conserving water and restoring soil quality (Kitchen et al. 2005). Thus, soil and water conservation practices are implemented with consideration to spatial and temporal variability across natural and agricultural landscapes (Lerch et al. 2005). This strategy can also minimize vulnerability of an aquifer to leaching of NO_3 and other agricultural chemicals, and surface water is also protected from transport of sediments and nutrients in runoff. Delineating spatial patterns of soil erosion in a field can be used in implementing precision conservation (Schumacher et al. 2005). In this regard, a spatial map indicating a field's erosion history, along with major processes and factors, could be helpful for implementation of site-specific conservation plans (Schumacher et al. 2005). The approach can be particularly useful in siting conservation buffers (e.g., riparian buffers, filter strips, grass waterways). When considering variable runoff patterns along field margins, the design of buffers and their effectiveness can be improved by precision conservation (Dosskey et al. 2005).

16.3.5 PRECISION TILLAGE

Soil compaction is a serious problem on mechanized farms and is exacerbated by heavy machinery. Its risks can be managed by selecting tire size, inflation pressure, rubber tracks, and controlled traffic (Godwin 2010). Precision tillage is aimed at minimizing the risks of soil compaction, saving energy, and conserving soil and water. It can be comprised of (1) subsoiling under the plant row prior to seeding, (2) guided or controlled traffic so that compaction is confined to specific parts of the field, and (3) improved tilth to facilitate seeding. Autoguidance via GPS satellite can improve production by keeping implements in the same traffic patterns year-to-year. Soil compaction sensors are used to efficiently measure soil compaction and cone index at different depths (Aguera et al. 2013) and appropriate measures are needed to alleviate it.

16.3.6 PRECISION SEEDING

Among a wide range of factors that affect crop stand (Santos et al. 2011) are the longitudinal distribution of seed and the placement depth, both of which depend on the seed. Thus, precision seeding implies placing an appropriate number of seeds at the desired depth and spacing. Thus, a seed-metering device is a key component of precision seeding drill (Zhai et al. 2014). Precision seeding reduces costs and increases the reliability of crop stand and agronomic production. In addition to reduced seed costs, other advantages include uniform stand and improved yield. However, equipment costs of precision drills are high. Precision seeding robots, equipped with an intelligent precision seeding device, can control accurate positions of seeds in the field (Xiu et al. 2010). Depending on the type of seed (rice, corn, wheat, rape, etc.), factors to be considered are hole diameter, rotating speed of cam, and vertical displacement to reduce seed injury (Yi et al. 2014).

Among depth-control devices, a promising option is a side gauge wheel and a poor choice is that of the rear presswheel (Karayel and Ozmerzi 2008). The use of a traction pneuma one suction type tillage-free fertilizing precision seeding machine is also considered an improved seeding device (Cai et al. 2013). These new drills save energy and reduce compaction risks (Yang et al. 2014), have multiple functions such as harrow, cultivator, and fertilizer/pesticide spreader, and can be used safely and reliably. In addition to sowing quality, a precision drill purchase must also consider cost and customer service (Ziegler et al. 2012). Clump planting and plant geometry are also a type of precision seeding that optimizes the use efficiency of water (Stewart and Lal 2012).

16.3.7 PRECISION ROW ORIENTATION

Maximizing exposure to light of the canopy can improve crop yield, moderate soil temperature, and reduce weed infestation. Increasing access of crops to light and decreasing access to weeds through row orientation and canopy management can improve agronomic productivity (Champion et al. 2008; Hozayn et al. 2012). Crop rows oriented at a right angle to sunlight direction may suppress weed growth through greater shading of weeds in interrow spaces and can be a useful weed control technique in wheat (*Triticum aestivum*) and barley (*Hordeum vulgare*; Borger et al. 2010). Row orientation and spacing are nonchemical weed management techniques that are also useful in organic farming (Bond and Grundy 2001).

The concept of row orientation is relevant to precision viriculture, which involves trellis design. The spatial and temporal variability of insolation affects the microclimate within the canopy. Thus, trellis design and row direction can affect quality and yield of grapes (Bergqvist et al. 2001; Weiss 2000; Weiss et al. 2003).

16.4 RECENT DEVELOPMENTS IN PA TECHNOLOGY

There are numerous developments in computer hardware and software, mapping techniques, methods of soil characterization, engineering, and communication, which have allowed better management of soil and water resources. In addition, systems analysis is used to integrate these tools for holistic management. For example, precision conservation has the potential to advance the sustainable use of natural resources and also enhance the resilience of agroecosystems to extreme climatic events. The rate of crop residue removal for biofuel feedstock and other competing uses can be assessed on the basis of site-specific factors so that the residue removal rate has a minimal environmental effect (Delgado et al. 2011).

There are some important advances in soil/site characterization and mapping. Optimization of N fertilization (avoiding over- and underfertilization) can be facilitated by sensing crop N status with fluorescence indicators. Fluorescence-based technology (real-time and *in vivo* assessment of chlorophyll inflorescence and phenolic compounds) can provide information on plant N status independent of soil, leaf area, biomass, and so forth, and is useful for PA (Tremblay et al. 2012). Field-scale assessment of soil fertility parameters can be done by using infrared spectral approaches to provide spatial soil properties data for PA (McCarty and Reeves 2006). Management zones can be established by using soil electrical conductivity (Corwin and Lesch 2005) and other fuzzy clustering techniques (Molin and de Castro 2008). Spatial heterogeneity of soil properties can be mapped by assessing apparent electrical conductivity (Mertens et al. 2008). Similarly, x-ray fluorescence spectrometry-based measurements can be used for precision management of soil P to optimize uptake and minimize transport into natural waters (Dao et al. 2011). Sampling intervals over a landscape for characterizing soil properties for using PA technology can be determined by the use of kriging techniques (Liu et al. 2010). Yield maps can also be used to optimize soil sampling (van Groenigen et al. 2000) by using transects or grids.

16.5 FUTURE RESEARCH AND DEVELOPMENT

Despite impressive scientific advances (see Section 16.4), there are numerous knowledge gaps that must be filled. In this context, additional research on PA is a challenging target for the future (Kirchmann and Thorvaldsson 2000).

1. *Economics and cost-effectiveness*: Environmental and economic benefits of PA technology have not been widely quantified, especially for horticultural crops (Alamo et al. 2012). Some of the innovative technologies (e.g., variable-rate irrigation equipment for managing irrigation spatially) can be adapted to commercial machines and equipment, but costs are high (Sadler et al. 2005). Equipment costs are also high for precision seeding (Ziegler et al. 2012). Thus, economic analysis of these and other emerging PA technologies must be done at the prevailing market prices. Indeed, economic factors may be a major constraint for resource-poor farmers and small landholders. The cost of soil sampling and analysis is another key factor (van Groenigen et al. 2000).

2. *Farm size*: Is PA technology applicable to small landholders managing <2 ha of farmland? It is argued that PA technology is scale-neutral, but this concept needs to be tested and validated under site-specific conditions. The adaptation of PA to small landholders is also confounded by the lack of relevant data and maps on soil, terrain, and climate variables. With reliance on information management, computers, GPS, and GIS, PA is expected to become more widespread for the small landholders of South Asia (Tiwari and Jaga 2012) and elsewhere in developing countries.

3. *Environmental protection*: The impact of PA on the environment must be quantified at different spatial (soilscape, landscape) and temporal (annual, decadal) scales. Specific measurements must be made with regard to (1) soil processes and properties such as erosion, salinization, soil organic carbon sequestration, densification, and nutrient imbalance, (2) water resources including pollution, contamination, eutrophication, and excessive withdrawal of groundwater, (3) biodiversity of above- and below-ground biota with regard to food chain and interconnectivity, (4) emission of GHGs, especially N_2O and methane (CH_4), and (5) aesthetic value of soil, water, vegetation, terrain, and other natural resources.

4. *Risk management*: PA technology has strong applications in management and reduction of environmental risks, including emission of GHGs, eutrophication of natural water, sedimentation of waterways and aquatic ecosystems, and soil pollution.

5. *Energy use and efficiency*: Direct and indirect use of energy-based inputs and production and use of biofuels are other priority themes.

6. *Organic farming*: Applications of PA to organic farming (e.g., weed and pest management) need additional research.

16.6 CONCLUSIONS

PA technology has strong applications in food and nutritional security, environmental quality, runoff and erosion control, and management of energy-based inputs (plant nutrients, pesticides, herbicides, tillage). This technology combines the goals of meeting the basic necessities of a growing population with the objectives of environmental stewardship and the aesthetic values of natural and managed ecosystems. It provides innovative and creative alternatives through adoption of the concepts of ecological intensification and sustainable intensification of agroecosystems.

There are several researchable themes, including economics, scale neutrality, application to small landholders, organic farming, and risk management. Adoption of innovative and creative measures can enhance the applicability of PA to diverse biophysical and socioeconomic regions.

REFERENCES

Aguera, J, M Perez-Ruiz, J Carballido, JA Gil, and JV Stafford. 2013. Soil compaction sensor for site-specific tillage: Design and assessment. *Precision Agriculture* 13:49–56.

Alamo, S, MI Ramos, FR Feito, and JA Canas. 2012. Precision techniques for improving the management of the olive groves of southern Spain. *Spanish Journal of Agricultural Research* 10 (3):583–595.

Bailey, JS, K Wang, C Jordan, and A Higgins. 2001. Use of precision agriculture technology to investigate spatial variability in nitrogen yields in cut grassland. *Chemosphere* 42 (2):131–140.

Bergqvist, J, N Dokoozlian, and N Ebisuda. 2001. Sunlight exposure and temperature effects on berry growth and composition of Cabernet Sauvignon and Grenache in the central San Joaquin Valley of California. *American Journal of Enology and Viticulture* 52 (1):1–7.

Bond, W, and AC Grundy. 2001. Non-chemical weed management in organic farming systems. *Weed Research* 41 (5):383–405.

Borger, CPD, A Hashem, and S Pathan. 2010. Manipulating crop row orientation to suppress weeds and increase crop yield. *Weed Science* 58 (2):174–178.

Cai, Z, P Fan, J Kang, L Sun, W Tian, B Yang, and H Zhao. 2013. Traction pneuma suction–type tillage-free-fertilization precision seeding machine, has frame fixed with mounting transmission shaft, where end of machine frame is mounted with pneuma absorbing mnomer profile modeling seeding device. Patent Number CN203675587-U (August 16, 2013, China).

Champion, GT, RI Froud-Williams, and JM Holland. 2008. Interactions between wheat (*Triticum aestivum* L.) cultivar, row spacing and density and the effect on weed suppression and crop yield. *Annals of Applied Biology* 133 (3):443–453.

Corwin, DL, and SM Lesch. 2005. Apparent soil electrical conductivity measurements in agriculture. *Computers and Electronics in Agriculture* 46 (1–3):11–43.

Dao, TH, YXX Miao, and FSS Zhang. 2011. X-ray fluorescence spectrometry-based approach to precision management of bioavailable phosphorus in soil environments. *Journal of Soils and Sediments* 11 (4):577–588.

Delgado, JA, and WC Bausch. 2005. Potential use of precision conservation techniques to reduce nitrate leaching in irrigated crops. *Journal of Soil and Water Conservation* 60 (6):379–387.

Delgado, JA, and JK Berry. 2008. Advances in precision conservation. *Advances in Agronomy* 98:1–44.

Delgado, JA, R Khosla, and T Mueller. 2011. Recent advances in precision (target) conservation. *Journal of Soil and Water Conservation* 66 (6):167A–170A.

Dosskey, MG, DE Eisenhauer, and MJ Helmers. 2005. Establishing conservation buffers using precision information. *Journal of Soil and Water Conservation* 60 (6):349–354.

Du, CW, and JM Zhou. 2009. Evaluation of soil fertility using infrared spectroscopy—A review. In *Climate Change, Intercropping, Pest Control and Beneficial Microorganisms*, ed. E. Lichtfouse, 453–483. Springer, Dordrecht, the Netherlands.

Godwin, RJ. 2010. The use of precision agriculture in soil management. *Trends in Agricultural Engineering*, 4th International Conference, September 7–10, 2010, Prague, Czech Republic, 2010:7–13.

Hochman, Z, PS Carberry, MJ Robertson, DS Gaydon, LW Bell, and PC McIntosh. 2013. Prospects for ecological intensification of Australian agriculture. *European Journal of Agronomy* 44:109–123.

Hozayn, M, TA El-Shahway, and FA Sharara. 2012. Implication of crop row orientation and row spacing for controlling weeds and increasing yield in wheat. *Australian Journal of Basic and Applied Sciences* 6 (3):422–427.

Irmak, A, JW Jones, WD Batchelor, and JO Paz. 2001. Estimating spatially variable soil properties for application of crop models in precision farming. *Transactions of the ASAE* 44 (5):1343–1353.

Jat, ML, MK Gathala, JK Ladha, YS Saharawat, AS Jat, V Kumar, SK Sharma, and R Gupta. 2009. Evaluation of precision land leveling and double zero-till systems in the rice-wheat rotation: Water use, productivity, profitability and soil physical properties. *Soil & Tillage Research* 105 (1):112–121.

Karayel, D, and A Ozmerzi. 2008. Evaluation of three depth-control components on seed placement accuracy and emergence for a precision planter. *Applied Engineering in Agriculture* 24 (3):271–276.

Khosla, R, K Fleming, JA Delgado, TM Shaver, and DG Westfall. 2002. Use of site-specific management zones to improve nitrogen management for precision agriculture. *Journal of Soil and Water Conservation* 57 (6):513–518.

Khosla, R, D Inman, DG Westfall, RM Reich, M Frasier, M Mzuku, B Koch, and A Hornung. 2008. A synthesis of multi-disciplinary research in precision agriculture: Site-specific management zones in the semi-arid western Great Plains of the USA. *Precision Agriculture* 9 (1–2):85–100.

Kirchmann, H, and G Thorvaldsson. 2000. Challenging targets for future agriculture. *European Journal of Agronomy* 12 (3–4):145–161.

Kitchen, NR, KA Sudduth, DB Myers, RE Massey, EJ Sadler, RN Lerch, JW Hummel, and HL Palm. 2005. Development of a conservation-oriented precision agriculture system: Crop production assessment and plan implementation. *Journal of Soil and Water Conservation* 60 (6):421–430.

Lan, Y, Y Huang, DE Martin, and WC Hoffmann. 2009. Development of an airborne remote sensing system for crop pest management: System integration and verification. *Applied Engineering in Agriculture* 25 (4):607–615.

Lerch, RN, NR Kitchen, RJ Kremer, WW Donald, EE Alberts, EJ Sadler, KA Sudduth, DB Myers, and F Ghidey. 2005. Development of a conservation-oriented precision agriculture system: Water and soil quality assessment. *Journal of Soil and Water Conservation* 60 (6):411–421.

Liu, GS, HL Jiang, SD Liu, XZ Wang, HZ Shi, YF Yang, XM Yang, HC Hu, QH Liu, and JG Gu. 2010. Comparison of kriging interpolation precision with different soil sampling intervals for precision agriculture. *Soil Science* 175 (8):405–415.

Lord, JC. 2005. From Metchnikoff to Monsanto and beyond: The path of microbial control. *Journal of Invertebrate Pathology* 89 (1):19–29.

McCarty, GW, and JB Reeves. 2006. Comparison of NFAR infrared and mid infrared diffuse reflectance spectroscopy for field-scale measurement of soil fertility parameters. *Soil Science* 171 (2):94–102.

Mertens, FM, S Paetzold, and G Welp. 2008. Spatial heterogeneity of soil properties and its mapping with apparent electrical conductivity. *Journal of Plant Nutrition and Soil Science-Zeitschrift Fur Pflanzenernahrung Und Bodenkunde* 171 (2):146–154.

Molin, JP, and CN de Castro. 2008. Establishing management zones using soil electrical conductivity and other soil properties by the fuzzy clustering technique. *Scientia Agricola* 65 (6):567–573.

Motomiya, AVD, JE Cora, and GT Pereira. 2006. Using indicator kriging for evaluating soil fertility indicators. *Revista Brasileira De Ciencia Do Solo* 30 (3):485–496.

Murthy, AR, IVM Krishna, and MSR Murthy. 2004. Precision conservation of natural resources for sustainable development. *IGARSS 2004: IEEE International Geoscience and Remote Sensing Symposium Proceedings* 7:4605–4608.

Naresh, RK, RS Rathore, P Kumar, SP Singh, A Singh, and UP Shahi. 2014. Effect of precision land leveling and permanent raised bed planting on soil properties, input use efficiency, productivity and profitability under maize (Zea mays)–wheat (*Triticum aestivum*) cropping system. *Indian Journal of Agricultural Sciences* 84 (6):725–732.

Sadler, EJ, RG Evans, KC Stone, and CR Camp. 2005. Opportunities for conservation with precision irrigation. *Journal of Soil and Water Conservation* 60 (6):371–379.

Santos, AJM, CA Gamero, RB de Oliveira, and AC Villen. 2011. Spatial analysis of longitudinal distribution of maize seeds in a precision grain drill. *Bioscience Journal* 27 (1):16–23.

Schumacher, JA, TC Kaspar, JC Ritchie, TE Schumacher, DL Karlen, ER Venteris, GW McCarty, TS Colvin, DB Jaynes, MJ Lindstrom, and TE Fenton. 2005. Identifying spatial patterns of erosion for use in precision conservation. *Journal of Soil and Water Conservation* 60 (6):355–362.

Silva, MC, FAC Pinto, DM Queiroz, J Gomez-Gil, and LM Navas-Gracia. 2012. Weed mapping using a machine vision system. *Planta Daninha* 30 (1):217–227.

Slaughter, DC, DK Giles, and D Downey. 2008. Autonomous robotic weed control systems: A review. *Computers and Electronics in Agriculture* 61 (1):63–78.

Steiner, U, K Burling, and EC Oerke. 2008. Sensor use in plant protection. *Gesunde Pflanzen* 60 (4):131–141.

Stewart, BA, and R Lal. 2012. Manipulating crop geometries to increase yield in dry areas. In *Soil Water and Agronomic Productivity*, eds. B.A. Stewart and R. Lal, 409–426. Taylor and Francis, Boca Raton, FL.

Tiwari, A, and PK Jaga. 2012. Precision farming in India—A review. *Outlook on Agriculture* 41 (2):139–143.

Tremblay, N, ZJ Wang, and ZG Cerovic. 2012. Sensing crop nitrogen status with fluorescence indicators. A review. *Agronomy for Sustainable Development* 32 (2):451–464.

van Groenigen, JW, M Gandah, and J Bouma. 2000. Soil sampling strategies for precision agriculture research under Sahelian conditions. *Soil Science Society of American Journal* 64 (5):1674–1680.

Weiss, SB. 2000. Vertical and temporal patterns of insolation in an old-growth forest. *Canadian Journal of Forest Resources* 30:1953–1964.

Weiss, SB, DC Luth, and B Guerra. 2003. Potential solar radiation in a vertical shoot positioned trellis at 38°N latitude. *Practical Winery and Vineyard* 16:16–23.

Xie, YW, JY Yang, SL Du, J Zhao, Y Li, and EC Huffman. 2012. A GIS-based fertilizer decision support system for farmers in Northeast China: A case study at Tong-le village. *Nutrient Cycling in Agroecosystems* 93 (3):323–336.

Xiu, YF, HB Lin, RX Wang, Q Li, and CJ Yi. 2010. The development of a wheat variable precision seeding robot based on GPS. *Intelligent Automation and Soft Computing* 16 (6):869–879.

Yang, D, X Chen, and Z Tang. 2014. Corn precision seeding machine, has machine frame connected with power transmission mechanism, cutter shaft component connected between fertilizing leg and fine sowing leg, and profile modeling wheel provided with fixing bracket. Patent Number: CN203675563-U. Patent Assignee: Anhui Jingtian Machinery Co Ltd. (February 12, 2014, China).

Yi, SJ, YF Liu, C Wang, GX Tao, HY Liu, and RH Wang. 2014. Experimental study on the performance of bowl-tray rice precision seeder. *International Journal of Agricultural and Biological Engineering* 7 (1):17–25.

Zhai, JB, JF Xia, Y Zhou, and S Zhang. 2014. Design and experimental study of the control system for precision seed-metering device. *International Journal of Agricultural and Biological Engineering* 7 (3):13–18.

Ziegler, K, E Gobel, O Schmittmann, and PS Lammers. 2012. Precision seed drill tests in Germany/Franconia 2009–2011. *International Sugar Journal* 114 (1367):8–15.

Index

Page numbers followed by f and t indicate figures and tables, respectively.

Q